Immunochemistry of viruses
The basis for serodiagnosis and vaccines

Immunochemistry of viruses

The basis for serodiagnosis and vaccines

edited by

M.H.V. Van Regenmortel
Institut de Biologie Moléculaire et Cellulaire (IBMC)
15 rue René Descartes, 67084 Strasbourg, France

and

A.R. Neurath
The Lindsley F. Kimball Research Institute,
New York Blood Center, 310 East 67th Street,
New York, NY 10021, U.S.A.

1985
Elsevier
Amsterdam · New York · Oxford

ISBN 0-444-80647-4

Published by:
Elsevier Science Publishers B.V. (Biomedical Division)
P.O. Box 211
1000 AE Amsterdam
The Netherlands

Sole distributors for the USA and Canada:
Elsevier Science Publishing Company, Inc.
52 Vanderbilt Avenue
New York, NY 10017
USA

Library of Congress Cataloging in Publication Data

Main entry under title:

Immunochemistry of viruses.

Includes bibliographies and index.
1. Viruses. 2. Immunochemistry. 3. Serum diagnosis.
4. Viral vaccines. 5. Virus diseases--Immunological aspects. I. Van Regenmortel, M.H.V. II. Neurath, A. Robert. [DNLM: 1. Immunochemistry. 2. Serodiagnosis.
3. Viral Vaccines. 4.Virus Diseases--immunology.
5. Viruses--immunology. QW 160 I33]
QR370.I46 1985 616′.0194 85-1512
ISBN 0-444-80647-4 (U.S.)

Printed in The Netherlands

Cover photograph: Immunoelectronmicroscopic localization of the hepatitis B virus (HBV) pre-S antigen on HBV particles. Partially purified HBV particles (provided by J. Gerin) were stained for the presence of pre-S antigen using a mouse pre-S-specific antibody and colloidal gold-labelled secondary antibody. The black colloidal gold particles surrounding the HBV particle and tubule indicate the selective presence of pre-S antigen on these forms. Note the absence of pre-S antigen on the 22-nm hepatitis B surface antigen particles. Magnification × 30 000 (photograph contributed by D.R. Milich and F.V. Chisari, Scripps Clinic and Research Foundation, La Jolla, CA). For further explanation see Chapter 19.

Preface

Until recently the vaccines and diagnostic reagents used in the management of virus diseases were obtained by the same empirical approach that successfully led to the conquest of most bacterial diseases. However, during the last ten years, our basic knowledge concerning the antigenic properties of viruses has expanded at an unprecedented rate, mainly as a result of advances in gene cloning and sequencing, in peptide synthesis and in hybridoma technology. The development of monoclonal antibodies using the technique of cell hybridization, as well as the determination of the amino acid sequence of viral gene products by deduction from nucleic acid sequence data, have made it possible to characterize the molecular structure of viral antigens with a degree of precision unattainable hitherto. The accumulated knowledge concerning these antigens is having a major impact on the development of diagnostic tests for viruses and on approaches for active immunization against viral diseases. Totally new strategies for the preparation of vaccines have been devised and new diagnostic tools of considerable specificity have been developed.

In the present volume, we attempt to review in a comprehensive manner the progress in virus immunoprophylaxis and diagnosis. The book, which represents the collaborative effort of many researchers, attempts to integrate new findings with preceding knowledge on the immunochemistry of distinct viruses. About half the chapters deal in general with the immunologic properties of viral antigens, with methods of viral diagnosis and with vaccines used in human and veterinary medicine; future trends in vaccine production based on recombinant DNA technology and synthetic peptides are also described. The other chapters each deal with the antigenic structure of distinct families of viruses that infect man, animals, insects and plants. The selection of virus families for inclusion was admittedly subjective and was dictated, in considerable measure, by the availability of a sufficient body of information concerning their antigenic structure. No attempt was made to review comprehensively all virus families.

The basic tenet of this book is that our expanding knowledge of virus immunochemistry, combined with an improved understanding of the immune response to viral infections provides a rational scientific base for the design of new virus vaccines and diagnostic reagents. By bringing together under one cover data on the antigenic properties of a wide range of different viruses, we hope to foster useful comparisons and to provide practising virologists with relevant information from outside their field of specialization. The book is intended for researchers and graduate students in the fields of virology, microbiology, immunology, infectious diseases and preventive medicine.

The editors wish to thank all contributors and express their special appreciation to Drs G. Air and G. Laver for contributing Chapter 15 on short notice.

M.H.V. Van Regenmortel
A.R. Neurath

November 1984

List of contributors

G.M. Air – *University of Alabama, Birmingham AL, U.S.A.*
R. Arnon – *Weizmann Institute of Science, Rehovot, Israel.*
F. Audibert – *Institut Pasteur, Paris, France.*
L.A. Babiuk – *Western College of Veterinary Medicine, Saskatchewan, Canada.*
F. Brown – *The Animal Virus Research Institute, Woking, Surrey, U.K.*
P.A. Brunell – *University of Texas Health Science Center, San Antonio TX, U.S.A.*
L. Chedid – *Institut Pasteur, Paris, France.*
B. Dietzschold – *The Wistar Institute, Philadelphia PA, U.S.A.*
D.D. Eckels – *Georgetown University School of Medicine, Washington DC, U.S.A.*
E.A. Emini – *Merck, Sharp and Dohme Research Laboratories, West Point PA, U.S.A.*
M.K. Estes – *Baylor College of Medicine, Houston TX, U.S.A.*
P.S. Gardner – *Central Public Health Laboratory, London, U.K.*
R.V. Gilden – *NCI-Frederick Cancer Research Facility, Frederick MD, U.S.A.*
D.Y. Graham – *Baylor College of Medicine, Houston TX, U.S.A.*
B.A. Jameson – *State University of New York, Stony Brook NY, U.S.A.*
D. Katz – *Israel Institute of Biological Research, Nes Ziona, Israel.*
S.B.H. Kent – *California Institute of Technology, Pasadena CA, U.S.A.*
A. Kohn – *Israel Institute of Biological Research, Nes Ziona, Israel.*
J.R. Lamb – *Medical Research Council, London, U.K.*
W.G. Laver – *John Curtin School of Medical Research, Canberra ACT, Australia.*
C. Leclerc – *Institut Pasteur, Paris, France.*
B. Mandel – *Long Island University, Brooklyn NY, U.S.A.*
R.J. Massey – *Igen Inc., Rockville MD, U.S.A.*
F.M. Moss – *University College Hospital, London, U.K.*
A.R. Neurath – *Lindsley F. Kimball Research Institute, New York NY, U.S.A.*
E. Norrby – *Karolinska Institute School of Medicine, Stockholm, Sweden.*
J.F. Obijeski – *Genentech Inc., South San Francisco CA, U.S.A.*

S. Oroszlan – *NCI-Frederick Cancer Research Facility, Frederick MD, U.S.A.*
C. Örvell – *Karolinska Institute School of Medicine, Stockholm, Sweden.*
E.J. Patzer – *Genentech Inc., South San Francisco CA, U.S.A.*
B.L. Petrie – *Baylor College of Medicine, Houston TX, U.S.A.*
U. Pettersson – *University of Uppsala, Uppsala, Sweden.*
G. Schochetman – *Igen Inc., Rockville MD, U.S.A.*
P.G. Spear – *University of Chicago, Chicago IL, U.S.A.*
E.G. Strauss – *California Institute of Technology, Pasadena CA, U.S.A.*
P.G. Strauss – *California Institute of Technology, Pasadena CA, U.S.A.*
T.W. Tinsley – *Natural Environment Research Council, Oxford, U.K.*
M.H.V. Van Regenmortel – *IBMC, Strasbourg, France.*
G. Wadell – *University of Umeå, Umeå, Sweden.*
T.J. Wiktor – *The Wistar Institute, Philadelphia PA, U.S.A.*
E. Wimmer – *State University of New York, Stony Brook NY, U.S.A.*
W.H. Wunner – *The Wistar Institute, Philadelphia PA, U.S.A.*
R.H. Yolken – *Johns Hopkins Hospital, Baltimore MD, U.S.A.*

Contents

Eds. M.H.V. van Regenmortel and A.R. Neurath
Immunochemistry of viruses. The basis for serodiagnosis and vaccines.

Structure of viral antigens 1

M.H.V. VAN REGENMORTEL[1] and A.R. NEURATH[2]

[1]IBMC, 15 rue Descartes, 67084 Strasbourg, France and [2]The Lindsley F. Kimball Research Institute of the New York Blood Center, 310 East 67th Street, New York, NY 10021, U.S.A.

The antigenic reactivity of viruses represents their capacity to undergo specific binding with antibodies and lymphocyte receptors. This reactivity resides in restricted parts of the viral surface that are known as antigenic determinants or epitopes. An epitope possesses a three-dimensional structure complementary to that of the binding site of an antibody molecule. The binding site, which is composed of a series of hypervariable loops forming a cleft at the distal end of the Fab regions of the immunoglobulin molecule, is able to accommodate protein epitopes consisting of about 5–7 amino acid residues. Although it is generally stated that the size of epitopes in proteins corresponds to about 6 amino acid residues, it seems that these dimensions simply reflect the size of the antibody tool used to gauge the antigen. There is in fact no ground for the belief that an antigenic patch at the surface of a protein need be restricted in size to that of the complementary area of a single antibody combining site.

In recent years, many attempts have been made to mimic the antigenic regions of proteins by synthesizing linear peptides. Since the antigenic reactivity of synthetic peptides of increasing size often showed a maximum value when a length of 6–7 residues had been reached, this finding also led to the view that protein epitopes correspond to linear sequences of up to seven residues (Crumpton, 1974; Atassi, 1975; Benjamini, 1977). Extensive investigations with several protein systems have shown, however, that such an interpretation is an oversimplification. When the surface of a globular protein is analysed by means of computer graphics or is represented by space-filling models, it can be clearly seen that only relatively small areas of the surface correspond to a linear array of contiguous residues in the protein sequence. Exposed surface patches of the protein are frequently made up of residues and side chains that are distant in the sequence and brought in close proximity by the folding of the chain. When protein surfaces are viewed in this light, attempts to mimic the complex surface topography of proteins by means of linear sequences of residues amount to searching for unidimensional solutions to a three-dimensional problem. It should come as no surprise then, that the antigenic reactivity of synthetic peptides measured by means of antibodies specific for the corresponding intact molecule is usually very low. Much remains to be learned about the conformational constraints that the surrounding residues impose on a short segment of polypeptide chain; our ignorance of these factors is responsible for our inadequate understanding of what constitutes a viral epitope.

1. Types of antigenic determinants

It has been customary to distinguish two types of epitopes – the so-called *sequential* and *conformational* determinants (Sela, 1969). Whereas a sequential determinant is viewed as a sequence of about seven residues in its unfolded random coil form, a conformational determinant is defined by a number of residues that are kept in a specific conformation, usually within the limitations of a particular macromolecular structure. It is generally assumed that antibodies directed to conformational determinants do not react with the unfolded peptides derived from the corresponding part of the native molecule. The distinction between conformation-dependent and independent determinants is perceived nowadays as somewhat artificial, since any sequence of residues within a native globular protein possesses a particular conformation. Operationally, a sequential determinant may be viewed as an epitope, present on a short peptide fragment, which elicits antibodies that recognize the same peptide region when it is inserted in the native molecule.

Another useful distinction, introduced by Atassi and Smith (1978), consists in differentiating between continuous and discontinuous determinants. A continuous determinant is defined as a contiguous sequence of residues exposed at the surface of a native protein and possessing distinctive conformational features. A discontinuous determinant consists of the juxtaposition in space of residues that are not contiguous in the primary structure. Distant residues may become contiguous through the folding of the peptide chain or by the juxtaposition of two separate peptide chains.

When discussing the antigenic structure of viruses, it is also useful to distinguish two other types of epitopes called *cryptotopes* and *neotopes*. Cryptotopes are hidden epitopes that become antigenically active only after fragmentation, depolymerization or denaturation of the antigen (Jerne, 1960). In the case of viral capsids, cryptotopes are found, for instance, on the surfaces of the subunits that are turned inward and become buried after polymerization (Van Regenmortel, 1966; 1982). Neotopes are epitopes that are specific for the quaternary structure of the virion and which are absent in the constituent monomeric subunits of the capsid (Van Regenmortel, 1966; Neurath and Rubin, 1971). Neotopes owe their existence either to conformational changes of the protein induced by intersubunit bonds or to the juxtaposition of residues from neighboring subunits.

Attempts to define the structure of epitopes solely in terms of surface residues in direct contact with the antibody molecule have led to difficulties, especially when certain epitopes were analysed by immunological cross-reactivity studies. When a single amino acid substitution in a viral mutant leads to an antigenic change, the usual interpretation is that the residue contributes to the structure of an epitope. However, without additional information, for instance about the three-dimensional structure of the molecule, one cannot be sure that the mutated residue is located within the boundaries of an epitope. It is now well established that substitutions occurring outside the surface domain of an epitope are able, by some kind of allosteric mechanism, to produce conformational changes that alter the antigenic reactivity of the epitope (Hurrell et al., 1977; White et al., 1978; Ibrahimi et al., 1979; Milton et al., 1980). It seems that the conformational uniqueness of epitopes is gained by long-range interactions at the level of secondary, tertiary and quaternary structure; the term *topographic determinant* (Urbanski and Margoliash, 1977) has been used to describe the situation where an epitope is modified by a distant substitution and therefore cannot be defined solely by the residues that are in direct contact with the antibody binding site (East et al., 1980). Since an excised epitope is unlikely to retain a memory of the conformational directives received from a distal part of the intact molecule before the cleavage occurred, there is little prospect of maintaining exactly the structure of an epitope after it has been isolated from the complete molecule. It is important, therefore, to consider how the analytical methods used to study epitopes influence the way in which they are defined by the investigator.

2. Methods used in the localization of viral epitopes

2.1. Fragmentation of the molecule

A series of peptides obtained by chemical or enzymatic cleavage of the protein are screened for antigenic reactivity by measuring their capacity to inhibit the reaction between the native antigen and its specific antibodies. Since the isolated peptides do not maintain the conformation they possessed in the parent molecule, large molar excesses of peptide over intact protein (e.g. 10^4:1) are usually needed to demonstrate the presence of antigenic reactivity in the peptides. This is usually ascribed to the fact that peptides exist in solution in a variety of random conformations in equilibrium with the native one, and that only conformations that approximate to the native form will bind to antibody (Sachs et al., 1972; Furie et al., 1975). However, it is also possible that the low level of residual antigenicity found in linear peptides simply reflects the fact that only some of the residues in any peptide contribute to the structure of epitopes that are mostly discontinuous rather than continuous.

2.2. Use of synthetic peptides

Once a particular protein fragment has been shown to possess antigenic reactivity, a more precise delineation of the epitope can be attempted by synthesizing a series of short peptides. Several viral epitopes have been defined in this way, for instance in the coat proteins of tobacco mosaic virus (Benjamini, 1977; Van Regenmortel, 1982) and turnip yellow mosaic virus (Quesniaux et al., 1983). When no information apart from the amino acid sequence of the viral protein is available, it is possible to predict the likely location of epitopes in the sequence by identifying regions of greatest local hydrophilicity. This prediction is based on the rule that hydrophilic regions are predominantly surface-oriented and thus more likely to be antigenic (Hopp and Woods, 1981; 1983). Using this approach, several laboratories have synthesized peptide analogues of the hepatitis B surface antigen that correspond to an antigenic region of the intact hepatitis antigen (Prince et al., 1982; Bhatnagar et al., 1982; Neurath et al., 1982). Since these linear synthetic peptides probably represented only part of the viral epitope and were probably folded incorrectly, it is not astonishing that the antigenic cross-reactivity between them and the native structure was rather low.

The location of epitopes in the sequence of viral proteins has also been achieved on the basis of computational methods used to predict protein secondary structure (Muller et al., 1982). Although these methods are not yet very satisfactory (Kabsch and Sander, 1983), confidence in the prediction is enhanced when a comparison of different viral serotypes reveals high sequence variability in the putative epitope region (Dreesmann et al., 1982). When α-helices are found to possess clear-cut hydrophobic and hydrophilic sides, there is also a good probability that the hydrophilic half is directed towards the solvent and that it corresponds to an epitope (Pfaff et al., 1982).

Synthetic peptides that correspond to the C- and N-termini of viral proteins (Walter et al., 1980; Altschuh et al., 1983; Quesniaux et al., 1983) possess higher than average antigenic cross-reactivity with the native structure. Peptides corresponding to the C-terminal sequences of tobacco mosaic virus and foot-and-mouth disease virus proteins were able, for instance, to induce antibodies that neutralized the infectivity of the two viruses (Anderer, 1963; Anderer and Schlumberger, 1965; Bittle et al., 1982). It seems likely that the surface orientation of the terminal residues reduces the conformational constraints of the peptide in the native structure and provides a better than average correspondence between the synthetic and native structures. It is known that many termini in proteins are located at the surface of the molecule and have a high relative flexibility (Thornton and Sibanda, 1983).

In recent years, the immunogenicity of synthetic peptides has received considerable attention. Antibodies raised against synthetic peptides have been used to determine which parts of the sequence are located at the protein surface (Lerner, 1982; Shinnick et al., 1983). It was originally believed that antibodies reactive with the intact protein would only be produced if the fragment used for immunization closely reproduced the conformation found in the protein. In the case of the hemagglutinin of influenza virus, it was found that a series of peptides scattered throughout the entire molecule elicited antibodies able to recognize the intact hemagglutinin; in contrast antihemagglutinin antibodies present in influenza virus antisera did not react with any of the synthetic peptides (Green et al., 1982). There is at the present time no satisfactory explanation for this apparent contradiction. A simple explanation based on the hypothesis that the immunogenic form of the peptides corresponds to a rare conformational occurrence was tested by estimating the frequency with which monoclonal antibodies directed against a peptide recognize the intact hemagglutinin (Niman et al., 1983). Most monoclonal antibodies raised against the peptide were found to react with the complete hemagglutinin, a finding which seemed to refute the view that a rare native-like conformation of the peptide was responsible for generating these antibodies. However, it is possible that the antipeptide monoclonal antibodies recognized the hemagglutinin because this molecule had been partly denatured by the immunoassay procedure used in the experiment. It has been shown, for instance, that monoclonal antibodies specific for neotopes of tobacco mosaic virus (TMV) were not reactive in immunoassay procedures in which the virus was directly adsorbed to the plastic surface of microtiter plates (Al Moudallal et al., 1984). Adsorption to the plastic appears to alter the conformation of the polypeptide chain so that the virus becomes antigenically similar to the dissociated subunits. These results obtained with TMV lead to the conclusion that the stochastic model of peptide immunogenicity has not been adequately tested in the experiments of Niman et al. (1983). It is also possible that cells of the immune system recognize only some peptide conformations, while others are not recorded as immunogenic. Further experiments are needed to establish whether truly native and correctly folded molecules are able to react with a large proportion of the monoclonal antibodies raised against a peptide.

It is clear, however, that many investigators have been successful in raising antipeptide polyclonal antisera that recognize apparently native proteins. This high rate of success may partly be attributed to the fact that the relative flexibility of peptides in solution can be mimicked by local disorder in short segments of the protein. It has been shown recently that the majority of continuous epitopes of TMV protein, myoglobin, lysozyme and myohemerythrin correspond to surface regions in these proteins that possess a high segmental mobility (Westhof et al., 1984; Tainer et al., 1984). It is possible that segmental motion in proteins provides the necessary degree of conformational similarity with isolated peptides, and that this ensures sufficient antigenic cross-reactivity between the intact molecule and an excised fragment.

Although linear peptides shorter than five residues usually are not reactive with antibodies directed to the intact protein (Atassi, 1975), even shorter synthetic analogues sometimes can be shown to possess specific antigenic reactivity (Anderer and Schlumberger, 1966). A low level of activity that is below detection in some immunoassays can be enhanced by increasing the potential hydrophobic or electrostatic interactions of the peptide. It was shown, for instance, that an inactive tripeptide analogue corresponding to residues 110–112 of TMV protein acquired specific binding activity by N-octanoylation (Benjamini et al., 1968). The addition of a lysine residue to a short peptide has also been shown to increase its binding to antibodies (Leach, 1983). These findings demonstrate that antigen-antibody binding is mediated only through the usual forces of protein-protein interactions that are brought into play when the contact residues of the antigen come into close apposition with complementary residues of the antibody binding site. In the case of short peptides, the conformational contribution to the binding may be nearly insignificant, and the addition, at a nonspecific location, of hydrophobic or charged residues may increase the reactivity of the peptide above the threshold value needed for detection (Van Regenmortel, 1984). These findings illustrate the difficulty of distinguishing between the specific and nonspecific binding activities of synthetic peptides that bind only poorly to antibodies directed to the parent molecule.

2.3. Study of antigenic cross-reactivity

This type of study, which requires a series of closely related viral proteins possessing a small number of amino acid exchanges, is most informative when carried out with monoclonal antibodies (Yewdell and Gerhard, 1981). The homogeneity of a monoclonal antibody preparation ensures that only one epitope is investigated at one time, and the influence of a residue exchange on antibody binding is therefore more readily detectable. When polyclonal antisera are used in this type of study, those antibodies that recognize an alteration in one particular epitope are often swamped by the large number of antibodies that continue to react in the normal way with unchanged epitopes of the virion. For instance, Underwood (1982), using a panel of 125 monoclonal antibodies raised against the hemagglutinin of influenza virus, identified in 16 virus strains several groups of variable amino acids that were given the status of likely hemagglutinin epitopes. Some of the proposed epitopes were the same ones that had been identified from studies of single point mutants of influenza virus. In these studies monoclonal antibodies had been used to select non-neutralized virus mutants in which the epitope recognized by the selecting antibody was altered by a single amino acid exchange (Wiley et al., 1981; Caton et al., 1982). Most single substitutions totally abolished the ability of the monoclonal antibody used to select the mutant to bind to the hemagglutinin. The same approach has been used in a study of the epitopes of the neuraminidase of influenza virus (Colman et al., 1983). However, in this case the epitopes seemed to form a nearly continuous antigenic surface across the top of the neuraminidase molecule, and it was not possible to recognize clearly distinct nonoverlapping antigenic areas.

The earlier view (Atassi, 1975; 1979a; 1979b) that a large portion of the surface of proteins is devoid of apparent antigenicity and that epitopes are restricted to a few highly accessible locations at the surface of the molecule appears today to be a gross oversimplification (White et al., 1978; Berzofsky et al., 1982; Leach, 1983; Van Regenmortel, 1984). When the antigenic reactivity of a protein is defined only in terms of short protein fragments or synthetic peptides, the three-dimensional complexity of the protein surface is ignored. It is becoming increasingly clear that a better understanding of the specificity of antigenic structures requires complementary information that can be obtained only through studies of the immunochemical cross-reactivity between related molecules. The advent of the hybridoma technology and improvements in solid-phase peptide synthesis should go a long way in providing answers to the central problem in viral immunochemistry: what are the immunogenic regions of viral antigens that give rise to neutralizing antibodies ?

3. Contribution of different viral components to antigenicity

Many viruses below a diameter of approximately 80 nm consist of nucleocapsids built up of a layer of protein molecules enclosing the infectious nucleic acid. Larger viruses usually have an additional membrane-like structure (the envelope) which usually encloses a nucleocapsid with helical or icosahedral symmetry. The envelope contains morphologically recognizable projections or spikes known as peplomers which are made up of glycoproteins. The main criteria used in grouping viruses are the type and size of the nucleic acid, the symmetry and size of the capsid, and the presence or absence of an envelope. The current state of virus classification is summarized in the report of Matthews (1982). The terms family, genus and species are used by most virologists in building up hierarchical groups. Plant virologists, however, are divided about the suitability of using Linnean terms for entities such as viruses that do not reproduce by sexual means, and some of them prefer to use the terms *virus* and *group* instead of species and genus.

The antigenic properties of viruses have also been used as a basis for building up groups. Serological groups are made up of viruses that cross-react with each other. In the Togaviridae family, for instance, serogroups are equated with genera: the genus Alphavirus is made up of serologically cross-reacting viruses that do not show cross-reactions with viruses in other genera such as flaviviruses or rubiviruses. As the 27 recognized plant virus groups in fact correspond to serogroups (there are no cross-reactions between the members of separate groups) the simplest way to build up a uniform classification encompassing all plant viruses would be to consider the clusters of virus strains as species and the 27 established groups of plant viruses as genera. The ending *virus* used in the name of existing plant virus groups is in agreement with the rule adopted by the International Committee on Taxonomy of Viruses for naming genera.

The immune reaction of vertebrate hosts establishes a selective pressure against virus infection which leads to viruses either evolving into a multitude of antigenic types (e.g. entero-, rhino- and adenoviruses) or modifying their antigenic sites continuously (e.g. myxoviruses). Only surface antigens are involved in these patterns of adaptation. In the case of influenza, for instance, only the hemagglutinin and neuraminidase undergo antigenic changes while the less accessible components, such as the matrix component and nucleocapsid, have stable antigenic properties. Some surface antigens are antigenically dominant and are found on all subtypes of a virus; they are known as group-specific antigens. Type-specific antigens, on the other hand, are those which are found to vary between different serotypes.

The glycoproteins found in viral envelopes consist of a hydrophilic moiety that can be removed by digestion with proteolytic enzymes, and a hydrophobic segment that remains associated with the lipid bilayer and anchors the protein in the membrane. The lipids of enveloped viruses are present in a bilayer form and consist mainly of phospholipids and cholesterol. These lipids are derived from pre-existing cellular membranes of the host and are devoid of antigenic activity. In virus groups such as the myxo- and rhabdoviruses, a layer of nonglycosylated matrix (M) proteins is present directly below the lipid bilayer.

Glycoproteins that are anchored to the lipids are known as integral membrane proteins. In addition, the envelope also contains peripheral proteins which do not have a hydrophobic anchoring part and are bonded instead to the integral membrane proteins.

The glycosylation site in viral proteins appears to be invariably the sequence Asn-X-Ser/Thr. Glycosylation is effected by enzymes of the host cell and leads to the incorporation of complex glycans consisting of mannose, galactose, *N*-acetylglucosamine, sialic acid and fucose (Kääriäinen and Pesonen, 1983).

Extensive information on the structure of viral glycoproteins is available for orthomyxoviruses (Chapter 14), togaviruses (Chapter 22), rhabdoviruses (Chapter 20) and retroviruses (Chapter 24). Because of their exposed surface location, glycoproteins interact preferentially with antibody and they are therefore the antigens most commonly involved in the neutralization of enveloped viruses.

The glycoproteins of virus envelopes such as hemagglutinins play a major role in the entry of the virus into the host cell. Fusion of the virus envelope with the host cell membrane leads to the release of the virus nucleocapsid into the cytoplasm. The ability of certain glycoproteins to agglutinate red blood cells is linked to the attachment and penetration functions of these viral proteins.

In a comparative study of the glycoproteins in the hemagglutinin of 18 influenza A strains, no variations in the glycosylation pattern were observed within a given serotype (Klenk et al., 1980). However, differences in the carbohydrate pattern of strains belonging to the same serotype can also be found. Since the oligosaccharide chains are specified by host enzymes, they are similar to those found in the glycoproteins of normal cells and are thus not immunogenic in the infected organism. However, a recent report (Skehel et al., 1984) has shown conclusively that the presence of oligosaccharide side chains can modify the antigenic structure in influenza hemagglutinin by a steric hindrance phenomenon. An exchange Asp→Asn at position 63 of the hemagglutinin of Hong Kong influenza virus generated a new glycosylation site. The presence of the additional oligosaccharide chain in the mutant was shown to mask an underlying epitope and to inhibit the binding of a monoclonal antibody specific for this region. The reactivity of many antibodies to protein epitopes on glycoproteins is influenced by attached carbohydrate chains (Alexander and Elder, 1984).

4. The immune response to viral antigens

The mechanisms by which higher organisms resist virus infections are either noninducible or inducible. The latter mechanisms are either nonspecific (e.g. induction of interferon synthesis) or specific, i.e. based on specific recognition of the virus and/or its gene products by cells of the immune system (Burns, 1975; Ogra et al., 1975). Viral gene products may be presented to these cells not only in the form of virus particles with only external structures exposed, but also as viral subunits, either in free form or incorporated into membranes of infected cells. Internal structural components of virions or gene products that do not become incorporated into virions may also be exposed on the surface of infected cells (Burns, 1975).

Antibodies, representing the earliest recognized defense mechanism of the host, may eliminate the infectious agents by the following mechanisms: (*a*) neutralization of extracellular virus; (*b*) complement-mediated lysis of virus-infected cells having exposed viral gene products, and (*c*) specific destruction of these cells by antibody-mediated cytotoxic cells.

As shown in the early 1970s, infected cells having exposed viral antigens may also become lyzed by specific cytotoxic T cells. Such lysis occurs only if the cytotoxic cells specifically recognize exposed viral gene products and if the effector and target cells are histocompatible (Sherman et al., 1983).

The synthesis of antiviral antibodies requires the collaboration between several cell types of the immune system as well as several antigen-specific and nonspecific helper factors. Such cooperation always involves B and T lymphocytes, each having distinct receptors for a viral antigen and recognizing either identical or distinct sites on the same antigen (Sherman et al., 1983.)

The interplay of defense mechanisms utilized in terminating a particular virus infection may differ from virus to virus. Some of these different mechanisms will be illustrated in the subsequent chapters.

5. References

Al Moudallal, Z., Altschuh, D., Briand, J.P. and Van Regenmortel, M.H.V. (1984) *J. Immunol. Methods* 68, 35–43.

Alexander, S. and Elder, J.H. (1984) *Science* 226, 1328–1330.

Altschuh, D., Hartman, D., Reinbolt, J. and Van Regenmortel, M.H.V. (1983) *Molec. Immunol.* 20, 271–278.

Anderer, F.A. (1963) *Biochim. Biophys. Acta* 71, 246–248.

Anderer, F.A. and Schlumberger, H.D. (1965) *Biochim. Biophys. Acta* 97, 503–509.

Anderer, F.A. and Schlumberger, H.D. (1966) *Biochim. Biophys. Acta* 115, 222–224.

Atassi, M.Z. (1975) *Immunochemistry* 12, 423–438.

Atassi, M.Z. (1979a) *CRC Crit. Rev. Biochem.* 6, 337–369.

Atassi, M.Z. (1979b) *CRC Crit. Rev. Biochem.* 6, 371–400.

Atassi, M.Z. and Smith, J.A. (1978) *Immunochemistry* 15, 609–610.

Benjamini, E. (1977) In *Immunochemistry of Proteins* (M.Z. Atassi, Ed.) Vol. 2, Plenum Press, New York. pp. 265–310.

Benjamini, E., Shimizu, M., Young, J.D. and Leung, C.Y. (1968) *Biochemistry* 7, 1261–1264.

Berzofsky, J.A., Buckenmeyer, G.K., Hicks, G., Gurd, F.R.N., Feldmann, R.J. and Minna, J. (1982) *J. Biol. Chem.* 257, 3189–3198.

Bhatnagar, P.K., Papas, E., Blum, H.E., Milich, D.R., Nitecki, D., Karels, M.J. and Vyas, G.N. (1982) *Proc. Natl. Acad. Sci. U.S.A.* 79, 4400–4404.

Bittle, J.L., Houghten, R.A., Alexander, H., Shinnick, T.M., Sutcliffe, J.G., Lerner, R.A., Rowlands, D.J. and Brown, F. (1982) *Nature* 298, 30–33.

Burns, W.H. (1975) In *Viral Immunology and Immunopathology* (A.L. Notkins, Ed.) Academic Press, New York. pp. 43–56.

Caton, A.J., Brownlee, G.G., Yewdell, J.W. and Gerhard, W. (1982) *Cell* 31, 417–427.

Colman, P.M., Varghese, J.N. and Laver, W.G. (1983) *Nature* 303, 41–44.

Crumpton, M.J. (1974) In *The Antigens* (M. Sela, Ed.) Vol. 2. Academic Press, New York. pp. 1–78.

Dreesman, G.R., Sanchez, Y., Ionescu-Matiu, I., Sparrow, J., Six, H.R., Peterson, D.L., Hollinger, F.B. and Melnick, J.L. (1982) *Nature* 295, 158–160.

East, I.J., Todd, P.E. and Leach, S.J. (1980) *Molec. Immunol.* 17, 519–525.

Furie, B., Schechter, A.N. Sachs, D.H. and Anfinsen, C.B. (1975) *J. Mol. Biol.* 92, 497–506.

Green, N., Alexander, H., Olson, A., Alexander, S., Shinnick, T.M., Sutcliffe, J.G. and Lerner, R.A. (1982) *Cell* 28, 477–487.

Hopp, T.P. and Woods, K.R. (1981) *Proc. Natl. Acad. Sci. U.S.A.* 78, 3824–3828.

Hopp, T.P. and Woods, K.R. (1983) *Molec. Immunol.* 20, 483–489.

Hurrell, J.G.R., Smith, J.A., Todd, P.E. and Leach, S.J. (1977) *Immunochemistry* 14, 283–288.

Ibrahimi, I.M., Prager, E.M., White, T.J. and Wilson, A.C. (1979) *Biochemistry* 13, 2736–2744.

Jerne, N.K. (1960) *Ann. Rev. Microbiol.* 14, 341–358.

Kääriäinen, L. and Pesonen, M. (1983) In *The Glycoconjugates* (M.I. Horowitz, Ed.) Vol. IV. Academic Press, New York. pp. 191–242.

Kabsch, W. and Sander, C. (1983) *FEBS Lett.* 155, 179–182.

Klenk, H.-D., Garten, W., Keil, W., Niemann, H., Schwarz, R.T. and Rott, R. (1980) In *Biosynthesis, Modification, and Processing of Cellular and Viral Polyproteins* (G. Koch and D. Richter, Eds.) Academic Press, New York. pp. 175–184.

Leach, S.J. (1983) *Biopolymers* 22, 425–440.

Lerner, R.A. (1982) *Nature* 299, 592–596.

Matthews, R.E.F. (1982) *Intervirology* 17, 1–199.

Milton de L., R.C., Milton, S.C.F., Von Wechmar, M.B. and Van Regenmortel, M.H.V. (1980) *Molec. Immunol.* 17, 1205–1212.

Muller, G., Shapira, M. and Arnon, R. (1982) *Proc. Natl. Acad. Sci. U.S.A.* 79, 569–573.

Neurath, A.R. and Rubin, B.A. (1971) *Viral Structural Components as Immunogens of Prophylactic Value* S. Karger, Basel. p. 87.

Neurath, A.R., Kent, S.B.H. and Strick, N. (1982) *Proc. Natl. Acad. Sci. U.S.A.* 79, 7871–7875.

Niman, H.L., Houghten, R.A., Walker, L.E., Reisfeld, R.A., Wilson, I.A., Hogle, J.M. and Lerner, R.A. (1983) *Proc. Natl. Acad. Sci. U.S.A.* 80, 4949–4953.

Ogra, P.L., Morag, A. and Tiku, M.L. (1975) In *Viral Immunology and Immunopathology* (A.L. Notkins, Ed.) Academic Press, New York. pp. 57–77.

Pfaff, E., Mussagay, M., Bohm, H.O., Schulz, G.E. and Schaller, H. (1982) *EMBO J.* 1, 869–874.

Prince, A.M., Ikram, H. and Hopp, T.P. (1982) *Proc. Natl. Acad. Sci. U.S.A.* 79, 569–573.

Quesniaux, V., Briand, J.P. and Van Regenmortel, M.H.V. (1983) *Mol. Immunol.* 20, 179–185.

Sachs, D.H., Schechter, A.N., Eastlake, A. and Anfinsen, C.B. (1972) *Proc. Natl. Acad. Sci. U.S.A.* 69, 3700–3794.

Sela, M. (1969) *Science* 166, 1365–1374.

Sherman, L.A., Vitiello, A. and Klinman, N.R. (1983) *Annu. Rev. Immunol.* 1, 63–86.

Shinnick, T.M., Sutcliffe, J.G., Green, N. and Lerner, R.A. (1983) *Annu. Rev. Microbiol.* 37, 425–446.

Skehel, J.J., Stevens, D.J., Daniels, R.S., Douglas, A.R., Knossov, M., Wilson, I.A. and Wiley, D.C. (1984) *Proc. Natl. Acad. Sci. U.S.A.* 81, 1779–1783.

Tainer, J.A., Getzoff, E.D., Alexander, H., Houghten, R.A. Olson, A.J., Lerner, R.A. and Hendrickson, W.A. (1984) *Nature* 312, 127–134.

Thornton, J.M. and Sibanda, B.L. (1983) *J. Mol. Biol.* 167, 443–460.

Underwood, P.A. (1982) *J. Gen. Virol.* 62, 153–169.

Urbanski, G.J. and Margoliash, E. (1977) *J. Immunol.* 118, 1170–1180.

Van Regenmortel, M.H.V. (1966) *Adv. Virus Res.* 12, 207–271.

Van Regenmortel, M.H.V. (1982) *Serology and Immunochemistry of Plant Viruses.* Academic Press, New York. pp. 302.

Van Regenmortel, M.H.V. (1984) In *Hybridoma Technology in Agriculture and Veterinary Research* (N.J. Stern and H.R. Gamble, Eds.) Rowman and Allenheld, Totowa N.J. pp. 42–80.

Walter, G., Scheidtmann, K.H., Carbone, A., Landano, A. and Doolittle, R.F. (1980) *Proc. Natl. Acad. Sci. U.S.A.* 77, 5197–5200.

Westhof, E., Altschuh, D., Moras, D., Bloomer, A.C., Mondragon, A., Klug, A. and Van Regenmortel, M.H.V. (1984) *Nature* 311, 123–126.

White, T.J., Ibrahimi, I.M. and Wilson, A.C. (1978) *Nature* 274, 92–94.

Wiley, D.C., Wilson, I.A. and Skehel, J.J. (1981) *Nature* 289, 373–378.

Yewdell, J.W. and Gerhard, W. (1981) *Ann. Rev. Microbiol.* 35, 185–206.

Eds. M.H.V. van Regenmortel and A.R. Neurath
Immunochemistry of viruses. The basis for serodiagnosis and vaccines.

Modulation of antibody response to viruses by delivery systems and muramyl peptides

2

C. LECLERC, F. AUDIBERT and L. CHEDID

Immunothérapie Expérimentale, Institut Pasteur, 28 rue du Dr. Roux, 75724 Paris Cedex 15, France

Several reviews (Neurath and Rubin, 1971; Cowan, 1973; Mandel, 1978; McNabb and Tomasi, 1981; Couch and Kasel, 1983) and some of the chapters of the present volume discuss in detail the roles of cell-mediated and humoral antiviral immunity. Therefore, we will limit ourselves to a brief description of the role of antibodies, and especially of those produced by the secretory immune system. Special attention will be given to adjuvants and carriers as quantitative and qualitative modulators of the host response to viruses.

The precise mechanisms by which the immune system protects the host against viral infections are still controversial. Moreover, the relative importance of immune lymphocytes, and especially of antiviral antibodies has been shown to depend critically upon the way in which viruses spread from one cell to another (Notkins, 1974). For instance, protection against influenza virus has been shown to be mediated either by B- or T-cell immunity, depending on the experimental system used (McMichael et al., 1982).

1. Humoral antiviral immunity

Passive transfer immunization has demonstrated that humoral immunity could be sufficient to protect against certain viral diseases. Thus, for instance, it has been reported that the passive transfer of herpes simplex virus (HSV) type 1-specific antibody could provide recovery from ocular HSV infection by inhibiting the spread of the virus within the brain after its entry into the central nervous system (Davies et al., 1979). More recently, passive immunization with monoclonal antibodies directed against glycoprotein D as well as other HSV glycoproteins was shown to be highly effective in protecting mice against challenge by a lethal dose of HSV (Norrild et al., 1979; Kapoor et al., 1982; Rector et al., 1982).

The neutralizing capacity of antiviral antibodies has been attributed to (*a*) complement-dependent virolysis; (*b*) aggregation of virus; (*c*) conformational alteration of the virus: the binding of one molecule of antibody to an antigenic subunit could stabilize that subunit in a specific nonfunctional conformation (Mandel, 1978); and (*d*) a role of antibody-dependent cell-mediated cytotoxicity (ADCMC) has also been suggested in certain viral diseases (Oakes and Lausch, 1981).

The protective activity of antiviral antibodies is related to their affinity, to their isotype and obviously to their specificity. Indeed, conventional vaccines, even when purified, elicit the production of antibodies of several specificities. Only some of these clonotypes play a role in the protection against the infectious agent. For instance, it has been shown that two surface glycoproteins are important in the case of influenza virus. A good level of passive protection could be obtained with anti-hemagglutinin antibodies or with antineuraminidase antibodies, although with the latter large amounts are required. In contrast, antibodies raised against the internal nonglycosylated proteins (matrix and nucleoprotein) are not protective (Couch and Kasel, 1983).

Most of these data were obtained from in vitro studies and their significance is still questionable. In this respect, it is interesting to mention the provocative results obtained by Rector et al. (1982). These workers have evaluated in vitro the biological activities of anti-HSV-1 monoclonal antibodies previously shown to protect mice in vivo against a lethal dose of HSV-1. These monoclonal antibodies were evaluated for their capacity to neutralize HSV-1, to promote complement-mediated lysis and antibody-dependent cellular cytotoxicity. None of these monoclonal antibodies were active in all three of the in vitro assays, although they were protective in vivo. Moreover, one of them had no activity in any of the three immunological reactions, even though it was highly effective in promoting recovery from HSV-1 induced ocular disease in vivo (Rector et al., 1982).

Therefore, in certain cases, in vitro evaluation of affinity, isotype and specificity of circulating antibodies is not sufficient to predict or rule out their protective activity in vivo. Indeed, there exists increasing evidence that local immunity plays an important role in resistance to many viral infections.

2. Secretory antiviral immunity

Local antibodies play a major role in resistance to many viral infections. For instance, in humans and in ferrets immunity to reinfection correlates better with secretory than with circulating antibodies (review in Shvartsman and Zykov, 1976). Moreover, Akerfeld et al. (1973) showed that to protect mice against experimental influenza infection with human γ-globulin, 2000–3000 mg/kg must be administered systemically compared with 50 mg/kg intranasally. Clinical studies have also demonstrated the effectiveness of administering hyperimmune serum intranasally in preventing (Shvartsman and Zykov, 1976) or curing influenza (Ritova and Zhdanov, 1962).

The role of secretory antibodies in poliomyelitis is also well established. Circulating poliomyelitis antibodies protect against paralytic symptoms of the infection but have little or no influence on the multiplication of the virus in the mucosal cells of the digestive tract. In contrast, secretory antibodies prevent the spreading of the virus in the host and the development of paralytic symptoms by neutralizing the virus at the portal of entry (Ogra and Karzon, 1969a; 1969b; 1971). Such antibodies are also of great epidemiological value because they restrict the dissemination of the virus to other hosts.

Local antibody responses play a role not only in mucosal but also in nonmucosal immunity. For instance, local specific IgA antibody formation has been repeatedly demonstrated in viral infections of the central nervous system (Waksman, 1983). Such antibodies may continue to be produced several years after an acute infection (Waksman, 1983).

Evidence for the existence of a common mucosal immune defence system is now substantial. Ingestion or inhalation of an antigen results in the sensitization of lymphoid cells in the gut-associated lymphoreticular tissue (GALT) and bronchial-associated lymphoreticular tissue and in the subsequent migration of IgA-secreting precursor B cells to distant mucosal sites (McGhee et al., 1978a; McDermott and Bienenstock, 1979). Thus, in several experimental models, intragastric immunization gives rise to antibodies in the gastrointestinal tract and also in mammary, lachrymal or salivary glands (see in McGhee et al., 1978a). McGhee et al. (1978b) have also demonstrated that in human volunteers fed *Streptococcus mutans*, specific IgA was detected in both saliva and tears.

Polymeric IgA is thus the predominant immunoglobulin in most exocrine secretions such as milk, saliva, tears and intestinal fluid. However, mucosal immunity can also be mediated by serum or locally derived IgG, as demonstrated by Couch and Kasel (1983) during influenza infection.

Because of the key role played by local immunity in antiviral infections, the influence of routes of injection has been actively studied. High levels of secretory anti-influenza antibodies have been produced in humans immunized through the respiratory tract (Waldman et al., 1968) whereas systemic immunization did not stimulate secretory immunity (Alford, 1969). Likewise, immunization with polio vaccines induces similar serum levels of IgM and IgG antibodies after oral or parenteral administration, whereas a significant secretory antibody response in the alimentary tract and nasopharynx occurs only after oral immunization (Ogra and Karzon, 1969a; 1969b).

However, it must be noted that effective oral immunization has been mostly obtained by using live viruses that colonize the gastrointestinal tract, whereas only limited success was achieved by using inactivated or killed viruses. As in the case of all vaccines except those containing live viruses, oral vaccines containing safer non-viable agents may necessitate the addition of appropriate potentiating agents such as delivery systems or adjuvants.

3. Delivery systems in future vaccines

Viral subunits, genetically engineered antigens or synthetic peptides have been administered in the following delivery systems: (*a*) recombinant vaccines expressing viral antigens; (*b*) inert delivery systems such as liposomes, and (*c*) natural and synthetic carrier molecules (which will be discussed later with synthetic vaccines).

3.1. Live recombinant vaccines

Infectious vaccinia virus recombinants expressing foreign genes from a variety of sources have been constructed as live vaccines for heterologous pathogenic agents (e.g. Panicali and Paoletti, 1982). Smith et al. (1983a) inserted the coding sequence for hepatitis B surface antigen (HBsAg) into the vaccinia virus genome. Cells infected with these vaccinia virus recombinants did synthesize and excrete HBsAg. Moreover, according to antigenicity, polypeptide composition, buoyant density and sedimentation rate, these particles were similar to those present in the serum of hepatitis B virus carriers. Following vaccination with one of these virus recombinants, rabbits developed typical local skin reactions and anti-HBsAg titers of more than 200 mIU/ml were detected in their serum. The same investigators have also reported successful production of vaccinia recombinants expressing a gene coding for influenza hemagglutinin (Smith et al., 1983b). Moreover, hamsters vaccinated with this recombinant virus were protected against respiratory infection with influenza virus (Smith et al., 1983b). More recently, Paoletti et al. (1984) have shown that mice immunized with vaccinia recombinants expressing the coat glycoprotein of HSV were protected from subsequent lethal challenge. Although very promising, more experiments are required to evaluate the efficacy of such vaccines and to delete the virulence of the vaccinia particles. Furthermore, it will be necessary to rule out the possibility of reversion of recombinants to new virulent forms. Thus, parallel efforts are being pursued to find inert delivery systems, such as liposomes and micelles.

3.2. Inert delivery systems

The use of water-soluble protein micelles has been shown to increase immunity against viral antigens such as HBsAg elicited by an HBsAg polypeptide (Skelly et al., 1981), or by a synthetic peptide (Sanchez et al., 1982).

More often, protein antigens have been encapsulated into liposomes (reviewed in Gregoriadis, 1976a; 1976b; Tyrrell et al., 1976). Such preparations elicited stronger immune responses for many reasons, including the 'depot' of antigen, entrapment by macrophages and better presentation. Moreover, Almeida et al. (1975) have shown by electron microscopy and serospecific agglutination that influenza virus surface proteins orient themselves in liposomes in a manner which appears similar to their disposition in the intact virus (Almeida et al., 1975; Tyrrell et al., 1976). Similar virosomes have been obtained with hepatitis B surface antigen (Manesis et al., 1978; 1979; Sanchez et al., 1980) and Gross virus-associated cell-surface antigen (Sakai et al., 1980).

Virosomes containing HBsAg (Sanchez et al., 1980), influenza-derived antigens (Thibodeau et al., 1981) or purified rabies glycoproteins (Thibodeau et al., personal communication) have been shown to enhance the humoral or cellular immune responses. These immunosomes were biologically much more active than the purified glycoprotein structure with its own viral lipids. Neutralizing antibody synthesis and resistance to virus challenge were increased after a single vaccine injection (Thibodeau et al. and Trudel et al., personal communications).

4. Adjuvants in present and future vaccines

Purified and synthetic vaccines will very often have to be enhanced since their reduction to simpler structures may eliminate the immunostimulating moieties present in their native form. In the future, host mediators such as interleukins will perhaps be used to amplify the antibody response. It can even be assumed that better performing copies of these endogenous molecules will one day be obtained by chemical synthesis.

Other immunological approaches for regulating the immune system also exist. One of the most promising is the utilization of the idiotype network. Injection of anti-idiotypic antibodies (anti-Id) can, under certain experimental conditions, induce antigen-specific helper T cells (Eichmann and Rajewsky, 1975; Julius et al., 1978; Benca et al., 1980; Miller et al., 1981; Gougeon et al., 1983). The expression of the corresponding Id is subsequently increased (Trenkner and Riblet, 1975; Cazenave, 1977; Urbain et al., 1977; Kelsoe et al., 1980). Anti-Id antibodies have also been shown to modulate the immune response to HBsAg. Injection of soluble anti-Id prior to HBsAg stimulation increased the number of anti HBs IgM-secreting spleen cells, whereas alum-precipitated anti-Id produced primarily anti-HBs IgG (Kennedy et al., 1983).

At present, experimental and clinical adjuvants are mainly mineral (aluminum hydroxide, mineral oil, saponin) or bacterial (Freund's complete adjuvant, *B. pertussis*) agents. After briefly reviewing the present status of adjuvants, we will describe in more detail muramyl peptides which are new synthetic molecules that have been recently administered in several vaccination models.

4.1. Adjuvants others than muramyl peptides

The most common adjuvants used in the laboratory are Freund's complete adjuvant (FCA) and incomplete adjuvant (FIA). The former cannot be applied for clinical or even for veterinary immunization. Mineral oil has been used for viral vaccines to prevent certain veterinary diseases such as foot-and-mouth and Newcastle disease infections (McKercher and Graves, 1977; reviewed in Edelman, 1980). Saponin has also been used extensively in such vaccines (Espinet, 1951; Dalsgaard, 1974; 1978; Dalsgaard and Jensen, 1977; Dalsgaard et al., 1977; Bomford, 1980a; 1980b). Nevertheless, both of these adjuvants, FIA and saponin, cannot be used in humans. Although the former has been assayed in clinical influenza trials (Hilleman, 1966; Davenport, 1968; Stuart-Harris, 1969) it has been abandoned since. The adjuvant which has been – and still is – the most used in humans is alum (Aprile and Wardlaw, 1966), administered sometimes with certain polyvalent vaccines containing *B. pertussis*, which is also considered to be an adjuvant. However, even these acceptable agents are not devoid of certain drawbacks. They can induce local reactions and other side effects such as an enhanced production of IgE (Nagel et al., 1977; 1979; Murphy et al., 1983). Development of new agents capable of potentiating antibody production, and hopefully devoid of side effects, must therefore be actively pursued.

4.2. Muramyl peptides

NAcMur-L-Ala-D-isoGln (known as muramyl dipeptide or MDP) represents a synthetic copy of part of the repetitive structure constitutive of most bacterial peptidoglycans which can substitute for whole mycobacteria in FCA (Ellouz et al., 1974; Kotani et al., 1975). Several hundred derivatives of MDP have now been synthesized (Azuma et al., 1979; Lefrancier and Lederer, 1981) and used to study the relationship between structure and immunostimulant activity (Chedid et al., 1978; Parant, 1979). Dissociation of one or several of the immunopharmacological properties of the original molecule was found to be possible by slight chemical modification (Chedid et al., 1978; Parant, 1979). MDP has been shown to be an adjuvant when given in saline with protein antigens (Audibert et al., 1976). It was also shown that the humoral response to an antigen given either intravenously or subcutaneously could be enhanced even when MDP was administered orally (Chedid et al., 1976). MDP was shown to increase nonspecific immunity to bacterial (Chedid et al., 1977; Parant, 1979), fungal (Matthews and Fraser-Smith, 1980) and parasitic challenge (Kierszenbaum and Ferraresi, 1979; Krahenbuhl et al., 1981). Interestingly, contrary to lipopolysaccharide, protection was obtained even after oral administration to adult or newborn mice (Parant et al., 1978). Nonspecific immunity could also be enhanced in immunocompromised hosts following cyclophosphamide therapy or splenectomy (Parant, 1979).

Lipophilic derivatives of MDP have been shown to be more active in certain systems and capable of inducing a cell-mediated response. These molecules may be very useful in the field of viral vaccines administered either in saline (Parant et al., 1980) or in metabolizable oil emulsion (Carelli et al., 1981; Yarkoni et al., 1981), and may also be included in liposomes (Siddiqui et al., 1978; Jolivet et al., 1981). Coupling MDP or its derivatives to carriers or to antigenic materials increases certain of their biological activities (Chedid et al., 1979; Reichert et al., 1980). Such procedures have often been used in synthetic vaccines.

The mechanism of action of MDP has been studied in several in vivo and in vitro systems. Its activity seems to be mainly mediated by macrophages (Leclerc and Chedid, 1982). Various immunopharmacological and neuropharmacological properties of MDP and its derivatives have been described (Chedid, 1983; Kotani and Takada, 1983; Leclerc et al., 1983; Leclerc and Chedid, 1984). For instance, early studies established that MDP is pyrogenic and displays various other side effects (Kotani et al., 1976; Masek et al., 1978; Dinarello et al., 1978; Riveau et al., 1980). It was possible by chemical modification, to eliminate these unwanted properties of MDP while retaining its immunopotentiating activity. The most promising compound is presently NAcMur-L-Ala-D-Gln-*α*-*n*-butyl-ester (also called murabutide), which is as active as MDP but devoid of any detectable side effects (Chedid et al., 1982; Lefrancier et al., 1982). This compound is currently undergoing clinical trials.

4.2.1. Enhancement of specific systemic immunity against conventional vaccines
Beneficial effects of MDP have been reported in experimental animals after administration of influenza, hepatitis and HSV vaccines. Experiments were performed using a subunit influenza vaccine which is poorly immunogenic in mice and hamsters. As reported by Audibert et al. (1977), administration of MDP with this antigen in an aqueous medium enhances the antibody response of mice. A similar adjuvant activity was also demonstrated in hamsters in which no antibodies were detectable after a single injection of this vaccine. In contrast, a single dose of influenza subunit vaccine with MDP produced a primary response and a stronger secondary response than the control (Webster et al., 1977).

High levels of specific antibodies were obtained in mice and guinea pigs following the administration of low dosages (1 μg) of hepatitis B antigen with murabutide. Moreover, even higher titers were obtained by combining murabutide with low dosages of aluminum hydroxide, without increasing the level of specific IgE (Audibert et al., 1982a). A single injection of hepatitis B vaccine with either of three 5.0 acetylated 6.0 acylated dipeptidyl furanose MDP derivatives also significantly stimulated production of circulating antibodies in mice (Durette et al., 1982).

MDP has also been shown to increase the immunogenicity of formalinized whole HSV (Thomson et al., 1983). Repeated doses of this vaccine without adjuvant provided only minimal protection against establishment of latent HSV-1 infection. The combination of whole virus vaccine with FCA produced almost complete protection as measured by primary herpetic lesions and survival. In this study, total protection from encephalitic death was also observed in groups vaccinated with whole virus mixed with MDP and administered in saline (Thomson et al., 1983).

4.2.2. Enhancement of specific systemic immunity against synthetic vaccines
Production of vaccines against certain viruses such as hepatitis B by conventional means is a difficult and costly task. It can be reasonably predicted that for this and other similar viral antigens, genetic engineering will offer a more advantageous source of antigenic material. However, especially for small peptides, synthesis may become a better choice because of lower cost and easier safety control. Before describing the production of synthetic viral immunogens it must be mentioned that anti-Id antibodies have also been used to produce substitutes for viral antigens.

Previously, Bluestone et al. (1981) had induced anti-H-2 antibodies in the absence of alloantigen exposure, by in vivo administration of anti-Id. Induction of parasitic immunity by anti-Id antibodies had also been reported (Sacks et al., 1982). These anti-Id antibodies were prepared against protective monoclonal antibodies raised against variable surface antigens (VAT) of three different clones of *Trypanosoma rhodesiense*. Administration of anti-Id to mice in the absence of antigen induced immunity against subsequent infection with trypanosomes bearing the homologous VAT (Sacks et al., 1982).

Anti-HBs IgG has also been induced without any viral exposure by administration of anti-Id antibodies to mice. These antibodies expressed an idiotype shared by anti-HBs of naturally infected humans who are considered to be immune to reinfection (Kennedy and Dreesman, 1984). These results indicate that anti-Id antibodies can induce protective immunity in the absence of a viral antigen.

A large number of studies using synthetic adjuvants and antigens have already been reported. The first successful vaccination against a bacterial pathogen using a totally synthetic antigen was reported by Goebel in 1939. An artificial antigen (the azobenzyl glycoside of cellobiuronic acid, contained in type III pneumococci) conjugated to horse serum globulin and administered to rabbits elicited specific antibodies which conferred passive protection to mice (Goebel, 1939). The following different lines of investigation have gradually led, during the last 20 years, to the development of synthetic peptide vaccines:

(*a*) Demonstration of the antigenicity of small natural fragments or synthetic peptides analogous to defined regions of the coat protein of tobacco mosaic virus (TMV) or of sperm whale myoglobin (Spitler, 1970; Atassi, 1975; Van Regenmortel et al., 1983).

(*b*) Demonstration that synthetic peptides can elicit antibodies which recognize native proteins such as collagen (Borek et al., 1969) or hen egg white lysozyme (Arnon et al., 1971; Arnon, 1980).

(*c*) Demonstration that synthetic peptides can induce antibodies which recognize and neutralize a plant virus such as TMV (Anderer, 1963) or a bacterial virus such as MS2 coliphage. Neutralizing antiviral antibodies were obtained by immunizing rabbits with a MS2 peptide attached to a synthetic carrier, multichain poly-DL-alanine-poly-L-lysine (A-L) (Langbeheim et al., 1976).

(*d*) Demonstration that synthetic peptides may induce protective antibodies against bacterial (Audibert et al., 1981; 1982b; Beachey et al., 1981; Chedid et al., 1983) and viral pathogens (Lerner et al., 1981; Dreesman et al., 1982; Bittle et al., 1982; Müller et al., 1982; Neurath et al., 1982; Prince et al., 1982).

Other synthetic antigens have also been shown to be capable of inducing immune responses against parasites (Godson et al., 1983) and hormones (Fraser, 1980; Lee et al., 1980; Carelli et al., 1982).

It is therefore possible to identify and synthesize antigenic determinants capable of eliciting antibodies that recognize the natural structure of a pathogen and are endowed with specific neutralizing activity. It is clear that, whatever their mode of preparation, most artificial peptides are poor immunogens that require the help of a carrier and of adjuvants. Indeed, the above-mentioned peptides have usually been coupled to a large carrier molecule, such as tetanus toxoid or keyhole limpet hemocyanin, and injected with FCA.

Recent investigations have demonstrated that it is possible to use synthetic adjuvants such as MDP instead of FCA for the induction of protective antibodies. In these models, MDP was used either free or conjugated to the peptide-carrier entity.

It has been shown that MDP could potentiate the production of protective antibody against diphtheria toxin when injected in saline together with a synthetic antigen coupled to a protein carrier (Audibert et al., 1982b).

More interestingly, immune responses have also been obtained using MDP conjugated to totally synthetic antigens. In earlier studies, MDP conjugated to P2-A--L, the synthetic conjugate for MS2 phage, induced titers of inactivating antibodies (Arnon et al., 1980) similar to those previously obtained with P2-A--L administered in FCA (Langbeheim et al., 1976). A totally synthetic diphtheria vaccine was also obtained by conjugating the synthetic diphtheria peptide to a synthetic carrier. This last immunogen induced protective antidiphtheria immunity when administered in a physiological medium (Audibert et al., 1982b). Immunization of mice with synthetic HBsAg peptide 122–137 associated with MDP in liposomes has also been reported (Dreesman et al., 1982).

A synthetic vaccine avoiding the use of carrier, liposomes and FCA was built by conjugating the decapeptide LH-RH (luteinizing hormone-releasing hormone) to MDP or MDP-Lys. Such a conjugate administered in saline to mice elicited immunological castration and anti-LH-RH antibodies (Carelli et al., 1982). The immunogenicity of this low molecular weight conjugate (2000 Da) was very marked: its activity was stronger than that of LH-RH coupled to proteins and emulsified in FCA.

In order to avoid the problems inherent to the use of a carrier, another approach consists of using polymerized synthetic peptides. Immunization with the respective homopolymers elicited antibodies recognizing diphtheria toxin (Audibert et al., 1982b), streptococcal M protein (Jolivet et al., 1983), foot-and-mouth disease VP-1 and HBsAg (Chedid et al., unpublished results). Except for the latter, all of these materials were able to elicit protection.

4.2.3. Enhancement of specific secretory immunity

As previously stated, oral administration of MDP and of a few other adjuvant-active derivatives has been shown to enhance the humoral response to protein antigen administered parenterally. Such derivatives are also capable of enhancing the secretory IgA response when administered with antigen by the appropriate route (Morisaki et al., 1983). Increased antibody titers, especially of the IgA isotype, in both serum and saliva were induced in rats that were given MDP and either *Streptococcus mutans* whole cells (WC) or purified cell walls (CW) by gastric intubation. Rats orally immunized with either WC or CW and MDP and challenged with virulent *S. mutans* exhibited significantly lower plaque and caries scores, as well as smaller numbers of viable organisms, than controls.

Increased salivary IgA responses in rats receiving MDP was also recently demonstrated by Taubman et al. (1983), who used a slightly modified model. He observed, in rats immunized orally with glucosyltransferase (GTF), an increased level of saliva IgA when MDP was injected orally, but not after subcutaneous injection. No circulating antibodies were detected, although subsequent infection of these primed animals with *S.mutans* bearing surface GTF triggered the production of specific humoral IgG antibodies. Both subcutaneous and oral MDP given with oral GTF before infection increased serum IgG levels.

Another derivative, 6-*O*-stearoyl-*N*-acetylmuramyl-L-alanyl-D-isoglutamine (S-MDP) has been claimed to enhance the immune response when administered orally with bovine serum albumin in liposomes (Genco et al., 1983).

Potentiation of the IgA response after oral administration of antigen with MDP could be mediated by cells present in Peyer's patches (PP) since MDP has been shown to promote their in vitro response to sheep red blood cells (Kiyono et al., 1982).

MDP is also effective when given through the respiratory system. When MDP was administered with ovalbumin by chronic aerosol, serum IgA antibody levels were increased as compared with non-MDP-treated controls administered with this soluble antigen. This IgA increase required parenteral priming with antigen and was not accompanied by an increase of circulating IgG levels (Butler et al., 1983). In other experiments, however, after the same treatment, an increase of both IgG and IgA levels could be observed in serum as well as in broncho-alveolar wash fluids (Richerson et al., 1982). In the same study it was shown that repeated and prolonged inhalation of antigen and MDP could produce a chronic granulomatous pneumonitis (Richerson et al., 1982). Although vaccination would require a much more limited and therefore safer exposure, it seems clear that such studies will produce a renewed interest in the administration of vaccines by the oral route. This is particularly true for the prevention of viral infections which are under the control of IgA immunity. It can be predicted that such vaccines will require delivery systems such as liposomes or attachment to a cholera toxin fragment for better targeting of the immunogen to the gut-associated lymphoreticular tissue cells. Recent studies have shown that utilization of the oral or intranasal route markedly reinforces the capacity of muramyl peptides to enhance the natural resistance against viral infections.

4.2.4. Enhancement of nonspecific immunity

Resistance to viral infection can be achieved by the enhancement of specific and nonspecific host responses. Nonspecific immunity can be increased by a considerable number of factors such as interferon, and also by mediators like interleukin 1, which triggers a great number of defence mechanisms such as fever (reviewed in Dinarello, 1984). The role of temperature in resistance to infections is well documented (Lwoff, 1959; Kluger et al., 1975; Kluger, 1979).

As stated above, muramyl peptides have been shown to increase the host resistance against bacterial, fungal and parasitic infections. More recently, similar results have been reported in relation to viral infections.

Nishimura et al. (1982) and Ikeda et al. (1983) have demonstrated that MDP and some lipophilic derivatives could protect mice against localized lesions caused by vaccinia or herpes simplex viruses. This protection could be obtained by parenteral or oral injection of MDP, MDP-Lys or B30-MDP. The mechanisms underlying this antiviral effect are still unclear but are not related to direct interferon production. However, α-interferon induction either by lipopolysaccharide or by Poly I:C was significantly enhanced in mice previously given MDP or lipophilic derivatives (Kotani et al., 1984). None of these muramyl peptides had any direct in vitro antiviral activity but seemed to increase adherent cell antiviral activity (Kotani et al., 1984). In these experiments, in vivo activity was obtained only by administering glycopeptides *before* the infection.

Even more interesting were the results recently presented by Dietrich et al. (1983) on the prophylactic and *therapeutic* antiviral effect of another MDP analog. This derivative, which is known as MTP-PE (*N*-acetylmuramyl-L-alanyl-D-isoglutaminyl-L-alanine (1′-2′-dipalmitoyl-*sn*-glycerol-3′-hydroxyphosphoryloxyethylamide)), increased the resistance of mice and guinea pigs to a broad spectrum of RNA or DNA viruses. It was effective even when administered in very small amounts by intranasal, oral, or intravaginal routes (Lukas et al., 1983).

Similar protection against influenza infection was observed by oral administration of nonlipophilic muramyl peptides after an intranasal challenge (unpublished results). Under the same conditions very little, if any, activity was found when mice were treated by the parenteral route. All these results indicate the importance of stimulating mucosal defence mechanisms in eliciting nonspecific immunity. Since muramyl peptides were also shown to be an active adjuvant with viral vaccines, it can be expected that such synthetic activators of both specific and nonspecific immunity will lead to improved antiviral vaccination.

Indeed, such compounds administered with a viral vaccine should induce, in addition to the protection conferred by specific antibodies, a broader spectrum of antiviral immunity and, therefore, a better defence against antigenic variation. This could be of major importance for viral diseases such as influenza. In a field where chemotherapy is very limited, such a 'therapeutic vaccination' would also be of great value against infections such as rabies.

5. References

Akerfeld, S., Lofberg, E. and Fuchs, G. (1973) *Biochem. Pharmacol.* 22, 2911–2917.

Alford, R.H. (1969) *J. Immunol.* 103, 20–24.

Almeida, J.D., Brand, C.M., Edwards, D.C. and Heath, T.D. (1975) *Lancet* 7941, 899–901.

Anderer, F.A. (1963) *Biochim. Biophys. Acta* 71, 246–248.

Aprile M.A. and Wardlaw, A.C. (1966) *Can J. Public Health* 57, 343–360.

Arnon, R. (1980) *Ann. Rev. Microbiol.* 34, 593–618.

Arnon, R., Maron, E., Sela, M. and Anfinsen, C.B. (1971) *Proc. Natl. Acad. Sci. U.S.A.* 68, 1450–1455.

Arnon, R., Sela, M., Parant, M. and Chedid, L. (1980) *Proc. Natl. Acad. Sci. U.S.A.* 77, 6769–6772.

Atassi, M.Z. (1975) *Immunochemistry* 12, 423–438.

Audibert, F., Chedid, L., Lefrancier, P. and Choay, J. (1976) *Cell. Immunol.* 21, 243–249.

Audibert, F., Chedid, L. and Hannoun, C. (1977) *C.R. Acad. Sci. Paris Ser. D* 285, 467–470.

Audibert, F., Jolivet, M., Chedid, L., Alouf, J.E., Boquet, P., Rivaille, P. and Siffert, O. (1981) *Nature* 289, 593–594.

Audibert, F., Przewlocki, G., Lefrancier, P., Choay, J. and Chedid, L. (1982a) *C.R. Acad. Sci. Paris Ser.* III 295, 611–614.

Audibert, F., Jolivet, M., Chedid, L., Arnon, R. and Sela, M. (1982b) *Proc. Natl. Acad. Sci. U.S.A.* 79, 5042–5046.

Azuma, I., Yamawaki, M., Uemiya, M., Saiki, I., Panio, Y., Kobayashi, S., Fukuda, T., Imada, I. and Yamamura, Y. (1979) *Gann* 76, 847–848.

Beachey, E.H., Seyer, J.M., Dale, J.B., Simpson, W.A. and Kang, A.H. (1981) *Nature* 292, 457–459.

Benca, R.J., Quintas, J., Kearney, J.F., Flood, P.M. and Schreiber, H. (1980) *Mol. Immunol.* 17, 823–831.

Bittle, J.L., Houghten, R.A. Alexander, H., Shinnick, T.M., Sutcliffe, J.G., Lerner, R.A., Rowlands, D.J. and Brown, F. (1982) *Nature* 298, 30–33.

Bluestone, J.A., Sharrow, S.O., Epstein, S.L., Ozato, K. and Sachs, D.H. (1981) *Nature* 291, 233–235.

Bomford, R. (1980a) *Clin. Exp. Immunol.* 39, 426–434.

Bomford, R. (1980b) *Clin. Exp. Immunol.* 39, 435–441.

Borek, F., Kurtz, J. and Sela, M. (1969) *Biochim. Biophys. Acta* 188, 314–323.

Butler, J.E., Richerson, H.B., Swanson, P.A., Kopp, W.C. and Suelzer, M.T. (1983) In *The Secretory Immune System* (J.R. McGhee, and J. Mestecky, Eds.) Vol. 409. Annals of the New York Academy of Sciences, New York, pp. 669–687.

Carelli, C., Audibert, F. and Chedid, L. (1981) *Infect. Immun.* 33, 312–314.

Carelli, C., Audibert, F., Chedid, L. and Gaillard, J. (1982) *Proc. Natl. Acad. Sci. U.S.A.* 79, 5392–5395.

Cazenave, P.A. (1977) *Proc. Natl. Acad. Sci. U.S.A.* 74, 5122–5125.

Chedid, L. (1983) *Microbiol. Immunol.* 27, 723–732.

Chedid, L., Audibert, F., Lefrancier, P., Choay, J. and Lederer, E. (1976) *Proc. Natl. Acad. Sci. U.S.A.* 73, 2472–2475.

Chedid, L., Parant, M., Parant, F., Lefrancier, P., Choay, J. and Lederer, E. (1977) *Proc. Natl. Acad. Sci. U.S.A.* 74, 2089–2093.

Chedid, L., Audibert, F. and Parant, M. (1978) *Pathol. Biol.* 26, 527–530.

Chedid, L., Parant, M., Parant, F., Audibert, F., Lefrancier, P., Choay, J. and Sela, M. (1979) *Proc. Natl. Acad. Sci. U.S.A.* 76, 6557–6561.

Chedid, L.A., Parant, M.A., Audibert, F.M., Riveau, G.J., Parant, F.J., Lederer, E., Choay, J.P. and Lefrancier, P.L. (1982) *Infect. Immun.* 35, 417–424.

Chedid, L., Jolivet, M., Audibert, F., Przewlocki, G., Beachey, E.H., Gras-Masse, H. and Tartar, A. (1983) *Biochem. Biophys. Res. Commun.* 117, 908–916.

Couch, R.B. and Kasel, J.A. (1983) *Annu. Rev. Microbiol.* 37, 529–549.

Cowan, K.M. (1973) *Adv. Immunol.* 17, 195–253.

Dalsgaard, K. (1974) *Arch. Gesamte Virusforsch.* 44, 243–254.

Dalsgaard, K. (1978) *Bull. Off. Int. Epiz.* 89, 11–12.

Dalsgaard, K. and Jensen, M.H. (1977) *Acta Vet. Scand.* 18, 367–373.

Dalsgaard, K., Jensen, M.H. and Sorensen, K.J. (1977) *Acta Vet. Scand.* 18, 349–360.

Davenport, F.M. (1968) *Ann. Allergy* 26, 288–292.

Davies, W.B., Taylor, J.A. and Oakes, J.E. (1979) *J. Infect. Dis.* 140, 534–539.

Dietrich, F.M., Lukas, B. and Schmidt-Ruppin, K.H. (1983) Abstract. 13th International Congress of Chemotherapy, August 28–September 2, Vienna.

Dinarello, C.A. (1984) *Rev. Infect. Dis.* 6, 51–95.

Dinarello, C.A., Elin, R.J., Chedid, L. and Wolff, S.M. (1978) *J. Infect. Dis.* 138, 760–767.

Dreesman, G.R., Sanchez, Y., Ionescu-Matiu, I., Sparrow, J.T., Six, H.R., Peterson, D.L., Hollinger, F.B. and Melnick, J.L. (1982) *Nature* 295, 158–160.

Durette, P.L., Dorn, C.P. Jr., Friedman, A. and Schlabach, A. (1982) *J. Med. Chem.* 25, 1028–1033.

Edelman, R. (1980) *Rev. Infect. Dis.* 2, 370–383.

Eichmann, K. and Rajewsky, K. (1975) *Eur. J. Immunol.* 5, 661–666.

Ellouz, F., Adam, A., Ciorbaru, R. and Lederer, E. (1974) *Biochem. Biophys. Res. Commun.* 59, 1317–1325.

Espinet, R.G. (1951) *Gac. Vet.* 13, 268–274.

Fraser, H.M. (1980) In *Immunological Aspects of Reproduction and Fertility Control* (J.P. Hearn, Ed.), MTP Press Ltd., Falcon House, Lancaster, U.K. pp. 143–171.

Genco, R.J., Linzer, R. and Evans, R.T. (1983) In *The Secretory Immune System* (J.R. McGhee and J. Mestecky, Eds.) Vol. 409. Annals of the New York Academy of Sciences, New York. pp. 650–668.

Godson, G.N., Ellis, J., Svec, P., Schlesinger, D.H. and Nussenzweig, V. (1983) *Nature* 305, 29–33.

Goebel, W.F. (1939) *J. Exp. Med.* 69, 53–65.

Gougeon, M.L., Bismuth, G., Leclerq, L., Somme, G. and Thèze, J. (1983) *Cell. Immunol.* 75, 103–110.

Gregoriadis, G. (1976a) *N. Engl. J. Med.* 295, 704–710.

Gregoriadis, G. (1976b) *N. Engl. J. Med.* 295, 765–770.

Hilleman, M.R. (1966) *Prog. Med. Virol.* 8, 131–182.

Ikeda, S., Negishi, T., Urushibara, T. and Nishimura, C. (1983) Proc. 103rd Annual Meeting of the Pharmaceutical Society of Japan, Tokyo. p. 354.

Jolivet, M., Sache, E. and Audibert, F. (1981) *Immunol. Commun.* 10, 511–522.

Jolivet, M., Audibert, F., Beachey, E.H., Tartar, A., Gras-Masse, H. and Chedid, L. (1983) *Biochem. Biophys. Res. Commun.* 117, 359–366.

Julius, M.H., Cosenza, H. and Augustin, A.A. (1978) *Eur. J. Immunol.* 8, 484–491.

Kapoor, A.K., Nash, A.A., Wildy, P., Phelan, J., McLean, C.S. and Field, H.J. (1982) *J. Gen. Virol.* 60, 225–233.

Kelsoe, G., Reth, M. and Rajewsky, K. (1980) *Immunol. Rev.* 52, 75–88.

Kennedy, R.C., Adler-Storthz, K., Henkel, R.D., Sanchez, Y., Melnick, J.L. and Dreesman, G.R. (1983) *Science* 221, 853–855.

Kennedy, R.C. and Dreesman, G.R. (1984) *J. Exp. Med.* 159, 655–665.

Kierszenbaum, F. and Ferraresi, R.W. (1979) *Infect. Immun.* 25, 273–278.

Kiyono, H., McGhee, J.R., Kearney, J.F. and Michalek, S.M. (1982) *Scand. J. Immunol.* 15, 329–339.

Kluger, M.J. (1979) *Fever, its Biology, Evolution and Function.* Princeton University Press, Princeton NJ. pp. 1–195.

Kluger, M.J., Ringler, D.H. and Anver, M.R. (1975) *Science* 188, 166–168.

Kotani, S. and Takada, H. (1983) *Yakagaku Zasshi* 103, 1–27.

Kotani, S., Watanabe, Y., Shimono, T., Narita, T., Kato, K., Stewart-Tull, D.E.S., Kinoshita, F., Yokogawa, K., Kawata, S., Shiba, T., Kusumoto, S. and Tarumi, Y. (1975) *Z. Immun. Forsch.* 149, 302–319.

Kotani, S., Watanabe, Y., Shimono, T., Harada, K., Shiba, T., Kusumoto, S., Yokogawa, K. and Taniguchi, M. (1976) *Biken J.* 19, 9–13.

Kotani, S., Takada, H., Tsujimoto, M., Ogawa, T., Mori, Y., Nishimura, C., Ikeda, S., Koga, T., Iribe, H., Kohashi, O., Tanaka, A., Nagao, S., Ogawa, H., Ozawa, S., McGhee, J.R. Michalek, S.M., Hamada, S., Morisaki, I., Kawata, S., Shiba, T. and Kusumoto, S. (1983) In *Immunology 83, Progress in Immunology V* (Y. Yamamura and T. Tada, Eds.) Academic Press, Tokyo. pp. 1359–1377.

Krahenbuhl, J.L., Sharma, S.D., Ferraresi, R.W. and Remington, J.S. (1981) *Infect. Immun.* 31, 716–722.

Langbeheim, H., Arnon, R. and Sela, M. (1976) *Proc. Natl. Acad. Sci. U.S.A.* 73, 4636–4640.

Leclerc, C. and Chedid, L. (1982) *Lymphokines* 7, 1–21.

Leclerc, C. and Chedid, L. (1984) *Int. J. Tissue Reactions* 6, 213–218.

Leclerc, C., Morin, A. and Chedid, L. (1983) In *Recent Advances in Clinical Immunology* (R.A. Thompson and N.R. Rose, Eds.) Vol. 3. Churchill Livingstone, Edinburgh. pp. 187–204.

Lee, A.C.J., Powell, J.E., Tregear, G.W., Niall, H.D. and Stevens, V.C. (1980) *Mol. Immunol.* 17, 749–756.

Lefrancier, P. and Lederer, E. (1981) *Prog. Chem. Organic Natural Products* 40, 1–47.

Lefrancier, P., Derrien, M., Jamet, X., Choay, J., Lederer, E., Audibert, F., Parant, M., Parant, F. and Chedid, L. (1982) *J. Med. Chem.* 25, 87–90.

Lerner, R.A., Green, N., Alexander, H., Liu, F.T., Sutcliffe, J.G. and Shinnick, T.M. (1981) *Proc. Natl. Acad. Sci. U.S.A.* 78, 3403–3407.

Lukas, B., Schmidt-Ruppin, K.H. and Dietrich, F.M. (1983) *Immunomodulation by Chemically Defined Adjuvants.* Proc. Intern. Symposium, Sapporo, Japan. p. 42.

Lwoff, A. (1959) *Bact. Rev.* 23, 109–124.

Mandel, B. (1978) *Adv. Virus Res.* 23, 205–268.

Manesis, E.K., Cameron, C.H. and Gregoriadis, G. (1978) *Biochem. Soc. Trans.* 6, 925–928.

Manesis, E.K., Cameron, C.H. and Gregoriadis, G. (1979) *FEBS Lett.* 102, 107–111.

Matthews, T.R. and Fraser-Smith, E.B. (1980) In *Current Chemotherapy of Infectious Diseases*, Proc. 11th Int. Congr. Chemoth. 2, 1734–1735.

Masek, K., Kadlecova, O. and Petrovicky, P. (1978) *Toxicon* 1, Suppl. pp. 991–1003.

McDermott, M.R. and Bienenstock, J. (1979) *J. Immunol.* 122, 1892–1898.

McGhee, J.R. Mestecky, J. and Babb, J.L. (1978a) *In Secretory Immunity and Infection.* Vol. 107. Plenum Press, New York. pp. 1–905.

McGhee, J.R., Mestecky, J., Arnold, R.R., Michalek, S.H., Prince, J. and Babb, J.L. (1978b) *Adv. Exp. Med. Biol.* 107, 177–184.

McKercher, P.D. and Graves, J.H. (1977) *Dev. Biol. Stand.* 35, 107–112.

McMichael, A.J., Askonas, B.A., Webster, R.G. and Laver, W.G. (1982) *Immunol. Today* 3, 256–260.

McNabb, C.P. and Tomasi, T.B. (1981) *Annu. Rev. Microbiol.* 35, 477–496.

Miller, G.G., Nadler, P.I., Asano, Y., Hodes, R.J. and Sachs, D.H. (1981) *J. Exp. Med.* 154, 24–34.

Morisaki, I., Michalek, S.M., Harmon, C.C., Torii, M., Hamada, S. and McGhee, J.R. (1983) *Infect. Immun.* 40, 577–591.

Müller, G.M., Shapira, M. and Arnon, R. (1982) *Proc. Natl. Acad. Sci. U.S.A.* 79, 569–573.

Murphy, M.D., Rasnack, J., Dickson, H.D., Dietch, M. and Brunell, P.A. (1983) *Pediatrics* 71, 200–205.

Nagel, J.E., Svec, D., Waters, T. and Fireman, P. (1977) *J. Immunol.* 118, 334–341.

Nagel, J.E., White, C., Lin, M.S. and Fireman, P. (1979) *J. Allergy Clin. Immunol.* 63, 308–314.

Neurath, A.R. and Rubin, B.A. (1971) In *Viral Structural Components as Immunogens of Prophylactic Value. Monographs in Virology* (J.L. Melnick, Ed.) Vol. 4. Karger, Basel.

Neurath, A.R., Kent, S.B.H. and Strick, N. (1982) *Proc. Natl. Acad. Sci. U.S.A.* 79, 7871–7875.

Nishimura, C., Ikeda, S., Negishi, T., Urushibara, T. and Okamoto, Y. (1982) *Proc. 102nd Annual Meeting of Pharmaceutical of Japan*, Osaka. p. 290.

Norrild, B., Shore, S.L. and Nahmias, A.J. (1979) *J. Virol.* 32, 741–748.

Notkins, A.L. (1974) *Cell. Immunol.* 11, 478–483.

Oakes, J.E. and Lausch, R.N. (1981) *Infect. Immun.* 33, 109–114.

Ogra, P.L. and Karzon, D.T. (1969a) *J. Immunol.* 102, 15–23.

Ogra, P.L. and Karzon, D.T. (1969b) *J. Immunol.* 102, 1423–1430.

Ogra, P.L. and Karzon, D.T. (1971) *Prog. Med. Virol.* 13, 156–193.

Panicali, D. and Paoletti, E. (1982) *Proc. Natl. Acad. Sci. U.S.A.* 79, 4927–4931.

Paoletti, E., Lipinskas, B.R., Samsonoff, C., Merser, C. and Panicali, D. (1984) *Proc. Natl. Acad. Sci. U.S.A.* 81, 193–198.

Parant, M. (1979) *Springer Semin. Immunopathol.* 2, 101–119.

Parant, M., Parant, F. and Chedid, L. (1978) *Proc. Natl. Acad. Sci. U.S.A.* 75, 3395–3399.

Parant, M.A., Audibert, F.M., Chedid, L.A., Level, M.R., Lefrancier, P.L., Choay, J.P. and Lederer, E. (1980) *Infect. Immun.* 27, 826–831.

Prince, A.M., Ikram, H. and Hopp, T.P. (1982) *Proc. Natl. Acad. Sci. U.S.A.* 79, 579–582.

Rector, J.T., Lausch, R.N. and Oakes, J.E. (1982) *Infect. Immun.* 38, 168–174.

Reichert, C.M., Carelli, C., Jolivet, M., Audibert, F., Lefrancier, P. and Chedid, L. (1980) *Mol. Immunol.* 17, 357–363.

Ritova, V.V. and Zhdanov, V.M., (1962) *Clin. Med.* 40, 117–139.

Richerson, H.B., Suelzer, M.T., Swanson, P.A., Butler, J.E., Kopp, W.C. and Rose, E.F. (1982) *Am. J. Pathol.* 106, 409–420.

Riveau, G., Masek, K., Parant, M. and Chedid, L. (1980) *J. Exp. Med.* 152, 869–877.

Sacks, D.L., Esser, K.M. and Sher, A. (1982) *J. Exp. Med.* 155, 1108–1119.

Sakai, F., Gerlier, D. and Doré, J.F. (1980) *Br. J. Cancer* 41, 227–235.

Sanchez, Y., Ionescu-Matiu, I., Dreesman, G.R., Kramp, W., Six, H.R., Hollinger, F.B. and Melnick, J.L. (1980) *Infect. Immun.* 30, 728–733.

Sanchez, Y., Ionescu-Matiu, I., Sparrow, J.T., Melnick, J.L. and Dreesman, G.R. (1982) *Intervirology* 18, 209–213.

Shvartsman, Y.A. and Zykov, M.P. (1976) *Adv. Immunol.* 22, 291–330.

Siddiqui, W.A., Taylor, D.W., Kan, S.C., Kramer, K., Richmond-Crum, S.M., Kotani, S., Shiba, T. and Kusumoto, S. (1978) *Science* 201, 1235–1237.

Skelly, J., Howard, C.R. and Zuckerman, A.J. (1981) *Nature* 290, 51–54.

Smith, G.L., Mackett, M. and Moss, B. (1983a) *Nature* 302, 490–495.

Smith, G.L., Murphy, B.R. and Moss, B. (1983b) *Proc. Natl. Acad. Sci. U.S.A.* 80, 7155–7159.

Spitler, L., Benjamini, E., Young, J.D., Kaplan, H. and Fudenberg, H.H. (1970) *J. Exp. Med.* 131, 133–148.

Stuart-Harris, C.H. (1969) *Bull. W.H.O.* 41, 617–621.

Taubman, M.A., Ebersde, J.L., Smith, D.J. and Stack, W. (1983) In *The Secretory Immune System* (J.R. McGhee and J. Mestecky, Eds.) Vol. 409. Annals of the New York Academy of Sciences, New York. pp. 637–649.

Thibodeau, L., Naud, P. and Boudreault, A. (1981) *Genetic Variation among Influenza Viruses*. Academic Press, New York. pp. 587–597.

Thomson, T.A., Hilfenhaus, J., Moser, H. and Morahan, P.S. (1983) *Infect. Immun.* 41, 556–562.

Trenkner, E. and Riblet, R. (1975) *J. Exp. Med.* 142, 1121–1132.

Tyrrell, D.A., Heath, T.D., Colley, C.M. and Ryman, B.R. (1976) *Biochim. Biophys. Acta* 457, 259–302.

Urbain, J., Wikler, M., Franssen, J.D. and C. Collignon (1977) *Proc. Natl. Acad. Sci. U.S.A.* 74, 5126–5130.

Van Regenmortel, M.H.V., Altschuh, D. and Briand, J.P. (1983) In *Protein Conformation as an Immunological Signal* (F. Celada, V.N. Schumaker and E.F. Sercarz, Eds.) Plenum Press, New York. pp. 181–190.

Waksman, B.H. (1983) In *The Secretory Immune System* (J.R. McGhee and J. Mestecky, Eds.) Vol. 409. Annals of the New York Academy of Sciences, New York. pp. 22–24.

Waldman, R.H., Kasel, J.A., Fulk, R.V., Togo, Y., Hornick, R.B., Heiner, G.C. and Dawkins, A.T. (1968) *Nature* 218, 594–595.

Webster, R.G., Glezen, W.P., Hannoun, C. and Laver, W.G. (1977) *J. Immunol.* 119, 2073–2077.

Yarkoni, E., Lederer, E. and Rapp, H.J. (1981) *Infect. Immun.* 32, 273–276.

Eds. M.H.V. van Regenmortel and A.R. Neurath
Immunochemistry of viruses. The basis for serodiagnosis and vaccines.

Monoclonal antibodies to viruses 3

RICHARD J. MASSEY and GERALD SCHOCHETMAN

Igen Inc., 1530 East Jefferson Street, Rockville, MD 20852, U.S.A.

Antibodies directed against viruses are important reagents for studying viruses and viral pathogenesis. Much of the early work on characterization of viruses employed polyspecific or polyvalent antisera composed of heterogeneous populations of antibodies with varying affinities. Compounding these difficulties is the fact that viruses are antigenically complex structures that are composed of different proteins, and that individual proteins consist of multiple antigenic determinants (epitopes). These limitations can be largely overcome by using groups of homogeneous antibody populations each capable of reacting with a single epitope. Monoclonal antibodies produced by cells derived from single cell clones possess these properties and represent the highest degree of monospecificity available today.

There have been numerous reports on the use of monoclonal antibodies to study a wide variety of viruses. The aim of this review is to provide an overview on the uses of monoclonal antibodies to viruses, rather than a detailed literature survey.

1. Antigenic relationships among viruses

The ability of monoclonal antibodies to discriminate antigenic differences, even at the level of single amino acid differences, makes them ideal reagents for investigating antigenic relationships among viruses. Antigenic polymorphism for virus-coded proteins has been described for most families of viruses. Sets of antigenic determinants which are shared by viruses belonging to the same family and others which are unique to a particular isolate have been identified. Additionally, certain classes of determinants may correlate with a particular biological property of a virus on which they are found. This aspect of monoclonal antibodies can be illustrated by recent studies with several viruses including rabies, influenza, measles and mouse mammary tumor virus (MMTV, a retrovirus which induces mammary adenocarcinomas in mice).

For MMTV, antigenic polymorphism has been shown with heterologous antiserum for the two envelope glycoproteins (Arthur et al., 1979; Marcus et al., 1979; Massey et al., 1980a) and a core protein (Teramoto and Schlom, 1978). The polyspecific antibody probes used in those studies were unable to clearly distinguish any one MMTV from other MMTVs. However, using monoclonal antibodies to one of the envelope glycoproteins three patterns of reactivity could be observed: (*a*) type-specific, i.e. unique for a single strain of MMTV; (*b*) class specific, i.e. reactive for determinants found only on a subset of MMTV strains; and (*c*) group-specific, i.e. reactive with determinants shared by all strains of MMTV (Massey et al., 1980b). The biological significance of this data relates to the observation that mice develop type-specific neutralizing and cytotoxic antibodies and group-specific precipitating antibodies (Schochetman et al., 1979; Massey et al., 1980a).

In the case of rabies virus, antigenic differences could not be detected between field and vaccine strains using polyspecific antisera. In contrast, antigenic differences could be demonstrated with monoclonal antibodies (Wiktor and Koprowski, 1978; Flammand et al. 1980). This could explain why rabies vaccination occasionally fails to protect against infection by field strains.

In the case of influenza virus, the use of monoclonal antibodies specific for the HA molecule of influenza virus A/Russia/1/77 made it possible to demonstrate antigenic differences among virus strains isolated during the same influenza epidemic (Webster et al., 1979).

Only one serotype of measles virus has been identified using polyspecific antisera (Kingsbury et al., 1978). The antigenic stability of this virus is illustrated by epidemiological observations showing that a single infection gives lifelong immunity. Monoclonal antibodies were made against five structural proteins of measles virus to determine the degree of antigenic variation within these proteins among nine strains of virus (Sheshberadaran et al., 1983). This study confirmed the monotypic nature of measles by demonstrating the occurrence of only limited variations in the epitopes on the hemagglutinin protein. Three variations were discovered in a comparative analysis of 11 monoclonal antibodies consisting of at least 9 different binding groups and 9 virus strains. Similar to these results for the hemagglutinin, the fusion protein also showed marked stability. These results suggested that the monotypic nature of the virus does not result from lack of genetic flexibility but instead suggests that variations in the hemagglutinin are limited by functional constraints. In contrast to the antigenic stability of the hemagglutinin, there was considerable variation in the antigenicity of the matrix protein. This finding emphasizes that variations in the antigenic nature of virus components occur even if these components are not subjected to any immune selection.

2. Topographical analysis and virus neutralization

Based on studies with polyvalent antisera, different mechanisms were described for antibody-mediated virus neutralization (Daniels, 1975; Cooper, 1979; Mandel 1978; 1979) and these were grouped into two general schemes. The first set of mechanisms involved a physical reduction in the number of virus particles either by complement-mediated viral lysis or by cross-linking virus particles which results in virus aggregation. The second set involved binding of antibody to viral antigens which affected one or more of the steps necessary in establishing a productive infection. Although the concept of a 'critical site' or target for neutralizing antibodies was developed (Dulbecco et al., 1956; Daniels, 1975), mechanisms of virus neutralization associated with this second pathway were not completely understood because viruses are composed of multiple epitopes and the polyvalent antisera contained antibodies with multiple specificities and binding affinities. The use of monoclonal antibodies simplified the study of the complex antigenic structure of viruses and enabled the determination of specific target sites and mechanisms for virus neutralization. In one example, monoclonal antibodies to epitopes on the major external viral glycoprotein (gp52) of the mouse mammary tumor virus (MMTV) were used in neutralization assays and for topographical analysis of antigenic determinants (Massey and Schochetman, 1981a). Based on the data presented in that study a 'critical site' or target for neutralizing antibody on gp52 was defined and a mechanism responsible for neutralization was identified. Monoclonal antibodies to MMTV gp52 showed that group-, class- and type-specific epitopes could be targets for virus neutralization. Competition data using a radioiodinated type-specific neutralizing antibody revealed that all monoclonal antibodies that had high avidities and neutralized virus infectivity also blocked the binding of that monoclonal antibody to the virus. In addition, a topographically distinct site from the type-specific determinant was found using other monoclonal antibodies. This site is defined by the observation that these monoclonal antibodies did not compete for binding of the type-specific neutralizing antibody and their epitopes did not function as targets for neutralizing antibodies. Therefore, two topographically distinct sites could be identified on an individual viral protein; one site functions as a target for neutralization and the second site binds antibody but does not affect neutralization.

The existence of the second antibody binding site demonstrated that binding of an antibody to the target antigen (gp52) was not sufficient to cause virus neutralization. This observation is not unique to MMTV and is a general phenomenon for many viruses including rabies virus (Flammand et al, 1980), reovirus (Spriggs et al., 1983), Venezuelan encephalomyelitis virus (Mathews and Roerhrig, 1982), tick-borne encephalitis virus, a flavivirus (Heinz et al., 1983), and paramyxoviruses (Yewdell and Gerhard, 1982; Nishikawa et al., 1983).

The ability of one site on a viral protein to function as a target for neutralization while another site does not derives from a specific mechanism(s) of neutralization different from simple virus aggregation. The observation (Massey and Schochetman, 1981a) that MMTV bound to cell surface receptors was no longer susceptible to neutralizing monoclonal antibody indicated that neutralization affected virus prior to its binding to cell surface receptors. This result was consistent with antibody preventing virus adsorption to cells. Therefore, neutralization could be mediated by antibodies binding to epitopes adjacent to those determinants which function in virus binding to cell surface receptors. This would lead to a steric hindrance and an inability of the virus to bind to the cell. It indicated that a restricted site on the MMTV gp52 was involved in cell receptor binding. This premise was supported by the previous demonstration of cell receptor specificity for individual MMTVs (Altrock et al., 1981). In this study, MMTV cell receptors on mouse and rat cells were shown to be related and were able to distinguish various strains of MMTVs. The mechanism of antibody-mediated steric hindrance involved in virus neutralization is analogous to the inhibition of enzyme-substrate interactions (Cinader et al., 1963). It was reported that the inhibitory effect on ribonuclease by a specific antibody to the enzyme depended on the molecular weight of both the substrate and the product. It was therefore concluded that the antibody-mediated inhibition was due to steric hindrance.

Other mechanisms of virus neutralization may occur, such as blocking of penetration or viral uncoating, which could be mediated by other types or combinations of monoclonal antibodies. Evidence for viral neutralization by blocking of viral penetration of a cell was derived from the observation that a monoclonal antibody to the transmembrane protein gp36 of MMTV also neutralized infectivity (Massey and Schochetman, unpublished results). Amino acid sequence analysis of gp36 demonstrated the highly hydrophobic nature of this protein (Arthur et al., 1982; Redmond and Dickson, 1983) and its analogy to fusion proteins in other viruses. Furthermore, MMTV can fuse cell membranes at low pH (Redmond et al., 1984). This information, together with the fact that anti-gp36 monoclonal antibody neutralizes infectivity but does not bind to cell-free MMTV, suggests that after MMTV binds to a cell surface receptor, the gp36 becomes accessible to anti-gp36 antibodies and that neutralization results from inhibition of viral and cell membrane fusion. This is an important step in infection of cells by enveloped viruses.

The demonstration of a topographically unique site involved in virus neutralization may define a general mechanism for virus neutralization and could also be used to explain how other viruses could escape inactivation by neutralizing antibodies. Certain viruses, such as the influenza A viruses, have the capacity for antigenic change leading to the emergence of new subtypes which are no longer neutralized by antibodies against the original strain. Major antigenic changes which occur in the hemagglutinin glycoprotein result in pandemics every 10 to 15 years (Kilbourne, 1975; Webster and Laver, 1975). During interpandemic periods, minor antigenic changes occur and these are believed to result from immune selection of spontaneously occurring antigenic variants by host immunity (Dowdle et al., 1974; Kilbourne, 1975; Webster and Laver, 1975). Based on studies with monoclonal antibodies to viruses, one would predict that only changes in those epitopes adjacent to the receptor binding site would affect antibody-mediated virus neutralization. In support of this proposal, a recent study demonstrated that only changes in a restricted set of epitopes on the hemagglutinin glycoprotein produced an epidemiologically significant variant which was incapable of being neutralized by a heterologous ferret serum directed against the parental strain (Webster and Laver, 1980). These changes would be expected to be topographically limited to those epitopes adjacent to the receptor binding site and must occur in each of the antigenic determinants that had elicited a neutralizing antibody response. Those variants, which underwent an antigenic shift so that all pre-existing targets for neutralization are altered, would have the potential for causing a pandemic.

In addition to isolating variants exhibiting a major antigenic shift, Webster and Laver (1980) isolated variants which were no longer neutralized by some monoclonal antibodies, but were still neutralized by the heterologous ferret sera. One would predict that for this second type of variant only a subset of epitopes adjacent to the receptor binding site had changed and this is what has been termed antigenic drift. These variants could potentially cause epidemics because the virus would again be infectious for individuals who had previously developed an antibody response only to the subset of epitopes which had changed.

It is apparent that the use of monoclonal antibodies for topographical analysis of epitopes on viral proteins represents an important tool for identifying critical sites involved in virus neutralization. More important than the identification of epitopes which undergo a high frequency of change would be epitopes which have a low frequency of change or would be highly conserved. An example of a highly conserved site may be the receptor binding site on some viruses (Fine et al., 1980). If this site is conserved among all epidemiologically significant variants, a virus which elicits a strong immunogenic response to this epitope would be an ideal candidate for a viral vaccine.

3. Inhibition of virus neutralization

The interactions of many viruses with neutralizing antibodies usually result in a small fraction of the viral population escaping neutralization. This observation is seen even in the presence of excess antibody (Andrewes and Elford, 1933; Burnet et al., 1937; Dulbecco et al., 1956; Majer, 1972). The biological significance of this phenomenon relates to the observations that infectious viruses can exist in animals in the presence of neutralizing antibody (Toolan, 1965; Gudnadottir and Palsson, 1965; Notkins et al., 1966; Porter and Larsen, 1967) and from its implications for chronic viral infections in animals and man (Oldstone, 1975). Until recently, the mechanisms for the generation of the nonneutralized fraction had not been completely analyzed. However, one of several mechanisms suggested by Dulbecco et al. (1956) was that antiserum contained uncharacterized molecules, presumably antibodies, which combined with viruses and inhibited the attachment of neutralizing antibody without inhibiting virus infectivity. This observation was supported by the demonstration that the nonneutralized fraction of virus could be neutralized by the addition of a secondary antibody (Mandel, 1958). This implied that the surviving virus had reacted with primary antibody but had remained infectious. To understand how one antibody can bind virus and not neutralize it but can interfere with the binding of a second antibody that does neutralize, requires knowledge both of the topography or spatial relationships of the combining sites for each antibody and the mechanism of antibody-mediated virus neutralization.

Monoclonal antibodies are ideal probes to study the topography of viral epitopes and the function of the epitopes as targets for neutralization. Based on topographical analysis of MMTV gp52 it was postulated that the neutralizing epitope was probably not the receptor binding site but was adjacent to it (Massey and Schochetman, 1981b). It was therefore predicted that epitopes existed within the neutralizing epitope domain but not within the virus receptor binding domain. It was postulated in that study that monoclonal antibodies to these hypothesized epitopes could compete for the binding of neutralizing monoclonal antibodies, but would themselves not neutralize virus infectivity. A consequence of the binding of these blocking antibodies would be the protection of virus particles from neutralization (Massey and Schochetman, 1981b). In this study a panel of monoclonal antibodies was tested for their ability to (*a*) compete for binding of a radiolabeled neutralizing monoclonal antibody to MMTV and (*b*) to neutralize virus infectivity. Monoclonal antibodies that gave discordant results, i.e. did not neutralize but competed for binding of neutralizing antibody would be potential blocking antibodies. Two out of a total of 13 monoclonal antibodies tested competed for the binding of the neutralizing antibody and did not neutralize virus infectivity. These two antibodies were further demonstrated to block virus neutralization by the neutralizing antibody. Although the nonneutralized fraction has been observed with other viruses and polyvalent antisera, to our knowledge this was the first direct demonstration that this occurs by way of blocking antibodies. The recent demonstration of blocking antibodies against the tick-borne encephalitis virus (Heinz et al., 1983) suggests that this may be a general phenomenon.

The use of monoclonal antibodies for studying the generation of the nonneutralized fraction represents an important approach in the analysis of virus-host cell interactions. This mechanism must now be considered in studying chronic viral infections in which infectious viruses persist in the presence of neutralizing antibodies. Examples of chronic viral infections in humans that fit into this category are chronic viral hepatitis, subacute sclerosing panencephalitis (measles virus) and Epstein-Barr virus infections. The results of this type of analysis should be applicable to therapeutic protocols for modifying such disease states. Finally, in designing future vaccination programs with purified antigens, fragments of antigens and adjuvants, the potential for generating blocking antibodies should be strongly considered.

4. Idiotype and anti-idiotype monoclonal antibodies for analysis of viral antigens

Unique antigenic determinants found in the variable region of antibody molecules are called idiotypes and the antibodies which react with them are called anti-idiotypes. Two types of anti-idiotype antibodies can be found; one type reacts with the antigen-combining site and the second type reacts with other determinants localized in the variable region of the heavy chain or light chain or both (framework determinants). The antigen-combining site is a unique conformation (paratope) which allows a 'lock and key' type interaction between the antibody and a specific antigen. This unique conformation is a derivative of the sequence of amino acids in the variable region of both the heavy and light immunoglobulin chains. Anti-idiotypic antibodies that specifically recognize the structure of the combining site of another antibody are thought to have a determinant that is similar, if not identical, to an epitope found on the antigen that induced the latter antibody.

Idiotypes have been studied extensively since they were first described (Kunkel et al., 1963). Idiotypes along with homologous anti-idiotypic antibodies are thought to be components of a network of complex reactions that regulates a given immune response. Such regulation at the level of idiotype recognition was postulated by Jerne (1974) to involve a series of idiotype-anti-idiotype reactions that are able to control both the afferent and efferent immune responses. Numerous studies have implicated idiotypic networks in regulating the immune response to a large variety of haptens and protein or carbohydrate antigens (Rodkey, 1980; Urbain et al., 1981).

More recently, anti-idiotype antibodies have been used to enhance the murine antibody response to transplantation antigens (Bluestone et al., 1981; Epstein et al., 1982) and to induce protective immunity in mice against lethal infection with trypanosomiasis (Sacks et al., 1982). More recent studies have employed anti-idiotypic antibodies to characterize idiotypes associated with antibodies to a variety of viruses.

An idiotype shared by human antibodies (anti-HBs) to hepatitis B surface antigen (HBsAg) has been described (Kennedy et al., 1982; Kennedy and Dreesman, 1983; Kennedy et al., 1983a). This shared idiotype was also found on anti-HBs antibodies produced in BALB/c mice and six other species, indicating an interspecies idiotypic cross-reaction (Kennedy et al., 1983b). Injection of anti-idiotype antibodies resulted in a significant increase in the population of spleen cells secreting IgM anti-HBs after the mice were challenged with HBsAg (Kennedy et al., 1983c). More importantly, injection with anti-idiotype antibodies alone induced cells producing antibodies having the interspecies idiotype (Kennedy et al., 1983c; Kennedy and Dreesman, 1984) and these antibodies could be found in the serum (Kennedy et al., 1984).

The rabies virus glycoprotein is responsible for many of the important biological properties of rabies virus, including the induction of neutralizing antibodies (Cox et al., 1977) and the attachment to cell surface receptors (Perrin et al., 1982). Although a complete functional epitope map for rabies virus is unknown, at least three major antigenic sites for type-specific virus neutralizing antibodies have been described using monoclonal antibodies (Lafon et al., 1983). Anti-idiotypic antibodies have been made to neutralizing monoclonal antibodies specific for rabies virus glycoprotein (Reagen et al., 1983). Two of the five anti-idiotypic antibodies, when injected into mice, led to a specific virus neutralizing antibody response. This indicated that anti-idiotypic antibodies can sufficiently mimic an epitope on a viral protein so as to induce a neutralizing antibody response.

Another example of antigenic mimicry of a viral paratope was achieved using a monoclonal anti-idiotypic antibody directed to a monoclonal antibody specific for the reovirus type 3 hemagglutinin (Kauffman et al., 1983). The reovirus hemagglutinin is the viral cell attachment protein (Lee et al., 1981). The anti-idiotype antibody bound to the cell surface receptor and inhibited the binding of type 3 reovirus to receptor-positive cells.

These studies show that anti-idiotypic antibodies will provide useful probes for studying the structure and regulation of viral receptors on cell surfaces. These antibodies should also prove useful for the purification of these receptors and may provide a means for blocking of virus binding and subsequent infection. More important would be the concept that monoclonal anti-idiotype antibodies could function as novel vaccines by inducing specific virus neutralizing antibodies.

5. References

Altrock, B.W., Arthur, L.O., Massey, R.J. and Schochetman, G. (1981) *Virology* 109, 257–266.

Andrewes, C.H. and Elford, W.J. (1933) *Br. J. Exp. Pathol.* 14, 367–383.

Arthur, L.O., Lovinger, G.G. and Schochetman, G. (1979) *J. Virol.* 32, 852–859.

Arthur, L.O., Copeland, T.D., Orozolan, S. and Schochetman, G. (1982) *J. Virol.*, 41, 414–422.

Bluestone, J.A., Epstein, S.L., Ozato, K., Sharrow, S.O. and Sachs, D.H. (1981) *J. Exp. Med.* 154, 1305–1318.

Burnet, F.M., Keogh, E.V. and Lush, D. (1937) *Aust. J. Exp. Biol. Med. Sci.* 15, 226–368.

Cinader, B. and Lafferty, K.L. (1963) In *Antibody to Enzymes-A Three-Component System* (B. Cinader, Ed.) Vol. 103 New York Academy of Sciences, New York. pp. 653–673.

Cooper, N.R. (1979) In *Comprehensive Virology* (H. Fraenkel-Conrat and R.R. Wagner, Eds.) Plenum, New York. pp. 123–170.

Cox, J.H., Dietzschold, B. and Schneider, L.G., (1977) *Infect. Immun.* 16, 754–759.

Daniels, C.A. (1975) In *Viral Immunology and Immunopathology* (A.L. Notkins, Ed) Academic Press, New York. pp. 79–97.

Dowdle, W.R., Colman, M.T. and Gregg, M.B. (1974) *Prog. Med. Virol.* 17, 91–135.

Dulbecco, R., Vogt, M. and Strickland, A.G.R. (1956) *Virology* 2, 162–205.

Epstein, S.L., Masakowski, V.R., Sharrow, S.O., Bluestone, J.A., Ozato, K. and Sachs, D.H. (1982) *J. Immunol.* 129, 1545–1552.

Fine, D.L., Arthur, L.O. and Schochetman, G. (1980) *Virology* 101, 176–184.

Flammand, A., Wiktor, T.J. and Koprowski, H. (1980) *J. Gen. Virol.* 48, 105–109.

Gudnadottir, M. and Palsson, P.A. (1965) *J. Immunol.* 95, 1116–1120.
Heinz, F.X., Berger, R., Tuma, W. and Kunz, C. (1983) *Virology* 130, 485–501.
Jerne, N.K. (1974) *Ann. Immunol. Paris* 125C, 373–389.
Kauffman, R.S., Noseworthy, J.H., Nepom, J.T., Finberg, R., Fields, B.N. and Greene, M.I. (1983) *J. Immunol.* 131, 2539–2541.
Kennedy, R.C., Sanchez, Y., Ionescu-Matiu, I., Melnick, J.L. and Dreesman, G.R. (1982) *Virology* 122, 219–221.
Kennedy, R.C. and Dreesman, G.R. (1983) *J. Immunol.* 130, 385–389.
Kennedy, R.C., Dreesman, G.R., Sparrow, J.T., Culwell, A.R., Sanchez, Y., Ionescu-Matiu, I., Hollinger, F.B. and Melnick, J.L. (1983a) *J. Virol.* 46, 653–655.
Kennedy, R.C., Ionescu-Matiu, I. Sanchez, Y. and Dreesman, G.R. (1983b) *Eur. J. Immunol.* 13, 232–235.
Kennedy, R.C., Adler-Storthz, K., Henkel, R.D., Sanchez, Y., Melnick, J.L. and Dreesman, G.R., (1983c) *Science* 221, 853–855.
Kennedy, R.C. and Dreesman, G.R. (1984) *J. Exp. Med.* 159, 655–665.
Kennedy, R.C., Melnick, J.L. and Dreesman, G.R. (1984) *Science* 223, 930–931.
Kilbourne, E.D. (1975) In *The Influenza Viruses and Influenza* (E.D. Kilbourne, Ed.) Academic Press, New York. pp. 483–538.
Kingsbury, D.W., Bratt, M.A., Choppin, P.W., Hanson, R.P., Hosaka, Y., Ter Meulen, V., Norrby, E., Plowright, W., Rott, R. and Wunner, W.H. (1978) *Intervirology* 10, 137–152.
Kunkel, H.G., Mannik, M. and Williams, R.C. (1963) *Science* 140, 1218–1219.
Lafon, M., Wiktor, T.J. and Macfarlan, R.I. (1983) *J. Gen. Virol.* 64, 843–851.
Lee, P.W.K., Hayes, P.C. and Joklik, W.K. (1981) *Virology* 108, 156–163.
Majer, M. (1972) *Curr. Top. Microbiol. Immunol.* 58, 69–84.
Mandel, B. (1958) *Virology* 6, 424–447.
Mandel, B. (1978) In *Advances in Virus Research*, Vol. 23 Academic Press, New York. pp. 205–268.
Mandel, B. (1979) In *Comprehensive Virology* (H. Fraenkel-Conrat and R.R. Wagner, Eds.) Vol. 15. Plenum, New York. pp. 37–121.
Marcus, S.L., Kopelman, R. and Sarkar, N. (1979) *J. Virol.* 31, 341–349.
Massey, R.J. and Schochetman, G. (1981a) *Virology* 115, 20–32.
Massey, R.J. and Schochetman, G. (1981b) *Science* 213, 447–449.
Massey, R.J., Arthur, L.O., Long, C.W. and Schochetman, G. (1980a) *J. Virol.* 33, 123–128.
Massey, R.J., Arthur, L.O., Nowinski, R.C. and Schochetman, G. (1980b) *J. Virol.* 34, 635–643.
Mathews, J.H. and Roehrig, J.T. (1982) *J. Immunol.* 129, 2763–2767.
Nishikawa, K., Isomura, S., Suzuki, S., Watanabe, E., Hamaguchi, M. Yoshida, T. and Nagai, Y. (1983) *Virology* 130, 318–330.
Notkins A.L., Mahar, S., Scheele, C. and Goffman, J. (1966) *J. Exp. Med.* 124, 81–97.
Oldstone, M.B.A. (1975) *Prog. Med. Virol.* 19, 84–119.
Perrin, P., Portnöi, P. and Sureau, P. (1982) *Ann. Virol. Paris* 133E, 403–422.
Porter, D.D. and Larsen, A.E. (1967) *Proc. Soc. Exp. Biol.* 126, 680–682.
Reagan, K.J., Wunner, W.H., Wiktor, T.J. and Koprowski, H. (1983) *J. Virol.* 48, 660–666.
Redmond, S.M.S. and Dickson, C. (1983) *EMBO J.* 2, 125–131.
Redmond, S., Peters, G. and Dickson, C. (1984) *Virology* 133, 393–402.
Rodkey, L.S. (1980) *Microbiol. Rev.* 44, 631–659.
Sacks, D.L., Esser, K.M. and Sher, A. (1982) *J. Exp. Med.* 155, 1108–1119.
Schochetman, G., Long, C. and Massey, R. (1979) *J. Virol.* 32, 131–139.
Sheshberadaran, H., Chen, S.-N. and Norrby, E. (1983) *Virology* 128, 341–353.
Spriggs, D.R., Kaye, K. and Fields, B. (1983) *Virology* 127, 220–224.

Teramoto, Y.A. and Schlom, J. (1978) *Cancer Res.* 38, 1990–1995.
Toolan, H.W. (1965) *Proc. Soc. Exp. Biol.* 119, 715–717.
Urbain, J., Wuilmart, C. and Cazanave, P.-A. (1981) *Contemp. Top. Mol. Immunol.* 8, 113–148.
Webster, R.G. and Laver, W.G. (1975) In *The Influenza Viruses and Influenza* (E.D. Kilbourne, Ed.) Academic Press, New York. pp. 269–314.
Webster, R.G. and Laver, W.G. (1980) *Virology* 104, 139–148.
Webster, R.G., Kendal, A.P. and Gerhard, W. (1979) *Virology* 96, 258–264.
Wiktor, T.J. and Koprowski, H. (1978) *Proc. Natl. Acad. Sci. U.S.A.* 75, 3938–3942.
Yewdell, J. and Gerhard, W. (1982) *J. Immunol.* 128, 2670–2675.

Eds. M.H.V. van Regenmortel and A.R. Neurath
Immunochemistry of viruses. The basis for serodiagnosis and vaccines.

Cellular immunity to viruses 4

J.R. LAMB[3], FIONA M. MOSS[1] and D.D. ECKELS[2]

[1] Imperial Cancer Research Fund, Human Tumour Immunology Unit, University College Hospital, University Street, London WC1E 6JJ, U.K., [2] Immunogenetics Research Section, The Blood Center of Southeastern Wisconsin, 1701 W. Wisconsin Avenue, Milwaukee, WI 53233, U.S.A. and [3] Medical Research Council, Tuberculosis and Related Infections Unit, Hammersmith Hospital, Ducane Road, London W12 OHS, U.K.

Understanding the complex nature of viruses, such as their intracellular location, expression of multiple antigenic determinants and the ability of different viruses to infect and replicate in a variety of cell types is necessary for appreciation of the intricacy of the immune responses to viral infections. Both the specific and nonspecific mechanisms by which the immune system responds to viral infection have therefore been the subject of intense investigation. However, while the role of antibody in neutralizing free virus and prevention of the spread of infection has been well documented, the importance of T cell responses in antiviral immunity is only now becoming more apparent and there is now convincing evidence that T cells are involved in the recovery from viral infection (McMichael et al., 1983).

The selective pressures induced by the interactions between the mammalian immune system and viruses have mutually influenced their evolution. For the immune system the result of this is seen by the polymorphism in the major histocompatibility gene complex (MHC) determinants, the multiplicity of MHC loci, and the selection for heterozygosity, as MHC antigens are so intimately involved in the recognition of viral antigen (Zinkernagel and Doherty, 1979). One example of the continual selective forces on viruses is variation in antigenic structure illustrated by the antigenic drift and shift of influenza strain A viruses (Ward, 1981).

The aim of this chapter is to concentrate on T and B cell recognition of both free viral determinants and those expressed on the surface of infected cells in association with products of the MHC. The intention is not to discuss immune responses in terms of acute, chronic, oncogenic, latent or slow viral infections, but to focus on the problems of viral antigen recognition in general terms. In many instances this will be illustrated by reference to immune responses to influenza viruses, as there has been detailed structural analysis of the antigenic structure of influenza and extensive studies on immune regulation of this virus both in man and animal models. Discussion of the nonspecific arc of immune responses to virus infection will be limited to the role of interferon.

Thus, in this review of cellular immunity to viruses emphasis is given to the recognition of viral antigen determinants.

1. Viral antigens recognized by the immune system

The ability of serum antibodies to recognize various determinants of free virus is well documented (Cowan, 1973). However, the advent of monoclonal antibodies has been the basis for defining variations among viral strains (Ward, 1981) and detailed antigenic structure of a number of viral systems such as murine leukemia (Nowinski et al., 1980) and rabies (Koprowski and Wiktor, 1980) viruses. It is perhaps influenza virus hemagglutinin (HA) that has been subject of the most detailed analysis (Gerhard et al., 1981; Caton et al., 1982). By monoclonal antibody testing on a panel of mutant influenza A viruses with defined amino acid substitutions, five immunodominant antigenic regions of the H1 subtype (PR/8) HA have been defined (Caton et al., 1982). These sites correspond to the four antigenic sites predicted from the three-dimensional structure of the HA of H3 subtype (A/Hong Kong) HA as determined by x-ray crystallography (Wiley et al., 1981; Wilson et al., 1981). In addition to defining epitope recognition, viral mutants are important in the analysis of the diversity of the B cell repertoire. A recent study by Staudt and Gerhard (1983) investigating the specificity of antibodies reactive with the four major B cell determinants of HA by paratypic and idiotypic analyses, has revealed that the same specificity recurred in separate mice only three times among 125 hybridomas tested. This observation emphasizes the enormous selective pressure exerted by influenza virus for the generation of diversity among B cells. The application of monoclonal antibodies, viral mutants and crystallography has provided invaluable reagents for the analysis of epitope recognition and the selective pressures on both the diversity of the immune response and viral expression.

Viral determinants as they are seen on the surface of infected cells present a further set of problems for both T and B cell recognition as compared to the intact virion. However, recognition of virally infected targets is of particular importance in protective immunity to viruses that replicate by budding. Firstly, early viral antigens expressed on the target cell surface may differ from those of the mature virion, and these determinants may be inaccessible in the intact virion. Secondly, it would appear that some viruses such as Sendai (Schrader and Edelman, 1977) are able to fuse sufficient antigen directly into the target cell membrane without de novo synthesis of viral protein. Clearly, this may influence the association of viral and target cell membrane antigens depending on whether or not the event occurs at the level of synthesis, postsynthetically or is merely a modification of the host protein. This is important as the antiviral response is concerned with the recognition of a complex determinant involving both viral and self antigens (Zinkernagel and Doherty, 1979). Viral determinants are expressed on infected cells in the context of MHC antigens so their structure may not reflect that predicted from the intact virion. Thus attempts to identify the viral determinants on infected cells critical in T cell recognition has proven difficult. This is well illustrated by analyses of target recognition of the cross-reactive population of influenza virus-immune cytotoxic T lymphocytes discussed in the next section.

2. T cell recognition of viral determinants

2.1. Specificity for viral antigens

Evidence would suggest that selective pressure has resulted in a highly specific antiviral B cell response in contrast with T cell responses, which appear to exhibit greater cross-reactivity (Zinkernagel and Rosenthal, 1981). In terms of antigen specificity the most extensively studied virus-immune T cells are cytotoxic T lymphocytes (Tc).

The majority of virus-induced Tc show viral specificity when compared with unrelated viruses, for example herpes simplex virus-induced Tc do not lyse targets infected with lymphocyte choriomeningitis virus (Pfizenmaier et al., 1977). However, when the fine specificity of Tc for closely related but serologically distinct viruses is analyzed, marked cross-reactivity can be observed. This is well illustrated by influenza virus-immune Tc.

A number of studies on the specificity of Tc responses to influenza A have revealed that there is no discrimination between serologically distinct type A subtypes (Braciale, 1977; Doherty et al., 1977; Zweerink et al., 1977; McMichael and Askonas, 1978). The existence of this cross-reactive subpopulation has been confirmed by the analysis of the specificity of interleukin 2-dependent Tc clones both in man (Fleischer, 1982) and in the mouse (Lin and Askonas, 1980; Braciale et al., 1981). Cross-reactive Tc have been reported for serologically defined viruses other than influenza such as reoviruses (Finberg et al., 1979) and vesicular stomatitis virus (VSV; Rosenthal and Zinkernagel, 1980). Although cross-reactive Tc appear to be the major population of T cells which respond to influenza there is also evidence for subtype-specific Tc (Braciale, 1977; Doherty et al., 1977; Ennis et al., 1977; Zweerink et al., 1977). HA has been proposed as the viral determinant recognised by this subpopulation of Tc from experiments using recombinant viruses (Ennis et al., 1977), cold target inhibition (Effros et al., 1977) and restimulation with purified HA (Braciale, 1979). Analysis at the clonal level by Braciale and his colleagues (1981) supports the evidence that HA is recognized by the subtype-specific Tc.

Attempts to define the fine antigen specificity of influenza-immune Tc have demonstrated that cyanogen bromide fragments of HA and, more recently, a defined peptide sequence of the HA_2 molecule are able to stimulate Tc responses (Wabuke-Bunoti et al., 1980; Wabuke-Bunoti and Fan, 1983). Comparable specificities have been reported for Sendai virus-reactive Tc (Guertin and Fan, 1980). Interestingly, at the clonal level Bennink and his colleagues (1982) have isolated influenza-induced Tc with specificity that maps with the polymerase gene (P3).

The identification of the viral determinant(s) recognized by cross-reactive influenza immune Tc has proved difficult. The highly conserved viral proteins matrix protein (MP), nucleoprotein (NP) and non-structural protein (NSP) have been proposed as the most likely candidates based on the assumption that the specifity of Tc should parallel the serology of the viral proteins. Early reports using heterologous antisera suggested that MP was expressed on the membrane of infected targets (Biddison et al., 1977). This observation, together with the inhibition of Tc target lysis by anti-MP antibody (Reiss and Schulman, 1980) support the role of MP as a viral target of Tc. However there are contradictory findings. The failure to inhibit Tc lysis with anti-MP antibodies (Hackett et al., 1980) and recent observations using monoclonal antibodies suggest that the glycoproteins HA and neuraminidase (NA), in contrast to MP, can be readily detected on infected cells (Yewdell and Gerhard, 1981). Interestingly, in the Tc response to VSV, the M protein which is located beneath the membrane of the infected cells may influence the expression of the membrane associated glycoprotein (G), since targets infected with temperature-sensitive mutants expressing G and not MP were not readily lysed (Zinkernagel et al., 1978). This result indicates a role for viral antigens other than those expressed on the membrane. As regards the conserved NP of influenza, Yewdell and his colleagues (1981) have shown that it can be identified on infected cells and as such may function as a recognition antigen for the cross-reactive Tc population. Indeed a murine Tc clone specific for the NP of PR/8 virus has recently been isolated (Townsend and Skehel, 1983). It now appears that HA may serve as the viral epitope recognised by both the subtype and cross-reactive Tc populations. The evidence for this is the following. Koszinowski et al. (1980) inserted HA and NA glycoproteins into liposomes which were then fused into targets using Sendai virus. They observed that these targets were lysed by Tc induced with a different subtype. This is supported by the observation that monoclonal anti-HA and anti-H-2 antibodies combined are able to inhibit the lysis by cross-reactive Tc (Askonas and Webster, 1980).

In concluding the discussion of Tc specificity it is worth mentioning that Tc are reported to have a protective capacity, since the transfer of cloned influenza virus-immune Tc into syngeneic recipients led to a reduction of viral titres in the lung (Lin and Askonas, 1981).

The viral antigen specificity of influenza virus-induced helper T cells (Th) parallels that reported for Tc in that both subtype specific (Gerhard et al., 1983; Lamb et al., 1982b) and cross-reactive (Anders et al., 1981; Lamb et al., 1982b; 1982c; Thomas et al., 1982) populations have been described. However, can any of the viral determinants be identified as determining the recognition specificity of these populations?

Studies in our laboratory suggest that MP-specific helper/proliferative T cells are a significant component of the T cell repertoire, since analysis of the viral specificity of a panel of human T cell clones revealed that 37% were reactive with MP (Lamb et al., 1982c). Furthermore, cloned MP-specific T cells were shown to collaborate with B cells in the production of anti-HA antibodies (Lamb et al., 1982d) as has been described for mouse Th (Russell and Liew, 1979). This phenomenon of 'intermolecular help' (Lake and Mitchison, 1976) may be important in the development of heterotypic immunity. The pattern of reactivity of the remaining T cell clones showed that 45% were specific for NA, 9% for NP and 9% for HA (Lamb et al., 1982c). Anders and her colleagues (1981) demonstrated that both subtype-specific and cross-reactive Th cells recognize determinants within the HA molecule, and this has recently been confirmed in our laboratory (Lamb et al., 1982b; Lamb and Green, 1983). Human T cell clones specific for HA (of A/Texas, H3) were analyzed for their viral antigen specificity using chemically synthesized peptides of the HA-1 molecule (Green et al., 1982). Clones 1.7 and 2.43 responded to the inducing antigen (HA) and to peptide 20 (termed p20, residues 306–329) but not to any other of the peptides (Table 4.1). In contrast, clone 2.61 recognized only peptide 11 (p11; 105–140) and HA, while 3.18 failed to respond to any of the peptides tested. Interestingly, both of these peptides are remote from the proposed antibody binding sites of subtype H3 (Wiley et al., 1981). However, the normal peripheral blood lymphocytes were able to respond to all the peptides, suggesting that there was no defect in the T cell repertoire (Table 4.1). In order to further define the recognition specificity, the ability of the clones to respond to representatives of the subtypes of influenza A was determined (Table 4.2). Clone 3.18 was only reactive to A/Texas and thus appears to recognize a determinant unique to the immunizing strain, whereas the other clones displayed variable patterns of cross-reactivity (Table 4.2). Clone 2.43 recognised all the subtypes, and clone 1.7 only the H3 subtypes, yet both were specific for p20. The p11-specific clone (2.61) was reactive with only the subtypes A/Texas and A/Victoria. Comparison of the amino acid substitutions between A/Texas and the other subtypes occurring within the sequence of p11 and p20 are summarized in Table 4.2. The results suggest that the epitope specific for clone 2.61 is associated with residues 123 and/or 127. Clone 1.7 appears to recognize a variant sequence of p20 occurring at residues 312–314, 316, 321 and 326–329. On the other hand, the recognition specificity of clone 2.43 is for an invariant sequence of p20, possibly including residues 306–311, 319–320 or 322–323. These findings confirm that the viral determinants recognized by influenza virus-immune T cells are strain specific, subtype specific or cross-reactive and reside in the HA molecule. Recent reports on the epitope recognition of murine Th clones specific for

TABLE 4.1

Fine antigen specificity of HA-induced T cell clones for the synthetic peptides of the HA-1 molecule

Synthetic peptides[a] of HA-1 molecule		Clone[c]					Antibody [d] binding sites
No.	Sequence	PBL[b]	1.7	2.43	2.61	3.18	
p^2	1–36	+	−	−	−	−	
p^4	39–65	+	−	−	−	−	
p^7	53–87	+	−	−	−	−	C disulphide bond 52-277
p^{10}	76–111	+	−	−	−	−	
p^{11}	105–140	+	−	−	−	−	
p^{16}	140–176	+	−	−	−	−	A
p^{17}	174–196	+	−	−	−	−	B
p^{19}	201–227	+	−	−	−	−	D
p^{21}	266–282	+	−	−	−	−	
p^{22}	282–303	+	−	−	−	−	
p^{20}	306–329	+	+	+	−	−	
HA		+	+	+	+	+	

[a–d]Data from Green et al. (1982), Lamb et al. (1982d), Lamb and Green (1983) and Wiley et al. (1981) respectively.

TABLE 4.2

Recognition of influenza virus strain A subtypes by HA-specific T lymphocyte clones

Virus subtypes		Clone [a]				Amino acid substitutions with:[b]	
		1.7	2.43	2.61	3.18	Peptide 11	Peptide 20
A/PR/8/34	H1	−	+	−	−	111–114, 116–119, 121–127, 129–135, 137, 138	312–314, 316–318, 321, 326–329
A/Japan/305/57	H2	−	+	−	−	110, 111, 113, 114, 116–119, 121–127, 130, 131, 133, 134, 138	312–314, 316, 321, 326–329
A/Aichi/2/68(x31)	H3	+	+	−	−	123, 127, 138	-
A/Victoria/3/57	H3	+	+	+	−	138	-
A/Texas/1/77	H3	+	+	+	+	-	-

[a] Data from Lamb and Green (1983).
[b] Sequence comparisons were derived from Gething et al. (1980), Winter et al. (1981) and Webster et al. (1982).

H1 have shown specificity for a nine amino acid sequence (111–119) with the presence of a glutamic acid residue at 115 being critical (Gerhard et al., 1983; Hackett et al., 1983a). It is important to emphasize that the epitopes recognized by the HA-reactive Th are distant from the antibody binding sites, suggesting that T and B cells show a preference for different determinants.

Although Lin and Askonas (1981) demonstrated that cloned influenza-immune T cells with the ability to mediate both cytotoxic and delayed type hypersensitivity (DTH) responses were able to limit virus replication in vivo, the role of DTH reactions in acute and chronic viral infections is not well understood. However, DTH does appear after inoculation of a wide range of viruses including lymphocytic choriomeningitis virus (Tosolini and Mims, 1971), Semliki Forest virus (Kraaijeveld et al., 1979) and influenza virus (Liew et al., 1979). DTH T cells (Td) from mice sensitized with live influenza are subtype cross-reactive, while noninfectious virus induces a HA-specific population (Leung et al., 1980a). Furthermore, Liew and Russell (1980) observed that suppressor T cells (Ts) of DTH were subtype specific and mediated their function at the level of Th, in contrast to Ts (Leung et al., 1980b), which acted on Tc. Ts in the regulation of anti-influenza immunity have not yet been investigated as thoroughly as either Tc or Th. However Ts undoubtedly exist (Ada et al., 1981). Fischer and his colleagues (Fischer et al., 1983) isolated an influenza-specific immune Th line that elaborated a nonspecific suppressor inducer factor. A Ts clone reactive against a MP-specific Th clone rather than a viral component has been described (Lamb and Feldmann, 1982).

To summarize, it seems that two categories of T cells, one specific, the other cross-reactive, are involved in viral antigen recognition. Nevertheless, even for the most studied of the viral systems (influenza) the precise nature of the determinants recognized by the immune system remain largely ill defined. Perhaps with synthetic viral peptides and the transfection of cloned viral genes creating better defined targets, together with the use of T cell clones, the important epitopes in viral antigen recognition will be identified.

2.2. Specificity for MHC-restricting determinants

An essential requirement for the activation and function of T cells is the recognition of MHC antigens in addition to foreign antigen (Zinkernagel and Doherty, 1979). The restriction of virus specific Tc effector function as first described in the mouse (Zinkernagel and Doherty, 1974) is through Class I (H-2K, D and L) molecules. In man, Tc-mediated lysis of cells infected by a number of viruses such as influenza (McMichael et al., 1977; Shaw and Biddison, 1979), mumps (Kress and Kreth, 1982) and Epstein-Barr (Rickinson et al., 1980) viruses is also restricted through Class I (HLA, A, B and C) antigens. The involvement of Class I molecules in both species has been confirmed by blockade of Tc target lysis with anti-Class I antibodies (Koszinowski and Ertl, 1975; McMichael et al., 1980). MHC Class I antigens are polymorphic and as such multiple determinants on an individual molecule can function as restricting elements for Tc antigen recognition. However, some data suggests that relatively few determinants are used. In a comparative structural analysis of a variety of HLA-A2 antigens that could be distinguished by influenza virus-immune Tc, but not serologically, it was suggested that residues 147–157 in the second heavy chain domain were critical for Tc restriction (Krangel et al., 1983). By comparing amino acid sequences of mutant and wild type H-2K^b molecules, it is possible to designate particular epitopes of H-2K^b determinants involved in Tc recognition of virus-infected cells (Zinkernagel, 1976). H-2K^b restricted Tc in a variety of viral systems do not lyse K^{bml} targets, yet the bml mutant only differs from the wild type by three amino acids (Pease et al., 1983). Structural evidence suggests that one H-2K^b mutant uses the same determinant (residues 147–157) as HLA-A2 (Krangel et al., 1983). However, Tc specific for a variety of viruses from H-2K^b mice exhibit different patterns of cross-reactivity, suggesting not only that multiple epitopes on individual Class I molecules are used for antigen presentation but that the role each plays in presentation varies for different viruses (Hurwitz et al., 1983). Similar heterogeneity of restriction associated with HLA-A2 has been observed with Epstein-Barr virus-specific Tc (Gaston et al., 1983). It is of interest that Hackett et al. (1983b) demonstrated that influenza Tc could be stimulated by virus inserted in liposomes in the absence of H-2 molecules. Further aspects of Class I restriction of viral Tc responses are reviewed in detail by Sherman et al. (1983).

The induction of Th requires the recognition of MHC Class II molecules on the antigen-presenting cell, namely H-2I region in the mouse (Klein, 1979) and HLA-D region-associated (D, DR, DC and SB) determinants (reviewed by Moller, 1982) in man. Influenza virus-specific Th analyzed at the clonal level were shown to be restricted by either D region (Lamb et al., 1982a; Eckels et al., 1982), DC (Lamb and Eckels, unpublished results) or SB (Eckels et al., 1983) region products. Additional evidence for the requirement of Class II molecules was the ability of anti-Class II antibody to inhibit the proliferative responses (Eckels et al., 1982). In this analysis, Eckels and his colleagues were unable to map viral antigen specificity with either haplotype of MHC Class II gene products. However, for influenza virus-immune Tc, Shaw et al. (1980) have suggested that responses to different influenza strains are preferentially restricted to one or other Class I specificity. The murine influenza Th clones described by Gerhard et al. (1983) are I region restricted. With respect to DTH, two subsets of T cells are involved, one of which is H-2K, D region restricted and the other I region restricted (Ada et al., 1981).

In summary, the polymorphism of MHC determinants and their intimate association with viral antigens function to regulate T cell responses in viral infection.

3. B cell recognition of viral determinants

Since the subtypes of influenza A, for example, are defined serologically it is apparent that B cells can recognize the virion. Indeed, neutralizing antibodies directed against the surface glycoproteins are critical for protection and recovery (Virelizier et al., 1979). The observation that lysis of measles virus-infected cells by the alternative pathway of complement required the presence of specific antibody, suggests that antibody can also recognize viral determinants as they are expressed on the surface of infected cells. It is perhaps worth digressing at this point to consider briefly the role of complement in immunity to viruses. The activation of the alternative pathway and the subsequent lysis of infected target cells is dependent on the presence of specific IgG and properdin. This is true for a number of viruses including herpes simplex I and II, measles and influenza. However, even though in human serum the classical complement pathway is activated by incubation of measles-infected cells with specific antibody, cytolysis occurs through the alternative pathway (Cooper and Oldstone, 1983).

As already discussed, Yewdell and his colleagues (Yewdell et al., 1981) have used monoclonal antibodies to demonstrate the presence of HA, NA and NP components on influenza virus-infected targets. If infected cells are used as the immunogen the antibody responses directed against HA and NA, compared with the response elicited using intact virion as the immunogen, are reduced by about 30% and 60% respectively (Wylie and Klinman, 1981). Conversely, NP appears to be recognized on targets but not on the intact virus. Therefore, some determinants expressed on the virion are inaccessible on the infected cells and these appear to be the cross-reactive determinants. In contrast to the major component of the Tc response, which is cross-reactive, the B cell response is more specific, suggesting that T and B cells recognise different epitopes. Analysis from our laboratory (Lamb et al., 1982b; Lamb and Green, 1983) of the fine specificity of HA-reactive T cell clones using chemically synthesized peptides of HA-1 demonstrates clearly that T and B cells recognize different epitopes on the same molecule. This has been confirmed by others (Gerhard et al., 1983; Hackett et al., 1983a).

A recent series of reports have demonstrated that B cell antigen recognition, like that of T cells, may require antigen to be presented in association with the relevant MHC determinants (e.g. Nisbet-Brown et al., 1981). A clear demonstration of this is the findings of Wylie et al. (1982) in which 57% of the antibody response of H-2^k mice immunized with H-2^k cell lines infected with influenza was reactive with the relevant infected cells, but not the virion or uninfected cells alone. Furthermore, the response was specific for HA and restricted to H-$2D^k$ Class I molecules. In contrast, the response of H-2^b B cells to influenza-infected H-2^b targets was H-$2K^b$ restricted. Thus B cell responses to influenza appear to be sensitive to variation in both the HA and MHC Class I determinants. Several important implications can be made from these findings, namely antiviral B cell activity in the case of influenza virus infection is not solely to neutralize the free virion. B cells, like T cells, may recognize antigen on virus-infected cells in the context of MHC determinants. While this draws one to the conclusion that T and B cell antigen recognition can operate through similar mechanisms, T and B cells may recognize different antigen epitopes associated with distinct MHC determinants.

4. Nonspecific mediators of antiviral immunity

Interferons (IFNs) are a group of polypeptides that function in the development of nonspecific antiviral activity (De Maeyer and De Maeyer-Guignard, 1979). Fibroblast (β-IFN) and leukocyte-derived (α-IFN) interferons are released when the cells are infected with virus. Although initially thought to be independent of the virus inducing its release, a recent report suggests that different viruses induce α-IFN from human macrophages with different activities (Bell et al., 1983). Immune or γ-IFN is produced by lymphocytes during an immune response, and appears to be induced more efficiently by DNA viruses than are the other IFNs. In support of this, a clinical study of Rasmussen et al. (1974) showed that patients with herpes simplex virus infections who made undetectable levels of γ-IFN had more frequent recurrence of lesions than those capable of producing γ-IFN. It is now well established that the stimulation of T cells with mitogen or antigen leads to the production of γ-IFN (Klein et al., 1982; Morris et al., 1982). Recently, Askonas and her colleagues (Morris et al., 1982) demonstrated that cloned murine influenza virus-immune Tc release γ-IFN only when stimulated with the correct influenza-infected target cell. This implies that the binding of specific antigen by the Tc clones is necessary for IFN release. The ability of Tc to release γ-IFN following contact with the appropriate target is of considerable importance in that it may help prevent the spread of infection. Furthermore, release of γ-IFN before actual lysis of the target would reduce the spread of virus from the infected cells and prevent the Tc itself from being infected. The local release of γ-IFN may also function in the activation of natural killer cells and so enhance the antiviral response. Interestingly, interleukin-2 alone appeared unable to induce IFN release from murine Tc clones (Bevan, M.J., personal communication). However, in contrast it was observed that both helper and suppressor T cell subsets in human peripheral blood were able to release γ-IFN after stimulation with human interleukin-2 (Kasahara et al., 1983). In our laboratory we have investigated IFN production by a panel of influenza virus-immune human T cell clones (all of the helper/inducer subset) following stimulation with autologous antigen-presenting cells and influenza virus (Jacobson, S., Lamb, J., Eckels, D. and Biddison, W., unpublished observations). Within the panel of clones some produced α-IFN, others γ-IFN and the third category appeared to release no detectable IFN. IFNs, through their interactions with the immune system, increase the expression of MHC Class I and II antigen (Lindahl et al., 1976; Rhodes et al., 1983) and activate immune effector cells, such as natural killer cells (Einhorn et al., 1978), thus having an indirect effect on antiviral immunity.

5. Conclusion

The previously established dogma that B cells recognize antigen per se, whereas T cell recognition of antigen is in the context of MHC determinants has supported the notion that these different lymphocyte populations recognize different epitopes. This is reflected in concepts of antiviral immunity, where it has been suggested that B cells are specific for free virus while T cells predominantly recognize virus-infected cells. However, it now appears that there are similarities between the major effector arcs of viral immunity, since both Tc and antibody for influenza virus, for example, are MHC restricted (Wylie et al., 1982). Yet, despite this similarity, they display differences in their selection of restriction elements. In addition, Tc have been shown to be highly MHC restricted (Krangel et al., 1983), while only 57% of antibody directed against virally infected targets was MHC restricted (Wylie et al., 1982), suggesting that T and B cells can use the same mechanism for antigen recognition, and that B cell antigen recognition may be through more than one mechanism.

B cell recognition of free virus or that expressed on infected cells is extremely sensitive to variation in antigenic structure as revealed by analysis of fine specificity on mutant strains of an influenza virus A subtype. While the major proportion of the Tc response is cross-reactive, failing to discriminate between serologically defined viral subtypes, the fine specificity of influenza virus-immune Th reflects that of B cells in that they appear to react with defined peptide sequences and are sensitive to single amino acid substitutions. Nevertheless, the epitopes they recognize are distinct from the antibody binding sites (Lamb et al., 1982b). However, within the repertoire of Th is a population that recognizes conserved viral determinants and as such displays patterns of cross-reactivity similar to Tc. These observations suggest that the effector mechanisms of viral immunity recognize a different group of determinants.

The selection of T and B cell repertoires may reflect the manner in which viral antigen interacts with the MHC determinants, such that interactions primarily influencing the MHC determinant through modification of structure or the generation of neo-MHC determinants would have a more powerful influence on Tc than antibody. With changes in the conformation or residue sequence of viral determinants the converse would be seen. Thus defining virus-target cell interactions and their influence on epitope expression is fundamental to understanding viral antigen recognition and the basis of cellular immunity to viruses.

6. References

Ada, G.L., Leung, K.-N. and Ertl, H. (1981) *Immunol. Rev.* 58, 1-24.
Anders, E.M., Katz, J.M., Jackson, D.C. and White, D.O. (1981) *J. Immunol.* 127, 669–672.
Askonas, B.A. and Webster, R.G. (1980) *Eur. J. Immunol.* 10, 151–156.
Bell, D.M., Roberts, N.J. and Hall, C.B. (1983) *Nature* 305, 319–321.
Bennink, J.R., Yewdell, J.W. and Gerhard, W. (1982) *Nature* 296, 75–77.
Biddison, W.E., Doherty, P.C. and Webster, R.G. (1977) *J. Exp. Med.* 146, 690–697.
Braciale, T.J. (1977) *Cell Immunol.* 33, 423–436.
Braciale, T.J. (1979) *J. Exp. Med.* 149, 856–869.
Braciale, T.J., Andrew, M.E. and Braciale, V.C. (1981) *J. Exp. Med.* 153, 910–923.
Caton, A.J., Brownlee, G.G., Yewdell, J.W. and Gerhard, W. (1982) *Cell* 31, 417–427.
Cooper, N.R. and Oldstone, M.B.A. (1983) *Immunol. Today* 4, 101–108.
Cowan, K.M. (1973) *Adv. Immunol.* 17, 195–253.
De Maeyer, E. and De Maeyer-Guignard, J. (1979) *Compr. Virol.* 15, 205–284.
Doherty, P.C., Effros, R.B. and Bennink, J. (1977) *Proc. Natl. Acad, Sci. U.S.A.* 74, 1209-1213.
Eckels, D.D., Lamb, J.R., Lake, P., Woody, J.N., Johnson, A.H. and Hartzman, R.J. (1982) *Human Immunol.* 4, 313–324.
Eckels, D.D., Lake, P., Lamb, J.R., Johnson, A.H., Shaw, S., Woody, J.N. and Hartzman, R.J. (1983) *Nature* 301, 716–718.
Effros, R.B., Doherty, P.C., Gerhard, W. and Bennink, J. (1977) *J. Exp. Med.* 145, 557–568.
Einhorn, S., Blomgren, H. and Strauder, H. (1978) *Int. J. Cancer* 22, 405–412.
Ennis, F.A., Martin, W.J. and Verbonitz, M.W. (1977) *J. Exp. Med.* 146, 893–898.
Fischer, A., Beverley, P.C.L., Feldmann, M. and Smith, S. (1983) *Immunology* 48, 177–186.
Finberg, R., Weiner, H.L., Field, B.N., Benacerraf, B. and Burakoff, S.J. (1979) *Proc. Natl. Acad. Sci. U.S.A.* 76, 442–446.
Fleischer, B. (1982) *J. Immunol.* 129: 1731–1736.
Gaston, J.S.H., Rickinson, A.B. and Epstein, M.A. (1983) *J. Exp. Med.* 158, 280–293.
Gerhard, W., Yewdell, J., Frankel, M.E. and Webster, R. (1981) *Nature* 290, 713–717.
Gerhard, W., Hackett, C. and Melchers, F. (1983) *J. Immunol.* 130, 2379–2385.
Gething, M.J., Bye, J., Skehel, J. and Waterfield, M.D. (1980) *Nature* 287, 301–306.
Green, N., Alexander, H., Olson, A., Alexander, S., Shinnick, T.M., Sutcliffe, J.G. and Lerner, R.A. (1982) *Cell* 28, 477–487.
Guertin, D.P. and Fan, D.P. (1980) *Nature* 283, 308–311.
Hackett, C.J., Askonas, B.A., Webster, R.G. and van Wyke, K. (1980) *J. Exp. Med.* 151, 1014–1025.
Hackett, C.J., Dietzschold, B., Gerhard, W., Ghrist, B., Knorr, R., Gillessen, D. and Melchers, F. (1983a) *J. Exp. Med.* 158, 294–302.
Hackett, C.J., Taylor, P.M. and Askonas, B.A. (1983b) *Immunology* 49, 255–263.
Hurwitz, J.L., Wettstein, P.J., Bennink, J.R. and Doherty, P.C. (1983) *J. Immunol.* 131, 471–474.
Kasahara, T., Hooks, J.J., Dougherty, S.F. and Oppenheim, J.J. (1983) *J. Immunol.* 130, 1784–1789.
Klein, J. (1979) *Science* 203, 516–521.
Klein, J.R., Raulet, D.H., Pasternack, M.S. and Bevan, M.J. (1982) *J. Exp. Med.* 155, 1198–1203.
Koprowski, H. and Wiktor, T. (1980) In *Monoclonal Antibodies* (R. Kennett, T. McKern and K. Bechtol, Eds.). Plenum Press, New York. pp. 335–351.
Koszinowski, U. and Ertl, H. (1975) *Nature New Biol.* 255, 522–524.
Koszinowski, U.H., Allen, H., Gething, M.J., Waterfield, M.D. and Klenk, H.D. (1980) *J. Exp. Med.* 151, 945–958.
Kraaijeveld, C.A., Marmsen, M. and Khader Boutahar-Trouw, B. (1979) *Infect. Immun.* 23, 219–223.

Krangel, M.S., Biddison, W.E. and Strominger, J.C. (1983) *J. Immunol.* 130, 1856–1862.
Kress, H.G. and Kreth, H.W. (1982) *J. Immunol.* 129, 844–849.
Lake, P. and Mitchison, N.A. (1976) *Cold Spring Harbour Symp. Quant. Biol.* 41, 589–595.
Lamb, J.R. and Feldmann, M. (1982) *Nature* 300, 456–458.
Lamb, J.R. and Green, N. (1983) *Immunology* in press.
Lamb, J.R., Eckels, D.D., Lake, P., Johnson, A.H. Hartzman, R.J. and Woody, J.N. (1982a) *J. Immunol.* 128, 233–238.
Lamb, J.R., Eckels, D.D., Lake, P., Woody, J.N. and Green, N. (1982b) *Nature* 300, 66–69.
Lamb, J.R., Eckels, D.D., Phelan, M., Lake, P. and Woody, J.N. (1982c) *J. Immunol.* 128, 1428–1431.
Lamb, J.R., Woody, J.N., Hartzman, R.J. and Eckels, D.D. (1982d) *J. Immunol.* 129, 1465–1470.
Leung, K.-N., Ada, G.L. and McKenzie, I.F.C. (1980a) *J. Exp. Med.* 151, 815–826.
Leung, K.-N., Ashman, R.B., Ertl, H.C.J. and Ada, G.L. (1980b) *Eur. J. Immunol.* 10, 803–810.
Liew, F.Y. and Russell, S.M. (1980) *J. Exp. Med.* 151, 799–814.
Liew, F.Y., Russell, S.M. and Brand, C.M. (1979) *Eur. J. Immunol.* 9, 783–790.
Lin, Y.-L. and Askonas, B.A. (1980) *Nature* 288, 164–167.
Lin, Y.-L. and Askonas, B.A. (1981) *J. Exp. Med.* 154, 225–234.
Lindahl, P., Gresser, I., Leary, P. and Tovey, M.G. (1976) *Proc. Natl. Acad. Sci. U.S.A.* 73, 1284–1287.
McMichael, A.J. and Askonas, B.A. (1978) *Eur. J. Immunol.* 8, 705–711.
McMichael, A.J., Ting, A., Zweerink, H.J. and Askonas, B.A. (1977) *Nature* 270, 524–526.
McMichael, A.J., Parham, P., Brodsky, F.M. and Pilch, J.R. (1980) *J. Exp. Med.* 152 (No. 2, Pt. 2), 195s-203s.
McMichael, A.J., Gotch, F.M., Noble, G.R. and Beare, P.A.S. (1983) *N. Engl. J. Med.* 309, 13–17.
Moller, G. (Ed) (1982) *Immunol. Rev.* Vol. 65. Munksgaard, Copenhagen. pp. 1–187.
Morris, A.G., Lin, Y.-L. and Askonas, B.A. (1982) *Nature* 295, 150–152.
Nisbet-Brown, E., Singh, B. and Diener, E. (1981) *J. Exp. Med.* 154, 676–687.
Nowinski, R.C., Stone, M.R., Tam, M., Lostrom, M.E., Burnette, W.N. and O'Donnell, P.V. (1980) In *Monoclonal Antibodies* (R. Kennett, T. McKern and K. Bechtol, Eds.) Plenum Press, New York. pp. 205–316.
Pease, L.R., Shulze, D.H., Pfaffenbach, G.M. and Nathenson, S.G. (1983) *Proc. Natl. Acad. Sci. U.S.A.* 80, 242–246.
Pfizenmaier, K., Jung, H., Starzinski-Powitz, A., Rollinghoff, M. and Wagner, H. (1977) *J. Immunol.* 119, 939–944.
Rasmussenn, L.E., Jordan, G.W., Stevens, D.A. and Merigan, T.C. (1974) *J. Immunol.* 112, 728–735.
Reiss, C.S. and Shulman, J.L. (1980) *Infect. Immun.* 29, 719–723.
Rhodes, J., Jones, D.H. and Bleehen, N.M. (1983) *Clin. Exp. Immunol.* 53, 739–743.
Rickinson, A.B., Wallace, L.E. and Epstein, M.A. (1980) *Nature* 283, 865–868.
Rosenthal, K.L. and Zinkernagel, R.M. (1980) *J. Immunol.* 124, 2301–2308.
Russell, S.M. and Liew, F.Y. (1979) *Nature* 280, 147–148.
Schrader, J.W. and Edelman, G.M. (1977) *J. Exp. Med.* 145, 523–539.
Shaw, S. and Biddison, W.E. (1979) *J. Exp. Med.* 149, 565–575.
Shaw, S., Shearer, G.M. and Biddison, W.E. (1980) *J. Exp. Med.* 151, 235–245.
Sherman, L.A., Vitiello, A. and Klinman, N.R. (1983) *Annu. Rev. Immunol.* 1, 63–86.
Staudt, L.M. and Gerhard, W. (1983) *J. Exp. Med.* 157, 687–704.
Thomas, D.B., Hackett, C.J. and Askonas, B.A. (1982) *Immunology* 47, 429–436.
Tosolini, F.A. and Mims, C.A. (1971) *J. Infect. Dis.* 123, 134–144.
Townsend, A.R.M. and Skehel, J.J. (1982) *Nature* 300, 653–657.
Virelizier, J.L., Allison, A.C. and Schild, G.C. (1979) *Brit. Med. Bull.* 35, 65–68.

Wabuke-Bunoti, M.A.N. and Fan, D.P. (1983) *J. Immunol.* 130, 2386–2391.

Wabuke-Bunoti, M.A.N., Fan, D.P. and Braciale, T.J. (1980) *J. Immunol.* 127, 1122–1125.

Ward, C.W. (1981) *Curr. Top. Immunol. Microbiol.* 94, 1–74.

Webster, R.G., Laver, W.G., Air, G.M. and Schild, G.C. (1982) *Nature* 296, 115–121.

Wiley, D.C., Wilson, I.A. and Skehel, J.J. (1981) *Nature* 289, 373–378.

Wilson, I.A., Skehel, J.J. and Wiley, D.C. (1981) *Nature* 289, 366–372.

Winter, G., Fields, S. and Brownlee, G.G. (1981) *Nature* 292, 72–75.

Wylie, D.E. and Klinman, N.R. (1981) *J. Immunol.* 127, 194–198.

Wylie, D.E., Sherman, L.A. and Klinman, N.R. (1982) *J. Exp. Med.* 155, 403–414.

Yewdell, J.W. and Gerhard, W. (1981) *Annu. Rev. Microbiol.* 35, 185–206.

Yewdell, J.W., Frank, M. and Gerhard, W. (1981) *J. Immunol.* 126, 1814–1819.

Zinkernagel, R.M. (1976) *J. Exp. Med.* 143, 437–443.

Zinkernagel, R.M. and Doherty, P.C., (1974) *Nature* 248, 701–702.

Zinkernagel, R.M. and Doherty, P.C. (1979) *Adv. Immunol.* 27, 51–177.

Zinkernagel, R.M. and Rosenthal, K.L. (1981) *Immunol. Rev.* 58, 131–155.

Zinkernagel, R.M., Althage, A. and Holland, J. (1978) *J. Immunol.* 121, 744–748.

Zweerink, H.J., Courtneidge, S.A., Skehel, J.J., Crumpton, M.J. and Askonas, B.A. (1977) *Nature New Biol.* 267, 354–356.

Eds. M.H.V. van Regenmortel and A.R. Neurath
Immunochemistry of viruses. The basis for serodiagnosis and vaccines.

Virus neutralization 5

BENJAMIN MANDEL

Department of Virology, The Public Health Research Institute of the City of New York Inc., New York, NY 10016 and Department of Biology, Long Island University, Brooklyn, NY 11201, U.S.A.

A pathogenic viral infection represents a triple-faceted phenomenon involving replication of virus, pathologic sequelae, and immunologic rebellion against the viral invader. An attempt to understand any one facet usually discloses its intimate relationships to the others. The early studies of Jenner and Pasteur indicated the existence of an inherent system capable of aborting or mitigating viral infections. However, recent studies have shown that in some instances this system may also be the root cause of pathologic manifestations. As early as the mid-1930s the interdependence of virology, immunology and pathology impressed Burnet who stated that, 'From the time of Jenner onward, immunological aspects have tended to dominate all studies of viruses pathologic for man and animals.'

Shortly after the demonstration that a bacterial toxin can induce the production of an antagonist (i.e. antibody), Sternberg (1892) demonstrated the analogous capability for viruses. The superficial resemblance of the chemical neutralization of an acid by a base to the inactivation of toxins and viruses by antibody, led to the application of the term 'neutralization' to these biological phenomena. With respect to viruses, neutralization connotes simply the nullification of the ability to infect as a result of interacting with antibody with, however, no implications as to mechanism.

Soon after Sternberg's demonstration of the immunogenic property of viruses and the in vitro neutralizing action of the respective antisera, other investigations showed these to be general phenomena. With the inclusion of bacterial viruses, the era of experimental viral immunology was inaugurated. Because of technical advantages, bacterial viruses became the system of choice for many of the early basic studies of neutralization. However, the study of animal virology, crude as it was in the 1920s and 1930s, was also extensively involved. A considerable body of data and concepts evolved from these studies as summarized by Burnet et al. (1937).

Interest in this subject was further stimulated by the studies and discussions of Dulbecco et al. (1956) and Fazekas de St. Groth and Reid (1958). Advances in the technology of animal virus propagation and quantitation prompted these and many other studies. Although a better comprehension is now at hand, fundamental problems of viral neutralization have yet to be solved. At present we are at the forefront of a new resurgence of interest stimulated by a major technological advancement – the method for obtaining monoclonal antibody. Recent investigations of viral neutralization have already shown the potential of this remarkable analytical tool.

1. Characteristics of the immune response

Although considerable information about the action and significance of antibody had accumulated by the mid-1930s, it was not until 1939 that it was learned that antibodies belong to the gamma globulin class of proteins. Subsequently it became evident that the term 'antibody' encompassed a family of diverse molecular species differing in physical, chemical and functional characteristics. In man, antibodies are divided into five classes: Immunoglobulin (Ig)G, IgM, IgA, IgD and IgE. Of these, the first three are involved in neutralization of viruses.

Immunoprotection against viral disease is afforded by a highly complex dual system consisting on the one hand of cellular elements, and on the other hand of soluble antibodies. Although the modus operandi of each is different, viz. cellular elements attack virus-infected cells, while antibodies neutralize extracellular virus, there are intricate interrelationships with regard to cooperation, production, regulation and function. Viral infections in which the cytopathology includes aberrations of plasma membrane proteins are controlled by cellular elements, while infections characterized by cytolysis with release of viral progeny are controlled by humoral or secretory antibody. It has become evident that this dichotomy of responsibility is not absolute; rather, one predominates with assistance from the other.

The qualitative and quantitative characteristics of the antibody response are determined by the dose of immunogen. The site of synthesis and/or localization of antibody depend on the site of entry of the immunogen and, in the case of viable virus, on the site(s) of replication and dissemination. Various experimental studies on the immune response have been conducted either using laboratory animals as models, or in human subjects receiving attenuated or inactivated vaccines or enduring naturally acquired infections. The results in all instances were in essential agreement, indicating the broad generality of the phenomena to be described.

1.1. Antibody response to low dose immunogenic stimulation

A single dose of immunogen in the unprimed host (e.g. $10^{6.4}$ PFU of poliovirus, representing about 0.05 μg of viral protein intravenously in a rabbit) induces IgM synthesis for about 5 days, after which antibody decays. Subsequent restimulation elicits essentially an equivalent response. Hence, feeble stimulation induces a response characterized by (*a*) a short period of synthesis, (*b*) IgM only, and (*c*) a failure to establish 'memory'.

1.2. Antibody response to high dose immunogenic stimulation

A single dose of immunogen in the unprimed host (e.g. $10^{7.4}$ PFU of poliovirus, representing about 0.5 μg of viral protein intravenously in a rabbit) induces production of IgM followed shortly by IgG and IgA. IgM titer rises for several days and then decays. In contrast, IgG synthesis persists for a variable but long time, possibly a full life span. The IgA response has not been intensively characterized but appears similar to IgG. Restimulation subsequent to IgM decay elicits an IgM response essentially the same as the initial response, whereas the IgG response differs from the initial response in that (*a*) the time lag before renewed synthesis is shortened, (*b*) the rate of synthesis is accelerated, and (*c*) the titer plateau is considerably elevated above the initial plateau.

The above studies (Svehag and Mandel, 1964) indicate that a durable high-level response in conjunction with 'memory' requires the transition from IgM to IgG synthesis, and that this requires a critical threshold of immunogenic stimulation. Probably during ongoing infection, naturally occurring or initiated by attenuated vaccines, there is sufficient output of progeny virus to attain the threshold before infection is aborted. With inactivated vaccines, the administered dose per se must provide the necessary critical mass of immunogen. The transient IgM response is of considerable diagnostic value since, it signifies an ongoing, or recently terminated, infection.

1.3. Sites of antibody synthesis and function

Serum antibodies are synthesized by terminally differentiated plasmacytes. These cells originate from primitive stem cells that differentiate sequentially into B-type lymphocytes producing IgM and then into plasmacytes yielding either IgG or IgA. Recent studies indicate that secretory IgA is synthesized in regional sites. For example, local stimulation of rat intestine elicits IgA-precursor cells in Peyer's patches followed by maturation and proliferation in the lamina propria of the rat gut (Husband, 1982). Cell cultures derived from mammary tissue of pregnant mice produced IgA during cultivation in vitro (Myrdal et al., 1983). Secretory IgA has been detected in milk, nasal secretions, tears, feces, saliva, the gut, and the central nervous system. In some disease states, e.g. IgA incompetence, IgG and IgM may substitute for IgA.

Almost any tissue or organ may serve as the site of viral replication in an animal host. However, in the vast majority of instances, invasion of the host is restricted to few access routes, and these are characteristically lined with mucous membranes. These membranes may be the primary and principal site of replication, or the primary site from which progeny virus disseminates to remote principal targets. A mechanism for antibody synthesis, or local extravasation, at the primary site clearly represents a major first-line of defense. To induce immunity through vaccination, it seems reasonable to mimic the natural infectious process by stimulating regional sites via local administration of vaccine.

2. Role of antibody as a deterrent to viral infection

Since antibody does not generally penetrate cells, its protective role as an independent agent is restricted to interaction with extracellular virus. Antibody can prevent infection when it encounters incoming virus, and it can ameliorate disease when it encounters viral progeny released from infected cells. Based on successful vaccination programs, it appears that acute viral diseases characterized by viremia and absence of a tendency toward latency, are most apt to be controlled by circulating antibody.

However, for infections displaying chronicity, prolonged intracellular habitat and latency, the role of antibody has not been clearly defined. Recent studies have focused on this problem as it applies to herpes simplex virus infections. Studies in athymic mice indicated restriction of viral replication at the site of inoculation (Kino et al., 1982), no restriction at the site of inoculation but reduction in secondary infection of ganglia and spinal cord (Kapoor et al., 1982), and no restriction at the primary site, no restriction of spread but reduced yield of virus from ganglia (Klein, 1980). These results suggest a positive role for humoral antibody principally in limiting the dissemination of virus.

A positive protective role for antibody has been described, for example, for rotavirus (Kapikian et al., 1983), respiratory syncytial virus (McIntosh et al., 1978), polioviruses (Ogra and Karzon, 1969; Lipton and Gonzalez-Scarano, 1978), coxsackievirus A24 (Langford et al., 1979), rubella virus (Schluederberg et al., 1978), rabies virus (Nicholson et al., 1979), Friend virus (Hunsmann et al., 1981) and Venezuelan equine encephalomyelitis virus (Mathews and Roehrig, 1982). To resolve the dilemma of recurrent infections by respiratory syncytial virus, Prince et al. (1983) evaluated the persistence of antibody in the upper (nose) and lower (lung) extremities of the respiratory tract of the cotton rat. Antibody persisted twice as long in the lungs, which they attributed to the source of antibody, viz. serum antibody in the lungs but local antibody in the nose.

3. The neutralization reaction

Although there were suggestions in the mid-1930s that viral infectivity is prevented by antibody that reacts with the target cell, there is at present no ambiguity about the direct interaction of viruses with homotypic antibody. With improvements in assay methods, studies of antigen-antibody interaction in the 1960s led to quantitative definitions of reaction parameters in terms of equilibrium, affinity, kinetics and thermodynamic evaluations.

Studies with chemically defined haptens indicated that the union of antigen and antibody is extremely rapid, necessitating sophisticated technology for measuring reaction rates. However, with viruses as antigens, assay by infectivity is sufficiently sensitive to permit the use of extremely low concentrations of reactants, with rates reduced to levels measurable in conventional time units. Studies of antigen-antibody interaction under diverse conditions – simple or complex antigens at low or high concentrations – have indicated the generality of the characteristics of antigen-antibody interaction involving broadly diverse antigens.

3.1. Chemical basis

Antigens and antibodies are bound by weak noncovalent short-range bonds of the electrostatic and van der Waals types. It is, therefore, essential that stereocomplementarity exist between the antigen and antibody. The higher the degree of complementarity, the greater the number of bonds. These forces are abetted by hydrophobic interactions resulting from mutual orientation of the reactants. Since the reaction is exothermic, binding is favored by decreasing temperature, whereas reaction rate is favored by increasing temperature because of increasing collision frequency. Finally, for most reactions the free energy change is negative, and the entropy change is positive, each change tending to drive the reaction forward.

Inasmuch as the binding forces are weak and noncovalent, the antigen-antibody complex may undergo spontaneous dissociation. This tendency varies considerably according to the binding properties of antibody and virus. With respect to antibody, binding affinity increases the longer an animal synthesizes that specific antibody. Valence–two, four, or ten–also contributes to the strength of the complex. With respect to the virus, the multiplicity of antigen sites per virion may afford opportunities for multiple monogamous bond formation. Such opportunities depend on the characteristics of the virion (e.g. topographical distribution and number of sites per unit of surface area, and presence of a lipoprotein envelope) and the characteristics of the antibody (e.g. bond affinity and valence). In order for dissociation to occur, all virion-antibody bonds must rupture synchronously, which is an improbable event.

Although the probability for spontaneous dissociation is of a low order of magnitude, reaction conditions can be modified to promote dissociation. Electrostatic bonds can be nullified by unusually low or high pH. Similarly, addition of salts such as divalent cations will decrease electrostatic as well as van der Waals forces. Other conditions, e.g. lowered surface tension by detergents, produce the same result. It has also been reported that some treatments may restore infectivity without concommitant dissociation (Kjellén, 1965), possibly resulting from a rearrangement thereby exposing critical sites.

Inasmuch as the antigen-antibody reaction is an equilibrium phenomenon, dilution of a reaction mixture at equilibrium should result in reversal in accordance with the equilibrium constant for this reaction. Reversal may be of sufficient magnitude to allow the virus to be detected by infectivity assay if the following conditions prevail: (*a*) dilution factor is sufficiently high; (*b*) antibody is of low affinity, and (*c*) dilution is done shortly after the reaction has reached equilibrium. The last requirement implies that after the initial formation of the antigen-antibody complex there is a tendency for the formation of additional bonds and/or existing bonds to acquire higher energy levels.

3.2. Kinetic analysis

Although the formation of the simplest virus-antibody complex is in fact a bimolecular reaction, when examined kinetically it conforms to a first-order reaction described by:

$$\frac{V_t}{V_0} = e^{-kt/D} \tag{1}$$

in which V_t and V_0 are the infectivity titers at times t and zero, k is a reaction rate constant, t is time (usually in minutes), and D is dilution factor of the serum. The neutralization reaction is, therefore, a pseudo-first-order system. The observed first-order kinetics imply that antibody is present in great excess over the amount consumed in the reaction.

When analyzed kinetically, the course of the reaction is graphically linear when plotted on semilogarithmic coordinates according to:

$$\log \frac{V_t}{V_0} = - \frac{k}{2.3D}\, t \tag{2}$$

This equation indicates that the slope of the line, represented by the coefficient of t, is inversely proportional to the value of D. Experimentally, it was shown (Dulbecco et al., 1956) that for a considerable part of the course of inactivation, equation (1) is applicable and that the rate of the reaction is proportional to the concentration of the antiserum (i.e. antibody). Of interest is the absence of a time lag preceding the start of inactivation suggesting that one molecule of antibody suffices to neutralize one virion. According to the hypothesis for multisite-dependent inactivation:

$$\frac{V_t}{V_0} = 1-(1-e^{-kt})^n \tag{3}$$

where n equals the number of 'hits' required for inactivation. In the case where n = 1, equation (3) is reduced to equation (1) and on this basis neutralization has been interpreted as a single-hit phenomenon. Although several critical and noncritical sites may be present on the viral surface, one molecule of antibody binding effectively to one critical antigenic site is sufficient regardless of the accretion of additional antibody molecules. It should be noted that this interpretation is not universally accepted as will be discussed in Section 6.

4. Neutralization test methods

The interaction of virus with antibody has biochemical and biological implications. Biochemically, the in vitro reaction has manifestations such as the formation of a precipitate, or the activation of the antibody molecule so that it acquires receptor activity for the complement system. Biologically, the virion is prevented from initiating infection. Procedures for examining these, as well as other manifestations have been standardized to produce data pertaining to antibody activity. It should be borne in mind that routine quantitative serological procedures do not measure directly the concentration of antibody molecules, but rather they provide estimates of antibody activity in accordance with an adopted set of laboratory conditions. The degree of activity is subject to variable conditions such as temperature, time, pH, salt concentration, etc. Comparative studies, therefore, require rigorous adherence to an adopted protocol of procedure in order for results to be meaningful.

The neutralization test consists of two phases: (*a*) setting up the virus-antiserum reaction in vitro and (*b*) assaying the result by infectivity titration. At present, infectivity titrations are performed preferably in cell cultures if possible, or if not, in susceptible animal hosts.

4.1. Quantal titration

The goal is to determine the minimal amount of antiserum that can reduce a small number of infectious units of virus (e.g. 100) to an average of 0.7 infectious units per volume of inoculum. This corresponds to the ID_{50} – the amount of virus for which there is a 0.5 probability of causing infection in one test subject. The procedure, commonly referred to as constant-virus-varying-serum, entails mixing a constant amount of virus with varying dilutions (preferably closely spaced) of antiserum. After allowing the reaction to proceed under standardized conditions, the concentration of residual infectious virus is assayed. Each mixture, without further dilution, is administered to several test subjects; the more test subjects used per mixture, the greater will be the statistical reliability. The 50% endpoint, i.e. that dilution of antiserum at which 50% of the test subjects display evidence of infection, is calculated by statistical interpolation according to methods such as that of Reed and

Muench (1938) or Kärber (1931). The result is expressed as a ratio, 1:X, where X is the dilution factor of the antiserum corresponding to the 50% endpoint for residual infectious virus.

4.2. Enumerative titration

An isolated viral lesion is an advanced stage of a progressing infection that evolved from a single cell. It has been shown that when the ratio of virus to cells is low, a lesion (i.e. pock, plaque, focus, etc.) originates from a single cell infected by a single infectious virion. Therefore, counting lesions is tantamount to counting viral particles that had 'succeeded' in initiating infection. 'Success' is the net result of numerous variables such as the proportion of functional particles, the homogeneity and susceptibility of the cells, and the biochemical and physical conditions of the assay system.

Enumerative assays of animal viruses are performed in monolayer cell cultures. Basically, the procedure involves two stages. First, the cells are propagated under conditions that will produce contiguous monolayers after several days' growth. The growth medium is then removed and a small specific volume of virus is dropped on the surface of the cell layer. The virus should be diluted to yield a suitable number of lesions: not so many that undercounting will result from coincidence errors, nor so few that large statistical variations will arise from dealing with small numbers. The infected monolayer is covered with an overlay medium that nourishes the cells and physically restricts the diffusion of released progeny virus only to the peripherally contiguous cells. This is achieved by incorporating a semisolid gelling agent such as agar into the overlay medium. The cultures are incubated long enough for the lesions to attain a visible size. Their presence is made more visible by incorporating a vital dye (e.g. neutral red) into the overlay medium. Lesions are counted and the titer is calculated in infectious units (IU)/ml according to the equation:

$$\mathrm{IU/ml} = \frac{n}{v \, . \, d} \tag{4}$$

where n is the number of lesions, v is the volume of the inoculum and d is the dilution of virus.

For assaying neutralizing activity, appropriately diluted virus is mixed with various dilutions of antiserum with the aim of finding the dilution that will cause a 50% reduction in lesion count. After allowing for virus-antibody interaction, the mixtures are assayed for residual virus as described above. Since it is unlikely that one of the virus-antiserum mixtures will show the exact 50% reduction, it is most convenient to estimate the dilution of antiserum corresponding to this reduction by graphic interpolation.

4.3. Kinetic titration

As indicated above (Section 3), the early course of neutralization is typical of a first-order reaction. The empirical neutralizing activity of an antiserum can be quantitatively characterized by the value of k, the reaction rate constant as defined in equation (1).

In essence, the kinetic assay of the neutralizing potency of an antiserum requires the evaluation of k. For this purpose the antiserum is diluted to a level that will allow measurement of the reaction at a moderate rate, e.g. a $\log_{10}$ in 5–10 minutes with a concentration of 10^6–10^7 IU/ml of virus. With an antiserum of unknown activity, preliminary trials will indicate an appropriate dilution. Since in the late stage the reaction may tend to deviate from the true course, it is necessary to collect several samples during the early stage. The reaction characteristics in this respect vary considerably for different virus-antibody systems. Hence, each system needs to be characterized individually. Inasmuch as the status of the reaction is being assessed at a specific time, it is essential that the collected sample be subjected immediately to a high degree of dilution in order to stop the reaction, and then be assayed promptly. The resultant data are plotted graphically as indicated by equation (2), viz. $\log_{10} V_t/V_o$ as a function of t. Suitable paired values for V_t/V_o and t can be read off the linear part of the plot and used to calculate k according to the equation:

$$k = \log_{10}\left(\frac{V_0}{V_t}\right).\frac{2.3D}{t} \qquad (5)$$

The kinetic assay procedure is sensitive to small variations in reaction conditions, and to small differences in antigenic specificity. By kinetic analysis Gould and Butler (1980) succeeded in detecting strain distinctions among several rubella virus isolates.

4.4. Statistical considerations

The quantitative index of the neutralizing activity of an antiserum is based on its effect on viral infectivity. Therefore, the statistical considerations are those that pertain to viral assay methods. The variability of the quantal and the enumerative assay methods are considerably different, as described by Dulbecco and Ginsberg (1980). A count of 100 lesions by the enumerative method can be expected to vary ±20% in 95% of repetitive trials. For the same level of confidence by the quantal method, about 100 replicate cell cultures would be required. Furthermore, the variability can be expected to be still greater if animals are used.

Since infectivity is the most sensitive measurable parameter for virus quantitation, it follows that neutralization should be the most sensitive method for measuring antibody activity.

5. Relevance of the neutralization test for serodiagnosis and for evaluation of vaccine efficacy

During viral replication in an immunocompetent host, antibodies will be elicited by structural as well as nonstructural viral proteins. Since neutralization involves only the structural proteins that are externally exposed on the surface of the virion, only antibodies specific for these proteins are of interest. It has become evident, however, that among these proteins only some are critical for neutralization. With the advent of monoclonal antibodies, a further subdivision has become necessary since a given protein may contain several antigenic determinants, only one or some of which are significant for neutralization. Consequently, it is evident that the presence of antibody for an unspecified antigen may not be relevant to the immune status of an individual, or to the protective efficacy of a vaccine.

Recent reports exemplify these considerations. Since individuals with antihemagglutinin antibody for rubella virus developed rubella infection, Schluederberg et al. (1978) recommended testing for neutralizing antibody as the only reliable indicator of immune status. Conversely, with influenza virus, vaccinees responded with neutralizing, but not HI antibody (Ennis et al., 1982). These investigators recommended the plaque neutralization test as their first choice for screening antibodies. Swenson et al. (1983) noted that antibody specific for the *d* subtype of hepatitis B virus failed to protect against infection by virus of the *ayw* subtype. Since the *a* component is common to all subtypes, it was considered to be the neutralization antigen. Among studies on hemagglutinating viruses, several reports have pinpointed the distinction between antibodies that inhibit hemagglutination and antibodies that neutralize (Foy et al., 1981; Chanas et al., 1982; Finberg et al., 1982; Peiris et al., 1982; Yewdell and Gerhard, 1982; Russell et al., 1983). Ukkonen and Penttinen (1981) found that IgG, but not IgM, was a reliable index of immune status with respect to mumps virus infection. During herpes infection, antibodies appear sequentially to early viral membrane antigens and then to late viral membrane antigens, but only the latter have neutralizing potential (Mikami et al., 1980).

The recent use of monoclonal antibodies has emphasized the diversity of viral determinants. Since all can evoke antibodies, but only some are relevant to neutralization, the choice of a serological procedure should be made with a view to the purpose of the search. For identification of a virus, or quantitative comparisons of antibody levels as diagnostic aids, any procedure of appropriate sensitivity, specificity, ease of performance, etc. is useful. For judging the immunoprotected state of an individual or the immunoprotective potential of a vaccine, the neutralization test is the most relevant indicator.

6. Mechanism of neutralization

The binding of antibody to virus may or may not result in neutralization depending on variables such as the antigenic site, the relative proportions of the reactants, the lineage of the cells, and the participation of other substances such as complement, rheumatoid factor, etc. (see reviews by Daniels, 1975; Mandel, 1978; 1979). At one extreme, neutralization may be the consequence of the mere cross-linking of virions to form an aggregate that registers as a single infectious unit. At the opposite extreme, virus particles with lipoprotein envelopes undergo physical degradation from the cooperative effects of antibody and complement. Between these extremes, antibody may thwart an early stage in the replication cycle by obstructing attachment, abrogating uncoating, or inhibiting a subsequent functional activity. Binding of antibody to a site that does not result in neutralization may, however, be mediated by the binding of a third component, e.g. complement, to the virus-antibody complex. Virus bound to nonneutralizing antibody is said to be sensitized.

The usual procedure for studying neutralization has been to prepare antiserum using intact virus as immunogen. Although it was known that such antisera contain antibodies of different classes and different avidities, there was no appreciation of the number of antibodies differing in specifity for different antigenic determinants (epitopes). This uncertainty has been resolved through the remarkable procedure introduced by Köhler and Milstein (1975) whereby antibody populations are derived from clonal populations of cells all of which are synthesizing antibody with precisely the same specificity for a single epitope. Such antibody has been called monoclonal antibody. The recent plethora of studies using monoclonal antibodies has revealed the multiplicity of epitopes associated with a given virus. It has become possible to define epitopes as neutralizing (critical) or nonneutralizing (noncritical) by use of monoclonal antibodies. In addition, it has become possible to allocate relative positions of the various epitopes by topological analyses. The presence of several epitopes in a single polypeptide is perhaps not unexpected in view of the minimum size of an antigenic determinant, e.g. a hexapeptide. Previous studies of the mechanism of neutralization based on the use of conventional (i.e. polyclonal) antisera defined two problems: (*a*) the minimum number of molecules of antibody per virion that is required, and (*b*) the precise effect on the virion that renders it noninfectious. Dulbecco et al. (1956) proposed on the basis of kinetic analysis of the neutralization reaction that one molecule sufficed. In opposition to this single-hit hypothesis, Della-Porta and Westaway (1978) presented arguments favoring a multi-hit mechanism. As to the second problem, Mandel (1976) presented data suggesting that one molecule of antibody induces a rearrangement in the quaternary structure of the viral capsid that renders it resistant to uncoating. In contrast, Della-Porta and Westaway (1978) proposed that neutralization is a consequence of steric blockade of viral attachment sites by antibody, several being required. Thus ended the polyclonal era.

The presumption that neutralization is a generic phenomenon representing several distinct neutralization pathways (Daniels, 1975; Mandel, 1979) has found support in recent studies in which monoclonal antibodies have been used, as will be described below.

6.1. The capsid rearrangement pathway

Studies on the neutralization of poliovirus by rabbit polyclonal antiserum (Mandel, 1971; 1976) showed that the capsid of this virus oscillates between two different conformations characterized by an isoelectric point (pI) of 7.0 or 4.5. Under physiological conditions the vast majority of particles are in the pI 7.0 state. However, when neutralized, particles are stabilized in the pI 4.5 state. It was also reported (Mandel, 1976) that neutralized virions resisted early-stage uncoating in an in vitro uncoating system. Hence, it was proposed that antibody induced a conformational reorganization of the viral capsid and thereby rendered it refractory to uncoating.

Recent studies have described the use of monoclonal antibodies for neutralization of poliovirus. Icenogle et al. (1983) studied the stoichiometry and kinetics of the neutralization reaction. Although the kinetic analysis indicated single-hit inactivation, it was found that 4 molecules of antibody per virion were required. The antibodies were bivalently bound and it was judged that each antibody cross-linked 2 pentameric subunits of the capsid. It was therefore proposed that by cross-linking 4 pairs of pentamers, uncoating was prevented. Electrophoretic analysis showed that the pI had not changed as a result of neutralization.

Emini et al. (1982; 1983c) isolated monoclonal antibodies for 7 different epitopes on poliovirus. All 7 were neutralizing and all were located on one of the capsid polypeptides. With one exception, these antibodies only partially prevented attachment and penetration. Attachment of neutralized poliovirus was previously described with the use of polyclonal antiserum (Mandel, 1967). When neutralized by the monoclonal antibodies, poliovirus showed conformational rearrangement as indicated by a change in pI from about 7.0 to about 4.1–3.6. Of interest, the one monoclonal antibody that neutralized and blocked attachment failed to induce a conformational transition. It was suggested, that the epitope for this antibody might be the attachment site (or a part thereof). It was also observed that an antibody that neutralized the Sabin strain of poliovirus in conjunction with a conformational change, failed to neutralize the Mahoney strain and did not affect the pI, although the antibody did bind to the virion. Finally, inability to demonstrate virus-specific transcriptase activity in cells that had reacted with neutralized virus was taken as evidence that uncoating of virus had not occurred. In another study (Emini et al., 1983d) the observation of Icenogle et al. was confirmed that neutralization required monogamous bivalent binding of antibody.

Neutralization of a quite different virus, the enveloped helical influenza virus, was described by Possee and Dimmock (1981) and Possee et al. (1982). Although polyclonal or monoclonal antibodies neutralized infectivity, all early stages including uncoating occurred in a near-normal manner. The uncoated nucleoprotein, when extracted from the exposed cells, was virtually devoid of transcriptase activity. The epitope involved in neutralization by the monoclonal antibody resided on the hemagglutinin glycoprotein. These investigators suggested that allosteric rearrangements induced by antibody in the hemagglutinin structure can be transmitted across the lipoprotein envelope to the inner located nucleoprotein, thereby affecting the viral transcriptase.

6.2. The aggregation pathway

Brioen et al. (1983) obtained a neutralizing monoclonal antibody to poliovirus that caused aggregate formation. By centrifugation analysis of a mixture of virus and antibody, various polymers of virions were obtained. The larger the polymer, the lower was the ratio of infectivity to virion. Of particular significance was their observation that virions that remained as monomers not only retained infectivity but also had not bound antibody. It was proposed that topological restrictions of antigenic sites excluded the possibility for monogamous bivalent binding for this specific epitope-paratope combination. These investigators concluded that there is '. . . no need to postulate any "intrinsic" modification of the individual virions although such modifications cannot be ruled out either'.

6.3. The attachment blockade pathway

Nowinski et al. (1981) isolated a neutralizing monoclonal antibody that reacted with an epitope on the gp70 glycoprotein of murine leukemia virus. Since this epitope, i.e. $gp70^f$, is proximal to the viral attachment site, it was suggested that neutralization is attributable to interference with attachment to cell receptors. A similar conclusion, based on more direct experimental evidence, was presented by Massey and Schochetman (1981) for mouse mammary tumor virus. Virus became neutralized if it was allowed to react with antibody first and was then exposed to cells, whereas virus escaped neutralization if exposed to cells first and then treated with antibody. Since in the latter situation virus had not yet penetrated and was therefore accessible to antibody, it was concluded that neutralization of this virus by this monoclonal antibody was due to obstruction of the attachment site on the virion. Because a monospecific antibody for the HN glycoprotein of paramyxovirus SV5 neutralized the virus, Merz et al. (1981) concluded that attachment was the antibody-sensitive stage. Hardy and Moore (1982) isolated an antibody from a polyclonal antiserum that cross-reacted with the 140S intact virion and the 12S subviral structure of foot-and-mouth disease virus. Although the antibody binds to a noncritical site, neutralization can be mediated by addition of antiglobulin antibody. It was proposed that the noncritical site is adjacent to the attachment site which, in turn, is blocked by the antiglobulin molecule.

6.4. Single-hit versus multi-hit requirement

The proposal that one molecule of antibody is sufficient to neutralize one virion (Dulbecco et al., 1956; Mandel, 1976) has been challenged by Della-Porta and Westaway (1978), who favor the need for several molecules critically distributed over the viral surface. Recent studies have provided evidence, mostly circumstantial, in support of a multi-hit mechanism. Peiris et al. (1982) observed a synergistic effect when neutralizing monoclonal antibodies specific for different epitopes reacted with West Nile virus. Also in support was their observation that antibody enhances infectivity at low multiplicity but neutralizes at high multiplicity. Also based on synergistic effect, Volk et al. (1982) envisioned a need for several antibody molecules to neutralize vesicular stomatitis virus. Icenogle et al. (1983) found that four molecules was the minimum number associated with a neutralized poliovirion, whereas kinetic analysis indicated that one was sufficient. Two explanations were considered: either one of four sites is critical, or each is critical and each reduced infectivity by a factor of 0.75. Based on mathematical considerations of the interaction of viruses with antibody as defined by Krummel and Uhr (1969), Icenogle et al. reconciled the stoichiometric and kinetic implications of their data consistent with a multi-hit mechanism. Similarly, Rappaport (1970) concluded that neutralization of bacteriophage required 4–5 molecules per virion.

6.5. Susceptibility of viral proteins to conformational rearrangement

Although antibody-induced conformational changes in proteins (e.g. enzymes) have been previously reported, only recently has this phenomenon been described for viruses. A change in the pI of poliovirus (Mandel, 1971; Emini et al., 1983c) and the inactivation of the transcriptase of influenza virus (Possee et al., 1982) were considered to be antibody-activated conformational rearrangements that rendered these viruses neutralized. Boudin and Boulanger (1981) observed the dissociation of the penton capsomer of adenovirus following interaction with antifiber antibody. Although these investigators did not relate this event to neutralization, it is a likely consequence. Similarly, Clegg et al. (1983) noted that a monoclonal antibody specific for the E1 component of the E1-P2 complex of Sindbis virus caused the complex to dissociate. Thorpe et al. (1982) concluded that 'sequence determinants alone do not mediate neutralization which may depend on antigenic determinants specified by complex conformational arrangements of the viral proteins' of poliovirus. Crainic et al. (1983) obtained neutralization-resistant variants of vaccine strains of poliovirus which were considered to be resistant to the necessary antibody-inducible conformational rearrangement although ability to bind antibody was not lost.

7. Conformational variability of viral proteins and its significance for their use as immunogens

Recent investigations have indicated the potential usefulness of viral immunogenic peptides as vaccines for foot-and-mouth disease (Bittle et al., 1982) and for poliomyelitis (Emini et al., 1983b). Use of a single immunogenic peptide, either derived from virions or synthetic, elicits antibody of restricted specificity. Emini et al. (1982; 1983a) and Crainic et al. (1983) have encountered neutralization-resistant variants of poliovirus when neutralization tests were done with monoclonal antibodies. Since the frequency of such variants is fairly high, e.g 1/10 000, the efficacy of a vaccine aimed at a single epitope may soon be circumvented.

The use of virus-specific peptides as immunogens may need to be evaluated in terms of (*a*) duration of the antibody response since it is a nonreplicating substance, (*b*) the ease whereby a virus can out-maneuver antibody by genetic variation (e.g. Laver et al., 1979, have shown that a single amino acid substitution in the hemagglutinin of influenza virus converted an epitope to one of different specificity), and (*c*) establishment of anamnesis (although Emini et al. (1983b) reported successful priming with a poliovirus peptide even in the absence of a detectable antibody response).

8. Conclusion

The role of antibody in immunoprotection has been established based on evidence that is indirect (e.g. consequences of gamma globulin dyscrasias), and direct (e.g. results of passive immunization with pooled hyperimmune gamma globulin). Following recognition of cell-mediated immune mechanisms, the role of antibody has been more precisely defined in terms of its independent action in neutralizing extracellular virus, and in terms of its participation as a mediator of cytotoxic cell activity. With the advent of monoclonal antibodies, it has become evident that a given virus particle represents a family of antigenic determinants, only some members of which have a role in neutralization. Consequently, in the immune response to viral infection, antibodies may be elicited to all antigenic determinants, but only some may contribute to the immune status of the individual. By use of the neutralization test, antibodies can be detected that have a protective function. It has been suggested that in vivo, nonneutralizing antibodies may also contribute through joint action with complement. Perhaps not relevant to the present discussion, but of interest nonetheless, is the observation that antibody may *promote* infectivity by providing a recognition link (Fc region) for cells that possess Fc receptors.

A considerable body of data has accumulated on the mechanism of neutralization. Various studies have delineated different mechanisms. Recent analyses involving 'simplified' systems (i.e. virus interacting with monoclonal antibody) have indicated the likelihood that neutralization may be the same end-result arrived at by different routes. This is perhaps to be expected since several neutralization targets are available: the virion itself, cell attachment, penetration, uncoating and initiation of viral specific macromolecular synthesis. For example, Merz et al. (1981) have described a monoclonal antibody for an epitope on the HN structure of paramyxovirus that neutralizes by preventing attachment, and a different monoclonal antibody for an epitope on the F structure that neutralizes by preventing uncoating. Heinz et al. (1983) described two monoclonal antibodies for tick-borne encephalitis virus, one of which neutralizes by blocking attachment, the other by an effect on the virion, possibly conformational derangement. Neutralization by steric obstruction is an easily accepted concept. In fact, the same result can be achieved with a monoclonal antibody directed at a viral receptor site on the cell surface (Campbell and Cords, 1983).

If the role of neutralizing antibody is accepted as a significant constituent of the immune apparatus, recent studies have established the following facts of immediate interest: (*a*) a virus particle comprises a group of epitopes; (*b*) antibody specific for only one, or some, may be relevant to immunity; (*c*) a meaningful antibody response requires stimulation with sufficient immunogen to produce IgG antibody directed at the appropriate epitope; (*d*) the route of vaccine administration may have to be considered with the goal of stimulating local or circulatory antibody according to the pathogenetic characteristics of the particular viral infection; and (*e*) evaluation of the immune status of an individual and evaluation of vaccine efficacy should be based on quantitation of neutralizing antibody per se, rather than antibody for an undetermined epitope.

9. References

Bittle, J.L., Houghten, R.A., Alexander, H., Shinnick, T.M., Sutcliffe, J.G., Lerner, R.A., Rowlands, D.J. and Brown, F. (1982) *Nature* 298, 30–34.

Boudin, M.-L. and Boulanger, P. (1981) *Virology* 113, 781–786.

Brioen, P., Dekegel, D. and Boeyé, A. (1983) *Virology* 127, 463–468.

Burnet, F.M., Keogh, E.V. and Lush, D. (1937) *Aust. J. Exp. Med. Sci.* 15, 231–368.

Campbell, B.A. and Cords, C.E. (1983) *J. Virol.* 48, 561–564.

Chanas, A.C., Gould, E.A., Clegg, J.C.S. and Varma, M.G.R. (1982) *J. Gen. Virol.* 58, 37–46.

Clegg, J.C.S., Chanas, A.C. and Gould, E.A. (1983) *J. Gen. Virol.* 64, 1121–1126.

Crainic, R., Couillin, P., Blondel, B., Cabau, N., Boué, A. and Horodniceanu, F. (1983) *Infect. Immun.* 41, 1217–1225.

Daniels, C.A. (1975) In *Viral Immunology and Immunopathology* (A.L. Notkins, Ed.) Academic Press, New York. pp. 79–97.

Della-Porta, A.J. and Westaway, E.G. (1978) *J.Gen. Virol.* 38, 1–19.

Dulbecco, R. and Ginsberg, H.S. (1980) In *Virology*, Harper and Row, New York. p. 883.

Dulbecco, R., Vogt, M. and Strickland, A.G.R. (1956) *Virology* 2, 162–205.

Emini, E.A., Jameson, B.A., Lewis, A.J., Larsen, G.R. and Wimmer, E. (1982) *J. Virol.* 43, 997–1005.

Emini, E.A., Dorner, A.J., Dorner, L.F., Jameson, B.A. and Wimmer, E. (1983a) *Virology* 124, 144–151.

Emini, E.A., Jameson, B.A. and Wimmer, E. (1983b) *Nature* 304, 699–703.

Emini, E.A., Kao, S.-Y., Lewis, A.J., Crainic, R. and Wimmer, E. (1983c) *J. Virol.* 46, 466–474.

Emini, E.A., Ostapchuk, P. and Wimmer, E. (1983d) *J. Virol.* 48, 547–550.

Ennis, F.A., Qi, Y.-H. and Schild, G.C. (1982) *J. Gen. Virol* 58, 273–281.

Fazekas de St. Groth, S. and Reid, A.F. (1958) *J. Immunol.* 80, 225–235.

Finberg, R., Spriggs, D.R. and Fields, B.N. (1982) *J. Immunol.* 129, 2235–2238.

Foy, H.M., Cooney, M.K., Allan, I.D., Frost, F., Blumhagen, J.M. and Fox, J.P. (1981) *J. Infect. Dis.* 143, 700–706.

Gould, J.J. and Butler, M. (1980) *J. Gen. Virol.* 49, 423–426.

Hardy, M.M. and Moore, D.M. (1982) *J. Gen. Virol.* 62, 287–295.

Heinz, F.X., Berger, R., Tuma, W. and Kunz, C. (1983) *Virology* 130, 485–501.

Hunsmann, G., Schneider, J. and Schulz, A. (1981) *Virology* 113, 602–612.

Husband, A.J. (1982) *J. Immunol.* 128, 1355–1359.

Icenogle, J., Shiwen, H., Duke, G., Gilbert, S., Rueckert, R. and Anderegg, J. (1983) *Virology* 127, 412–425.

Kapikian, A.Z., Wyatt, R.G., Levine, M.M., Yolken, R.H., VanKirk, D.H., Dolin, R., Greenberg, H.B. and Chanock, R.M. (1983) *J. Infect. Dis.* 147, 95–106.

Kapoor, A.K., Nash, A.A., Wildy, P., Phelan, J., McLean, C.S. and Field, H.J. (1982) *J. Gen. Virol.* 60, 225–233.

Kärber, G. (1931) *Arch. Exp. Pathol. Pharmakol.* 162, 480–483.

Kino, Y., Hayashi, Y., Hayashida, I. and Mori, R. (1982) *J. Gen. Virol.* 63, 475–479.

Kjellén, L. (1965) *Virology* 27, 580–588.

Klein, R.J. (1980) *J. Virol.* 49, 401–405.

Köhler, G. and Milstein, C. (1975) *Nature* 256, 495–497.

Krummel, W.M. and Uhr, J.W. (1969) *J. Immunol.* 102, 772–785.

Langford, M.P., Stanton, G.J., Barber, J.C. and Baron, S. (1979) *J. Infect. Dis.* 139, 653–658.

Laver, W.G., Air, G.M., Webster, R.G., Gerhard, W., Ward, C.W. and Dopheide, A.A. (1979) *Virology* 98, 226–237.

Lipton, H.L. and Gonzalez-Scarano, F. (1978) *J. Infect. Dis.* 137, 145–151.

Mandel, B. (1967) *Virology* 31, 238–247.

Mandel, B. (1971) *Virology* 44, 554–568.

Mandel, B. (1976) *Virology* 69, 500–510.

Mandel, B. (1978) *Adv. Virus Res.* 23, 205–268.

Mandel, B. (1979) In *Comprehensive Virology* (H. Fraenkel-Conrat and R.R. Wagner, Eds.) Vol. 15. Plenum Press, New York. pp. 37–121.

Massey, R.J. and Schochetman, G. (1981) *Virology* 115, 20–32.

Mathews, J.H. and Roehrig, J.T. (1982) *J. Immunol.* 129, 2763–2767.

McIntosh, K., Masters, H.B., Orr, I., Chao, R.K. and Barkin, R.M. (1978) *J. Infect. Dis.* 138, 24–32.

Merz, D.C., Scheid, A. and Choppin, P.W. (1981) *Virology* 109, 94–105.

Mikami, Y., Inoue, M., Kodama, H., Inage, F., Onuma, M. and Izawa, H. (1980) *J. Gen. Virol.* 47, 221-226.

Myrdal, S.E., Snoddy, S., Ashkar, S., Nixon, D.W. and Binkley, F. (1983) *J. Immunol.* 131, 869–876.

Nicholson, K.G., Cole, P.J., Turner, G.S. and Harrison, P. (1979) *J. Infect. Dis.* 140, 176–182.

Nowinski, R.C., Pickering, R., O'Donnell, P.V., Pinter, A. and Hammerling, U. (1981) *Virology* 111, 84–92.

Ogra, P.L. and Karzon, D.T. (1969) *J. Immunol.* 102, 1423–1430.

Peiris, J.S.M., Porterfield, J.S. and Roehrig, J.T. (1982) *J. Gen. Virol.* 58, 283–289.
Possee, R.D. and Dimmock, N.J. (1981) In *ICN-UCLA Symposium* (D.P. Nayak and C.F. Fox, Eds.) Vol XXI. Academic Press, London. pp. 473–480.
Possee, R.D., Schild, G.C. and Dimmock, N.J. (1982) *J. Gen. Virol.* 58, 373–386.
Prince, G.A., Horswood, R.L., Camargo, E., Koenig, D. and Chanock, R.M. (1983) *Infect. Immun.* 42, 81–87.
Rappaport, I. (1970) *J. Gen. Virol.* 6, 25–32.
Reed, L.J. and Muench, H. (1938) *Am. J. Hyg.* 27, 493–497.
Russell, P.H., Griffiths, P.C., Goswani, K.K.A., Alexander, D.J., Cannon, M.J. and Russell, W.C. (1983) *J. Gen. Virol.* 64, 2069–2072.
Schluederberg, A., Horstmann, D.M., Andiman, W.A. and Randolph, M.F. (1978) *J. Infect. Dis.* 138, 877–883.
Sternberg, G.M. (1892) *Trans. Assoc. Am. Physicians* 7, 68–86.
Svehag, S.-E. and Mandel, B. (1964) *J. Exp. Med.* 119, 1–19.
Swenson, P.D., Escobar, M.R., Carithers, R.L. Jr. and Sobieski, T.J. III (1983) *J. Clin. Microbiol.* 18, 305–309.
Thorpe, R., Minor, P.D., Mackay, A., Schild, G.C. and Spitz, M. (1982) *J. Gen. Virol.* 63, 487–492.
Ukkonen, P. and Penttinen, K. (1981) *J. Infect. Dis.* 144, 496–497.
Volk, W.A., Snyder, R.M., Benjamin, D.C. and Wagner, R.R. (1982) *J. Virol.* 42, 220–227.
Yewdell, J.W. and Gerhard, W. (1982) *J. Immunol.* 128, 2670–2675.

Eds. M.H.V. van Regenmortel and A.R. Neurath
Immunochemistry of viruses. The basis for serodiagnosis and vaccines.

Immunoelectron microscopy of viruses

6

ALEXANDER KOHN and DAVID KATZ

Israel Institute for Biological Research, P.O.Box 19, Ness Ziona 70450, Israel

It is now generally agreed that the electron microscope is an indispensable tool for the identification, classification and diagnosis of viruses. Its role in virology is equivalent to that played by the light microscope in bacteriology. The visualization of virus-antibody interactions by immunoelectron microscopy adds the dimension of antigenic specificity to the morphological identification of viruses and immunoelectron microscopy (IEM) is, therefore, a more specific technique than electron microscopy (EM).

Different types of virus-antibody interactions can be observed by IEM, and several techniques have been developed to identify viruses in thin tissue sections by means of labeled antibodies, or in suspensions by simple positive or negative staining techniques.

In this chapter we shall describe the development of IEM techniques with an emphasis on methodology. The different methods will be discussed with examples from plant, animal and human virology.

1. Immunoelectron microscopy (IEM) of viruses in suspension

The negative staining procedure introduced by Brenner and Horne in 1959 greatly improved the visualization of viral particles, and led to a considerable increase in our knowledge of viruses. As a result, morphology became an important criterion for differential diagnosis, especially when similar clinical signs were caused by viruses differing in their morphology (Almeida et al., 1974; Doane and Anderson, 1977; Almeida, 1980; Kjeldsberg, 1980; Field, 1982; Palmer and Martin, 1982).

When viruses that cause similar symptoms also have a similar morphology, only the combined use of EM and serology permits a specific identification of an etiological agent. The interaction of a virus with its homologous antibodies was first observed in the electron microscope by Anderson and Stanley (1941) in their study of tobacco mosaic virus (TMV). In subsequent years, IEM techniques were extensively used for the detection and identification of many noncultivable or elusive viruses such as those causing warts (Almeida and Goffe, 1965; Pfister et al., 1981), rubella (Best et al., 1968), hog cholera (Ritchie and Fernelius, 1967), hepatitis A and B (Bayer et al., 1967; Feinstone et al., 1973), St. Louis encephalitis (Luby et al., 1980) and non-A non-B hepatitis (Marciano-Cabral et al., 1981; Yoshizawa et al., 1981) as well as viruses such as coronaviruses (Kapikian et al., 1980), parvoviruses (Paver et al.,1973), rotaviruses (Flewett and Boxall, 1976), caliciviruses (Chiba et al., 1979), rhinoviruses (Kapikian et al., 1972) and astroviruses (Konno et al., 1982), and agents such as the Norwalk and Norwalk-like agents (Kapikian et al., 1980; Oshiro et al., 1981), the Snow Mountain agent (Dolin et al., 1982) and the Otofuke agent (Tanigushi et al., 1979).

Technical details of the IEM methods and their use for different systems have been described in numerous reviews (Almeida and Waterson, 1969; Doane and Anderson, 1977; Milne and Luisoni, 1977a; Almeida, 1980; van Regenmortel, 1981; 1982; Field, 1982; Chernesky, 1983; Katz and Kohn, 1984).

The IEM techniques can be subdivided according to the observed results of the virus-antibody interaction into:

(*a*) Antibody-induced clumping of virions,

(*b*) Specific serological trapping of virions on grids or other solid surfaces, and

(*c*) Virus coating (decorating) with antibodies.

These 'clumping', 'trapping' and 'decorating' phenomena can be brought about by different methods, some of which are listed in Table 6.1.

1.1. Direct 'classical' IEM

The classical IEM method has different variations, all of them involving one basic procedure–the incubation of antibody with the virus, and the visualization of the resulting agglutination (clumping) of virions by EM. Different incubation times, temperatures and centrifugation steps have been used in different systems. Many plant virologists use the simple 'leaf dip' technique for viruses that are abundant in the tissue, provided there is no serious interference by cell debris (Ball, 1971; Milne and Luisoni, 1977a). For animal viruses that are less abundant in clinical specimens, a centrifugation step is sometimes needed (Almeida and Waterson, 1969; Milne and Luisoni, 1977a; Chernesky, 1983).

The formation of clumps in the 'classical' IEM technique is dependent on optimal reagent proportions as in the immunoprecipitation technique. Antibody or antigen excess inhibits lattice formation, and this prevents the formation of clumps.

TABLE 6.1

Immunoelectron microscopical (IEM) methods

	Technique group			
	'Classical'	Agar filtration	Pseudoreplica IEM	Immunosorbent electron microscopy (ISEM)
Name of technique	1. Direct with or without centrifugation (Almeida and Waterson, 1969) 2. Indirect (Valters et al., 1975)	1. Agar diffusion filtration (ADF) (Kelen et al., 1971) 2. Indirect ADF (Berthiaume et al., 1981b) 3. Serum in agar (SIA) (Anderson and Doane, 1973)	1. Immunopseudoreplica (El Mekki et al., 1979) 2. Microwell pseudoreplica EM (MW-PR-IEM) (Katz et al., 1983)	1. Antibody-coated grid technique (AB-CGT) (Derrick, 1973) 2. Protein A-coated grid technique (PA-CGT) (Shukla and Gough 1979) 3. Protein-coated bacteria technique (PA-CBT) (Katz et al., 1980) 4. Antigen-coated grid technique (AG-CGT) (Milne and Luisoni, 1975)
Morphology of virus-antibody interaction	Clumping	Clumping	Clumping	1–3. Single virion trapped on solid phase. 4. Single virion decorated with antibodies.

Thus many single virions are lost in the supernatant following centrifugation. If the reaction suspension is not centrifuged, the single virions may adsorb to the grid and become coated with antibody, in which case it may be difficult to differentiate between them and antibody-free unrelated virions.

The prozone effect was reported by Almeida and Waterson (1969) for polyoma and wart viruses and by Locarnini et al. (1974) and Dienstag et al. (1976) for hepatitis A viruses. Elkana and Guttman-Bass (1983) re-evaluated the 'classical' IEM technique for poliovirus and concluded that its efficiency was also influenced by the configuration of the virus antibody complexes. Odenwald et al. (1978) used the prozone effect for serotyping avian infectious bronchitis viruses, and determined the relative degree of aggregation for homologous and heterologous systems when the amount of virus was kept constant and the dilution of antiserum was varied.

1.2. Indirect IEM

The dependence of the 'classical' IEM technique on equivalent concentrations of antibodies and viruses is inconvenient, especially when the concentration of virions in the clinical specimen is not known (Almeida and Waterson, 1969). To obviate the use of several antibody dilutions, Valters et al. (1975) used an indirect technique in which predetermined concentrations of anti-immunoglobulins were employed for agglutinating antibody-coated single adenovirus particles. A similar technique was used by Edwards et al. (1975) for the identification of viruses associated with respiratory diseases. Both reports demonstrated that indirect IEM was significantly more sensitive than direct IEM.

According to our experience with the indirect method, large aggregates often obscure the specific virus-antibody clumps, a difficulty that was also reported by Edwards et al. (1975).

1.3. Agar filtration techniques

Prior to IEM, viruses can be concentrated and partly purified by agar filtration. It is also possible to visualize the virions present in immunodiffusion precipitates, but this method is not very sensitive and sometimes fails completely (Field, 1982; Barron, 1971).

1.3.1. Agar diffusion filtration (ADF)

Kelen et al. (1971) incubated drops (0.025 ml) of diluted hepatitis B virus antigen or antibodies with the same volume of diluent for 30 minutes at 37°C in microplates. A volume of 0.01 ml of the mixture was then applied on 0.8% agar on a glass slide. Formvar-coated grids were placed face down on the drops and removed after the liquid had diffused completely into the agar. This agar diffusion filtration (ADF) technique was as sensitive as complement fixation, and more sensitive than cross-over electrophoresis.

Other investigators used the ADF technique for serotyping alphaviruses, flaviviruses, bunyaviruses and rhabdoviruses (Fauvel et al., 1977) and for detecting rotaviruses (Payne et al., 1981) and other gastroenteritis viruses (Vaucher et al., 1982).

1.3.2. Serum in agar (SIA)

Another agar filtration IEM method is the 'serum in agar' (SIA) technique, devised by Anderson and Doane (1973). One percent melted agar was cooled and mixed with diluted single or pooled antisera and poured into the wells of a microtiter plate. These plates could be stored for several months at 4° C. For use in typing enteroviruses, formvar-carbon-coated grids were put, face up, in the wells and 0.02 ml of specimens deposited on them. Following diffusion and drying of agar for 30–60 minutes, each grid was negatively stained for 1 minute. The grids were then removed, blotted on filter paper and observed by EM. The specificity of the SIA technique for enteroviruses was equal to that obtained in the neutralization test, and it had the advantage that results were not dependent on optimal serum dilutions. However, the specificity for individual enterovirus serotypes was not absolute since cross-reactions sometimes occurred when the antisera were not diluted enough. Anderson and Doane (1973) reported that the sensitivity of this SIA technique was about 100 times higher than that of the regular EM procedure.

Lamontagne et al. (1980) found the SIA technique superior to classical IEM for identifying bovine and canine adenoviruses and infectious bovine rhinotracheitis viruses. Trepanier et al. (1981) examined fecal samples by various methods and found that the SIA technique was less sensitive than the indirect ADF technique. The SIA method, being serologically specific, obviously will detect only viruses for which antibodies are present in the agar. This has a drawback in the case of viral gastroenteritis where more than one virus may be involved in the etiology of the disease. To broaden the specificity, therefore, Berthiaume et al. (1981a) used a modified SIA method in which a specific antiserum was replaced by commercial pooled human gamma globulin that contained antibodies to a number of viruses. Other enteric viruses such as rota-, adeno-, astro- and picorna-like viruses could be detected with this method. A modification of the ADF technique based on the use of ferritin-labeled antiglobulins has been used by Berthiaume et al. (1981b) to facilitate the detection of rotaviruses in stools.

1.3.3. Pseudoreplica methods in IEM

The pseudoreplica technique, first devised by Sharp (1960), is chronologically the first agar filtration method ever used for EM. Because of its sensitivity this technique was mainly exploited for quantitation and enumeration of virus particles. Viruses that are concentrated on gels can be efficiently transferred onto grids by a 'pseudoreplication' principle.

Pseudoreplication is performed by forming a thin layer of formvar on the surface of the gel. The layer is then floated off onto a 'negative' staining solution and transferred to a grid with the virions entrapped in the formvar film. This step circumvents a whole range of problems that are usually encountered with the use of precoated grids (Katz and Kohn, 1984). Modifications of the pseudoreplica method have been described by Smith and Melnick (1962), Palmer et al. (1975), El Mekki et al. (1978) and Palmer and Martin (1982).

Lee et al. (1978) and Macris et al. (1982) used pseudoreplica for the rapid diagnosis of cytomegalovirus from urines and oral swabs. In a study of arboviruses, (El Mekki et al., 1978), the pseudoreplica technique was compared to the SIA method, and found to be about 64 times more sensitive (the minimal number of virus particles detected by SIA was 10^6 ID_{50}, and by pseudoreplica 1.5×10^4).

Modifications of the pseudoreplica method developed by Barker et al. (1969) and Hirschman et al. (1969) were found to be unsuitable for handling clinical specimens (Kelen et al., 1971). El Mekki et al. (1979) devised an immunopseudoreplica method and a lectin (concanavalin A) pseudoreplica method that were relatively simple to perform. Using Sindbis virus as a model, they compared the two methods with the regular pseudoreplica and with the SIA methods. They found the lectin pseudoreplica to be 100 times more sensitive than pseudoreplica and SIA, and 10 times more sensitive than the immunopseudoreplica. For other glycoprotein-containing viruses, the lectin pseudoreplica was at least 10 times more sensitive than the regular pseudoreplica.

In our laboratory, we have found the methods of El Mekki et al. (1979) unsuitable for rapid handling of several specimens. We developed, therefore, a method combining the advantages of the SIA and the pseudoreplica technique which we named the microwell pseudoreplica IEM (MW-PR-IEM) technique (Katz et al., 1983). In our procedure, agarose gels with antiserum are prepared in flat-bottomed perforated microwells of a polyvinylchloride microtiter plate. The plates with the hardened gels are sealed with parafilm and stored in a humid environment at 4°C until used. For use, the agarose pellets are removed from the wells and used for pseudoreplication. The films are floated off in the negative staining solution, which is contained in a set of syringes with partly cut off needle holders. Such a set of containers permits a simultaneous processing of several specimens. The MW-PR-IEM was twice as sensitive as that of PA-CGT or PA-CBT, 3 times more sensitive than the AB-CGT, 6 times more sensitive than the 'classical' IEM, 10 times more sensitive than the SIA and 40 times more sensitive than the ADF techniques.

1.4. Immunosorbent electron microscopy (ISEM)

In 1973, Derrick introduced an IEM technique which was similar to solid phase immunoassays such as solid phase radioimmunoassay (SPRIA) or enzyme linked immunosorbent assay (ELISA) (Catt and Tregear, 1967; Engvall and Perlman, 1971). The technique is suitable for examination of clinical specimens, since salts and interfering debris are removed during the washing steps.

The different IEM techniques in which one of the reagents is bound to a solid phase have been reviewed recently (Katz and Kohn, 1984). We refer to all solid IEM methods under the name immunosorbent electron microscopy (ISEM) (Roberts and Harrison 1979; Roberts et al., 1982). Depending on the nature of the solid support and on the adsorbed reagent, there are four different variations of the technique:

(*a*) the antibody-coated grid technique (AB-CGT; Derrick, 1973),
(*b*) the antigen-coated grid technique (AG-CGT; Milne and Luisoni, 1975),
(*c*) the protein A-coated grid technique (PA-CGT; Shukla and Gough, 1979), and
(*d*) the protein A-coated bacteria technique (PA-CBT; Katz et al., 1980).

Initially ISEM methods were used only by plant virologists, but in recent years, they have also been applied successfully to animal viruses. The most convenient procedures for the four ISEM techniques are summarized in Table 6.2.

1.4.1. Antibody-coated grid technique (AB-CGT)

In the AB-CGT, the grid is coated with antibodies which serve for the specific trapping of viruses. For the antibody coating step, antisera are incubated on grids for 5 minutes and, after washing, the virus suspension is deposited on grids for 15 minutes (Milne and Luisoni, 1977a). For staining, some authors (Derrick and Brlansky, 1976) use alcoholic uranyl acetate (positive staining), while others (Milne and Luisoni 1977a; 1977b) prefer aqueous uranyl acetate (negative staining).

AB-CGT was useful for the detection of viruses in plant sap without the need for centrifugation or other purification steps (Milne and Lesemann, 1978; Roberts and Harrison, 1979; Lesemann et al., 1980; Nicolaieff and van Regenmortel, 1980; Milne, 1980). Each author introduced some minor modifications of the technique, but all agreed what for best results it was important to find the optimal serum dilution for each virus-antibody system. This factor was neglected in the earlier work of Derrick (1973) and Milne and Luisoni (1977b), because they assumed that in a solid phase type of assay the antibody concentration would not be very critical. Subsequently, it was found that grids coated with a high concentration of antiserum trapped fewer virions. This prozone-like effect was attributed to the interference of serum proteins adsorbed to the grids, a phenomenon which is known also to occur in other solid phase immunoassays.

Nicolaieff and van Regenmortel (1980) and van Regenmortel (1982) showed that the AB-CGT could also be used for distinguishing between different viral serotypes. They found that the number of virions trapped on grids coated with a single antiserum was proportional to the serological distance between the virus strains.

To overcome the nonspecific trapping of virions that is sometimes observed in AB-CGT, Milne and Luisoni (1977b) suggested the use of a 'decoration' step to confirm the specificity of the reaction (Roberts and Harrison, 1979; Giraldo et al.,

TABLE 6.2

The most common immunosorbent electron microscopy procedures

Procedure	First layer of active reagent on solid support				Washing steps		Second layer of reagents		
	Solid support	Name of reagent	Recommended adsorbtion procedure [f]		Amount (drops)	Buffer	Name of reagent	Recommended incubation [f]	
			Time (min)	Temp. (°C)				Time (min)	Temp. (°C)
AB-CGT	EM-Grid [b]	Antiviral [d] capture antibody	5–15	RT	6	PBS-BSA	Virus suspension	60–180	RT
AG-CGT	EM-Grid	Virus suspension	1–15	RT	30	PBS	Decoration antiviral antibody [d]	15	RT
PA-CGT	EM-Grid	Protein A [e]	5–15	RT	6	PBS	Antiviral capture antibody [d,g]	5–15	RT
PA-CBT	Centrifuged [a] pellet of 3×10^8 *S. aureus* [c] protein A-containing bacteria per sample	Protein A naturally present on bacterial cell walls	—		—		Antiviral capture antibody [d,g] (1 ml/pellet of bacteria)	15	RT/37°C

[a] Centrifugation steps may be performed in an Eppendorf bench centrifuge (1 ml volumes).
[b] In our experience, carbon-coated commercial grids (Polaron Equipment Ltd., Watford, U.K.) were superior for the first three methods while carbon-fronted formvar-coated grids were used for the *S. aureus* bacteria.
[c] Killed or live bacteria can be used.
[d] Optimal antiserum dilution have to be determined for each technique in preliminary experiments.
[e] Protein A may be diluted in PBS to concentrations of 1–10 μg/ml.
[f] For adsorbtion incubation or staining, the grid on drop method is recommended, and 20–50 μl drops are use

Washing steps		Third layer of reagents			Negative staining [f,j]		
Amount (drops)	Buffer	Name of reagent	Recommended incubation[f]				Result in EM
			Time (min)	Temp. (°C)	Reagent	Time (min)	
6	PBS	Decoration antiviral antibody [d,h]	5–15	RT	2% UE	3	Single virions on grids [k]
20 50	PBS+ water	Double decoration anti-antibody [d,h]	15	RT	2% UW	2	Decorated single virions trapped on grids [k]
6	PBS-BSA	Virus suspension [i]	60–180	RT	2% UE	3	Single virions distributed on grids
3 × 1 ml	PBS/ PBS-BSA	Virus suspension (1 ml/pellet of bacteria)	40–60	RT/ 37°C	1% PT	1	Virions attached to surface of bacteria

[g] Antibody must be of a protein A binding type of immunoglobulins.
[h] The decoration step is optional.
[i] Centrifuge virus suspension before staining and resuspend in 0.1 ml.
[j] Blot excess fluid and do not dry grids before staining. UE, Uranylacetate in 47% ethanol; UW, uranyl acetate in water; PT, phosphotungstic acid (pH 7.3).
[k] The diameter of decorated viruses may increase 2–3 times.

1982). The 'decoration' step corresponds to the coating of virions with a layer of antibody molecules and increases somewhat the sensitivity of the method, enhancing the threshold of detection for small viruses. Kerlan et al. (1981) used still another layer of antiglobulin antibodies, and claimed that this 'double decoration' step was useful for detecting virions at low magnification. It has also been shown that the specificity of AB-CGT can be enhanced by using proteins such as bovine serum albumin for washing antibody-coated grids and for diluting the virus suspension (Katz and Kohn, 1984; Katz and Straussman, 1984). Some of the failures reported with AB-CGT may be due to blocking by soluble viral coat antigens (Milne 1980), pH sensitivity of certain viruses (Cohen et al., 1982) and inadequate support films (Pegg-Feige and Doane 1983). The AB-CGT has been used for detecting rotaviruses in fecal specimens (Nicolaieff et al., 1980; Rubenstein and Miller 1982) and was reported to be 9 times more sensitive than the classical EM procedure, and equally sensitive as ELISA.

Giraldo et al. (1982) detected BK virus in crude and purified infected tissue culture fluids using a combination of AB-CGT and decoration. Before staining, the specimens were fixed with 0.8% glutaraldehyde. Compared to direct EM they detected about 20 times more virus particles in crude and purified virus preparations.

1.4.2. Antigen-coated grid technique (AG-CGT)

In this technique, the virus is first adsorbed to the grid, as in standard EM, and is then identified by 'decoration' with a specific antibody (Milne and Luisoni, 1975; Luisoni et al., 1975). The technique was successfully used by Yanagida and Ahmad-Zadeh (1970) and Yanagida (1972) for the identification of antigenic precursors and gene products of the bacteriophage T4, by Wrigley et al. (1977) for studying the binding antibody of hemagglutinin and neuraminidase of influenza virus, by Vernon et al. (1981) for localizing herpes virus nucleocapsid polypeptides, and by Norrby (1969) for identifying antigenic determinants on the surface of adenovirus. The AG-CGT has also been used for determining the serological distance between viruses and for detecting viral antibodies (Milne and Luisoni, 1977a; 1977b; Milne and Lesemann, 1978; Katz and Kohn, 1984). When virus-coated grids are floated on antiserum, the virions seem to be able to migrate to form clumps, a phenomenon that has been called SMOG (specific migration on grid) (Almeida et al., 1981). Variations of the AG-CGT using colloidal gold as a marker have been described as 'antigen-controlled immunodiagnosis' (ACID) by Muller and Baignet (1980), and as 'gold-labelled antibody decoration' (GLAD) by Pares and Whitecross (1982).

1.4.3. Protein A-coated grid technique (PA-CGT)

The PA-CGT is a modification of the AB-CGT (Shukla and Gough, 1979) based on the use of protein A, a cell wall component of *Staphylococcus aureus* possessing high affinity for the Fc part of mammalian immunoglobulins. Protein A-coated grids are used for anchoring IgG molecules through their Fc part, leaving the Fab part free to interact with viral antigens. In the case of sugarcane mosaic virus, the PA-CGT was found to be 25–340 times more sensitive than the AB-CGT (Shukla and Gough, 1979; Gough and Shukla, 1980).

Since the PA-CGT technique is less dependent on optimal antiserum dilutions (Milne, 1980) it allows the visualization of week heterologous reactions that require high serum concentrations for detection (Lesemann and Paul, 1980).

Using PA-CGT, Nicolaieff et al. (1980) detected rotaviruses in 71% of the clinical specimens, whereas the standard EM detected only 20%. Grids coated with protein A and rabbit antiserum trapped 5–10 times more virions than grids coated with rabbit antiserum alone. Obert et al. (1981) compared the PA-CGT to standard EM, crossed immunoelectrophoresis (CIEP) and ELISA for rotavirus detection. In these tests ELISA and PA-CGT detected twice as many positive samples as the other techniques. With PA-CGT about 80 samples a day could be processed by one person.

In similar comparative studies of PA-CGT and ELISA for the detection of rotaviruses in fecal specimens, Kjeldsberg and Montensson-Egmund (1982) found PA-CGT to be less sensitive than ELISA but more sensitive than EM, while Svensson et al. (1983) found that PA-CGT was 30 times more sensitive than EM and 10 times more sensitive than ELISA. However, the advantage of EM was that other viruses, such as adeno-, astro- and caliciviruses could be detected simultaneously. ELISA, on the other hand, is more suitable for handling a large number of specimens a day.

The use of PA-CGT for serotyping animal viruses has been described by Polatnik and Wool (1981) and Svensson and von Bonsdorff (1982). Most adenovirus serotypes could be identified by PA-CGT, with the exception of serotypes 7 and 11 which cross-reacted. In the study of Svensson et al. (1983), the PA-CGT was not absolutely specific, since adenoviruses, when present at high concentrations, were occasionally seen on the grids. The authors suggested that by using pools of the relevant antisera, all the virions likely to be found in stools could be identified on the basis of morphological differences only. Chernesky (1983) reported that a 'decoration' step helped to detect cytomegalovirus by PA-CGT.

We have compared PA-CGT with AB-CGT and standard EM using Sindbis virus and rabbit antiserum as a model (Katz and Straussman, 1984; Katz and Kohn, 1984). When optimal conditions for each system were used, the PA-CGT was only 1.5 times more sensitive than the AB-CGT, and 10 times more sensitive than regular EM. The specificity of PA-CGT depended on the use of 0.1% BSA-containing buffer for the washing steps. The protein A-coated grids were washed with this buffer after the adsorption of the antiserum, and the virus suspension was also diluted in it. When an incubation time of more than 3 hours was used, a decrease in the number of virions trapped on the grids was observed.

1.4.4. Protein A-coated bacteria technique (PA-CBT)
In this technique, *Staphylococcus aureus* cells possessing protein A on their surface are coated with specific antiviral antibodies and are used to trap the corresponding viruses (Katz et al., 1980).

Nicolaieff et al. (1982) found that the sensitivities of PA-CGT and PA-CBT were similar for plant viruses. PA-CBT was, however, less reproducible because of such technical problems as clumping and heterogeneity. Calculations showed that the efficiency of trapping of virions by protein A-coated bacteria approached 100%, though only 2% of the total trapped virions could be seen on the periphery of the cells. Lee et al. (1983) used a modified PA-CBT to detect virus particles present in immune complexes by electron microscopy. In their modification viruses are first trapped on protein-A-containing *Staphylococcus aureus* bacteria and after centrifugation are eluted at low pH. The eluted virions are then processed for EM by the pseudoreplica technique.

2. Thin section immunoelectron microscopy

The visualization of virions in thin sections by EM can be done by direct observation of nonspecifically stained sections with one of the standard EM stains, or by the use of specific immunoglobulins coupled to various markers. However, the direct EM methods lack the specificity that can be achieved by IEM.

The literature on the preparation and use of EM for thin sections of tissues is very extensive (Pease, 1964; Siegal, 1964; Morgan and Rose, 1967; Higashi, 1973; Hsiung, 1979; Andres et al., 1978; Griffiths et al., 1983).

The pioneer work in this field was done by Palade (1952), who explored the fixatives to be used and recommended buffered OsO_4. The study of the development of viral infections in tissues of infected animals was pioneered by Reissig and Melnick (1955) for herpes B virus in kidneys of infected monkeys, and by Morgan et al. (1956) for adenovirus cultured in HeLa cells.

2.1. Embedding

The preparation of thin sections requires proper fixation methods, embedding, cutting and immunostaining with antibodies conjugated to various markers. Embedding and cutting are common to EM and IEM, while staining is not. Since fixation and embedding are time consuming, numerous methods were developed to shorten this procedure to a few hours (Coulter, 1967; Estes and Apicella, 1969; Benscome and Tsutsumi, 1970; Johannessen, 1973). Hayat and Giaquinta (1970) have reduced the time to 3.5 hours, and Doane et al. (1974) to 2 hours. In this method the specimens are fixed for 15 minutes at 4°C in 2.5% glutaraldehyde, rinsed 3 times in phosphate buffer and postfixed for 15 minutes with 1% OsO_4. Dehydration in acetone in 5 steps is followed by embedding in Spurr medium (1969) (epoxy resin) and polymerized at 95°C for 1 hour. Recently an improved method for tissue embedding using polyethylene glycol has been described by Bosman (1982). Details of sectioning and deposition on grids may be found in the work of Morgan and Rose (1967).

2.2. Conjugates

In direct EM the thin sections are stained with uranyl acetate, lead citrate or phosphotungstic acid (Morgan and Rose, 1967). For immunoelectron microscopy, the staining of thin sections is done with conjugated antibody using markers such as ferritin (Hsu, 1981), ferritin bound to lectins (Hixson et al., 1979), horseradish peroxidase (Singer, 1959; Kraehenbuhl and Jamieson 1976), peroxidase-anti-peroxidase (Hansen et al., 1979; Hsu and Ree, 1980), cytochrome C (Kraehenbuhl et al., 1971; Singer, 1974), hemocyanin (Gonda et al., 1981) or colloidal gold bound to protein A (Weiland, 1981; Burrage et al., 1983). The staining can be direct, with the conjugated antibody, or indirect (sandwich type) where the first antibody is not labeled, and the second antibody is directed against the immunoglobulin of the species in which the first antiserum was produced (Patterson, 1975; Locarnini et al., 1977).

2.2.1. Ferritin

Ferritin conjugates have been extensively used in EM of thin sections since 1959 (Borek and Silverstein, 1961). The subject up to 1969 has been extensively reviewed by Howe et al. (1969). The glutaraldehyde fixed tissue may be frozen in liquid nitrogen, sectioned in a cryomicrotome, deposited on the grid and stained with ferritin-antibody conjugate (Painter et al., 1973).

In order to avoid nonspecific reactions due to the binding of immunoglobulins to Fc receptors that may be present on the surface of the cells examined, ferritin may be bound to the $F(ab')_2$ piece only, rather than to the whole molecule. A more sophisticated approach was described by Hammerling et al. (1969). He has produced hybrid $F(ab')_2$ in which one arm was directed against the gamma globulin of the species in which the antiserum to the antigen was prepared, and the second arm was an antibody to ferritin. A conjugate of ferritin with plant lectins with different sugar specificities (concanavalin A, wheat germ agglutinin, Ricinus agglutinin) was used by Hixson et al. (1979) to characterize virus-specific oligosaccharide determinants on cellular plasma membranes infected with various strains of mouse mammary tumor virus.

2.2.2. Enzymes

Immunoenzymatic techniques for IEM were introduced by Nakane and Pierce in 1967. Details of the immunoperoxidase technique are given by Kurstak et al.

(1977), Smart et al. (1981), Dmochowski et al. (1974) and Fink et al. (1979). The peroxidase staining method has been applied to a variety of viruses (Table 6.3).

Peroxidase-antiperoxidase IEM (Sternberg, 1979) staining permitted the determination of subcellular distribution of herpes simplex virus type I in rabbit cornea cells from 2 to 19 hours after infection (Hansen et al., 1979).

A possible immunoperoxidase artifact has been described by Dourmashkin et al. (1982), who observed plasma membrane staining with anti-influenza antibody, after infection of Vero cells with influenza virus. Since such staining also occurred on fixed cells or with fixed virus, they concluded that the staining was nonspecific and due to diffusion of diaminobenzidine reaction product away from the enzymic site, i.e. from the surface of the viral envelope to the cell plasma membrane.

An interesting technique for the demonstration of redistribution of virus-induced antigens (VGA) on the surface of infected cells has been described by Kumon et al.

TABLE 6.3

Diagnostic uses of ferritin and peroxidase staining in thin section IEM

Marker	Virus	Viral structure	Reference
Ferritin	Influenza PR8	Protein	Morgan et al., 1961a
			Morgan et al., 1961b
	Influenza A WSV	Envelope proteins	Green et al., 1981
	Influenza Lee B	Neuraminidase	Blough, 1967
	Parainfluenza 2	Nucleocapsid	Howe et al., 1967
	Mumps	Nucleocapsid	Duc-Nguyen and Rosenblum, 1967
	Sendai		Yasuda et al., 1981
	NDV	Envelope proteins	Meiselman et al., 1967
	Vesicular stomatitis	Envelope proteins	Green et al., 1981
		Glycoproteins	Bergmann et al., 1981
	Herpes 1 and 2		Dowler and Veltri, 1982
	Adeno, type 12	T antigen	Kalnins et al., 1967
			Stich et al., 1967
			Levinthal et al., 1967
Peroxidase	Influenza	Hemagglutinin, neuraminidase	Bohn, 1980
			Kohama et al., 1981
	Measles	Hemagglutinin, hemolysin	Armstrong et al., 1982
			Budka et al., 1982
	Avian leucosis	Pr 76	Stanislawski et al., 1980
	Jap. encephalitis	Virions (*Culex*)	Ito, 1977
	Herpes		Meyers-Elliot et al., 1980
			White and Taxy, 1983
	Warts		Viac et al., 1978
	SV40	T antigen	Leduc et al., 1969
	Rotavirus		Chasey, 1980

(1978) and Fumio (1979), who conjugated anti-influenza IgG to the T4 phage, applied this conjugate to MDCK cells infected with influenza virus B and observed in a scanning electron microscope the polar capping of VGA within 4 hours of infection.

2.3. Clinical uses

The demonstration of viruses in thin sections makes it possible not only to follow the course of the infection, but also to define the etiology of the disease. In recent years, examination of tissues by EM has been mainly applied to brain, liver and various tumors (e.g. skin, lymph nodes). IEM, however, has been used in selected cases only. Studies on viral infections of the human brain have been reviewed by Mirra and Takei (1976). Viruses such as papova, measles herpes, CMV and parainfluenza have been shown to be present in thin brain sections.

Indirect immunoperoxidase staining was used to detect hepatitis b and s antigens and delta antigen in liver and kidney biopsies (Gerber et al., 1974; Huang and Neurath, 1979; Canese et al., 1979; Slusarczyk et al., 1980).

3. Discussion

Immunoelectron microscopy (IEM) provides a means for direct detection and identification of viruses in clinical specimens such as body fluids, excretions and solid tissues. Since IEM is based both on viral morphology and antigenicity, it gives few false positive results. The sensitivity of IEM can be increased by steps such as centrifugation, agar filtration and 'decoration'. The widely used 'clumping' technique permits the detection of many fastidious noncultivable viruses which are involved in the etiology of some diseases. The classical IEM techniques, although simple, require a centrifugation step which may remove a considerable number of viral aggregates, and thus decrease the sensitivity of the method (Narang and Codd, 1981). Other drawbacks of the 'classical' clumping technique are its dependence on optimal proportions of virus and antibody, and the fact that the presence of rheumatoid factors or complement may lead to false positive results (Almeida and Waterson, 1969; Stannard et al., 1980). The prozone effect may be avoided if the indirect clumping method is used (Valters et al., 1975), but even so the method is not very practical, since the production of heavy precipitates of immune complexes requires an additional centrifugation step.

Centrifugation is avoided in the gel filtration methods of Kelen et al. (1971) and Anderson and Doane (1973), the various modifications of the pseudoreplica technique (Sharp, 1960), and when the solid phase approach is used. Grids coated with antibodies, either directly or via protein A, enhance selective 'trapping' of virions in the presence of impurities, and the selectivity and specificity of virus detection are increased by the process of washing.

The method in which the antigen is adsorbed to the grid (AG-CGT) is less sensitive than the other ISEM techniques, but is suitable for antibody detection.

The problems encountered in IEM of thin sections are related to nonspecific staining and to the need of prolonged fixation. Prolonged fixation may often reduce the antigenicity of the structure that is investigated and may limit the access of the reagents into the fixed cells (Smith et al., 1974; Brown and Thormar, 1976). The nonspecific staining may be due to several factors (Thomson et al., 1967; Hansen et al., 1979; Gardner et al., 1978). Tissues that possess endogenous peroxidase may obscure the specific reaction of peroxidase conjugates. A number of methods for the removal of endogenous peroxidase have been described (Straus, 1971; Fink et al., 1979).

Nonspecific reactions can also be caused by the binding of the antibodies to unrelated antigens. This source of error, common to all immunologic reactions, can be eliminated either by a proper dilution of the serum, by the use of immunoadsorbents to enhance the specificity of the antiserum, or by the use of monoclonal antibodies.

Although the existence of a large number of IEM techniques probably indicates that no technique is perfect, the multiplicity of published methods ensures that the investigator will probably find one that is suited to his particular problem.

4. Acknowledgements

We acknowledge with thanks the assistance of Mrs. Janina Gitelman in bibliographical search and proof reading.

5. References

Almeida, J.D. (1980) *Yale J. Biol. Med.* 53, 5–18.
Almeida, J.D. and Goffe, A.P. (1965) *Lancet* 2, 1205–1207.
Almeida, J.D. and Waterson, A.P. (1969) *Adv. Virus Res.* 15, 307–338.
Almeida, J.D., Gay, F.W. and Wreghitt, T.G. (1974) *Lancet* 2, 748–750.
Almeida, J.D., Skelly, J., Howard, C.R. and Zuckerman, A.J. (1981) *J. Virol. Methods* 2, 169–174.
Anderson, N. and Doane, F.W. (1973) *Can. J. Microbiol.* 19, 585–589.
Anderson, N. and Stanley, W.M. (1941) *J. Biol. Chem.* 139, 339–344.
Andres, G.A., Hsu, K.C. and Seegal, B.C. (1978) In *Handbook of Experimental Immunology* (D.M. Weir, Ed.) Vol. 2. pp. 37. 1–37.54.
Armstrong, M.A., Fraser, K.B., Dermott, E. and Shirodaria, P.V. (1982) *J. Gen. Virol.* 59, 187–192.
Ball, E.M. (1971) In *Methods in Virology* (K. Maramorosch and H. Koprowski, Eds.) Vol. 5. Academic Press, New York. pp. 445–450.
Barker, L.F., Smith, K.O., Gehle, W.D. and Shulman, N.R. (1969) *J. Immunol.* 102, 1529–1532.
Barron, A.L. (1971) In *Methods in Virology* (K. Maramorosch and H. Koprowski, Eds.) Vol. 5. Academic Press, New York. pp. 347–373.
Bayer, M.E., Blumberg, B.S. and Werner, B. (1968) *Nature* 218, 1057–1059.
Benscome, S.A. and Tsutsumi, V. (1970) *Lab. Invest.* 23, 447–450.

Bergmann, J.E., Tokuyasu, K.T. and Singer, S.J. (1981) *Proc. Natl. Acad. Sci. U.S.A.* 78, 1746–1750.
Berthiaume, L., Alain, R., McLaughlin, B., Payment, P. and Trepanier, P. (1981a) *J. Gen. Virol.* 55, 223–227.
Berthiaume, L., Micusan, V., Alain, R. and Trepanier, P. (1981b) *J. Virol. Methods* 2, 367–373.
Best, J.M., Banatvala, J.E., Almeida, J.D. and Waterson, A.P. (1967) *Lancet* 2, 237–239.
Blough, H.A. (1967) *Virology* 31, 514–522.
Bohn, W. (1980) *J. Histochem. Cytochem.* 26, 293–297.
Borek, F. and Silverstein, A.M. (1961) *J. Immunol.* 87, 555–561.
Bosman, F.T. (1981) *Histochemistry* 73, 195–200.
Brown, H.R. and Thormar, H. (1976) *Acta Neuropathol.* 36, 259–267.
Budka, H., Lassmann, H. and Popow-Kraupp, T. (1982) *Acta Neuropathol.* 56, 52–62.
Burrage, T.G., Tignor, G.H. and Smith, A.L. (1983) *J. Virol. Methods* 7, 337–350.
Canese, M.G., Rizzetto, M., Arico, S., Crivelli, O., Zanetti, A.R., Macchiorlatti, E., Ponzeto, A., Leone, L., Mollo, F. and Verme, G. (1979) *J. Pathol.* 128, 169–176.
Catt, K. and Tregear, G.W. (1967) *Science* 158, 1570–1572.
Chasey, D. (1980) *J. Gen. Virol.* 50, 195–200.
Chernesky, M.A. (1983) In *Medical Virology* (M.L. de la Maza and E.M. Petrson, Eds.) Part II. Elsevier, Amsterdam. pp. 1–26.
Chiba, S., Sakuma, Y., Kogasawa, R., Akihara, M., Horino, K., Nakao, T. and Fukui, S. (1979) *J. Med. Virol.* 4, 249–254.
Cohen, B.J. (1978) *J. Med. Virol.* 3, 141–149.
Cohen, J., Loebenstein, G. and Milne, R.G. (1982) *J. Virol. Methods* 4, 323–330.
Coulter, H.D. (1967) *J. Ultrastr. Res.* 20, 346–355.
Derrick, K.S. (1973) *Virology* 56, 652–653.
Derrick, K.S. and Brlansky, R.H. (1976) *Phytopathology* 66, 815–820.
Dienstag, J.L., Krugman, S., Wong, D.C. and Purcell, R.H. (1976) *Infect. Immun.* 14, 1000–1003.
Dmochowski, L., Hoshino, M., Shigematsu, T., Horaki, S. and Priori, E.S. (1974) In *Viral Immunodiagnosis* (E. Kurstak and R. Morisset, Eds.) Academic Press, New York pp. 183–213.
Doane, F.W., Anderson, N., Chao, J. and Noonan, A. (1974) *Appl. Microbiol.* 24, 407–410.
Doane, F.W. and Anderson, N. (1977) In *Comparative Diagnosis of Viral Diseases* (E. Kurstak and C. Kurstak, Eds.) Vol. II. Academic Press, New York pp. 505–539.
Dolin, R., Reichman, R.C., Roessner, K.D., Tralka, T.S., Schooley, R.T., Gary, W. and Morens, D. (1982) *J. Infect. Dis.* 146, 184–189.
Duc-Nguyen, H. and Rosenblum, E.N. (1967) *J. Virol.* 1, 415–429.
Dourmashkin, R., Patterson, S., Shah, D. and Oxford, J.S. (1982) *J. Virol. Methods* 5, 27–34.
Dowler, K.W. and Veltri, R.W. (1982) *Fed. Proc.* 41, 1751.
Edwards, E.A., Valters, W.A., Boehm, L.G. and Rosenbaum, M.J. (1975) *J. Immunol. Methods* 8, 159–168.
Elkana, Y. and Guttman-Bass, N. (1983) *J. Virol. Methods* 6, 53–58.
El Mekki, A.A., van der Groen, G. and Pattyn, S.R. (1978) In *Ebola virus haemorrhagic fever* (S.R. Pattyn, Ed.) Elsevier, Amsterdam pp. 261–267.
El Mekki, A.A., van der Groen, G. and Pattyn, S.R. (1979) *Ann. Soc. Belge Med. Trop.* 59, 279–283.
Engvall, E. and Perlmann, P. (1971) *Immunochemistry* 8, 871–874.
Estes, L.W. and Apicella, J.V. (1969) *Lab. Invest.* 20, 159–163.
Fauvel, M., Artsob, H. and Spence, L. (1977) *Amer. J. Trop. Med. Hyg.* 26, 798–807.
Feinstone, S.M., Kapikian, A.Z. and Purcell, R.H. (1973) *Science* 182, 1026–1028.
Field, A.M. (1982) *Adv. Virus Res.* 27, 2–69.
Fink, B., Loepfe, E. and Wyler, R. (1979) *J. Histochem. Cytochem.* 27, 686–688.
Flewett, T.H. and Boxall, E. (1976) *Clinics Gastroenterol.* 5, 359–385.
Fumio, U. (1979) *J. Electron Microsc.* 28, 83–92.

Gardner, P.S., Grandien, M. and McQuillin, J. (1978) *Bull. W.H.O.* 56, 105–110.
Gerber, M.A., Hadziyannis, S. and Vissoulis, C. (1974) *Am. J. Pathol.* 75, 489–502.
Giraldo, B., Beth, E., Lee, J., de Harven, E. and Chernesky, M. (1982) *J. Clin. Microbiol.* 15, 517–521.
Gonda, M.A., Benton, C.V., Massey, R.J. and Schultz, M.A. (1981) *Scanning Electron Microsc. U.S.* 45–62.
Gough, K.H. and Shukla, D.D. (1980) *J. Gen. Virol.* 51, 415–419.
Green, R.F., Meiss, H.K. and Rodriguez-Boulan, E. (1981) *J. Cell. Biol.* 89, 230–239.
Griffiths, G., Simons, K., Warren, G. and Tokuyasu, K.T. (1983) In *Biomembranes, Methods in Enzymology* (S. Fleischer and B. Fleischer, Eds.) Vol. 96, Pt.I. Academic Press, Orlando pp. 446–484.
Hämmerling, U., Aoki, T., Wood, H.A., Old, L.J., Boyse, E.A. and de Harven, E. (1969) *Nature* 223, 1158–1159.
Hansen, B.L., Hansen, G.N. and Vestergaard, B.F. (1979) *J. Histochem. Cytochem.* 27, 1455–1461.
Hayat, M.A. and Giaquinta, R. (1970) *Tissue Cell* 2, 191–195.
Higashi, N. (1973) *Prog. Med. Virol.* 15, 331–379.
Hirschman, R.J., Shulman, N.R., Baker, L.F. and Smith, K.O. (1969) *J. Am. Med. Assoc.* 208, 1667–1670.
Hixson, D.C., Bowen, J.M., Ohisuku, Y., Scalon, M. and Dmochowski, L. (1979) *Cancer Res.* 39, 199–206.
Howe, C., Morgan, C., de Vaux St-Cyr, C., Hsu, K.C. and Rose, H.M. (1967) *J. Virol.* 1, 215–237.
Howe, C., Morgan, C. and Hsu, K.C. (1969) *Prog. Med. Virol.* II, 307–353.
Hsiung, G.D., Fong, C.K.Y. and August. M.J. (1979) *Prog. Med. Virol.* 25, 133–159.
Hsu, K.C. (1981) *Scanning Electron Microsc. U.S.* 17–26.
Hsu, S.M. and Ree, H.J. (1980) *Am. J. Clin. Pathol.* 74, 32–40.
Huang, S.N. and Neurath, A.R. (1979) *Lab. Invest.* 40, 1–17.
Ito, T. (1977) *J. Kansai Med. Univ.* 29, 591–625.
Johannessen, J.V. (1973) *Kidney Int.* 3, 46–50.
Kalnins, V.I., Stich, H.F., Gregory, C. and Yohn, D.S. (1967) *Cancer Res.* 27, 1874–1886.
Kapikian, A.Z., Almeida, J.A. and Stott, E.J. (1972) *J. Virol.* 10, 142–146.
Kapikian, A.Z., James, H.D., Kelly, S.J. and Vaughn, A.L. (1973) *Infect. Immun.* 7, 111–116.
Kapikian, A.Z., Dienstag, J.L. and Purcell, R.H. (1980) In *Manual of clinical immunology* (N.R. Rose and H. Freedman, Eds.) 2nd Edn. American Society for Microbiology, Washington pp. 70–83.
Katz, D. and Kohn, A. (1984) *Adv. Virus Res.* 29, In press.
Katz, D. and Straussman, Y. (1984) *J. Virol. Methods* In press.
Katz, D., Straussman, Y., Shahar, A. and Kohn, A. (1980) *J. Immunol. Methods* 38, 171–174.
Katz, D., Shahar, A. and Straussman, Y. (1983) *European Group for Rapid Viral Diagnosis.* Meeting, Turku, Finland.
Kelen, A.E., Hathaway, A.E. and McLeod, D.E. (1971) *Can. J. Microbiol.* 17, 993–1000.
Kerlan, C., Mille, B. and Dunez, J. (1981) *Phytopathology* 71, 400–404.
Kjeldsberg, E. (1980) *Path. Res. Pract.* 167, 3–21.
Kjeldsberg, E. and Mortensson-Egnund, K. (1982) *J. Virol. Methods* 4, 45–53.
Kohama, M.T., Cardenas, J.M. and Seto, J.T. (1981) *J. Virol. Methods* 3, 293–302.
Konno, T., Suzuki, H., Ishida, N., Chiba, R., Mochizuki, K. and Tsunoda, A. (1982) *J. Med. Virol.* 9, 11–18.
Kraehenbuhl, J.P., de Grandi, P.B. and Campiche, M.A. (1971) *J. Cell. Biol.* 50, 432–445.
Kraehenbuhl, J.P. and Jamieson, J.D. (1976) In *Methods in Immunology and Immunochemistry* (C.A. Williams and M.W. Chase, Eds.) Vol. 5. Academic Press, New York pp. 482–495.
Kumon, H., Uno, F., Saburi, Y. and Tawara, J. (1978) *J. Electron Microsc.* 27, 215–222.
Kurstak, E., Tijssen, P. and Kurstak, C. (1977) In *Comparative diagnosis of viral diseases* (E. Kurstak and C. Kurstak, Eds.) Vol. 2, Part B. Academic Press, New York. pp. 430–436.

Lamontagne, L., Marsolais, G., Marois, P. and Assaf, R. (1980) *Can. J. Microbiol.* 26, 261–264.
Leduc, E.H., Wicker, R., Avrameas, S. and Bernhard, W. (1969) *J. Gen. Virol.* 4, 609–614.
Lee, F.K., Nahmias, A.J. and Stagno, S. (1978) *N. Engl. J. Med.* 299, 1266–1270.
Lee, F.K., Nahmias, A.J., Nahmias, D.G.A. and McDougal, J.S. (1983) *J. Virol. Methods* 7, 167–181.
Lesemann, D.E. and Paul, H.L. (1980) *Acta Hortic.* 110, 119–128.
Lesemann, D.E., Bozarth, R.F. and Koenig, R. (1980) *J. Gen. Virol.* 48, 257–264.
Levinthal, J.D., Cerattini, J.C., Ahmad-Zadeh, C. and Wicher, R. (1967) *J. Cancer* 2, 85–102.
Locarnini, S.A., Ferris, A.A., Scott, A.C. and Gust, I.D. (1974) *Intervirology* 4, 110–118.
Locarnini, S.A., Ferris, A.A., Lehmann, N.I. and Gust, I.D. (1977) *Intervirology* 8, 309–318.
Luby, J.P., Murphy, F.K., Gilliam, J.N., Kang, C.-Y. and Frank, R. (1980) *Am. J. Trop. Med. Hyg.* 29, 265–268.
Luisoni, E., Milne, R.G. and Boccardo, G. (1975) *Virology* 68, 86–96.
Macris, M.P., Nahmias, A.J., Bailey, P.D., Lee, F.K., Visintine, A.M. and Brann, A.W. (1981) *J. Virol. Methods* 2, 315–320.
Marciano-Cabral, F., Rublee, K.L. and Carithers, R.L., Jr. (1981) *Hepathology* 1, 575–582.
Meiselman, N., Kohn, A. and Danon, D. (1967) *J. Cell Sci.* 2, 71–76.
Meyers-Elliott, R.H., Pettit, T.H. and Maxwell, W.A. (1980) *Arch. Ophthalmol.* 98, 897–904.
Milne, R. (1980) *Acta Hortic.* 110, 129–135.
Milne, R.G. and Lesemann, D.E. (1978) *Virology* 90, 299–304.
Milne, R.G. and Luisoni, E. (1975) *Virology* 68, 270–274.
Milne, R.G. and Luisoni, E. (1977a) In *Methods in Virology* (K. Maramorosch and H. Koprowski, Eds.) Vol. 6. Academic Press, New York. pp. 265–281.
Milne, R.G. and Luisoni, E. (1977b) *Virology* 80, 12–20.
Mirra, S.S. and Takei, Y. (1976) *Prog. Neuropathol.* 3, 69–88.
Morgan, C. and Rose, H.M. (1967) In *Methods in Virology* (K. Maramorosch and H. Koprowski, Eds.) Vol. 3. Academic Press, New York. pp. 575–616.
Morgan, C., Howe, C., Rose, H.M. and Moore, D.H. (1956) *J. Biophys. Biochem. Cytol.* 2, 351–360.
Morgan, C., Hsu, K.C., Rifkind, R.A., Knox, A.W. and Rose, H.M. (1961a) *J. Exp. Med.* 114, 825–832.
Morgan, C., Rifkind, R.A., Hsu, K.C., Holden, M., Seegal, B.C. and Rose, H.M. (1961b) *Virology* 14, 292–296.
Müller, G. and Baigent, C.L. (1980) *J. Immunol. Methods* 37, 185–190.
Nakane, P.K. and Pierce, G.B. (1967) *J. Cell. Biol.* 33, 307–318.
Narang, H.K. and Codd, A.A. (1981) *J. Clin. Microbiol.* 13, 982–988.
Nicolaieff, A. and van Regenmortel, M.H.V. (1980) *Ann. Virol.* 131E, 95–110.
Nicolaieff, A., Obert, G. and van Regenmortel, M.H.V. (1980) *J. Clin. Microbiol.* 12, 101–104.
Nicolaieff, A., Katz, D. and van Regenmortel, M.H.V. (1982) *J. Virol. Methods* 4, 155–166.
Norrby, E. (1969) *Virology* 37, 565–576.
Obert, G., Gloeckler, R., Burckard, J. and van Regenmortel, M.H.V. (1981) *J. Virol. Methods* 3, 99–107.
Odenwald, W.F., Johnson, R.B., Morquardt, W.W. and Hetrick, F.M. (1978) *J. Clin. Microbiol.* 7, 568–575.
Oshiro, L.S., Haley, C.E., Roberto, R.R., Riggs, J.L., Crougham, M., Greenberg, H. and Kapikian, A. (1981) *J. Infect. Dis.* 143, 791–795.
Painter, R.G., Tokuyashi, K.T. and Singer, S.J. (1973) *Proc. Natl. Acad. Sci. U.S.A.* 70, 1649–1653.
Palade, G.E. (1952) *J. Exp. Med.* 95, 285–298.
Palmer, E.L. and Martin, M.L. (1982) *An atlas of mammalian viruses.* CRC Press, Boca-Raton. pp. 1–154.

Palmer, E.L., Martin, M.L. and Gary, G.W. (1975) *Virology* 65, 260–265.
Pares, R.D. and Whitecross, M.I. (1982) *J. Immunol. Methods* 51, 23–28.
Patterson, S. (1975) *J. Immunol. Methods* 9, 115–122.
Paver, W.K., Ashley, C.R., Caul, E.O. and Clarke, S.K.R. (1973) *Lancet* 1, 237–240.
Payne, C.M., Ray, C.G. and Yolken, R.H. (1981) *J. Med. Virol.* 7, 299–313.
Pease, D.C. (1964) *Histological techniques for electron microscopy.* 2nd Edn. Academic Press, New York.
Pegg-Feige, K. and Doane, F.W. (1983) *J. Virol. Methods* 7, 315–319.
Pfister, H., Nuernberger, F., Gissman, L. and Zur Hausen, H. (1981) *Int. J. Cancer* 27, 645–650.
Polatnick, J. and Wool, S. (1981) *Can. J. Comp. Med.* 45, 275–278.
Reissing, M. and Melnick, J.L. (1955) *J. Exp. Med.* 101, 341–352.
Ritchie, A.E. and Fernelius, A.L. (1967) *Vet. Rec.* 81, 417–418.
Roberts, I.M. and Harrison, B.D. (1979) *Ann. Appl. Biol.* 93, 289–297.
Roberts, I.M., Milne, R.G. and van Regenmortel, M.H.V. (1982) *Intervirology* 18, 147–149.
Rubenstein, A.S. and Miller, M.F. (1982) *J. Clin. Microbiol.* 15, 938–944.
Sharp, D.G. (1960) *Proc. Int. Congr. Electron Microscopy*, 4th Congr. 1958, Berlin. pp. 542–548.
Shukla, D.D. and Gough, K.H. (1979) *J. Gen. Virol.* 45, 533–536.
Siegal, B.M. (1964) *Modern development in electron microscopy.* Academic Press, New York.
Singer, I.I. (1974) In *Viral immunodiagnosis* (E. Kurstak and R. Morisset, Eds.) Academic Press, New York. pp. 101–123.
Singer, S.J. (1959) *Nature* 183, 1523–1524.
Slusarczyk, J., Michalak, T., Nazarewicz-de Mezer, T., Krawczynski, K. and Nowoslawski, A. (1980) *Am.J. Pathol.* 98, 29–43.
Smart, Y., Jerrome, D. and Millard, P.R. (1981) *J. Immunol. Methods* 45, 95–105.
Smith, J.W., Meijer, C.J.L.M., Decary, F. and Feltkamp-Vroom, T.M. (1974) *J. Immunol. Methods* 6, 93–98.
Smith, K.O. and Melnick, J.L. (1962) *Virology* 17, 480–490.
Spurr, A.R. (1969) *J. Ultrastruct. Res.* 26, 31–43.
Stanislawski, L., Mongiat, F. and Haguenau, F. (1980) *Intervirology* 14, 261–271.
Stannard, L.M., Almeida, J.D. and Tedder, R.S. (1980) *J. Med. Virol.* 6, 153–164.
Sternberg, L.A. (1979) *Immunochemistry.* 2nd Edn. Wiley and Sons, New York. pp. 104–169.
Stich, H.F., Kalnins, V.I., MacKinnon, E. and Yohn, D.S. (1967) *J. Ultrastruct. Res.* 19, 556–562.
Strauss, W. (1971) *J. Histochem. Cytochem.* 19, 682–688.
Svensson, L. and Bonsdorff, C.H. (1982) *J. Med. Virol.* 10, 243–253.
Svensson, L., Grandien, M. and Patterson, C.-A. (1983) *J. Clin. Microbiol.* 18, 1244–1249.
Taniguchi, K., Urasawa, S. and Urasawa, T. (1979) *J. Clin. Microbiol.* 10, 730–736.
Thomson, R.O., Walker, P.D., Batty, I. and Baillie, A. (1967) *Nature* 215, 393–394.
Trepanier, P., Alain, P., Micusan, V., McLaughlin, B. and Berthiaume,L. (1981) *Microbiol. Immunol.* 25, 1019–1024.
Valters, W.A., Boehm, L.G., Edwaeds, E.A. and Rosenbaum, M.J. (1975) *J. Clin. Microbiol.* 1, 472–475.
Van Regenmortel, M.H.V. (1981) In *Comprehensive Virology* (H. Fraenkel-Conrat and R.R. Wagner, Eds.) Vol. 17. Plenum Publishing Corp., New York. pp. 183–243.
Van Regenmortel, M.H.V. (1982) *Serology and immunochemistry of Plant Viruses.* Academic Press, New York. pp. 302.
Vaucher, Y.E., Ray, C.G., Minnich, L.L., Payne, C.M., Beck, D. and Lowe, P. (1982) *J. Infect. Dis.* 145, 27–36.
Vernon, S.K., Ponce de Leon, M., Cohen, G.M., Eisenberg, R.J. and Rubin, B.A. (1981) *J. Gen. Virol.* 73, 39–46.
Viac, J., Schmitt, D. and Thivolet, J. (1978) *J. Invest. Dermatol.* 70, 263–266.
Weiland, F. (1981) *Ann. Virol.* 132, 549–556.

White, C.L. and Taxy, J.B. (1983) *Hum. Pathol.* 14, 135–139.
Wrigley, N.G., Laver, W.G. and Downie, J.C. (1977) *J. Mol. Biol.* 109, 405–421.
Yamada, G., Takanashi, T., Mizuno, M., Sakamoto, Y., Kaga, K., Kobayashi, T. and Nagashima, H. (1980) *Acta Med. Okayama* 34, 175–188.
Yanagida, M. (1972) *J. Mol. Biol.* 65, 501–517.
Yanagida, M. and Ahmad-Zadeh, C. (1970) *J. Mol. Biol.* 51, 411–421.
Yoshizawa, H., Itoh, Y., Iwakari, S., Kitajima, K., Tanaka, A., Nojiri, T., Miyakawa, Y. and Mayumi, M. (1981) *Gastroenterology* 81, 107–113.
Yasuda, Y., Hosaka, Y., Fukami, Y. and Fukai, K. (1981) *J. Virol.* 39, 273–281.

Eds. M.H.V. van Regenmortel and A.R. Neurath
Immunochemistry of viruses. The basis for serodiagnosis and vaccines.

Precipitation, agglutination and complement fixation techniques

7

M.H.V. VAN REGENMORTEL

IBMC, 15 rue Rene Descartes, 67084 Strasbourg, France

This chapter discusses some of the oldest and most widely used techniques in virus serology. Although it may seem superfluous to review once more procedures that have been used in virology for so many years, a study of the literature shows that the principles of precipitation, agglutination and complement fixation still continue to be applied frequently to the development of new and sophisticated methods of immunochemical analysis. These procedures are, therefore, by no means outdated. Some basic principles underlying these methods will be reviewed briefly and a number of recent developments will be described. These modern applications will demonstrate that the old methods are still worthy of interest and that their considerable potential has not yet been fully exhausted.

1. Precipitation techniques

1.1. Precipitin tests

Precipitation tests are based on the direct visualization of the virus-antibody interaction in the form of a precipitate, and can be used only with viruses or viral proteins that are available in relatively large amounts. This is the case for most plant viruses, and it is in the field of plant virology that precipitin reactions have found the greatest number of applications. Precipitin tests in liquid are commonly used, for instance, for quantifying the degree of serological cross-reaction between plant viruses, especially elongated viruses that diffuse poorly in agar gel.

The various factors that influence the results of tube precipitin tests have recently been analyzed (van Regenmortel, 1982). It is a common misconception that precipitation results from the formation of a lattice of antigen and antibody molecules that grows in size until it is so large that the complex becomes insoluble. That this view is an oversimplification is shown by the fact that precipitation does not occur in solutions of low ionic strength, although binding of antigen to antibody occurs in the normal way in the absence of salt (Kleczkowski, 1965). The precipitation reaction appears to involve two different mechanisms. The first stage of the reaction is serologically specific and leads to the formation of relatively small immune complexes. Thermodynamic considerations have shown that very large complexes are unlikely to form (Steensgaard and Frich, 1979). The second stage of the reaction corresponds to the formation of a hydrophobic phase and is not immunologically specific (Steensgaard, 1984).

One of the main applications of liquid precipitin tests is the characterization of plant viruses by means of antiserum precipitin titers. It has been found convenient to express the degree of serological cross-reactivity between two viruses by the number of two-fold dilution steps that separate homologous from heterologous precipitin titers. This number, called the serological differentiation index or SDI, is only reliable if it corresponds to an average value calculated from several different antisera (van Regenmortel, 1975; 1982). Average SDI values between two viruses obtained in reciprocal tests, i.e. when antisera are used against each of the two viruses, have been shown to agree closely, and it is possible in this way to compute a reliable index of serological cross-reactivity between viruses. In the tobamovirus group, the extent of this cross-reactivity was found to correlate well with the variations in amino acid sequence of viral coat proteins (Paul et al., 1980).

1.2. Immunodiffusion

Immunodiffusion tests are precipitin tests that are carried out in gels instead of free liquid (Crowle, 1973). Several different types of immunodiffusion techniques are commonly used in virology, for instance radial immunodiffusion tests (Grandien and Norrby, 1975; Slack and Shepherd, 1975), radial hemolysis in gel (Grillner and Blomberg, 1976; Steinmann and Marzock, 1983) and double diffusion in plates. This last technique, known as the Ouchterlony method, is still one of the most widely used techniques in virus serology. Its main advantage is that it can provide a visual demonstration of the relationships that exist between viral antigens. The position and curvature of precipitin lines in the gel also give an indication of the size of the reacting antigen while the number of precipitin lines formed under conditions of serological equivalence indicates the number of different antigens present in a preparation (Nilsson, 1983). The minimum range of antigen concentration detected by double immunodiffusion lies in the vicinity of 2–20 μg/ml (van Regenmortel, 1982). Although less sensitive than many of the more recently developed techniques such as solid phase immunoassays, the simplicity and visual impact of immunodiffusion analysis cause it to be still widely used in all fields of virology.

When two antigens diffuse from neighboring wells in a gel toward the same antibody source, different precipitation patterns can be observed at the position where the lines meet. Three basic patterns are found – coalescence, spur formation and crossing of lines – and these provide information on the extent of immunological cross-reactivity existing between the antigens under study. It should be stressed that useful information is obtained only when two antigens are tested against a single antiserum. When two antisera prepared against two dissimilar but cross-reacting antigens are tested against only one of the antigens, the mechanics of the diffusion process ensure that a pattern of coalescence of precipitin lines is always obtained (van Regenmortel, 1982). Since such a merging of lines is often referred to as a 'reaction of identity' this label sometimes induces investigators to believe that antigens that are only distantly related are antigenically identical (Sim and Watson, 1973; Reinganum and Scotti, 1976).

In order to serologically compare two related viruses, it is often useful to cross-absorb antisera with the heterologous antigen. This is best done by a procedure known as intragel cross-absorption in which the antigen preparation used for absorption is allowed to diffuse in the gel to form a concentration gradient (van Regenmortel, 1982). The antiserum is later allowed to diffuse from the same well. This leads to a ring of precipitation of all cross-reacting antibodies around the well. Unabsorbed antibodies that recognize only the homologous antigen will diffuse freely and form a second precipitin line, thereby demonstrating the presence of unique antigenic determinants in individual serotypes.

Large viruses that do not diffuse in 0.5–1% agar gel can be studied by immunodiffusion after they have been degraded by dissociating agents. Purcifull and Batchelor (1977) have listed 35 elongated plant viruses that were studied by immunodiffusion after dissociation with sodium dodecyl sulfate. When the extent of cross-reactivity between viruses is measured with dissociated subunits instead of intact virions, the apparent degree of similarity between them is often much greater (Shepard et al., 1974; Rybicki and Von Wechmar, 1981). This is probably due to the fact that additional internal sequence homologies are uncovered when a larger portion of the viral antigens is immunochemically expressed.

1.3. Immunoelectrophoresis

Immunoelectrophoresis combines the phenomena of electrophoretic mobility in a gel with immunodiffusion. In the standard immunoelectrophoretic analysis, the antigen mixture is first separated into its components by electrophoresis in agar or agarose gels. Antiserum is then placed in a trough parallel to the path of electrophoretic migration and the immunodiffusion reaction is allowed to proceed. This method is useful for differentiating between closely related viruses and for verifying the structural integrity of viral antigens (Crowle, 1973; Schmidt and Lennette, 1973; van Regenmortel, 1982).

In counter-immunoelectrophoresis the two serological reactants are placed in opposite wells in the gel. The antigens, which are negatively charged at pH 7–8, are made to migrate toward the anode while the antibodies simultaneously move toward the cathode by electroendosmotic flow (Waldstrom, 1983). Precipitin lines usually appear in less than an hour. The main advantages of this technique are its rapidity and sensitivity (Grauballe et al., 1977). As in all gel diffusion procedures, the sensitivity of antigen detection can be enhanced by staining the precipitin lines, for instance with Coomassie blue (Crowle, 1973; Middleton et al., 1976).

In rocket immunoelectrophoresis, the antigen is allowed to migrate in an electric field in a layer of antibody-containing agarose (Axelson et al., 1983). This migration leads to the formation of a rocket-shaped precipitation pattern, the height of which is proportional to the antigen concentration in the sample. As the electrophoresis progresses, all antigen molecules at the moving front will be eventually complexed with antibody and the precipitation pattern will then remain stationary. Depending on the pH and the degree of electroendosmosis shown by the particular gel medium, the antibodies may also migrate, and this may result in cigar-shaped rather than rocket-shaped precipitates. The technique has been used for measuring the concentration of viral polypeptides (Oxford and Schild, 1977; Robinson et al., 1977) and whole virions (Havranek, 1978). Because of its miniaturization, the technique was able to detect a plant virus quantitatively in as little as 25 mg of infected leaf tissue (Hagen et al., 1982).

A variation of the rocket immunoelectrophoresis procedure is the crossed immunoelectrophoresis method (Grubb, 1983). The technique involves first the electrophoretic separation of a mixture of antigens in agarose gel and then a second electrophoresis, at right angles to the first, into a gel containing antibodies to the different antigens. Crossed immunoelectrophoresis is a powerful method for analyzing very complex antigen-antibody systems and has been used, for instance, for distinguishing the many antigenic components of herpes simplex virus (Vestergaard et al., 1977). A large number of newly developed electroimmunoprecipitation techniques have been reviewed recently (Axelson, 1983).

1.4. Immunoblotting

In immunoblotting techniques the antigens are first separated by gel electrophoresis and then transferred by blotting to sheets of nitrocellulose or diazobenzyloxymethyl paper before being revealed with antibody (Renart et al., 1979; Towbin et al., 1979). The blotting procedure achieves a superior discrimination between different proteins than the older immunofixation procedures (Ritchie and Smith, 1976; Kohn and Roches, 1978). In immunofixation, proteins that have been separated by electrophoresis are revealed by overlaying the gel surface with antiserum; the immunoprecipitates are then stained after removing other proteins by extensive washing.

In the so-called Western-blotting procedure, proteins in sodium dodecyl sulfate-containing gels are blotted onto nitrocellulose by a transverse electrophoretic transfer and are then detected with antibody and labelled protein A (Burnette, 1981). Multiple replicas of one polyacrylamide gel on several sheets of paper can be obtained and each replica can then be analyzed with a different antiserum (Legocki and Verma, 1981). One disadvantage of the technique is that the proteins remain denatured to an unknown degree after electrophoretic separation and blotting; as a result monoclonal antibodies that are specific for the native conformation of an antigen may fail to detect the relevant molecule. Polyclonal antisera are less likely to cause such a problem.

The immunoblotting technique has been used successfully for detecting the presence of common epitopes in the proteins of different baculoviruses (Smith and Summers, 1981; Knell et al., 1983) and for investigating the presence of cross-reactions among human and animal enteroviruses (Mertens et al., 1983). Viral proteins and glycoproteins have also been detected on Western blots by the use of lectins and protein A conjugated to peroxidase (Ramirez et al., 1983) as well as by means of monoclonal antibodies (Goldstein et al., 1982). In plant virology, immunoblotting has been found useful for detecting viruses in plant sap and for studying distant antigenic relationships within virus groups (O'Donnell et al., 1982; Rybicki and Von Wechmar, 1982).

1.5. Immunoprecipitation

The immunoprecipitation technique refers to the precipitation of radiolabeled antigens by specific antibody followed usually by an analysis of the disrupted precipitate by polyacrylamide gel electrophoresis. This is one of the most commonly used methods for identifying the virus-encoded polypeptides produced in infected cells after biosynthetic labeling (Crawford et al., 1980; Suh et al., 1980; Dougherty and Hiebert, 1980; Grose and Friedrichs, 1982). As the amount of antigen is always very small, the mass of the precipitate is usually increased by adding either a second anti-immunoglobulin antibody or protein A-containing heat-killed staphylococci (Kessler, 1981). A detailed analysis of the experimental conditions needed for successful immunoprecipitation has recently been published (Goding, 1983). Compared to immunoblotting, the advantage of immunoprecipitation is that the antibodies interact with native antigens. Monoclonal antibodies that are specific for the native conformation of the antigen may thus react in immunoprecipitation but fail to do so in immunoblotting.

2. Agglutination tests

Agglutination refers to the clumping of cells or particulate antigens of similar size. Fewer antibody molecules are needed to produce an agglutination visible to the naked eye than to produce a visible precipitate, and agglutination reactions are thus more sensitive than precipitin tests.

2.1. Viral hemagglutination

Many viruses possess the property to agglutinate the erythrocytes of certain animal species. This phenomenon called hemagglutination can be inhibited by letting the virions first react with antibody molecules. This is the basis of the hemagglutination inhibition (HAI) tests which are used either to identify virus isolates by means of specific antisera or to measure antibody levels by means of standard virus suspensions. The HAI test is one of the most widely used procedures for the serological diagnosis of viral infections (Mayr et al., 1977; Lennette and Schmidt, 1979). Unfortunately, HAI tests are hampered by the fact that many sera contain various nonspecific inhibitors of agglutination that may block agglutinating sites on the erythrocytes. Some of the methods used for removing these inhibitors have been listed by van Regenmortel (1981).

2.2. Passive hemagglutination

Passive hemagglutination, also called indirect hemagglutination, utilizes red blood cells to which either virions or antibodies have been coupled. Agglutination of the coated cells is brought about by the addition of the complementary serological reactant. Antigens have been coupled to red blood cells using reagents such as tannic acid, bisdiazotized benzidine, carbodiimides, difluorodinitrobenzene and chromic oxide (Sundick and Rose, 1980; Wang, 1982). Prior to the coupling reaction, the erythrocytes are stabilized with compounds such as tannic acid, glutaraldehyde or sulfosalicyl acid (van Regenmortel, 1982). The passive hemagglutination reaction with virus-coated cells can be inhibited by prior incubation of the antiserum with the homologous virus. This inhibition test lends itself to the measurement of very small quantities of viral antigen (Anderer and Ströbel, 1972; Ghanekar et al., 1979).

2.3. Latex test

In this test either the antigen or the antibody is adsorbed onto polystyrene latex particles. Optimal attachment of the reactant to the latex occurs only within a narrow range of pH values and reagent concentration (Litwin, 1977). When antisera of low titer are used to coat the latex particles, better results are obtained by coupling antibody to the latex by an intermediate layer of protein A (Torrance, 1980). The latex test has been widely used in plant virology (van Regenmortel, 1982) and has also recently been applied to the detection of animal viruses such as rotaviruses

(Sanekata et al., 1981; Haikala et al., 1983). The test is extremely rapid and should be performed on glass plates as the stickiness of plastic may prevent agglutination. The test requires no instrumentation and can be performed outside the laboratory. However, proper controls are necessary to identify the false positives that occur as a result of nonspecific agglutination.

2.4. Immune adherence hemagglutination assay

In this technique, three successive reactions take place. First the antigen and antibody are allowed to form complexes; complement is then added and finally type O human erythrocytes are added to agglutinate the complement-bearing complexes (Ito and Tagaya, 1966; Mayumi et al., 1971). The procedure which is normally carried out in microtiter plates is more sensitive and easier to perform than classical complement fixation (CF) tests. The immune adherence hemagglutination assay is less affected by variations in test conditions than CF and is also less subject to interference by the anticomplementary activity of certain antisera (Lennette, 1979). It has been successfully used for detecting antibodies to a wide variety of viruses as well as for measuring the concentration of viral antigens (Lennette and Lennette, 1978; Lennette and Schmidt, 1979). The many advantages of the immune adherence hemagglutination assay will no doubt lead to an increased use of this technique for many diagnostic tasks that are presently carried out by standard CF tests.

3. Complement fixation techniques

The complement fixation (CF) test is one of the most widely applicable serological tests and it has been used extensively in the diagnosis of virus infections of man and animals (Lennette and Schmidt, 1979). Complement is a multicomponent system of enzymes present in an inactive state in any serum. The system becomes activated when a specific site on the Fc portion of certain antibody molecules becomes exposed as a result of antigen binding and is then able to bind the C1q component of complement (Osler, 1976). When the antibody is directed against the membrane antigens of red blood cells, the binding of C1q leads to a cascade of enzymatic reactions which ends in the lysis of the cells and the release of hemoglobin. When the antibody is directed against a virus particle, complement also becomes irreversibly bound upon formation of the virus-antibody complex and this leads to the effective removal of complement from the antiserum.

The CF test is a two-stage procedure involving a viral test system and an indicator system consisting of sheep erythrocytes and their corresponding antibodies. The complement present in both the test antiserum and the anti-sheep erythrocyte serum is destroyed by heating. A small amount of guinea pig complement, sufficient to cause complete hemolysis of the sheep erythrocytes, is added to the test antiserum.

The first part of the test consists in the overnight incubation of viral antigen, antibody and complement. Depending on the amount of antigen and antibody present, different amounts of complement will be used up. In the second part of the reaction, the erythrocytes and their antibodies are introduced as an indicator. Any unused complement left in the test system will be quantitated by measuring the amount of lysis produced in the indicator system. The amount of hemoglobin released from the lysed cells is inversely proportional to the extent to which the test antigen and antibody have interacted and fixed complement.

The CF test requires a careful standardization of all reagents and experimental conditions (Levine, 1978). The test is mostly employed to assay the amount of viral antibody present in sera and titrations are usually performed in disposable microtiter plates. Standard procedures have been described by Lennette and Schmidt (1979) and Palmer (1980).

Quantitative CF tests are usually performed in tubes and are very useful for detailed antigenic analysis of viruses. In these tests, the percentage of complement fixed is calculated from photometric measurements of the amount of hemoglobin released during hemolysis. For a given antiserum dilution, a plot of the percentage of complement fixed by increasing quantities of viral antigen takes the form of a bell-shaped curve. A detailed procedure for quantitative CF tests has been described by van Regenmortel (1982).

The attachment of at least two IgG molecules to a multivalent antigen is required for activating the complement system. However, the antigenic activity of monovalent fragments of viral proteins can be measured by their ability to inhibit CF by the intact antigen. Such inhibition of CF is a sensitive method for detecting antigenic activity in short peptides and for localizing the epitopes of viral proteins (Quesniaux et al., 1983). Although a cleaved fragment of an antigen may be able to react with only a small fraction of the total antibodies present in an antiserum, it may nevertheless cause complete inhibition of the CF reaction. The total abolition of CF by a single antigenic fragment should not be construed to mean that this peptide represents the only epitope of the viral protein (van Regenmortel, 1982).

The main difficulty encountered with CF tests is the anticomplementary activity of certain sera and antigen preparations which gives them the ability to fix complement in the absence of any serological reaction (Pauri et al., 1981). Anticomplementary activity in serum may result from the presence of immunoglobulin aggregates or preformed antigen-antibody complexes, and can be removed by a preliminary incubation with guinea pig complement, followed by heating at 60°C (Mayr et al., 1977). Guinea pig complement is a readily available source of complement which can be used in conjunction with many mammalian antisera. However, it cannot be used with avian antisera since the first component (C1) of guinea pig complement is not activated by avian antibody. It is possible to overcome the incompatibility between mammalian C1 and chicken antibody by using chicken C1 and a guinea pig serum reagent containing all the components of complement except C1 (Stolfi et al.,

1971). This method was found suitable for measuring the complement-fixing ability of chicken antiviral antibody.

It is also possible to measure the amount of complement fixed by the test system in the absence of a hemolytic indicator system. This can be done, for instance, by an enzyme immunoassay using a labeled antibody specific for the C3 component of complement (Hinchliffe and Robertson, 1983) or by the hemagglutination assay described in Section 2.4.

4. References

Anderer, F.A. and Ströbel, G. (1972) *Eur. J. Immunol.* 2, 278–282.

Axelson, N.H. (1983) Handbook of Immunoprecipitation-in-Gel Techniques. *Scand. J. Immunol.* (Suppl. 10) 17, 1–383.

Axelson, N.H., Bock, E., Larsen, P., Blirup-Jensen, S., Just Svendsen, P., Pluzek, K.J., Bjerrum, O.J., Bog Hansen, T.C. and Ramlau, J. (1983) Handbook of Immunoprecipitation-in-Gel Techniques. *Scand. J. Immunol.* (Suppl. 10) 17, 87–96.

Burnette, W.N. (1981) *Anal. Biochem.* 112, 195–203.

Crawford, L.V., Pim, D.C. and Lane, D.P. (1980) *Virology* 100, 314–325.

Crowle, A.J. (1973) *Immunodiffusion.* 2nd Edn. Academic Press, New York.

Dougherty, W.G. and Hiebert, E. (1980) *Virology* 104, 174–182.

Ghanekar, A.M., Reddy, D.V.R., Iizuka, N., Amin, P.W. and Gibbons, R.W. (1979) *Ann. Appl. Biol.* 93, 173–179.

Goding, J.W. (1983) *Monoclonal Antibodies: Principles and Practice.* Academic Press, New York. 276 pp.

Goldstein, L.C., McDougall, J., Hackman, R., Meyers, J.D., Thomas, E.D. and Nowinski, R.C. (1982) *Infect. Immun.* 38, 273–281.

Grandien, M. and Norrby, E. (1975) *J. Gen. Virol.* 27, 343–353.

Grauballe, P.C., Genner, J., Meyling, A. and Hornsleth, A. (1977) *J. Gen. Virol.* 35, 203–218.

Grillner, L. and Blomberg, J. (1976) *J. Clin. Microbiol.* 4, 11–15.

Grose, C. and Friedrichs, W.E. (1982) *Virology* 118, 86–95.

Grubb, A.O. (1983) Handbook of Immunoprecipitation-in-Gel Techniques. *Scand. J. Immunol.* (Suppl. 10) 17, 113–124.

Hagen, T.J., Taylor, D.B. and Meagher, R.B. (1982) *Phytopathology* 72, 239–242.

Haikala, O.J., Kokkonen, J.O., Leinonen, M.K., Nurmi, T., Mantyjarvi, R. and Sarkkinen, H.K. (1983) *J. Med. Virol.* 11, 91–97.

Havranek, P. (1978) *Phytopath. Z.* 93, 97–104.

Hinchliffe, P.M. and Robertson, L. (1983) *J. Clin. Pathol.* 36, 100–103.

Ito, M. and Tagaya, I. (1966) *Jpn. J. Med. Sci. Biol.* 19, 109–126.

Kessler, S.W. (1981) *Methods Enzymol.* 73, 442–459.

Kleczkowski, A. (1965) *Immunology* 8, 170–181.

Knell, J., Summers, M.D. and Smith, G.E. (1983) *Virology* 125, 381–392.

Kohn, J. and Riches, P. (1978) *J. Immunol. Methods* 20, 325–331.

Legocki, R.P. and Verma, D.P.S. (1981) *Anal. Biochem.* 111, 385–392.

Lennette, E.T. (1979) In *Diagnosis of Viral Infections* (D.A. Lennette, S. Specter and K.D. Thompson, Eds.) University Park Press, Baltimore. pp. 73–83.

Lennette, E.T. and Lennette, D.A. (1978) *J. Clin. Microbiol.* 7, 282–285.

Lennette, E.H. and Schmidt, N.J. (1979) *Diagnostic Procedures for Viral, Rickettsial and Chlamydial Infections.* 5th Edn. American Public Health Association Inc., Washington.

Levine, L. (1978) In *Handbook of Experimental Immunology* (D.M. Weir, Ed.) 3rd Edn. Blackwell, Oxford.

Litwin, S.D. (1977) In *Methods in Immunology and Immunochemistry* (C.A. Williams and M.W. Chase, Eds.) Vol. 14. Academic Press, New York. p. 115.

Mayr, A., Bachmann, P.A., Bibrack, B. and Wittmann, G. (1977) *Virologische Arbeitsmethoden*. Vol. II. Serologie. Gustav Fischer, Stuttgart.

Mayumi, M.K., Okochi, K. and Nishioka, K. (1971) *Vox Sang.* 20, 178–181.

Mertens, T., Pika, U. and Eggers, H.J. (1983) *Virology* 129, 431–442.

Middleton, P.J., Petric, M., Hewitt, C.M., Szymanski, M.T. and Tam, J.S. (1976) *J. Clin. Pathol.* 29, 191–197.

Nilsson, L.Å. (1983) Handbook of Immunoprecipitation-in-Gel Techniques. *Scand. J. Immunol.* (Suppl. 10) 17, 57–68.

Niwayama, S., Sato, S., Ochiai, H., Motoda, S., Iwase, I. and Hayashi, K. (1981) *J. Virol. Methods* 2, 261–268.

O'Donnell, I.J., Shukla, D.D. and Gough, K.H. (1982) *J. Virol. Methods* 4, 19–26.

Osler, A.G. (1976) *Complement. Mechanisms and Functions.* Prentice-Hall, Englewood Cliffs NJ. pp. 193.

Oxford, J.S. and Schild, G.C. (1977) *J. Gen. Virol.* 38, 187–193.

Palmer, D.F. (1980) In *Manual of Clinical Immunology* (N.R. Rose and H. Friedman, Eds.) 2nd Ed. Amer. Soc. Microbiol., Washington. pp. 35–47.

Paul, H.L., Gibbs, A. and Wittmann-Liebold, B. (1980) *Intervirology* 13, 99–109.

Pauri, P., Bagnarelli, P. and Clementi, M. (1981) *J. Virol. Methods* 3, 329–335.

Purcifull, D.E. and Batchelor, D.L. (1977) *Univ. Fl. Agric. Exp. Stn. Tech. Bull.* 788, 39.

Quesniaux, V., Briand, J.P. and van Regenmortel, M.H.V. (1983) *Molec. Immunol.* 20, 179–185.

Ramirez, P., Bonilla, J.A., Moreno, E. and Leon, P. (1983) *J. Immunol. Methods* 62, 15–22.

Reinganum, C. and Scotti, P.D. (1976) *J. Gen. Virol.* 31, 131–134.

Renart, J., Reiser, J. and Stark, G.R. (1979) *Proc. Natl. Acad. Sci. U.S.A.* 76, 3116–3120.

Ritchie, R.F. and Smith, R. (1976) *Clin. Chim. Acta* 22, 497.

Robinson, O.R. Jr., Shibley, G.P. and Sevoian, M. (1977) *Infect. Immun.* 18, 60–67.

Rybicki, E.P. and Von Wechmar, M.B. (1981) *Virology* 109, 391–402.

Rybicki, E.P. and Von Wechmar, M.B. (1982) *J. Virol. Methods* 5, 267–278.

Sanekata, T., Yoshida, Y. and Okada, H. (1981) *J. Immunol. Methods* 41, 377–385.

Schmidt, J. and Lennette, H. (1973) *Prog. Med. Virol.* 15, 244–308.

Shepard, J.F., Secor, G.A. and Purcifull, D.E. (1974) *Virology* 58, 464–475.

Sim, C. and Watson, D.H. (1973) *J. Gen. Virol.* 19, 217–233.

Slack, S.A. and Shepherd, R.J. (1975) *Phytopathology* 65, 948–955.

Smith, G.E. and Summers, M.D. (1981) *J. Virol.* 39, 125–137.

Steensgaard, J. (1984) *Immunol. Today* 5, 7–10.

Steensgaard, J. and Frich, J.R. (1979) *Immunology* 36, 279–291.

Steinmann, J. and Marzock, H.J. (1983) *J. Immunol. Methods* 59, 221–227.

Stolfi, R.L., Fugmann, R.A., Jensen, J.J. and Sigel, M.M. (1971) *Immunology* 20, 299–306.

Suh, M., Kessous, A., Poirier, N. and Simard, R. (1980) *Virology* 104, 303–311.

Sundick, R.S. and Rose, N.R. (1980) In *Methods in Immunodiagnosis*. 2nd Ed. (N.R. Rose and P. Bigazzi, Eds.) John Wiley and Sons, New York. pp. 109–127.

Torrance, L. (1980) *Ann. Appl. Biol.* 96, 45–50.

Towbin, H., Staehelin, T. and Gordon, J. (1979) *Proc. Natl. Acad. Sci. U.S.A.* 76, 4350–4354.

Van Regenmortel, M.H.V. (1975) *Virology* 64, 415–420.

Van Regenmortel, M.H.V. (1981) In *Comprehensive Virology* (H. Fraenkel-Conrat and R.R. Wagner, Eds.) Vol. 17. Plenum Press, New York. pp. 183–243.

Van Regenmortel, M.H.V. (1982) *Serology and Immunochemistry of Plant Viruses.* Academic Press, New York. p. 302.

Vestergaard, B.F., Bjerrum, O.J., Norrild, B. and Grauballe, P.C. (1977) *J. Virol.* 24, 82–90.
Waldstrom, T. (1983) Handbook of Immunoprecipitation-in-Gel Techniques. *Scand. J. Immunol.* (Suppl. 10) 17, 97–102.
Wang, A.C. (1982) In *Antibody as a Tool. The Applications of Immunochemistry* (J.J. Marchalonis and G.W. Warr, Eds.) John Wiley and Sons, New York. pp. 139–161.

Eds. M.H.V. van Regenmortel and A.R. Neurath
Immunochemistry of viruses. The basis for serodiagnosis and vaccines.

Immunofluorescence techniques and their application 8

P.S. GARDNER

Division of Microbiological Reagents and Quality Control, Central Public Health Laboratory, 175 Colindale Avenue, London NW9 5HT, U.K.

Immunofluorescence (IF) together with electron microscopy were the first of the now many generations of rapid diagnostic techniques that have been developed. However, the basis of these new generations of rapid techniques have, without doubt, been derived from IF. This does not imply that the technique is obsolete but that there are now areas of the diagnostic field where all are agreed that IF is still the most sensitive and reliable method (WHO Bulletin, 1977; Almeida et al., 1979; WHO Report, 1981) while in other areas enzyme-linked immunosorbent assay (ELISA), radioimmunoassay (RIA) and electron microscopy (EM) are more appropriate. The aim of this chapter is to attempt to redefine the place of IF as a diagnostic tool in relation to its sister techniques. The major reservation most virologists have about IF is its labour intensity and the need for the training of skilled staff for using this technique. It is intended to examine these criticisms closely and see if changes are possible in the labour intensiveness of the method. In these days of fast moving technical progress it is remarkable that IF has remained relatively unchanged as a major diagnostic tool over the last fifteen years.

1. Principles of immunofluorescence

The basic principle of IF is that when a fluorochrome dye such as fluorescein isothiocyanate has been irradiated with light of certain wavelengths, that light is absorbed and changes the molecule of dye from a resting level of energy to that of an excited energy level. Once it has reached a state of excitation it immediately reverts to its original resting level of energy by releasing energy. This is done by emitting light of longer wavelength than that used to stimulate the molecule of dye originally. It is this longer wavelength emission which constitutes fluorescence.

There are two principal dyes concerned in IF – fluorescein isothiocyanate and rhodamine isothiocyanate. The wavelength of maximum absorption for fluorescein isothiocyanate is 490 nm, in the blue part of the spectrum, and that for emission is 517 nm, in the green part of the spectrum; the corresponding figures for rhodamine are 575 nm (absorption) and 595 nm (emission). These figures immediately summarise one of the earliest problems in IF: the separation of absorption wavelength from emission wavelength. Early on, with ordinary glass filter systems, the closeness of these two wavelengths prevented their separation and as a consequence such filters could not be used at their optimum wavelength of absorption. With the introduction of interference filters it became possible to separate two similar wavelengths and achieve stimulation at the wavelength of maximum absorption of the fluorochrome and at the same time it was possible to achieve, with the appropriate barrier filter, maximum emission. This principle is illustrated in Figs. 8.1 and 8.2.

To achieve stimulation of the fluorochrome an efficient light source is required. The burner most frequently used is the HBO 200 mercury vapor lamp. It has a life of approximately 200 hours and shows high peaks of emission at 365 nm, 435 nm, 546 nm and 578 nm. These peaks, with reliable filters, make this burner suitable for the stimulation of both fluorescein isothiocyanate and rhodamine isothiocyanate near or even at the wavelength of maxium absorption.

With efficient interference filtration systems it is possible to use lower wattage burners such as HBO 50 with good results. Two other light sources for the stimulation of fluorochromes are possible:

(*a*) Xenon mercury arc lamps, which are very reliable for fluorescein isothiocyanate but poor for rhodamine isothiocyanate.

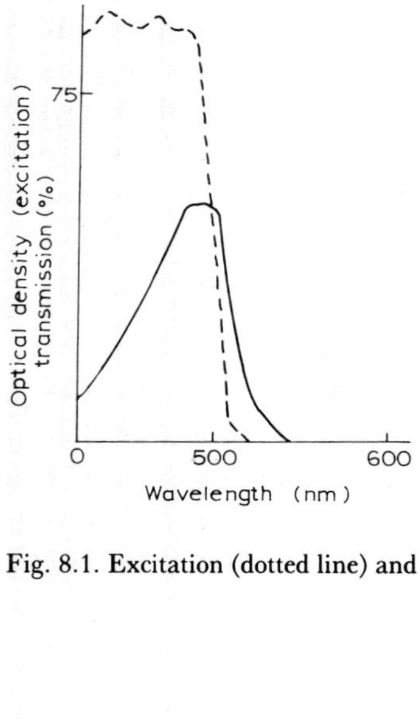

Fig. 8.1. Excitation (dotted line) and absorption (solid line) curves for FITC using interference filter.

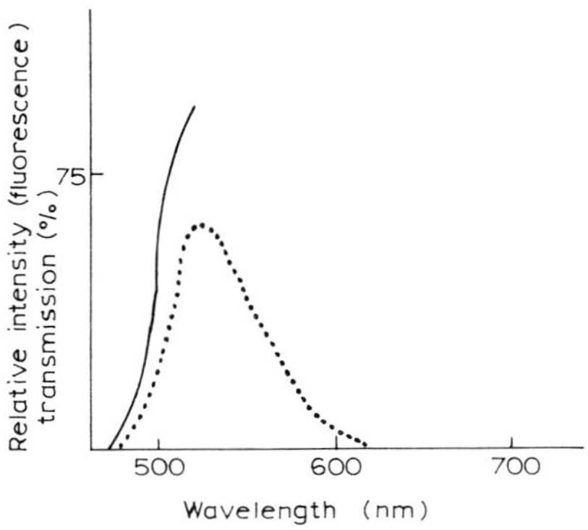

Fig. 8.2. Emission (dotted line) and transmission (dotted line) curves for FITC using a barrier filter.

(*b*) Laser beams, which can excite any fluorochrome at its peak wavelength of absorption, although so far this has not been applied to routine fluorescence microscopy (Berquist, 1973).

Although transmitted light microscopes can be used and are still used in conjunction with darkground illumination, there can be no doubt that the introduction of the incident light fluorescence microscope with improved interference systems has been of the greatest value to the routine diagnostic virologist and is the only system that can be recommended. Briefly, light, having been transmitted through the appropriate excitation filter, impinges on a chromatic beam splitter which reflects light below 510 nm onto the specimen; light above 510 nm passes through the chromatic beam splitter and is wasted. The specimen, having been irradiated with light below 510 nm, emits light of above 510 nm; this again impinges on the same chromic beam splitter and is transmitted through a barrier filter to the eye. Light below 510 nm is now reflected by the chromatic beam splitter and is lost. It is not possible to discuss in detail here the various principles involved but for those interested the whole subject of fluorescence microscopy for the virologist has been adequately covered elsewhere (Gardner and McQuillin, 1980).

2. Reagents and their assessment

There are three basic reagents involved in IF, viz. buffers and diluents, antisera and conjugates. The last two may be the same reagent if the direct test is used (see below). These reagents have been described fully by Gardner and McQuillin (1980) in respect of conventional antisera, buffers and conjugates. However, since that time events have overtaken the text book and newer methods of antisera production have appeared.

The major requirement in the manufacture of viral antisera is the production of large quantities of the material which should require little purification. This statement is equally applicable whether the indirect or direct test is being used. However, if the indirect test is to be employed, the number of species in which antisera are produced should be very small (preferably one) since it is preferable to use as small a number of antispecies conjugates as possible.

The manufacturing procedures of these antisera over the last decade illustrates well how production methods have evolved.

Calves are the natural hosts of a number of viruses which are antigenically similar to those infecting humans. Respiratory syncytial (RS) virus and parainfluenza virus type 3 are good examples of this phenomenon. Because of this it was easy to produce high titer antisera to these viruses in calves, well above any nonspecific activity. However, this method could not be used for all viruses and had the serious disadvantages that many animals have their own microbiological infections which are reflected as unwanted antibodies in the final reagent; gnotobiotic animals for the production of all diagnostic antisera would be impractical. Moreover reagents so produced would inevitably still show batch variations.

A new development has been the use of specific globulins extracted from egg yolks following the inoculation of the hens with the appropriate virus. The virus-specific globulins so produced can be obtained in large amounts and with little nonspecific activity. The product is fairly reproducible although the hen may need an occasional booster dose of virus (Gardner and Kaye, 1982). However, chickens have their own microbial flora and the egg globulin could theoretically reflect this.

The latest development in this field has been the production of monoclonal antibodies (Russell et al., 1981; Gardner, 1982). Monoclonal antibodies have the advantage of being completely specific. They may also be produced in unlimited supplies ensuring the absence of batch variations. However, the narrow specificity of the antibody may be a disadvantage because for rapid diagnosis by IF, as well as by ELISA and other rapid techniques which are often used to detect antigen in clinical material, any minor virus variation could lead to the failure to detect a small number of 'wild' virus strains. This may be overcome by using cocktails of appropriate monoclonal antibodies.

Before monoclonal antibodies are adopted for diagnostic work, it is essential to ensure that they are fully evaluated so that they will not fail to detect the majority if not all 'wild' viruses or their antigens. Large scale clinical trials will be needed to achieve this before a particular monoclonal antibody or a cocktail of monoclonal antibodies is approved for routine diagnostic use.

Conjugates will either be the specific viral antibody conjugated with a fluorochrome dye (direct test) or an antispecies conjugate similarly labeled (indirect test), and will need as careful assessment as unlabeled specific antibodies destined for viral IF. Almeida et al. (1979) and the European Group for Rapid Viral Diagnosis have laid down precise protocols for the assessment of antibodies and conjugates.

3. Procedures for the quality control of reagents

3.1. Virus antisera or antibodies

(*1*) Infected and control cells should be tested to determine the optimal dilution for detecting antigen. This dilution must be at least 4-fold greater than that revealing any non specific fluorescence with uninfected cells. The control cell cultures should include a continuous human line, a continuous monkey kidney line, a diploid cell strain and a primary monkey kidney line.

(*2*) A limited titration for final evaluation should be made on positive clinical material appropriate for the virus under test.

(*3*) Antisera will be tested at their optimum dilution on negative pharyngeal secretions and, where possible, on other negative human specimens.

(*4*) Antisera should be tested on cell cultures infected with a representative collection of viruses, e.g. respiratory syncytial virus and parainfluenza 1, 2, 3, 4a and 4b, mumps, influenza A and B, measles, herpes virus, cytomegalovirus and at least two enteroviruses. There should be no nonspecific reaction with heterologous antigens at the optimum dilution of the test antiserum.

3.2. Antispecies conjugates

(*1*) Dilutions of antiglobulin conjugate with appropriate antiserum should be tested on positive tissue cultures and positive human specimens to find the optimum dilution for use. A preliminary optimum dilution can be obtained on tissue cultures, but this has to be modified for its final use for material taken directly from the patient.

(*2*) They are then tested at their optimum dilution for any nonspecific reaction on cell lines likely to be used in the laboratory, including those mentioned above for antisera.

(*3*) Antiglobulins should then be tested at their optimum dilution on negative human material, including nasopharyngeal secretions, lung material and biopsy material etc. for nonspecific activity.

These criteria will only apply under the conditions of the test system and it is therefore of great advantage to users of the reagents to use the staining technique, including a counterstain, as suggested by Almeida et al., (1979).

4. The techniques

There are three basic techniques – the direct fluorescent antibody technique, the indirect antibody technique and the complement staining technique. All three techniques have been fully described elsewhere (Gardner and McQuillin, 1980), and that textbook should be consulted for practical details. However, the decision as to which method to use needs some consideration.

The direct test consists of the clinical specimen containing the antigen fixed on a slide and the sole diagnostic reagent is the specific antiserum which has been directly conjugated.

The indirect test also consists of the clinical specimen containing the antigen fixed on a slide but is first treated with a specific antiserum; after suitable washing the preparation is treated with the appropriate antispecies conjugate.

The direct test is quicker and usually suitable if one is only handling a single antigen, e.g. for the diagnosis of rabies. On the other hand it is tedious for all antisera to be conjugated if a moderate number of different antigens are being investigated. Under these circumstances it is better to use an antispecies conjugate having obtained specific antisera from one or at the most two animal species.

The complement method makes use of the principle that complement is bound when multivalent antigens bind to antibody. The presence of bound complement can be dectected, for instance, with an antibody to C3 complement.

The indirect and the complement methods are considered to be much more sensitive than the direct method by a factor which may be as much as ten (Pressman, Yagi and Hiramoto, 1958).

A number of practical points should be emphasised.

(*a*) Especially during the investigation of clinical specimens a counterstain such as naphthalene black should be used. In the indirect method the conjugate can be reconstituted by using a solution of phosphate buffered saline containing naphthalene black at a concentration of 1 mg/ml. This is solely to diminish autofluorescence of the specimen, a counterstain should never be used to mask a bad reagent.

(*b*) Never omit a final wash in distilled water before drying, as phosphate buffered saline crystals will be deposited and make the specimen unreadable.

(*c*) For routine diagnosis there is never any requirement to mount a preparation under a cover slip. Immersion oil (nonfluorescent) can be put directly on the specimen, after which the specimen is observed and then discarded. Mounted specimens are only necessary for teaching purposes or for exchange of specimens with other laboratories.

Occasionally it is necessary to stain a specimen by the double staining technique. For example if one observes dull fluorescent cells in a secretion it may be that they are coated with antibody. It would therefore be useful to stain the antigen with rhodamine and the antibody with fluorescein. This technique has been described by Gardner and McQuillin (1978) and practical details should be sought in that publication.

5. The application of immunofluorescence to diagnosis

The object of this chapter is to place IF in its proper place as a diagnostic tool. To be a modern diagnostic tool in virology the method has to be able to provide a rapid answer at a time when the diagnosis can influence the management and treatment of a patient or a community. There are basically two ways of achieving a rapid viral diagnosis: (*a*) detection of antigen at the site of lesion, and (*b*) detection of the early production of specific IgM. Although the measurement of the latter can be effected by IF, the majority of modern systems have been developed from sensitive ELISA and RIA techniques which are more suitable for this purpose and therefore the application of IF in this context will not be pursued further.

However, it is in the field of antigen tracing that IF has its most important diagnostic role to play. A number of WHO bulletins (1977 and 1978) and a WHO report (1981) have been issued on rapid virus diagnosis and the conclusion reached was that IF was the most sensitive method available for the diagnosis of respiratory virus infections. It strongly recommended this technique and has run a substantial number of workshops to teach this technique to laboratory workers, especially those in developing countries. Although appreciating that training was necessary and that at the moment IF is a labour-intensive technique, WHO still considered it to be a practical method even for developing countries. Their faith has been borne out by the trials which have been run involving the cross checking of slides from field laboratories by WHO reference centers.

Apart from the respiratory tract, for those laboratories not in possession of an electron microscope, the fluorescent antibody technique is suitable for the investigation of skin and eye lesions as well as for encephalitis when brain biopsy material was available.

Another field in which diagnosis by IF is strongly recommended by WHO (1973) is for rabies. This was based on the original work of Goldwasser et al. (1958), who first showed the increased sensitivity over the then conventional staining for Negri bodies. This specialised subject has been fully reviewed by Gardner and McQuillin (1980).

The taking and preparation of specimens from the respiratory tract and those from skin and eye have been fully described by Gardner and McQuillin (1980).

It can only be reiterated that a technique can only be as successful as the quality of the specimen received allows. If a specimen is carefully taken and prepared, the technique will prove to be as sensitive as, and in certain cases more sensitive than, viral isolation procedures (Gardner and McQuillin, 1980).

Secretions for the diagnosis of respiratory virus infections are taken at the bedside by passing a catheter through the nasal passages into the nasopharynx with suction being applied during the whole manoeuvre; secretions are collected in a mucous trap and brought on ice to the laboratory as soon as possible. Cells are separated from mucus by gentle washing in buffered saline to remove mucus, and drops of a concentrated suspension of these cells are then spotted onto previously prepared slides, air dried and fixed in acetone for 10 minutes at 4°C. The slides can then be stained but spare unstained slides can be stored below −20°C for future reference purposes. When confidence in the technique has been established, there is no need for parallel isolation procedures and the urgency of transporting and processing the specimen is less than if one needed to obtain viable virus growth on tissue culture (Gardner and McQuillin, 1980).

As virus antigenicity lasts longer than virus viability, the technique can be used to detect antigen rather later in the illness than viable virus can be isolated. The technique can also be used for the investigation of autopsy material. The latter may be prepared as impression smears of lung material, as tracheal and bronchial secretions or as frozen section of lungs (Gardner and McQuillin, 1980).

Skin and corneal scrapings for the investigation of herpes simplex or other virus infections of the skin may be taken and cells teased apart in drops of buffered saline on slides before air drying and fixing as described above.

To some extent the purpose of rapid virus diagnosis can be defeated by using the fluorescent antibody technique to detect virus and confirm the cause of cytopathogenicity in tissue cultures, the cells being scraped off the tube and resuspended in buffered saline before spotting onto slides for specific staining. True rapid diagnosis can only be achieved by detecting antigen in clinical material and not by waiting for the growth of virus in tissue culture.

5.1. Application to rapid diagnosis

All the frequently occurring respiratory virus infections which bring children into hospital can be diagnosed by IF, these include respiratory syncytial virus, influenza virus types A and B, the adenovirus group, and parainfluenza virus types 1, 2 and 3. The rhinoviruses cannot be diagnosed by IF as the number of different antigenic types is so great. However, rhinoviruses are a relatively minor cause of admission to hospital (Martin et al., 1978). Fig. 8.3a shows an example of rapid virus diagnosis being made on secretions taken and sent to the laboratory within 3 hours of these patients being admitted to hospital. Fig. 8.3b demonstrates the appearance of herpes simplex in a skin lesion. It is important for ease of diagnosis to obtain a vesicular lesion and scrape the base for cells; crusted lesions will give a certain amount of nonspecific fluorescence and make reading and interpretation more difficult.

5.2. Deaths from respiratory virus

Infection is still an all too common cause of death in the early months of childhood, some of these deaths resembling or even appearing to be identical with the sudden infant death syndrome (SIDS) on histological examination of the lungs (Ferris et al., 1973). The majority of these deaths are due to infection with respiratory syncytial virus, but other viruses may also play a role, e.g. influenza virus type A, parainfluenza virus type 3 and members of the adenovirus group. IF is an effective and often the only way to determine the cause of respiratory deaths. It is also a means of bringing some relief of the guilt feeling that parents of SIDS must suffer at this time, blaming themselves for something they could not forsee or prevent.

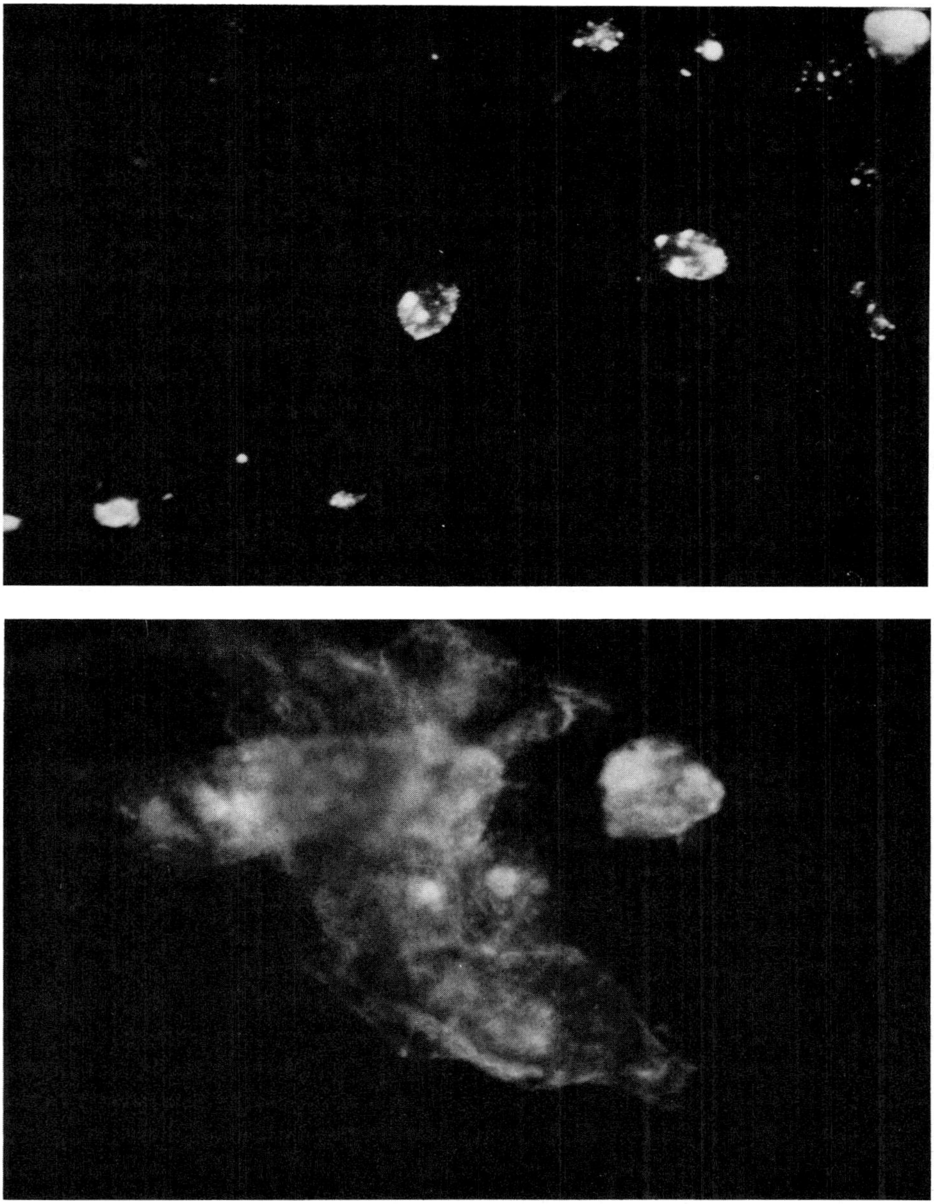

Fig. 8.3. Top, respiratory syncytial virus in cells of nasopharyngeal secretions of a child with acute bronchiolitis. Magnification ×2200. Bottom, herpes simplex virus in cells taken from a vesicular skin lesion and showing intranuclear fluorescence. Magnification ×2200.

5.3. Detecting viruses during convalescence

During the course of a respiratory illness antibodies develop rapidly in nasal secretions, mainly specific IgA which coat infected cells. This makes it difficult, after the first few days of illness, to isolate the virus which can, however be demonstrated by IF. It is this ability to detect antibody-coated cells which makes this technique more sensitive than virus isolation and also other techniques, such as RIA and ELISA (Gardner et al., 1970).

5.4. Hospital cross-infection

Microbial cross-infection is a constant threat to hospitalized patients and viruses constitute one of the major problems. Much has been written on the subject by Gardner et al. (1971), Weightmann et al. (1974) and Sims et al. (1975) working in Newcastle-upon-Tyne and IF is a potent tool for quickly identifying infected infants on admission to hospitals. As an example, influenza A frequently causes a febrile convulsion in the toddler which leads to admission of the child, who by then is no longer convulsing and has little pyrexia. He will be well enough to wander around the ward, excreting infected material over other patients, including those who are immunosuppressed; and this may have devastating results. The time is long overdue when these newer method should be applied for the benefit of all inpatients and for increasing the safety of all those who are ill and debilitated.

5.5. Immunosuppression

Immunosuppression forms an integral part of treatment of many malignancies, including acute lymphatic leukemia, which is particularly common among young children. Children and even adults who are immunosuppressed are prone to virus infections with severe consequences. One particularly devastating infection in the leukemic child is measles. The great majority of those who acquire this infection during the course of their leukemia will die. The child will eventually succumb to a giant cell pneumonia, usually after being severely ill for some weeks. Occasionally the illness is difficult to diagnose because of either the absence of rash or the appearance of one which is atypical. Diagnosis is extremely easy since at all stages of the illness until death the secretions are loaded with virus-laden giant cells. This is illustrated in Fig. 8.4.

Other virus infections such as respiratory syncytial virus may continue as an illness spread over many months and even in the older child it may prove fatal (Craft et al., 1977). In all these continuing infections in the immunosuppressed patient, virus-infected cells can be found in nasopharyngeal secretions. However, even after a considerable time there is no evidence of coating of cells by the patient's antibody as has already been discussed in the case of a normal infected patient.

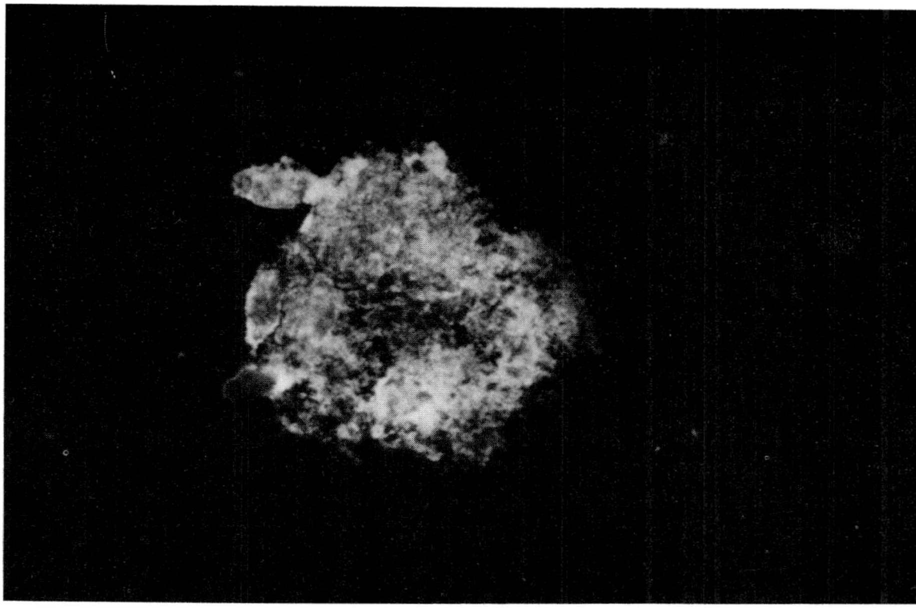

Fig. 8.4. Measles virus in a giant cell of nasopharyngeal secretions of a five year old child with acute leukaemia with a terminal measles pneumonia. Magnification ×2200.

5.6. Other aspects

Unusual presentation of clinical symptomes may cause difficulties in diagnosis. It has been mentioned above that influenza virus type A may often cause the signs and symptoms of febrile convulsions in toddlers (Stokes et al., 1977). Another example is infection with influenza virus type B, which can cause severe abdominal pain leading to an unnecessary laparotomy which could be avoided by using rapid diagnostic procedures (Kerr et al., 1975).

Some viruses are difficult to grow on tissue culture in the laboratory. The coronaviruses fall into this category, but can be identified in secretions by IF as demonstrated by McIntosh and his colleagues (1978).

Those laboratories which are remote from a virus laboratory can also be helped by the use of the fluorescent antibody technique. Slides can be prepared and fixed at the local laboratory for diagnosis elsewhere. These slides are stable and can be sent by post or public transport. That such a service is feasible has been demonstrated in the field where a hospital which was situated 100 miles from the virus laboratory had an effective 24-hour service using the post and a final telephone call (Downham et al., 1974).

There are obvious shortcomings to the fluorescent antibody technique and it would be as well to examine these critically. The basic weakness is that the technique as it stands needs someone to read the result down a microscope. This means that it becomes labour intensive and requires trained personnel to observe and recognise positive cells. Paradoxically this is also the strength of the technique, as the human eye can appreciate various nuances which a machine will fail to detect. In other words the observer working according to certain rules makes this technique the most sensitive in use so far. Only a single cell needs to be observed, but it must be of the right type and contain the virus antigen in the correct distribution: there is no point in calling a squamous cell positive if nuclear or even cytoplasmic fluorescence is seen in such a cell in nasopharynx secretions reputed to be infected with respiratory syncytial virus. The cells need to be ciliated columnar cells and the fluorescence must be restricted to the cytoplasm and be distributed as particles or granules of various sizes. Moreover, only the eye can detect the exact color of the fluorescence, which has a very characteristic apple-green appearance. Small deviations from the normal color should put the observer on the alert.

A trial between the author's laboratory and the State Bacteriological Institute in Stockholm was carried out to compare the sensitivity of ELISA and the fluorescence antibody technique for detecting respiratory virus antigens in nasopharyngeal secretions. The London laboratory examined prepared slides sent to it by the Stockholm laboratory, who examined the same material by ELISA. The two viruses which were sought were Influenza virus type A and respiratory syncytial virus. Out of a total of 73 specimens 21% were identified as Influenza virus type A by IF and 25% as respiratory syncytial virus. In the same series 19% of influenza virus type A and 19% of respiratory syncytial virus was detected by ELISA. It would therefore appear that there is little difference in the detection rate of influenza virus type A by the two methods but about 20% difference in sensitivity for respiratory syncytial virus. These results will be reported more fully elsewhere. Although on a small scale these results confirm the sensitivity advantage of IF, they are contradicted by the finding reported by Hornsleth and his coworkers in Denmark (1982).

6. Developments

There is no doubt that a major disadvantage of IF is its work intensiveness, and the possibility of automating a fluorescence system would therefore be most desirable. Progress has been made in this area by Halonen and his colleagues in Turku (1983) with their preliminary work on time-resolved fluoroimmunoassay. In this method, a long-life fluorescent probe is excited by a short light pulse and specific fluorescence is measured after a selected interval of time. During this interval the non specific background fluorescence, because of its shorter decay time, is lost and only the specific fluorescence is detected. Europium, a rare earth metal, has a long decay time of 100–1000 μs and moreover demonstrates in its conjugated form a large difference between excitation and emission wavelengths; this also helps with the elimination of background noise.

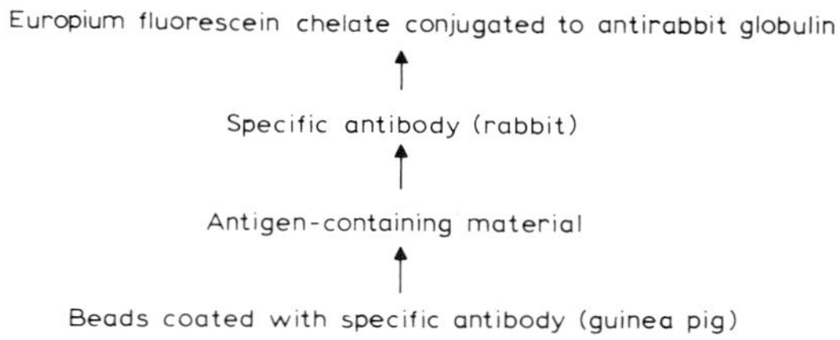

Fig. 8.5. The methodology of time-resolved fluoroimmunoassay.

Fig. 8.5 illustrates the similarity of this test with other immunoassays. The specimens are read in a single-photon cavity fluorometer equipped with a xenon flash lamp. Excitation is produced with a wavelength of 340 nm and the length of time of the excitating pulse is 1 μs.

After an interval of 400 μs single-photon emissions are counted for 500 μs at 613 nm wavelength. After a further interval of 100 μs a new cycle commences. There are 1000 cycles in the overall counting time of 1 s. These workers claim that their results compare very favourably with RIA and conventional immunofluorescence for routine virus diagnosis on clinical specimens. This method has as yet to be investigated by other workers to confirm its reproducibility and sensitivity and must then be examined, in terms of modern jargon, for 'cost effectiveness'. One has to remember that even at the height of massive epidemics, respiratory infection patients admitted to hospital only generate approximately 20–30 new specimens a day, a total easily handled by the conventional IF method. Expensive apparatus with a larger capacity for handling increased numbers of specimens may not be justified on the basis of sample numbers although it may be justified by its sensitivity and accuracy.

Methods have also been looked at for improving sensitivity in other ways, such as by using alternative conjugates. Two possible methods have been proposed. The first is the use of staphylococcal protein A, a protein binding to Fc portions of many immunoglobulins. However, this does not appear to be practical, especially nowadays when many of the antiglobulins in use, derived from egg yolk, do not bind at all to staphylococcal protein A.

7. Conclusions

The question that now needs to be asked is whether IF, which has been an essential evolutionary method in rapid virus diagnosis, still has an important diagnostic role to play. It would appear that the question can be answered clearly and unhesitantly in the affirmative.

Historically, Coons and collaborators (1941; 1950) perfected his technique in the 1940s and 1950s and slowly this was taken up for diagnostic use. The main utilisation of the method was for the respiratory group of viruses (Liu, 1956; Hers et al.,

1968; Gardner and McQuillin, 1968; McQuillin et al., 1970; Gardner et al., 1971) for herpes infections (Biegeleisen et al., 1959; Gardner et al., 1968) and for measles (McQuillin et al., 1976). It was in fact the first method to employ a solid phase. All new developments so far seem to be based on the principle of a solid phase, which either consists of the specimen or of an antibody which will capture the antigen in the specimen for diagnosis; the rest of the test is solely a variation of the indicator system. Immunofluorescence, therefore, has been the guideline for modern viral diagnosis. Should it now retire gracefully into the background having served a honourable place in the annals of virology, or has it still a diagnostic function?

So far it has remained the only rapid method for the diagnosis of respiratory virus infections and is strongly recommended by WHO (WHO bulletins, 1977 and 1978; WHO report, 1981). Other methods have so far not reached its reliability or sensitivity.

Although it has been claimed that IF requires much training, specialised apparatus and excessive labour, a considerable number of laboratories use this method for rapid diagnosis. In the UK alone, over 90 laboratories were using this method in 1981. That laboratory workers can learn and practise these techniques quickly and accurately is shown by a number of collaborative trials which have been run (MRC Report, 1978; Orstavik et al., 1980; 1984) recently. On a smaller scale, WHO laboratories which have recently taken up immunofluorescence are advised to send duplicate specimens to WHO reference laboratories until they are confident in their procedures. This has also worked successfully.

The development in automated readings on the lines originated by the Turku workers (Halonen et al., 1983) can only lead to progress and improvement in the technique. However, older workers in all branches of laboratory medicine must regret the passing of the days when they actually used their eyes to record results and relied on their own judgement rather than on rows of automated numbers. Those practicing IF will have that extra pang of regret as one of the last pieces of beauty that can be enjoyed in scientific work is automated.

8. References

Almeida, J.D., Atanasiu, P., Bradley, D.W., Gardner, P.S., Maynard, J., Schuurs, A.W., Voller, A. and Yolken, R.H. (1979) *WHO Offset Publication* 47.

Bergquist, W.R. (1973) *Scand. J. Immunol.* 2, 37–44.

Biegeleisen, J.Z. Jr., Scott, L.V. and Lewis, V. Jr. (1959) *Science* 129, 640–641.

Coons, A.H., Creech, H.J. and Jones, R.N. (1941) *Proc. Soc. Exp. Biol. Med.* 47, 200–202.

Coons, A.H. and Kaplan, M.H. (1950) *J. Exp. Med.* 91, 1–13.

Craft, A.W., Reid, M.M., Bruce, E., Kernahan, J. and Gardner, P.S. (1977) *Arch. Dis. Childhood* 52, 752–757.

Ditchburn, R.K., McQuillin, J., Gardner, P.S. and Court, S.D.M. (1971) *Br. Med. J.* 3, 671–673.

Downham, M.A.P.S., Elderkin, F.M., Platt, J.W., McQuillin, J. and Gardner, P.S. (1974) *Arch. Dis. Childhood* 49, 467–471.

Ferris, J.A.J., Aherne, W.A., Locke, W.S., McQuillin, J. and Gardner, P.S. (1973) *Brit. Med. J.* 2, 439–442.

Gardner, P.S. (1982) *Proc. R. Soc. Edinburgh* 81B, 277–291.

Gardner, P.S. and Kaye, S. (1982) *J. Virol. Methods* 4, 257–262.

Gardner, P.S. and McQuillin, J. (1968) *Br. Med. J.* 3, 340–343.

Gardner, P.S. and McQuillin, J. (1978) *J. Med. Virol.* 2, 77–87.

Gardner, P.S. and McQuillin, J. (1980) *Rapid Virus Diagnosis, Application of Immunofluorescence.* 2nd Edn. Butterworths, London.

Gardner, P.S., McQuillin, J., Black, M.M. and Richardson, J. (1968) *Br. Med. J.* 4, 89–92.

Gardner, P.S., McQuillin, J. and McGuckin, R. (1970) *J. Hyg. Camb.* 68, 575–580.

Gardner, P.S., McQuillin, J., McGuckin, R. and Ditchburn, R.K. (1971) *Br. Med. J.* 2, 7–12.

Goldwasser, R.A., Kissling, R.E., Carski, T.R. and Hosty, T.S. (1958) *Proc. Soc. Exp. Biol. Med.* 98, 219–223.

Halonen, P., Meurman, O., Lovgren, T., Hemmila, I. and Soini, E. (1983) In *Current Topics in Microbiology and Immunology.* Vol. 104 (P.A. Bachmann, Ed.) Springer-Verlag, Berlin. pp. 133–146.

Hers, J.F.Ph., van der Kuip, L. and Masurel, N. (1968) *Lancet* 1, 510–511.

Hornsleth, A., Friis, B., Andersen, P. and Brenoe, E. (1982) *J. Med. Virol.* 10, 273–281.

Kerr, A.A., Downham, M.A.P.S., McQuillin, J. and Gardner, P.S. (1975) *Lancet* 1, 291.

Liu, C. (1956) *Proc. Soc. Exp. Biol. Med.* 92, 883–887.

Martin, A.J., Gardner, P.S. and McQuillin, J. (1978) *Lancet* 2, 1035–1038.

McIntosh, K., McQuillin, J., Reed, S. and Gardner, P.S. (1978) *J. Med. Virol.* 2, 341–346.

McQuillin, J., Bell, T.M., Gardner, P.S. and Downham, M.A.P.S. (1976) *Arch. Dis. Childhood* 51, 411–419.

McQuillin, J., Gardner, P.S. and McGuckin, R. (1970) *Lancet* 2, 690–695.

MRC Subcommitee on RSV Vaccines (1978) *Br. Med. J.* 2, 796–798.

Orstavik, I., Grandien, M., Halonen, P., Arstila, P., Mordhurst, C.H., Hornsleth, A., Popow-Kraupp, T., McQuillin, J. and Gardner, P.S. (1980) *Lancet* 2, 32.

Orstavik, I., Grandien, M., Halonen, P., Arstila, P., Mordhurst, C.H., Hornsleth, A., Popow-Kraupp, T., McQuillin, J., Gardner, P.S., Almeida, J., Bricout, F. and Marques, A. (1984) *Bull. W. H. O.* In Press.

Pressman, D., Yagi, Y. and Hiramoto, R. (1958) *Int. Arch. Allergy* 12, 125–136.

Russell, W.C., Gunvanti, P., Precious, B., Sharp, I. and Gardner, P.S. (1981) *J. Gen. Virol.* 56, 393–408.

Sims, D.G., Downham, M.A.P.S., Webb, J.K.G., Gardner, P.S. and Weightman, D. (1975) *Acta. Paediat. Scand.* 64, 541–545.

Stokes, M.J., Downham, M.A.P.S., Webb, J.K.G., McQuillin, J. and Gardner, P.S. (1977) *Arch. Dis. Childhood* 52, 129–133.

Weightmann, D., Downham, M.A.P.S. and Gardner, P.S. (1974) *J. Hyg. Camb.* 73, 53–60.

WHO Expert committee on rabies (6th report) (1973) *Technical Report Series* 523.

WHO Bulletin (1977) *WHO* 55, 33–37.

WHO Bulletin (1978) *WHO* 56, 241–244.

WHO Scientific Group (1981) *Technical Report Series* 661.

Eds. M.H.V. van Regenmortel and A.R. Neurath
Immunochemistry of viruses. The basis for serodiagnosis and vaccines.

Solid phase immunoassays for the detection of viral diseases

9

ROBERT H. YOLKEN

The Johns Hopkins Hospital, 600 N. Wolfe Street, CMSC 1109, Baltimore, MD 21205, U.S.A.

Traditionally, the identification of viruses in biological specimens has been accomplished by the cultivation and isolation of the infecting virus in tissue or organ culture, or in laboratory animals. Such viral cultivation methods have been extremely valuable for the identification of viral pathogens capable of infecting humans. In addition, viral cultivation is the mainstay for the diagnosis of viral infections in individual patients. However, there are a number of limitations to the use of viral cultivation for clinical diagnosis. The most important limitation is the fact that the time required for the cultivation of many viruses is too long to provide useful information to clinicians caring for patients with suspected viral infections. Thus, important clinical decisions must be made before the results of viral studies are available. This limitation is particularly troublesome in situations in which the use of antiviral chemotherapy is under consideration, since many chemotherapeutic agents are sufficiently toxic to require a definite viral diagnosis prior to administration to humans (McIntosh et al., 1980; Yolken, 1980). In addition, the availability of a rapid method for measuring the level of viral infection in response to therapy would be extremely useful in the development of strategies for the optimal use of antiviral chemotherapy.

For these reasons, there has been a great deal of interest in the development of more rapid techniques for the identification of viral infections. There have been numerous approaches undertaken in an effort to establish efficient diagnostic systems for this purpose. Some of the systems are based on the visualization of viral inclusions by light microscopy or, more importantly, by the direct identification of viral particles by means of electron microscopy (Flewett et al., 1973; Kapikian et al., 1980; Almeida, 1981). Some of these techniques are described in detail in Chapter

6. Electron microscopic techniques have been very effective in detecting new agents of gastrointestinal disease, hepatitis and hemorrhagic fever (McCormick et al., 1982) and have been very useful in the diagnosis of infections caused by these agents. However, microscopic techniques require the availability of a trained observer capable of distinguishing viral particles from each other and from artifacts. For this reason, there has been a good deal of interest in developing assays which do not rely on viral morphology for diagnosis. The antigenic properties of viruses and sub-viral components allow for the widespread use of immunological techniques for the direct identification of viral antigens in cultures and in clinical specimens. Regardless of their specific format, all immunoassays share the property that they are based on the binding of specific antibody to an antigen which might be present in the specimen and the subsequent identification of the antigen-antibody complex by means of a defined detection system.

The method of identification of the immune reaction can be quite diverse depending upon the nature of the immunoassay system. In all cases, the performance characteristics of immunoassays depend to a major extent on the nature of the antigen-antibody interactions (Pesce et al., 1978; Ekins, 1980; Yolken, 1982; Lechtonen and Earola, 1982). The kinetics of antigen-antibody reaction follow the law of mass action:

$$K_a = \frac{[AgAb]}{[Ag]\,[Ab]}$$

where [AgAb], [Ag] and [Ab] are the molar concentrations of the antigen-antibody complex, unbound antigen and unbound antibody respectively, and K_a is the affinity constant of the antigen-antibody reaction.

It can be seen that, at equilibrium, the amount of immune complex formed is proportional to the concentration of antibody, the concentration of antigen and the affinity constant of the antibody. In terms of concentration of antibody it should be noted that the important factor is the concentration of antibody which is actually capable of reacting with the antigen. This concentration will depend upon the concentration of total immunoglobulin in the system as well as the fraction of the total immunoglobulin which is actually directed at the antigen. With antibodies of equal affinity, an antibody preparation which contains a higher proportion of antibody directed at a particular antigen will bind to more antigen and result in a more sensitive immunoassay system than a preparation with a lower proportion of specific antibody. In terms of antigen concentration, it should be noted that this factor refers to the amount of free antigen which is capable of reacting with antibody. Thus, antigen which is in the form of endogenous immune complexes or otherwise inaccessible to antibody will not be capable of forming antigen-antibody complexes and thus will not be measured in the subsequent immunoassay procedure (Watanabe et al., 1977). In terms of the affinity constant, for polyclonal antibodies this factor represents an average of the affinities of different clones of immunoglobulins. Since most antigens derived from viruses are macromolecular and are capable of reacting with a considerable number of antibody molecules, it is also possible that cooperative binding can occur. The mathematical relationships defining the interaction of individual immunoglobulin molecules of different affinity to macromolecular antigens are complex and not totally elucidated. However, it is possible that the binding of the antigen molecule to one molecule of immunoglobulin can increase the affinity of unbound sites on the antigen for additional molecules of antibody. The occurrence of variable amounts of cooperative binding might explain the differences in assay sensitivity noted in systems using different sources of antiviral antibodies and differences noted in the reactivities of monoclonal and polyclonal antibodies (see below) (Frankel et al., 1979). In addition, it should be noted that the law of mass action describes the state of antigen-antibody interaction at equilibrium, i.e. when the rate of antigen-antibody complex formation equals the rate of antigen-antibody complex dissociation. The law of mass action does not describe the kinetics of reaching the equilibrium state. In the case of antigen-antibody reactions occurring in solutions, the rate of reaching equilibrium is quite rapid, being limited mainly by the diffusion constants of the molecules in the solution. However, the immobilization of one of the components of the immunoreactants onto a solid phase surface can markedly alter the interaction of immunoreactants and result in a marked retardation of attaining equilibrium. Similarly, the linkage of the components to marker molecules can also alter the antigen-antibody kinetics. This can occur by means of an increase in the molecular weight of the antibody-marker complex, or by the alteration of the charge or degree of hydration of the labelled antibody.

Also, while the law of mass action accurately describes the formation of immune complexes in pure solutions, the measurement of viral antigens in body fluids involves the measurement of antigen-antibody reactions in reaction environments in which antigen-antibody reaction might be suboptimal. Thus, materials in clinical specimens which inhibit antigen-antibody reactions can be expected to decrease the formation of the antigen-antibody complexes with a resulting decrease in the sensitivity of the corresponding immunoassay system. Similarly, the added immunoglobulin can react nonspecifically with the materials in the clinical specimen (Yolken et al., 1979; Kato et al., 1979). This will lead to a nonspecific signal which can obscure the signal generated by specific antigen-antibody interaction.

The sensitivity of an immunoassay system for the measurement of an antigen in body fluid will be dependent upon a number of factors, as discussed above. The average affinity and concentration of specific antibody as well as the magnification inherent in the immunoglobulin label will form one limit of immunosensitivity. Furthermore, the affinity of the antibody for nonspecific reactants and their concentration will form an additional limit. Clearly, the coefficient of variation of the reaction will also serve as a limiting factor since a signal cannot be considered to represent the presence of an antigen unless it is measurably greater than the signal generated by nonspecific reactions. Efforts to increase the sensitivity and specificity of immunoassay systems should thus be directed at the optimization of all of these factors. It can thus be predicted that an increase in antibody concentration or antibody affinity or a decrease in nonspecific reactions and the coefficient of variation will result in immunoassay systems capable of detecting smaller amounts of antigens in clinical specimens.

The concentration of antibody in the immunoassay system is also limited by the properties of solid phase. Most widely used solid phases consist of plastic surfaces in the form of tubes, microtiter plates or beads (Catt et al., 1967; Scharpe et al., 1976; Purcell et al., 1976). Immunoglobulin molecules will generally bind to plastic surfaces by means of nonspecific hydrophobic interactions in such a way that a significant portion of the immunologic activity of the bound immunoglobulin is retained following solid phase attachment (Cantarero et al., 1979; Lehtonen et al., 1980). The principal advantage of the use of hydrophobic binding to solid phase supports is convenience, since all that is required for binding is reaction of a dilute solution of immunoglobulin to the solid phase surface in a neutral or alkaline buffer. In addition, such supports are relatively inexpensive and widely available. However, the use of nonspecific binding to plastic has a number of significant disadvantages. The most important is that the capacity of plastic supports for immunoglobulin molecules is quite limited, ranging from 0.5 to 1 μg immunoglobulin/cm^2 of surface area (Pesce et al., 1977). For standard 96-well microtiter plates, this degree of binding to microtiter plates corresponds to approximately 0.5 μg immunoglobulin per microtiter well. This represents approximately 0.05% of the amount of IgG present in 0.1 ml of a typical antiserum. This fact explains why increasing the concentration of antibody used to coat the solid phase beyond this amount often does not result in an increase in assay sensitivity compared to the use of more dilute solutions. A second serious limitation of nonspecific solid phase binding is that the binding can be variable and elution of immunoreactants from the solid phase can occur during the course of the immunoassay procedure (Kricka et al., 1980). In addition to further decreasing the antibody concentration available to bind antigen, the eluted antibody can decrease the sensitivity of the immunoassay system still further by competing for free antigen with antibody which is still bound to the solid phase (Yolken, 1982). This phenomenon is probably responsible for the prozone or ‘hook’ effect noted when high concentrations of solid phase immunoreactants are utilized. Due to these limitations, there has been a great deal of interest in the development of more efficient solid phase systems for the binding of immunoreactants. Approaches to improvements in solid phase systems have generally taken two forms. One approach is to utilize the hydrophobic binding of immunoglobulins to surfaces with greater binding capacity for immunoglobulins than that which is inherent in unmodified plastics (Park, 1978). Some of the most convenient surfaces available are derived from paper materials (Towbin and Gordon, 1984), and in particular nitrocellulose-derived surfaces, which bind up to 100 μg/cm^2. The use of papers or other high capacity supports has the advantage of binding large amounts of protein material in short periods of time. The binding capacity is sufficiently high so that coating antibody is not required but rather antigens from clinical specimens can be applied directly to the solid phase surface. This method has the advantage that the binding can be performed under conditions which will result in dissociation of endogenous immune complexes. This method can thus potentially allow for the detection of antigens which would be inaccessible to antibodies in standard immunoassay systems due to the inability of labeled antibodies to penetrate the immune complexes. However, the use of such high capacity solid phase surfaces has a number of disadvantages. One is that the high binding capacity, in addition to providing for increased absorption of specific immunoreactants, also leads to an increase in nonspecific binding. This nonspecific binding can result in a marked decrease in the effective sensitivity of the immunoassay systems. It is thus important that efforts should be made to minimize the amount of nonspecific binding to the solid phase surface. These efforts include the blocking of binding sites by means of neutral proteins and extensive washing to remove nonspecific material. An alternative approach to the binding of material to the solid phase involves the covalent attachment of immunoreactants to a solid phase surface (Hendry and Herrmann, 1979). The use of covalent attachments has a number of advantages over the use of nonspecific absorption. These include a high binding capacity and a negligible rate of elution from the solid surface during the immunoassay procedure. However, unlike paper and other supports which rely on hydrophobic interactions, solid phase supports utilizing covalent attachment can be washed under dissociating conditions which minimize nonspecific hydrophobic binding. This can thus result in an assay system with minimal nonspecific activity. For example, the addition of 0.1% sodium dodecylsulphate into wash solutions virtually eliminates nonspecific hydrophobic binding to the solid phase and results in the virtual absence of detectable signal in the absence of antigen-antibody interaction. Originally, solid phase supports utilized cyanogen bromide to achieve a covalent attachment of the immunoreactants to the solid phase surfaces. However, the use of such attachment procedures is limited by the difficulty in controlling the cyanogen bromide activation reaction as well as the potential environmental hazard involved in the use of cyanide derivatives. More recently alternate means of linking proteins to the solid phase surfaces have been devised. One of the most promising involves the use of polymers substituted with amino or carboxy moieties or with agarose polymer hydrolyzed with mild acid to expose glycoside bonds (Stults et al., 1983). Protein antigens or antibodies can be covalently linked to such surfaces by the use of cross-linking agents such as gluteraldehyde or carbodiimide derivatives. Such polymers can be made in virtually any form including microtiter plates, gels or sticks. The widespread availability of practical solid phase systems which provide for the covalent attachment to immunoreactants should improve the reproducibility and performance characteristics of immunoassay systems.

1. Preparation of immunoreactants

As discussed above, the reaction parameters of solid phase immunoassays depend to a great extent on the binding characteristics of the immunoreactants which are utilized in the assay system. For this reason, it is important that the antibodies used in the assay system have favorable affinity characteristics and are highly specific. In the case of antisera prepared by animal hyperimmunization with purified viral components, it is important to use reagents without detectable antibody directed at heterologous viruses or host components contaminating the immunizing material. The performance characteristics of the antisera can be enhanced by purification by means of affinity chromatography with purified viral components. This type of purification not only can decrease the rate of nonspecific binding of the sera but can also increase the proportion of immunoglobulin in the reagent which is directed at the target antigen, thus increasing the effective concentration of antibody in the reaction system. However, the utility of this approach is limited by difficulties in obtaining sufficient supplies of purified viral components for the performance of the binding and elution procedures.

Another method of improving the quality of immunoreactants is offered by the recently developed techniques related to the production of monoclonal antibodies (Kennett et al., 1981). These techniques allow for the production of unlimited supplies of reagents with defined reactivities. The ability to produce monoclonal antibodies offers a number of striking advantages for the production of reagents for immunoassay systems. Monoclonal antibody technology allows for the production of a large amount of antibody with defined antigen binding characteristics. Since the antibody-producing cells can be grown in large quantities and can be distributed to a large number of laboratories, monoclonal antibody technology offers a possibility of producing highly standardized and reproducible reagents. In addition, since each clone produces antibody with a single range of immunological reactivity, monoclonal antibodies can be generated which will react with viral components but not with host antigen which might be contaminating the immunogen preparation. This is particularly advantageous in the case of viral proteins which cannot be completely purified from host cell components by standard physicochemical techniques. Finally, the proportion of immunoglobulin produced by a clone which is directed at the desired antigen is extremely high. In the case of antibody produced in vitro the proportion can be close to 100%, although in practice it is often somewhat less due to the production of some nonfunctional antibody. In the case of monoclonal antibodies amplified in ascites fluid, proportions of antibody directed at the desired antigen of 20–50% can be obtained. From the discussion of the law of mass action above, it can be seen that this increased proportion of antibody directed at the antigen can result in a marked increase in the theoretical sensitivity of solid phase immunoassay systems, especially when the antibody is utilized as the solid phase immunoreactant.

In light of these advantages of monoclonal antibodies, the question might be raised as to whether monoclonal antibodies will totally replace the use of polyclonal animal-derived antibodies in immunoassay systems. Unfortunately, however, monoclonal antibodies have a number of limitations which have hindered their ascension to predominance in the immunoassay field. One of the most important has been the difficulty in producing monoclonal antibodies of sufficient affinity to be useful in the solid phase systems. As can be seen from Table 9.1, a decrease in antibody affinity can counterbalance advantages expected from the increase in antibody concentration. The reason for the difficulty in obtaining high affinity monoclonal antibodies for some antigens is not known but might have to do with the kinetics of the immunization process. In addition, since monoclonal antibodies are directed at single antigenic determinants, the reaction of single clones of antibody cannot involve the cooperative binding effects to macromolecular antigens (Friguet et al., 1983) and thus cannot make use of kinetic advantages which might be inherent in such cooperative binding. Another theoretical limitation of monoclonal antibodies is that they might be 'too specific' in that they would recognize antigenic determinants which would be present on the immunizing strains of viral antigen but not be found widely in viruses infecting humans. In such cases immunoassays utilizing the monoclonal antibodies could not be utilized for sensitive detection systems. While this has been a problem in some cases, several investigators have demonstrated that antibodies to

TABLE 9.1

Kinetics of antigen-antibody interactions

Antibody	K_a	Antibody concn. (M)	% Antibody bound
High affinity[a]	10^{-8}	10^{-7}	90
		10^{-8}	50
		10^{-9}	9
		10^{-10}	1
Low affinity[a]	10^{-6}	10^{-7}	9
		10^{-8}	1
		10^{-9}	0.1
		10^{-10}	0.01
Monoclonal		10^{-7}	100
Affinity purified		3×10^{-8}	30
Hyperimmune		10^{-8}	10
Post-infection		10^{-9}	1

[a]Theoretical values.
[b]Assuming a maximum binding capacity of 10 μg/ml and a molecular weight of 100 000.

common antigens of rotaviruses, adenoviruses, herpes simplex virus, cytomegaloviruses and respiratory syncytial virus can be generated which will react with a wide range of antigens expressed by the viruses in tissue culture fluids and in clinical specimens (Yolken, 1983). These studies indicate that it is possible to select monoclonal antibodies which react with common antigens which are expressed during infection. A third practical problem with monoclonal antibodies is that many of the monoclonal antibodies have physical properties which make them difficult to utilize in solid phase immunoassay systems (Kammer, 1983). For example, many monoclonal antibodies which we have encountered are difficult to label with radioactive, fluorescent or enzymatic markers utilizing standard labeling techniques. In addition, monoclonal antibodies can be difficult to attach to a solid phase surface either by nonspecific absorption or covalent attachment techniques. While the exact nature of these problems is not clear, it is possible that some of the binding and labeling procedures interfere with the antigen-combining sites of the monoclonal antibody by means of steric interactions. Additional efforts should be directed at establishing means of producing monoclonal antibodies of high affinity on a consistent basis and of devising means of modifying the monoclonal antibodies so that they can be efficiently used in solid phase immunoassay systems.

2. Immunoassay formats

Solid phase immunoassays can be performed in a number of different formats. Many of the currently available systems utilize a noncompetitive or sequential binding format. In such assays, antibody is bound to a solid phase surface either by nonspecific absorption or covalent attachments as discussed above. The solid phase can be utilized immediately or stored under appropriate conditions until use. The clinical specimen or tissue culture fluid is added to the solid phase and, following sufficient periods of time for reaction, is removed by means of a washing procedure. A second antibody is then added which will react with antigenic sites which are present on the added antigen but which have not been occupied by binding to the solid phase antibody. In some formats, this antibody is directly linked with the label. In such cases, the reaction is completed by washing unbound antibody and measuring the amount of label bound to the solid phase surface (Fig. 9.1). Alternatively, the second antibody can be added in an unlabeled form and its presence can be quantitated by the addition of another immunoreactant which will bind specifically with the second antibody (indirect format) (Fig. 9.2). The second antibody can also be labeled with a nonenzymatic moiety and the reaction can be quantitated by the addition of reactant which will recognize that moiety specifically (e.g. avidin-biotin immunoassay). In all cases unreacted labeled antibody is removed by washing and the amount of label bound to the solid phase is determined by the appropriate instrumentation.

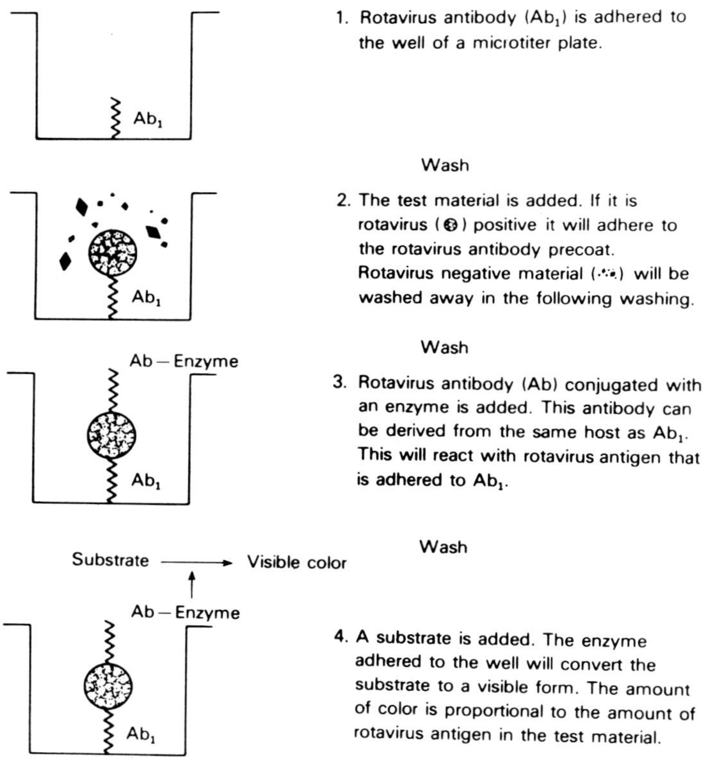

Fig. 9.1. Direct ELISA for antigen measurement.

One practical modification of the above sandwich formats involves the simultaneous addition of antigen and labeled antibody to the solid phase (Yolken and Leister, 1982). If the antigenic determinants recognized by the solid phase antibody and the labeled antibody are sufficiently distinct, there will be little competitive interaction and this binding of antigen to the solid phase antibody and labeled antibody can thus occur simultaneously. This format is thus more rapid and simpler to perform than the sequential formats described above. However, this format will be somewhat less sensitive if there is any competition between the solid phase and labeled immunoreactants or if there is material in the clinical specimen which interferes with the labeled immunoglobulin.

The amount of signal generated by specimens is compared to appropriate negative controls and its positivity or negativity is determined. While there is no standard method for the determination of positivity, the quantitative nature of the solid phase assay allows the determination of a normal distribution of signals generated by known negative specimens. Thus positivity can usually be defined in terms of whether the signal of an unknown specimen falls within a fixed number of standard

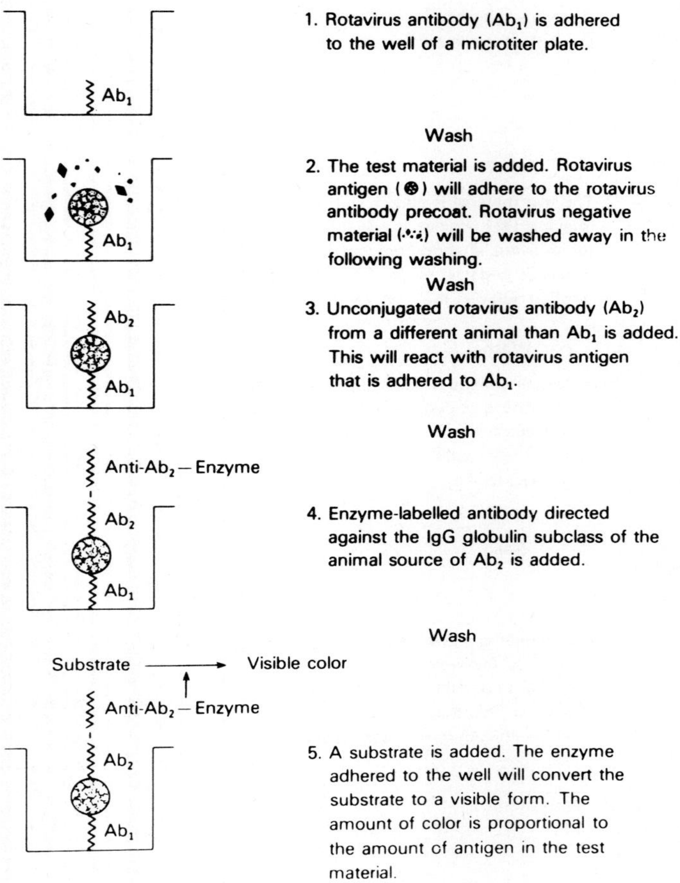

Fig. 9.2. Indirect ELISA for rotavirus antigen measurement.

deviations from the mean of the negative specimens. In terms of nonspecific reactivity, one of the significant advantages of solid phase immunoassays is that they allow for the performance of quantitative control reactions. These are generally performed by the substitution of nonimmune reagents for immune reagents in the immunoassay procedure (Yolken and Stopa, 1979) (Fig. 9.3). In such cases a specific activity can be defined as a difference between the reactivity of specimens with immunoreagents directed at a specific antigen as compared to the reactivity generated when nonimmune immunoreactants are utilized. When the nonimmune immunoreactants are closely matched with the immune ones in terms of animal species and of immunoglobulin class, these specific activity measurements can accurately distinguish specific antigen-antibody reactions from nonspecific interactions of immunoglobulin molecules with nonantigenic materials in the specimens. This

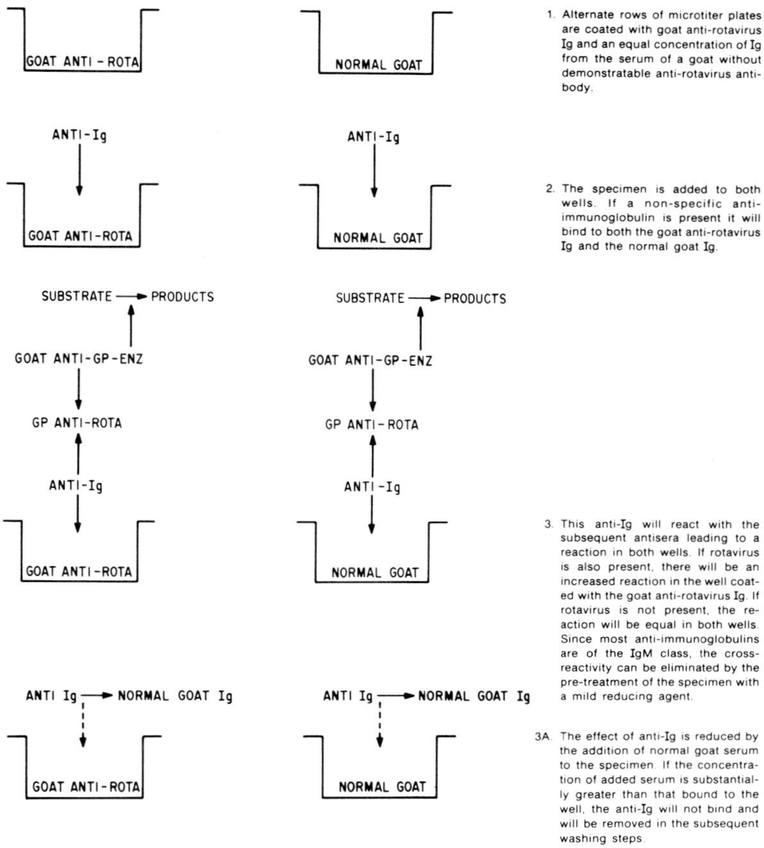

Fig. 9.3. Cross-reactivity due to anti-Ig.

type of reaction can avoid false positive reactions due to immunoglobulin binding materials which might be present in clinical specimens such as rheumatoid factors and immunoglobulin-binding bacteria.

While immunoassays for viral antigens are often performed in the noncompetitive formats described above, accurate assays can also utilize a competitive format, as depicted in Fig. 9.4. In one type of competitive assay, a specimen is allowed to react with labeled antibody in liquid phase. The antigen-antibody mixture is then transferred to a solid phase coated with specific antigen. While the antigen can be bound nonspecifically, it is often advantageous to utilize one of the covalent binding systems described above. Antigen in the initial specimen will then be manifested by a measurable decrease in the binding of labeled antibody to the solid phase antigen as compared to negative control specimens. Competitive assay systems offer a number of potential advantages. One is that they involve fewer reaction steps than the

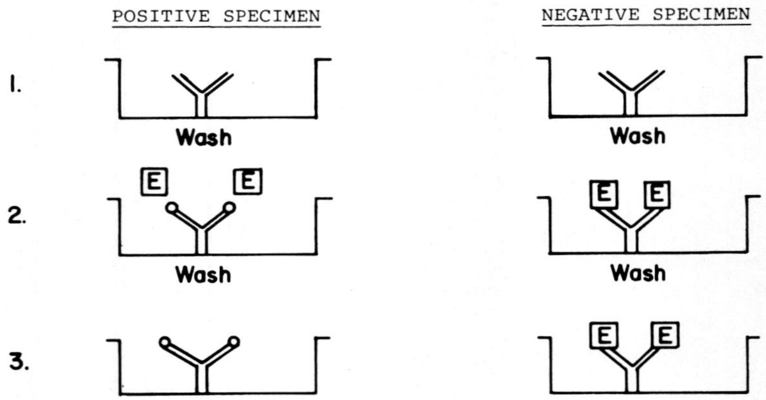

1. Antibody (Y) is bound to the solid phase. Unbound antibody is removed in the washing step.
2. The specimen is added. If it contains antigen (O) it will bind to the solid phase antibody. Enzyme-labelled antigen (E) is then added. This will react with antibody sites not occupied by antigen from the specimen.
3. Unbound enzyme-labelled antigen is removed by washing and the amount of bound enzyme-labelled antigen is quantitated by the addition of the appropriate substrate. The amount of substrate product is inversely proportional to the amount of antigen in the test specimen.

Fig. 9.4. Competitive immunoassay format.

noncompetitive assays described above. In addition, they require the use of only a single labeled antibody. This is advantageous in situations in which it is difficult to produce antibodies in different animal species to a single viral antigen. In addition, the fact that the reaction of labeled antibody and the unknown antigen takes place in the liquid phase allows for the more rapid reaching of the equilibrium conditions and a potentially faster rate of reaction than the solid phase sandwich assays described above. In addition, competitive assays can offer a good deal of specificity since they are based on the ability of an unknown specimen to interfere with the binding of a specific antigen to a purified antibody. However, competitive assays also suffer from a number of disadvantages. The most important is that they require a consistent supply of viral antigens which can be bound reproducibly to a solid phase surface. This can be difficult to obtain for some viral proteins. However, the ability to produce viral proteins by recombinant DNA techniques and the availability of covalent solid phase binding systems, should make this a more practical approach in the future. Another disadvantage of competitive assays is that there might be materials in clinical specimens which will degrade antibody in the liquid phase and will thus result in a decrease in signal (Viscidi et al., 1983). This decrease could then be construed as a false positive reaction. However, this problem can be overcome by the use of appropriate control reactions and the use of reaction media which inhibit the breakdown of antibody by endogenous enzymes and other interfering substances.

3. Immunoglobulin labels

Immunoassays are often distinguished on the basis of the labels which are utilized for the measurement of the antigen-antibody reactions. The labels which originally achieved widespread usage in quantitative solid phase immunoassays were radioisotopes, the most widely used of which is ^{125}I. The use of radioactivity as the immunoreactant label has a number of important advantages. Radioactivity can be quantitated in an accurate manner by means of widely available gamma-counting instrumentation. In addition, since small amounts of radioactivity can be measured by such instrumentation, radioimmunoassays offer the potential for a good deal of sensitivity, often in the range of 1–10 ng of labeled antibody. However, radioactivity offers some disadvantages for widespread use in immunoassay systems. The most important limitation is that immunoglobulins labeled with radioactivity are often unstable both in terms of decay of the isotope and in terms of alterations in immunoglobulin structure and function due to the release of gamma emitter. This fact necessitates the constant relabeling and restandardization of the immunoassay reagents. Although this is not a problem in situations in which a large number of specimens are tested on a regular basis, as in the case of the testing of blood specimens for hepatitis B surface antigen, it is somewhat problematic in situations in which the need for immunoassays cannot be anticipated on a regular basis. In addition, standard oxidative methods for radiolabeled immunoglobulins, while they are fairly easy to perform, are somewhat difficult to standardize and the procedure of linking immunoglobulin to radioactive label involves the generation of free reaction iodine, which is a potential biohazard.

For these reasons there has been a large amount of interest in the development of nonisotopic labels for immunoassay systems. Enzymes are nonisotopic markers which offer a number of advantages for use in immunoassay systems. Since enzyme-immunoglobulin conjugates can be formulated which are relatively stable and not hazardous, the potential problems of safety and instability inherent in the use of radioactive labels are avoided. In addition, since a single molecule of enzyme can react with a large number of substrate molecules, enzymes offer the potential for a large degree of magnification of the antigen-antibody reaction. Furthermore, since enzyme substrates can be utilized which generate visibly colored products, the results of the reactions can be followed with the naked eye or measured quantitatively by the use of color-measuring instrumentation.

Enzyme immunoassays have been devised for the detection of a large number of viral antigens in clinical specimens. Such antigens include hepatitis A, hepatitis B, adenovirus, rotavirus, respiratory syncytial virus, influenza virus, parainfluenza virus, and herpes simplex virus (Yolken, 1982). In all cases the sensitivity of the assay systems approximated that of other rapid detection systems such as radioimmunoassay or immunofluorescence. In cases in which the virus can be cultivated in standard tissue culture systems, the sensitivity of the solid phase enzyme immunoassay is often less than that of cultivation. However, this decreased sensitivity is often counter-balanced by the decreased cost and time required to attain an accurate diagnosis. In addition, the solid phase enzyme immunoassay has been very useful for the detection of viruses, such as rotavirus and enteric adenoviruses, which are difficult to cultivate in commonly available tissue culture systems.

The sensitivity of enzyme immunoassays is determined both by the kinetics of antigen-antibody interactions, as discussed above, and by the kinetics of the enzyme-substrate interaction. Thus, the performance characteristics of enzyme immunoassays are dependent to a great extent on the enzymes which are utilized as the antibody marker. For this reason, it is important that the enzymes utilized have high turnover rates and that they be capable of being efficiently conjugated with immunoglobulin molecules by reproducible methods. Although a large number of enzymes could theoretically be utilized for this purpose, most available immunoassays utilize alkaline phosphatase, horseradish peroxidase, or β-galactosidase as the marker enzyme. Such enzymes have turnover rates which allow for the detection of nanogram quantities of labeled immunoreagents in practical periods of time. However, it should be noted that there are additional enzymes which might display even more favorable kinetics. One such enzyme is β-lactamase (Yolken and Wee, 1984). This enzyme, which is produced in high quantities by many strains of bacteria, can convert penicillin compounds to the corresponding penicillinoic acids with a high degree of efficiency. The penicillinoic acids can be measured by a number of colorimetric, fluorescent or radioactive systems. One particularly advantageous substrate system involves the use of starch-iodine solutions to measure the generation of penicillinoic acids. In this system the presence of β-lactamase is manifested by the conversion of a deeply blue starch iodine to a colorless solution of iodide salt. The decoloration can be easily visualized by the naked eye or quantitated in a microplate colorimeter. In addition, the results can be recorded by a standard office photocopying machine, thus allowing for the recording of the immunoassay results in situations in which expensive instrumentation is not available. β-Lactamase offers the advantage of high turnover rates as well as being an inexpensive, safe and widely obtainable substrate. The successful adaptation of this and other high turnover enzymes could markedly improve the utility of enzyme immunoassay systems. In addition, it is possible that enzymes can be reacted in a cascade format similar to that which naturally occurs in complement and blood coagulation systems (Chuck and Stadtman, 1983). In cascade formats, the reaction of one enzyme catalyzes the activation of a second enzyme or group of enzymes. In such formats the activation of each molecule of enzyme can result in the generation of very large amounts of substrate with a corresponding increase in the potential sensitivity of the assay systems. For example, cascade formats can be devised which make use of proteolytic proenzymes. In such systems one enzyme can convert an inactivated proteolytic enzyme to an active one which in turn cleaves a large number of amino acids from a macromolecular substrate. The cascade systems can be further expanded by the incorporation of additional enzyme substrate systems. For example, a three enzyme cascade can be formulated utilizing urokinase and streptokinase. The successful adaptation of cascade systems could markedly increase the sensitivity of enzyme immunoassay systems. Another approach to increasing the sensitivity of enzyme substrate systems involves the use of substrate systems which generate high energy products which can be detected in lower concentrations than substrates which are measured by visible light. Thus, enzyme substrate systems have been devised which generate fluorescent or chemiluminescent products on enzyme reaction. The use of such substrates can result in the detection of attomole (10^{-18} M) quantities of enzyme and can thus be utilized to detect small amounts of enzyme-linked immunoreactant (Shalev, 1980). However, it should be noted that both with the use of high energy substrates and the cascade systems described above the sensitivity of the enzyme immunoassay system can be no greater than the limits of the antigen-combining capacity as determined by the kinetics of antigen-antibody interaction. The use of higher energy systems thus will not necessarily lead to assay systems which are more sensitive than ones utilizing standard enzyme substrate reactions in terms of concentration of antigen which can be detected (Yolken and Leister, 1981). However, the assay system utilizing sensitive enzyme substrate systems will reach the limits of detectability more rapidly than that utilizing standard systems, thus resulting in a more rapid assay system. In addition, the higher energy substrates will allow for the use of much smaller quantities of immunoreactants than those utilizing standard systems. This can be useful in situations in which small volumes of samples are available and in situations in which it is desirable to conserve limited supplies of reagents.

However, it should be noted that the use of enzymes as immunoglobulin labels is associated with a number of inherent disadvantages. The very nature of the enzyme-substrate reaction introduces a number of variables into the quantitation of the signal generated in the assay system. These variables include reaction time, temperature and ionic strength as well as the presence of enzyme inhibitors which might be contaminating wash solutions or immunoreactants. For this reason, enzyme immunoassays can be more difficult to quantitate than assay systems which do not require a secondary enzyme-substrate reaction. Another difficulty involved in the use of enzymes as markers is that enzymes are relatively large molecules with respect to other immunoreactants. Thus, the direct labeling of enzymes to immunoglobulins can be difficult to control and perform in a reproducible fashion (Avrameas et al., 1978). The principal limitation of such conjugations is related to the fact that macromolecular complexes of undetermined size are formed during conjugation and that many of the antigen binding and substrate binding sites of the complexes will be inactivated during the conjugation procedure. The inactivation of antigen binding sites is particularly deleterious since this results in a decrease in the concentration of effective antibody in the reaction system and a subsequent decrease in the formation of antigen-antibody complexes. The conjugation of antibodies to enzymes is generally performed by the linkage of amino groups in the immunoglobulin to either amino groups or carbohydrate residues in the enzyme molecule. However, the difficulty in performing these reactions in a reproducible fashion has led to some interest in devising alternative methods of conjugation. One of the most successful involves the linkage of enzymes to immunoglobulin molecules by means of the sulfhydryl group in the hinge region of the immunoglobulin molecule by the use of maleimide derivatives (Yoshitake et al., 1982). This is generally performed by first generating the F(ab) fragment of the immunoglobulin by means of papain digestion of intact immunoglobulin. Since each F(ab) fragment has a hinge region in a predictable location removed from the antigen binding site of the fragment, such labeling procedures result in the formation of a homogenous conjugate with virtually complete retention of antigen binding capacity. An alternative approach involves the labeling of immunoglobulin molecules with small molecules which would not be expected to interfere with the antigen binding capacity or diffusion characteristics of the immunoglobulin. In such cases, the reaction is quantitated by the subsequent addition of a second labeled immunoreactant, which is capable of binding to the small molecule. While a number of such low molecular weight markers can be utilized, one which is particularly suited for this purpose is biotin, a small (molecular weight 254) vitamin which can be covalently linked to immunoglobulin by a number of methods. Suitable linkage methods involve the covalent linkage of biotin to the amino, carboxyl or sulfhydryl portions of the immunoglobulin molecule. In this way biotinillated antibodies can be formulated which retain virtually all of their antigen binding capacity (Yolken et al., 1983). Reactions with biotinillated immunoreactants are quantitated by reaction with avidin, a protein which has an extremely high affinity for biotin and biotinillated molecules. The avidin can be labeled directly with enzyme by means of covalent linkage techniques. However, since such techniques can result in a decrease in the functional activity of the avidin, it is often more efficient to form soluble complexes of avidin and biotinillated enzyme. The use of such systems has the advantage of allowing the reproducible generation of labeled antiviral reagents for use in enzyme immunoassay systems. However, it does not alleviate the problems inherent in the use of enzyme substrate reactions for quantitative systems.

Due to the limitations of enzymatic methods there has been renewed interest in the labeling of immunoglobulins with nonisotopic markers which can be directly measured by practical instrumentation. Molecules which generate fluorescence have been long used in immunoassay procedures. However, while standard fluorescent markers such as fluorescein and rhodamine are extremely useful in microscopic immunofluorescent reactions, they generally do not emit sufficient energy to be useful in quantitative systems which are sufficiently sensitive to allow for the detection of viral antigens in clinical specimens. However, a number of higher energy fluorescence-emitting markers have recently been devised. Some of the most useful are rare earth metals which emit fluorescent energy over a relatively long period of time. This property allows such markers to be utilized in the technique of time-resolved fluorescence (Soini and Hemmila, 1979). In this system, the label is stimulated with a short pulse of light and the resulting fluorescent energy is determined at a fixed period following the pulse emission. This method has the advantage of being highly sensitive and being relatively free of interfering reactions due to fluorescing materials in clinical specimens. Time-resolved fluorescent reactions have thus been devised for the detection of a number of viral antigens in clinical specimens. The principal limitation of the use of such labels is that they are absolutely dependent on a dedicated instrumentation and that the optimal conditions for the generation of the fluorescent conjugates have not been conclusively established. The optimalization and general availability of these and other systems which rely on the direct measurement of nonisotopic labeled immunoglobulins would markedly widen the scope of solid phase immunoassay technology for the detection of viral antigens.

4. Conclusion

Solid phase immunoassays have been devised for the detection of large numbers of viral antigens in clinical specimens. The widespread usage of some of these assay systems both in clinical and research laboratories attests to their utility under a wide range of experimental and clinical conditions. While solid phase immunoassays have reached an advanced stage of development, it can be anticipated that recent advances in the ability to generate viral antigens and antibodies will lead to striking improvements in immunoassay technology. Similarly, identification and optimalization of efficient methods for labeling immunoreactants with molecules that generate large amounts of signal will similarly lead to an increase in immunoassay sensitivity. These advances, in conjunction with the developing field of nucleic acid hybridization, should allow for a striking expansion of the range of immunoassays which are available for the detection of viral antigens and will lead to the application of quantitative solid phase detection techniques for the diagnosis of a wide range of viral diseases.

5. Acknowledgement

This work was supported by the National Institute of Allergy and Infections Diseases contract NO1-AI-22680 and the Roth Memorial Fund.

6. References

Almeida, J.D. (1981) *Am. Soc. Microbiol.* 231.

Avrameas, S., Ternynck, T., and Guesdon, J.L. (1978) *Scand. J. Immunol.* 8, 7.

Cantarero, L.A., Butler, J.E. and Osborne, J.W. (1979) *J. Immunol, Methods* 30, 209.

Catt, K. and Tregear, G.W. (1967) *Science* 158, 1570.

Chuck, P.B. and Stadtman, E.R. (1983) *Methods Enzymol.* 64, 297.

Ekins, R. (1980) *Nature* 284, 14.

Feinsteon, S.M., Kapikian, A.Z. and Purcell, R.H. (1973) *Science*, 182, 1026.

Flewett, T.H., Bryden, A.S. and Davies, H. (1973) *Lancet* 2, 1497.

Frankel, M.E. and Gerhard, W. (1979) *Mol. Immunol.* 16, 101.

Friguet, B., Djavadi-Ohaniance, L., Pages, J., Bussard, A. and Goldberg, M. (1983) *J. Immunol. Methods* 60, 351.

Hendry, R.M. and Herrmann, J.E. (1979) *J. Immunol. Methods*, 30, 209.

Kammer, K. (1983) *Immunology* 48, 7.

Kapikian, A.Z., Diestag, J.L. and Purcell, R.H. (1980) In *Manual of Clinical Immunology*, (N. Rose and H. Friedman, Eds.) 2nd Edn. ASM, Washington DC. p. 70.

Kato, K., Umeda, U., Suzuki, F., Hayashi, D. and Kosask, A. (1979) *FEBS Lett* 99, 172.

Kricka, L.J., Carter, T.J.N., Burt, S.M., Kennedy, J.H., Holder, R.L., Halliday, M.I., Telford, M.E. and Wisdom, G.B. (1980) *Clin. Chem.* 26, 741.

Lechtonen, G.P. and Eerola, E. (1982) *J. Immunol. Methods* 54, 233.

Lehtonen, O.P. and Viljanen, M.K. (1980) *J. Immunol. Methods* 34, 61.

McIntosh, K., Wilfert, C.W., Chernesky, M., Plotkin, S. and Mattheis, J.M. (1980) *J. Infect. Dis.* 142, 793.

Park, H. (1978) *J. Immunol. Methods* 20, 349.

Pesce, A.J., Ford, D.J. and Gaizutis, A. (1978) *Scand. J. Immunol.* 8, 1.

Pesce, A.J., Ford, D.J., Gaizutis, M. and Pollak, V.E. (1977) *Biochim. Biophys. Acta* 492, 399.

Purcell, R.H., Wong, D.C., Mortisugu, Y., Dienstag, J.L. Routenberg, J.A. and Boggs, J.D. (1976) *J. Immunol.* 116, 349.

Scharpe, S.L., Cooreman, W.M., Bloome, W.J. and Laekeman, G.M. (1976) *Clin. Chem.* 22, 733.

Shalev, A., Greenberg, A.H. and McAlpine, P.J. (1980) *J. Immunol. Methods* 38, 125.

Soini, E. and Hemmila, I. (1979) *Clin. Chem.* 25, 353.

Stults, N.L., Lin, P., Hardy, M., Lee, Y.C., Uchida, Y., Tsukada, Y. and Sugimori, T. (1983) *Anal. Biochem.* 135, 392.

Towbin, H. and Gordon, J. (1984) *J. Immunol. Meth.* 72, 313.

Viscidi, R., Laughon, B.E., Yolken, R.H., Bo-Linn, P., Moench, T., Ryder, R. and Bartlett, J.G. (1983) *J. Infect. Dis.* In Press.

Watanabe, H. and Holmes, I.H. (1977) *J. Clin. Microbiol.* 6, 319.

Yolken, R.H. (1980) *Yale J. Biol. Med.* 53, 85.

Yolken, R.H. (1981) *Clin. Immunol. News* 3, 25.

Yolken, R.H. (1982) *Rev. Infect. Dis.* 4, 35.

Yolken, R.H. (1983) In *Current Topics in Microbiology 104 and Immunology: New Developments in Diagnostic Virology* (P.A. Bachmann, Ed.) Springer-Verlag, New York.

Yolken, R.H. and Leister, F.J. (1981) *J. Clin. Microbiol.* 13, 738.

Yolken, R.H. and Leister, F.J. (1982) *J. Infect. Dis.* 146, 43.

Yolken, R.H., Leister, F.J., Santosham, M. and Whitcomb, L.S. (1983) *J. Immunol. Methods* 56, 319.

Yolken, R.H. and Stopa, P.J. (1979) *J. Clin. Microbiol.* 10, 703.

Yolken, R.H. and Wee, S.B. (1984) *J. Clin. Microbiol.* In Press.

Yoshitake, S., Imagawa, M., Ishikawa, E. et al., (1982) *J. Biochem.* 92, 1413.

Eds. M.H.V. van Regenmortel and A.R. Neurath
Immunochemistry of viruses. The basis for serodiagnosis and vaccines.

Synthetic peptide vaccines 10

R. ARNON

Department of Chemical Immunology, The Weizmann Institute of Science, Rehovot 76100, Israel

Ever since the times of Jenner and Pasteur, vaccination has been accepted as a part of our way of life. Existing vaccines, which mostly consist of killed or live attenuated microbial agents, purified viral proteins or their subunits, and bacterial toxoids, have diminished the incidence, morbidity and mortality of a large number of infectious diseases. However, there are still several major problems that await solution, for example the lack of sufficient material from natural sources, as in the case of hepatitis B virus, or safety considerations due to the potential risks of some vaccines. Another problem is the constant evolution of some viruses into new strains, with different serological specificities, thus requiring the continuous development of new vaccines.

It is of some interest that real advances in the development of antiviral vaccines have always been the result of a major breakthrough. The first breakthrough occurred at the end of the 17th century when Jenner introduced vaccination with cowpox virus to protect against smallpox. This was the first attempt at active immunization. Almost a century passed before the second breakthrough: Pasteur's use of attenuation of a virus by passage in a new host, which led to the rabies vaccine. Following this development, the major problem became the growing of viruses, and the next breakthrough was the use of the embryonated egg for this purpose, leading to the development of several other vaccines such as yellow fever and influenza. The cell culture technique was the next breakthrough, and resulted in the production of many vaccines, including polio and measles.

It is clear that thanks to existing vaccines various kinds of infectious or communicable diseases have been diminished in their incidence and importance. However, concomitantly new diseases, caused by other viruses or by different strains of the same viruses, are replacing them, and this is the reason why new vaccines are required continuously. Further breakthroughs in this field are now required, probably through the exploitation of molecular biology, or possibly a synthetic approach such as the one discussed in this chapter.

The basis for the utilization of synthetic antigens for vaccination has been laid during the years when they were used for studies of protein antigenicity (Sela, 1969). Once it became apparent that synthetic peptides were capable of eliciting antiprotein immune responses (Arnon et al., 1971), it was predicted that adequate molecular engineering should make it possible to design synthetic materials that would elicit antiviral immunity, leading eventually to multivalent synthetic vaccines (Arnon, 1972). The rapid development of DNA cloning and sequencing techniques which in turn facilitate protein sequencing has made the synthetic approach more feasible and practical than before. Primary sequences are available today for many proteins, enabling the synthesis of any selected region, using either chemical laboratory techniques, which are advisable in the case of short peptides, or genetic engineering procedures in the case of longer protein segments, subunits, or even whole proteins. Moreover, the synthetic approach should allow the carrier, as well as the adjuvant, to be chosen, the goal being the production of multivalent vaccines with built-in adjuvanticity.

1. Identification of antigenic determinants

Before attempting to synthesize a vaccine one should have some knowledge about the immunological properties of the antigens involved. This subject was discussed in detail in the review by Arnon and Geiger (1977). In the course of an immune response the antigens exhibit two distinct reactivities, which may not necessarily coincide. The first is immunogenicity, i.e. the capacity of the antigenic material to elicit an immune response manifested either by antibody production or by cell-mediated immunity. The second characteristic, antigenic specificity, refers to the antigen's capacity to react in a specific manner with antibodies or with immune lymphocytes, regardless of how these have been raised. The regions in the antigen molecule that come into direct contact with the active site of the antibodies or with the cell receptors are defined as antigenic determinants.

Macromolecules may carry a large number and variety of possible antigenic determinants that dictate their antigenic specificity. However, only a limited number of the potential antigenic sites are immunodominant and important in terms of immunogenicity. In viral antigens, even a smaller number of determinants are involved in inducing neutralizing immunity. The parameters found to be most important for the immunological properties are the molecular size of the immunogen, the spatial conformation, and the accessibility of the antigenic determinants. The molecular size is important primarily for immunogenicity. Consequently, small molecules are often nonimmunogenic, but can become immunogenic after being coupled to carriers. This is therefore a common procedure employed in the preparation of synthetic vaccines. The accessibility of the determinant is a crucial factor since the interaction of any individual antigenic determinant with either the antibodies or specific receptors of the immunocompetent cells necessitates its exposure on the antigenic molecule. As a result, in many native proteins, antigenic determinants consist of segments that are at the 'corners' of the folded polypeptide chain and are fully or partly exposed on the surface of the protein molecule.

The spatial conformation of proteins plays a decisive role in determining the antigenic specificity. There is considerable evidence that a drastic change in antigenic properties occurs upon denaturation of native proteins or unfolding of their polypeptide chains, or even as a result of more subtle conformational alterations. This led to the conclusion that most of the antigenic determinants of proteins are conformational determinants, i.e. they result from the steric conformation and may include residues that are remote in the unfolded peptide chain but are brought together in the native molecule. Antibodies specific towards such determinants will not necessarily react with isolated peptides derived from the molecule, whereas antibodies directed against sequential determinants will usually recognize stretches of amino acid sequences in the protein and react with short peptide fragments of it.

Identification of antigenic determinants is usually accomplished by fragmentation of the native protein, either by chemical cleavage or by controlled proteolysis. The resultant fragments are then screened for immunologically active components that can bind to antibodies and interfere with their interaction with the intact antigen. Alternatively, monoclonal antibodies with neutralizing activity could be useful in identifying relevant antigenic sites. Crystallographic studies of the three dimensional structure of proteins serve as an additional useful tool for prediction of potential antigenic determinants. Cumulative data achieved by those methods have resulted in the identification of antigenic determinants in several protein antigens such as tobacco mosaic virus protein (Benjamini et al., 1964; Altschuh et al., 1983), sperm whale myoglobin (Atassi, 1975) or hen egg-white lysozyme (Shinka et al., 1962; Arnon and Sela, 1969). Moreover, such information might indicate which segments in the molecule possess the 'ingredients' that are necessary for an antigenic determinant to give rise to a neutralizing immune response.

2. Protein-specific synthetic antigens

Once antigenic determinants in a protein have been identified, it is possible to synthesize them chemically and investigate their immunological properties. This approach has been employed in the case of several protein antigens and has demonstrated clearly that synthetic antigenic fragments can induce antibodies that react with the intact protein, in its native form. The spatial conformation was found to play an important role in this activity, as can be illustrated by the following examples.

Collagen is the main protein constituent of connective tissue. The collagen of vertebrates is composed of three polypeptide chains, two identical α_1 chains and one α_2 chain, which are supercoiled in the form of a characteristic triple helix, with nonhelical sequences, known as telopeptides, at the N- and C-termini. Immunization with the native collagen leads to antibodies against the telopeptides, with almost no response to the helical region which is very highly conserved in evolution. A synthetic periodic polypeptide, (Pro-Gly-Pro)$_n$, which was shown to have a collagen-like triple helical structure (Traub and Yonath, 1966) was found to be immunogenic in guinea pigs (Borek et al., 1969). Immunization with this copolymer elicited antibodies that, in addition to their reactivity with the homologous polymer, also reacted with native collagens of several species (Fuchs et al., 1974) leading to an autoimmune phenomenon. This cross-reaction is by virtue of the triple helical conformation, which is common to the synthetic material and the various collagens.

In the case of *hen egg-white lysozyme*, two antigenic regions have been identified: one contains the C- and N-terminal segments linked by a disulfide bond, while the second is a larger region, comprising residues 57–104 and containing two disulfide bonds (Shinka et al., 1962). The latter was further split to yield a smaller immunologically active region, containing residues 60–83 with a single disulfide bond between cysteines 64 and 80, which was denoted 'loop' (Arnon and Sela, 1969). This region, which is exposed on the surface of the molecule, was demonstrated to be a conformation-dependent antigenic determinant (Maron et al., 1971, Pecht et al., 1971). This loop determinant was later chemically synthesized and attached to a synthetic carrier, and the resultant completely synthetic conjugate elicited antibodies that were reactive with native lysozyme and could still recognize the conformation-dependent determinant in the intact molecule (Arnon et al., 1971).

Another example is the *carcinoembryonic antigen* (CEA) of the colon, a glycoprotein of about 200 k Da. A peptide corresponding to the 11 amino acid residues of the NH_2-terminal portion of the molecule has been synthesized and chemically attached to a synthetic carrier and to bovine serum albumin. Both macromolecular conjugates provoked antipeptide antibodies that reacted not only with the homologous antigens but also with the intact CEA molecule (Arnon et al., 1976), and could serve in an immunodiagnostic assay of CEA in sera of cancer patients.

As seen from the above-mentioned examples, synthetic antigens that contain immunoreactive regions of protein molecules can give rise to immune responses towards the intact proteins. There is no reason that prevents this approach from being used for viral and bacterial components for the purpose of inducing antibodies reactive with the intact organisms and leading to their neutralization. A prerequisite for this approach is a detailed knowledge of the structure or at least the amino acid sequence of their relevant antigenic proteins. In recent years some attempts have been made in this direction, indicating that this goal is indeed within reach.

3. Synthetic vaccines for antibacterial and antitoxin immunity

Streptococcus pyogenes M protein is one example of a surface protein which may serve as the basis for a future synthetic vaccine. It has been recognized for a long time that the resistance of group A Streptococci to phagocytosis is largely attributable to the presence of a type-specific antigen, the M protein, on the surface of virulent organisms. In the immune host, specific antibodies against the M protein neutralize the antiphagocytic effect and allow the rapid elimination of any invading streptococci organisms having the same serotype as the M protein. Studies on the primary structure of the M protein of type 24 have indicated that the molecule is composed of repeating covalent structures, each of which contains antigenic determinants important for protection (Beachy et al., 1978). Consequently a 35 amino acid residue peptide corresponding to the sequence of a cyanogen bromide fragment (CB7) of the M protein has been synthesized and tested as an antigen. Rabbits immunized with this peptide showed both cellular and humoral immune responses against the intact M protein. Moreover, passive immunization of mice with sera of these rabbits provided protection against a challenge infection with type 24 streptococci (Beachy et al., 1981). These results could lead to the development of M protein vaccines that could protect against the streptococcal infections that cause debilitating diseases such as acute rheumatic fever, glomerulonephritis and rheumatic heart disease.

Another example is the diphtheria toxin, which consists of a single polypeptide chain of 62 kDa with two disulfide bridges. Vaccination against diphtheria is achieved by immunization with the detoxified toxin, or toxoid, which leads to antibodies that also combine with the native toxin and neutralize its toxicity. Audibert et al. (1981) have shown that active immunization against diphteria can be achieved by a synthetic peptide attached to a protein carrier. This peptide, a tetradecapeptide consisting of residues 188–201 in the animo acid sequence of diphtheria toxin, represents the loop fragment sustained by the disulfide bridge near the NH_2 terminus of the molecule. Conjugates of this peptide, or the hexadecapeptide 186–201, linked covalently to a protein carrier or to a synthetic copolymer, elicited antibodies in guinea pigs that not only bind specifically to the toxin but also neutralize its dermonecrotic activity and lethal effect. The same results were achieved when the conjugate contained the synthetic adjuvant muranyl dipeptide attached to the carrier, instead of the commonly used Freund's adjuvant. In this case a completely synthetic soluble molecule elicited antitoxin activity (Audibert et al., 1982).

A more recent example is the use of the synthetic approach to induce an immune response against cholera toxin (Jacob et al., 1983). This toxin is composed of two subunits, A and B. Subunit A activates adenylate cyclase which triggers the biological activity, and subunit B is responsible for binding to cell receptors, and expresses most immunopotent determinants. Antibodies to the B subunit are capable of neutralizing the biological activity of the intact toxin. The B subunit (choleragenoid) is a pentamer, each of the chains containing 103 amino acid residues of known sequence (Lai, 1980). Six peptides corresponding to various regions in the B subunit have been synthesized. The conjugates of all six peptides attached to tetanus toxoid led to the induction of peptide-specific antibodies, and four antisera out of the six also reacted, to different extents, with the intact B subunit and native cholera toxin. The strongest cross-reactivity with the native toxin was obtained with antibodies against the region 50–64, which was the only peptide that reacted with antibodies raised against the intact toxin. The antibodies raised against this synthetic peptide, as well as against the sequence 8–20 of the B subunit, were capable of neutralizing the biological activity of cholera toxin. This neutralization was manifested both on the biological level of the toxin activity, namely its capacity to cause skin permeability and to induce fluid secretion into ligated intestinal loops, and on the biochemical level, namely the induction of adenylate cyclase activity (Jacob et al., 1984).

4. Antiviral response induced by synthetic antigens

4.1. Tobacco mosaic virus

The first report of a peptide fragment that induced an antiviral response was by Anderer (1963a) describing his work on tobacco mosaic virus. This virus is composed of a 5% RNA core surrounded by a coat of 2200 identical protein subunits (TMVP) that are packed in a characteristic helical pattern. The interaction of TMV with antiserum raised against the intact virus results in neutralization of the virus infectivity. Anderer observed that peptides isolated from enzymatic digests of TMV protein had the capacity to partially inhibit the quantitative precipitation of TMV by its rabbit antibodies. Apparently only a few peptides in the digest were active, most of which originated from the C-terminus of the viral coat protein. These results led to the conclusion that the only immunogenically active region of the TMVP that led to antibodies capable of interacting with intact virus was the C-terminal hexapeptide. This situation might be due to the exposure of this peptide on the outer surface of the virus, as a result of the polypeptide chain folding and the packing of the protein subunits around the RNA core (Bloomer et al., 1978).

Based on the above findings, Anderer (1963b) isolated the C-terminal hexapeptide Thr-Ser-Gly-Pro-Ala-Thr from a TMVP enzymatic digest, attached it to bovine serum albumin and used the conjugate for the immunization of rabbits. The resultant antibodies precipitated with the homologous antigen and also with the intact virus, although to a much lesser extent. But, most importantly, when mixed in large excess with tobacco mosaic virus, the antiserum partially inhibited the formation of viral lesions on tobacco leaves. Although the hexapeptide used in this study was obtained from an enzyme digest of the native protein, it can still be considered within the realm of synthetic materials, since such short peptides are easily prepared by chemical synthesis. The limitation of this system, however, is the low efficacy and the incompleteness of neutralization of the virus infectivity by the antipeptide antibodies.

4.2. MS-2 coliphage

A model system in which efficient viral neutralization was obtained by antibodies against a synthetic peptide is the MS-2 coliphage (Langbeheim et al., 1976). The MS-2 phage is an RNA-containing virus with icosahedral symmetry, containing 180 identical 13 kDa coat protein subunits, in the viral capsid. Each subunit consists of a single polypeptide chain with 129 animo acid residues of known sequence. Antisera elicited by the total phage or even by the monomeric coat protein are efficient in neutralizing the intact phage, thus reducing the number of its plaques. A synthetic peptide corresponding to the sequence 88–108 of the coat protein and denoted P_2 was found to be an efficient inhibitor of antibody-induced viral neutralization and was therefore used for immunization. Antibodies raised by a synthetic conjugate of this peptide, denoted P_2-A--L, in Freund's adjuvant were indeed capable of neutralizing the viability of MS-2 almost as efficiently as antibodies raised against the intact coat protein. Furthermore, subsequent studies (Arnon et al., 1980) showed that when the synthetic adjuvant MDP was covalently linked to P_2-A--L, the resulting conjugate elicited antibodies in rabbits that were very effective in neutralizing MS-2. This conjugate was effective when administered in physiological medium, showing that a completely synthetic antigen with built-in adjuvanticity could lead to an antiviral response.

4.3. Hepatitis B virus

Numerous attempts have been made to produce a synthetic vaccine against hepatitis B virus. This virus cannot be grown in culture, and the only material presently available for vaccine preparation is the surface antigen (HBsAg) obtained from human blood of chronic carriers. As a result, the presently existing vaccine against hepatitis B is not readily available and is very expensive. Hence, this is probably one of the most important systems for which the synthetic approach might provide a solution not easily attained by classical vaccination methods. Five research groups synthesized peptides from different regions in the molecule which could induce antibodies reactive with the intact surface antigen, HBsAg.

Lerner and his colleagues synthesized thirteen peptides corresponding to amino acid sequences predicted from the nucleotide sequence of the HBsAg DNA, and covering most of its 220 residues (Lerner et al., 1981). The free or carrier (KLH)-linked synthetic peptides were injected into rabbits, and 7 of the 13 peptides elicited an antipeptide response. Antisera against four of the six soluble peptides that were longer than 10 amino acids were reactive with the native antigen and specifically precipitated the viral particles. The most effective peptide corresponded to the sequence 95–109. Melnick and his colleagues (Dreesman et al., 1982) have used another approach. Rather than synthesizing a whole series of peptides, they chose one region which, according to their prediction, should be exposed on the intact particle and consequently could represent a potential antigenic determinant. They chose the region of the animo acid sequence 117–137, which contains two cysteine residues that could be closed to form an internal disulfide 'loop'. Although there is no information as to whether such a disulfide bond exists in the native molecule, they predicted that it could contribute to the antigenic impact of the synthetic peptide. Two cyclic peptides from this region were synthesized, both of them unrelated to any of Lerner's peptides. These peptides, without linkage to a carrier, but used in conjunction with various adjuvants, elicited anti-native HBsAg antibody responses in mice after a single injection.

Hopp and Woods (1981) synthesized a peptide corresponding to residues 138–149 of HBsAg. The peptide was selected on the basis of a theoretical model which predicted a putative dominant HBsAg epitope corresponding to the sequence of 141–146. The 138–149 peptide exhibited cross-reaction with anti-HBsAg and, when coupled to aldehyde-stabilized human erythrocytes and injected into mice, it induced the formation of anti-HBsAg antibodies with and without the use of Freund's adjuvant (Prince et al., 1982). The fourth report concerning the hepatitis B virus is that of Vyas and his colleagues (Bhatnagar et al., 1982). These authors synthesized seven peptide analoges of HBsAg, linked them to KLH, and immunized rabbits with the conjugates. Three of the analogs, obtained from the same region of the molecule, induced anti-HBsAg antibodies. This indicated that the nonapeptide sequence 139–147, which is similar to the peptide identified by Prince et al. (1982), represents the total or an essential part of the a determinant of HBsAg that is common to all serotypes of the virus. An additional report by Neurath and his colleagues (1982) showed that a synthetic peptide corresponding to the region 135–155, elicited a heterogenous population of antibodies, one of which cross-reacted with HBsAg.

In all the reports mentioned above, there was no indication whether the antibodies produced by any of these peptides could neutralize the virus or whether their conjugates provided any protection against infection. The fact that the only species relevant to human hepatitis, in which protection can be investigated, is the chimpanzee, makes this task quite difficult. Recently, however, Gerin et al. (1983) reported that the peptide 110–137 stimulated a brisk type-specific response in chimpanzees which led to a partial protective effect. In view of these promising results efforts are now being continued in this direction.

4.4. Influenza virus

The influenza virus provides a very suitable model for studying the synthetic approach to vaccination. Since detailed information is available on the structure, function and antigenic properties of its surface protein components, it is possible to synthesize peptides with a high potential immune reactivity. Moreover, since this is an animal virus for which the viral strains infective in humans are also infectious for mice, the immune response against the synthetic peptides can be evaluated not only for in vitro reactivity, but also for its protective effect in vivo. The main protein component which is responsible for the serological specificity of the various strains of influenza, as well as for the viral neutralization by antibodies, is the hemagglutinin (Schulman, 1975). This molecule was therefore used in several laboratories including our own to investigate the value of synthetic peptides as viral antigens.

The first peptide we synthesized, before the three dimensional structure was known, contained 18 amino acid residues corresponding to the sequence 91–108 of the influenza hemagglutinin molecule. A conjugate of this peptide with tetanus toxoid elicited antibodies in both rabbits and mice that reacted immunochemically with the synthetic peptide, as well as with the intact influenza virus of several strains of type A. These antibodies were capable of inhibiting the hemagglutination by the relevant strains. They also interfered with the in vitro growth of the virus in tissue culture. But, most important, mice immunized with the peptide-toxoid conjugate were partially protected against challenge infection with the virus (Müller et al., 1982). Furthermore, since the region 91–108 is a conserved sequence present in the hemagglutinin of several influenza strains, it was of interest to determine whether immunization with the same conjugate would elicit protection against more than one hemagglutinin strain. The finding that this is indeed the case, is an indication that the synthetic approach might lead to multivalent vaccines for cross-strain protection.

Subsequent studies in other laboratories using synthetic peptides corresponding to various regions in the hemagglutinin molecule also led to antibodies that were cross-reactive with the hemagglutinin (Green et al., 1982; Atassi and Webster, 1983), but with no indication of protective capability.

According to the three-dimensional structure of the influenza hemagglutinin four main antigenic sites have been located (Wiley et al., 1981). The region 140–146, which forms antigenic site A, is a 'loop' of seven amino acids unusually protruding from the surface of the molecule. It is also exposed in the trimeric structure assumed by the hemagglutinin on the viral surface. One would expect that peptides corresponding to this region where 'natural' immunogenic determinants of the hemagglutinin molecule are located, would elicit a better and more specific immune response against the virus than the 91–108 peptide. It was therefore of interest to synthesize peptides covering the 'loop' region for immunization purpose. Jackson et al. (1982) have used a synthetic peptide representing the sequence 123–151 of the hemagglutinin of strain X-31, which includes the 'loop' region, for immunization of rabbits. They did not find significant binding between the antipeptide antibodies and X-31 virus. However, antibodies raised against the intact X-31 virus did exhibit some binding to the synthetic peptide.

In recent studies in our laboratory we have confirmed these results. We have found that octapeptides corresponding to the sequence 139–146 induced antibodies which did not cross-react with the intact virus although antibodies against the virus recognized these 'loop' peptides. However, a longer peptide, corresponding to the sequence 138–164, induced antibodies that recognized the intact virus, and were also capable of interfering with its hemagglutination activity. Moreover, immunization with a conjugate of this peptide provided partial protection against a challenge infection with a viral strain containing the corresponding sequence (Shapira et al., 1984). These findings are indicative of the importance of the length and conformation of the synthetic peptides for their immunological reactivity.

4.5. Foot-and-mouth disease virus

In the case of foot-and-mouth disease (FMD) virus, for which an efficient vaccine is available, the main advantage of a synthetic vaccine would be its safety, since a few outbreaks of foot-and-mouth disease have been associated with failures of the virus inactivation process. Hence, an effort in this direction has been made, using both cloning of the virus genome and chemical synthesis of fragments of its proteins. The virus particles consist of one molecule of the infectious single-stranded RNA, and 60 copies of each of four structural polypeptides VP1–VP4. Recently neutralizing immunity against FMD virus has been elicited using synthetic fragments of the capsid proteins produced by genetic engineering (Kleid et al., 1981). When inoculated into six cattle and two swine, the protein synthesized by the *Escherichia coli* elicited high levels of neutralizing antibodies and protection against challenge with the virus.

VP1 was shown to be of antigenic importance in the virus, although in the isolated form it is a weak immunogen. Nevertheless, chemically synthesized fragments of this protein have been recently prepared and evaluated as immunogens for inducing response against the intact virus. Pfaff et al. (1982) prepared a hexadecapeptide corresponding to residues 144–159 and showed that, when injected coupled to KLH, it elicited antibodies that specifically recognized and neutralized FMD virus. Bittle et al. (1982) prepared peptides corresponding to three different regions in the VP1. After coupling to KLH, peptides of two regions produced high levels of serotype-specific virus neutralizing antibodies in cattle, guinea pigs and rabbits. All these antisera reacted in the immunoprecipitation technique with the virus native structural proteins and their precursor molecules, which indicates that their specificity resembles that of antiviral sera. Moreover, a single inoculation of the synthetic peptide 141–160 elicited sufficient neutralizing antibody to protect guinea pigs against subsequent challenge with FMD virus, although its protection efficiency was only 1–10% of that of the inactivated virus particle when calculated on an equal weight basis. Interestingly the neutralizing antibody titer elicited by the single peptide 141–160 was several orders of magnitudes greater than the best results obtained with the whole VP1 protein, which was shown to elicit antibodies specific mainly toward the denatured form of the protein. It is thus possible that an entire protein isolated from a virus does not fold properly when deprived of the scaffolding provided by the other structural proteins, whereas a single relatively short synthetic peptide may have better prospects as an appropriate immunogen. These results are therefore an important support to the notion of totally synthetic vaccines.

4.6. Poliovirus

Very recent studies indicate that poliovirus vaccine may also be amenable to the synthetic approach. Chow and Baltimore (1983) have used both genetic engineering methods and chemically synthesized peptides. They have constructed several plasmids in which the VP 1 gene was inserted, and studied the expression of VP1 in both *E. coli* and mammalian cells. Preliminary data indicate, however, that the capsid proteins, when expressed, are highly toxic to the host organism. In order to define the antigenic determinants of VP1 which induce neutralizing antibodies, they have injected synthetic peptides corresponding to various regions of protein. Using anti-peptide and antivirion antibodies, four regions of VP1 have been identified which apparently elicit neutralizing responses. The main activity is in the region 80–100 in the VP1 sequence, which elicits neutralizing antibodies on both rabbits and rats. These species are not susceptible to poliovirus, and thus it was not possible to determine whether the synthetic conjugates provide in vivo protection against infection.

Emini et al. (1983) have also used synthetic peptides for induction of antipoliovirus neutralizing antibodies. They have synthesized five peptides corresponding to the hydrophilic regions of the VP1, as they assumed that such regions are more likely to be exposed on the virion's surface. Four of these peptides were indeed recognized by neutralizing monoclonal antibodies raised against the intact virion. After conjugation to a carrier protein, one of these peptides corresponding to the sequence 93–103 of the VP1 induced in rabbits a significant level of neutralizing antibodies. Moreover, all five peptides were capable of priming the immune system of the rabbits so that when followed by a subsequent single inoculation of a sub-effective dose of the virus, they provided a significant level of long-lasting virus-neutralizing IgG antibodies. These results indicate that synthetic peptides are suitable for mimicking neutralizing epitopes of poliovirus and for specifically priming the immune systems as a strategy in vaccination.

5. Conclusions

The cumulative information described above illustrates the tremendous potential of synthetic peptides in the preparation of some vaccines. This approach, if it materializes, will offer several advantages. Firstly, it could allow the attachment of several peptides, representing the relevant antigenic portions of different disease-causing agents, to the same carrier molecule, thus providing multivalent synthetic vaccines. Secondly, this approach might eliminate the formation of antibodies against many irrelevant antigens that contaminate the essential immunogen, and thereby reduce undesired side effects. Thirdly, it might permit the production of vaccines with built-in adjuvanticity that would be efficient when administered in physiological medium. This would eliminate the hazard inherent in most of the presently available adjuvants that are utilized for augmentation of the efficacy of the existing vaccines. The key question is, of course, whether synthetic peptides can form the basis for *practical* vaccines. An answer to this question will hopefully be found in the not too distant future.

6. References

Altschuh, D., Hartman, D., Reinbolt, J. and van Regenmortel, M.H.V. (1983) *Mol. Immunol.* 20, 271.

Anderer, F.A. (1963a) *Z. Naturforsch. Teil. B* 18, 1010.

Anderer, F.A. (1963b) *Biochim. Biophys. Acta* 71, 1246.

Arnon, R. (1972) In *Immunity of Viral and Rickettsial Diseases* (A. Kohn and A.M. Klinberg, Eds.) Plenum Press, New York. pp. 209–222.

Arnon, R. and Geiger, B. (1977) In: *Immunochemistry* (L.E. Glynn, Ed.) John Wiley and Sons, New York. p. 307.

Arnon, R. and Sela, M. (1969) *Proc. Natl. Acad. Sci. USA* 62, 163.

Arnon, R., Maron, R., Sela, M. and Anfinsen, C.B. (1971) *Proc. Natl. Acad. Sci. USA* 68, 1450.

Arnon, R., Bustin, M., Calef, E., Chaitchik, S., Haimovich, J., Novik, N. and Sela, M. (1976) *Proc. Natl. Acad. Sci. USA* 73, 2123

Arnon, R., Sela, M., Parant, M. and Chedid, L. (1980) *Proc. Natl. Acad. Sci. USA* 77, 6769.

Atassi, M.Z. (1975) *Immunochemistry* 12, 423.

Atassi, M.Z. and Webster, R.G. (1983) *Proc. Natl. Acad. Sci. USA* 80, 840

Audibert, F., Jolivet, M., Chedid, L., Alouf, J.E., Boquet, P., Rivaille, P. and Shiffert, O. (1981). *Nature* 289, 593.

Audibert, F., Jolivet, M., Chedid, L., Arnon, R. and Sela, M. (1982) *Proc. Natl. Acad. Sci. USA* 79, 5042.

Beachy, E.H., Seyer, J.M. and Kang, A.H. (1978) *Proc. Natl. Acad. Sci. USA* 75, 3163.

Beachy, E.H., Seyer, J.M., Dale, J.B., Simpson, W.A. and Kang, A.H. (1981) *Nature* 292, 457.

Benjamini, E., Young, J.D., Shimizu, M. and Leung, C.Y. (1964) *Biochemistry* 3, 1115.

Bhatnagar, P.K., Papas, E., Blum, H.E., Milich, D.R., Nitecki, D., Karels, M.J. and Vyas, G.N. (1982). *Proc. Natl. Acad. Sci. USA* 79, 4400.

Bittle, J.L., Houghtern, R.A., Alexander, H., Shinnick, T.M., Sutcliffe, J.G., Lerner, R.A., Rowlands, D.J. and Brown, F. (1982). *Nature* 298, 30.

Bloomer, A.C., Champness, J.N., Bricogne, G. and Staden, R. (1978) *Nature* 276, 362.

Borek, F., Kurtz, J. and Sela, M. (1969) *Biochem. Biophys. Acta* 188, 314.

Chow, M. and Baltimore, D. (1983) *Abstracts of Cold Spring Harbor Meeting on Modern Approaches to Vaccination.* p. 47

Dreesman, G.R., Sanchez, Y., Ionescu-Matiu, I., Sparrow, J.T., Six, H.R., Peterson, D.L., Hollinger, F.B. and Melnick, J.L. (1982). *Nature* 295, 158.

Emini, E.A., Jameson, B.A. and Wimmer, E. (1983) *Nature* 304, 699.

Fuchs, S., Maoz, A. and Sela, M. (1974) *Israel J. Chem.* 12, 681.
Gerin, J.L., Alexander, H., Shih, J.W., Purcell, R.H., Dapolito, T., Engle, R., Green, N., Sutcliffe, J.G., Shinnick, T.M. and Lerner, R.A. (1983) *Proc. Natl. Acad. Sci. USA* 80, 2365.
Green, N., Alexander, H., Olson, A., Alexander, S., Shinnick, T.M., Sutcliffe, J.G. and Lerner, R.A. (1982) *Cell* 28, 477.
Hopp, T.P. and Woods, K.R. (1981) *Proc. Natl. Acad. Sci. USA* 78, 3824.
Jackson, D.C., Murray, J.M., White, D.O., Fagan, C.N. and Tregear, G.W. (1982) *Virology* 120, 273.
Jacob, C.O., Sela, M. and Arnon, R. (1983) *Proc. Natl. Acad. Sci. USA* 80, 7611.
Jacob, C.O., Arnon, R., Sela, M., Pines, M. and Hurwitz, S. (1984) *Nature*, In Press.
Kleid, D.G., Yansura, D., Small, B., Dowbenko, D., Moore, D.M., Grubman, M.J., McKescher, P.D., Morgan, D.O., Robertson, B.H. and Bachrach, H.L. (1981) *Science*, 214, 1125.
Lai, C. (1980) *CRC Critical Reviews in Biochemistry* 9, 1171.
Langbeheim, H., Arnon, R. and Sela, M. (1976) *Proc. Natl. Acad. Sci. USA* 73, 4636.
Lerner, R.A., Green, N., Alexander, H., Liu, F.T., Sutcliffe, G.J. and Shinnick, T.M. (1981) *Proc. Natl. Acad. Sci. USA* 78, 3403.
Maron, E., Shiozawa, C., Arnon, R. and Sela, M. (1971) *Biochemistry* 10, 763.
Muller, G., Shapira, M. & Arnon, R. (1982) *Proc. Natl. Acad. Sci USA* 79, 569.
Neurath, R.A., Kent, S.B.H. and Strick, N. (1982) *Proc. Natl. Acad. Sci. USA* 79, 7871.
Pecht, I., Maron, E., Arnon, R. and Sela, M. (1971) *Eur. J. Biochem.* 19, 369.
Pfaff, E., Mussgay, M., Bohn, H.O., Schulze, G.E. and Schaller, H. (1982) *EMBO J.* 1, 869.
Prince, A.M., Ikram, H. and Hopp, T.P. (1982) *Proc. Natl. Acad. Sci. USA* 79, 579.
Schulman, J.L. (1975) In *The Influenza Viruses and Influenza* (E.D. Kilbourne, Ed.) Academic Press, New York p. 373.
Sela, M. (1969) *Science* 166, 1365.
Shapira, M., Jibson, M., Muller, G. and Arnon, R. (1984) *Proc. Natl. Acad. Sci. USA* 81, 2461.
Shinka, S., Imanishi, M., Kuwahara, O., Fujio, H. and Amano, T. (1962) *Bikins J.* 5, 181.
Traub, W. and Yonath, A. (1966) *J. Mol. Biol.* 16, 404.
Wiley, D.C., Wilson, I.A. and Skehel, J.J. (1981) *Nature* 289, 373.

Eds. M.H.V. van Regenmortel and A.R. Neurath
Immunochemistry of viruses. The basis for serodiagnosis and vaccines.

Vaccines prepared from translation products of cloned viral genes

11

ERIC J. PATZER and JOHN F. OBIJESKI

Department of Vaccine Development, Genentech Inc., 460 Point San Bruno Boulevard, South San Francisco, CA 94080, U.S.A.

Perhaps the greatest accomplishment of medical science in the twentieth century has been the control and in some cases the eradication of infectious diseases through the use of safe and efficacious vaccines. Vaccination became a powerful prophylactic tool of modern medicine when it was discovered that the immune system could mount an antibody response to an attenuated virus and that this response would protect the host from subsequent infections by the virulent virus. The concept of using an attenuated virus became the theoretical basis for most of today's vaccines. Initially virus attenuation was classically accomplished by selecting a naturally occurring noninfectious virus (for the host being treated) that was immunologically very similar to the infectious agent (such as vaccinia for variola), or by growing the infectious virus in foreign tissues (rabies) or cell lines (polio) that resulted in an apparently stable mutation of the virus to a noninfectious but immunologically protective form. Contemporary 'live' vaccines have been attenuated by repeated passaging in cell culture until the virus produces only a mild or inapparent infection. Live vaccines have the advantage of usually producing lifelong immunity to disease. However, they tend to be less stable and there is always the risk of spreading the infection to unvaccinated or immunosuppressed individuals. Killed or inactivated vaccines are prepared by chemically inactivating the infectious virus under conditions which preserve the antigenicity of the virus.

This classical approach to producing vaccines by various methods of attenuating viruses has remained unchanged until recent times. The strategy was a hit-or-miss procedure which did not require a detailed understanding of the virus on a molecular level. With the advent of recombinant deoxyribonucleic acid (rDNA) techniques and their application to viruses for vaccine research, there has been an explosion of information about the molecular structure and replication of many viruses. rDNA technology in conjunction with several other emerging technologies, e.g. monoclonal antibodies, solid phase synthesis of peptides and prediction of protein conformation on the basis of amino acid sequence, has provided a powerful battery of techniques that in many cases has allowed the identification of specific sites on the virion surface that elicit neutralizing antibodies (Minor et al., 1983). Knowledge of these sites allows one to design a subunit vaccine that utilizes one of the virion proteins or regions of a particular protein in the absence of any other viral proteins or the viral nucleic acid. The advantages of this approach are: (*a*) that there are no potentially infectious agents contained in the vaccine if the inactivation procedure is incomplete, (*b*) there is less chance of complications from the vaccine due to nonessential viral components in the vaccine, (*c*) a purified protein or polypeptide will usually be more stable than virus particles during storage, and (*d*) many times larger quantities of an antigen can be produced by rDNA techniques than by classical vaccine methods.

1. Cloning and expression of viral genes

1.1. Cloning of viral genes

Once the causative agent of a disease has been isolated, an important prerequisite for the production of rDNA-derived vaccines is the identification and, if possible, characterization of the structural protein(s) responsible for eliciting a neutralizing and protective antibody response in the host. The important structural protein can be identified immunologically by using either antibodies from immune individuals or neutralizing monoclonal antibodies. These virus-specific antibodies, whether poly- or monoclonal, provide an important tool for the development of subunit vaccines by identifying the essential virus structure necessary for stimulating a protective host response. After identification of the essential surface protein(s) the viral gene encoding this protein(s) can be isolated by molecular cloning techniques. Since virus genomes are relatively small in comparison to those of more complex parasites, it is usually less complicated to identify the genomic region containing the correct gene.

Unlike other organisms, viruses have genomes of either ribonucleic acid (RNA) or deoxyribonucleic acid (DNA). Cloning of structural genes from RNA viruses such as poliovirus, rabies virus, influenza virus, measles virus and foot-and-mouth disease virus (FMDV) can be accomplished by converting the viral RNA into complementary cDNA using the enzyme reverse transcriptase (Goodman and MacDonald, 1979). Three general methods have been used for producing cloned viral sequences from RNA viruses which take advantage of the unique characteristics of different virus groups. RNA from viruses such as poliovirus and FMDV which contain positive-strand RNA genomes that are polyadenylated at their 3'-terminus can be used directly as a template for reverse transcriptase. With negative-strand RNA viruses, such as rabies virus and influenza virus, it is usually more convenient to use the polyadenylated viral mRNA species as a template for reverse transcriptase. Another approach that is used with double-stranded segmented RNA genome viruses, such as reovirus, rotavirus or blue tongue virus, is to add polyadenosine residues to the 3'-termini of both RNA strands before cDNA synthesis by reverse transcriptase.

More specifically, the RNA genome or viral mRNA is first transcribed into DNA using reverse transcriptase and a DNA primer molecule, such as oligo-dT, which hybridizes to the polyadenosine tail. After first strand synthesis, the RNA is removed and a complementary second DNA strand is synthesized using DNA polymerase I (Klenow fragment) from *Escherichia coli.* A single strand specific nuclease is used to cleave the 'hairpin' found at one end of the double stranded DNA molecule produced during synthesis. The final step in the synthesis of the viral gene is the addition of a homopolymeric group of deoxycytosine to the 3'-terminus of each DNA strand using the enzyme terminal transferase.

To synthesize larger quantities of the viral gene, the DNA copy is inserted into a plasmid, such as pBR322, which is capable of self-replication as an extrachromosomal element in *E. coli* (Bolivar et al., 1977). Since pBR322 codes for the antibiotic resistance markers tetracycline (Tc) and ampicillin (Ap), bacteria harboring pBR322 can be selected for by growth in the presence of Tc and Ap. The plasmid can be linearized with restriction endonuclease enzymes at many sites in the molecule, including the two antibiotic resistance markers, Tc and Ap. Endonucleases are enzymes which cleave DNA at specific nucleotide sequences. Each enzyme is absolutely specific for a particular nucleotide sequence, and precisely defined DNA fragments can thus be generated by endonuclease digestion of a given DNA molecule. For example, pBR322 may be cleaved at a number of nucleotide sequences including the *PstI* recognition site within the Ap resistance gene. If the viral gene is inserted within the Ap resistance gene, a recombinant plasmid can be generated which contains an inactivated Ap resistance gene. The viral gene is inserted within the Ap resistance gene by first adding homopolymeric deoxyguanosine residues (with terminal transferase) to the 3'-termini of pBR322 generated by *PstI* cleavage and then hybridizing with C-tailed viral DNA under conditions which allow base pairing between the C and G residues. This mixture is used to transform *E. coli* cultures to antibiotic resistance by allowing the recombinant plasmid (with inserted viral DNA sequences) to enter semipermeabilized bacterial cells. The recombinant plasmid is still capable of conferring resistance to Tc, but not Ap. Plating of the transformed bacterial culture on antibiotic selection media allows colonies resistant to one antibiotic (Tc) to be assayed for insertion of viral sequences into the plasmid by testing for resistance to the second antibiotic (Ap). By these procedures large quantities of DNA corresponding to the viral RNA genome can be synthesized and subsequently used for the expression of the selected protein.

Molecular cloning of genes from DNA viruses such as herpes simplex virus (HSV) follows the same scheme as described for RNA viruses, but does not require synthesizing a cDNA copy. Since many restriction endonucleases are now available, viral DNA genomes can be cleaved at many restriction sites containing specific nucleotide sequences. Plasmids which contain these same restriction sites can also be constructed. Cleavage of the viral DNA genome and the plasmid with the same restriction enzymes followed by ligation of the viral fragments and the plasmids with T4 DNA ligase allows the insertion of a variety of DNA virus genes into specific sites within plasmids. The same antibiotic resistance selections, as described previously, can be utilized in the molecular cloning of DNA virus genomic fragments. A library of DNA fragments representing the entire viral genome may be cloned using this procedure. Bacteriophage vector systems have also been used successfully to establish virus gene banks.

After having cloned the entire DNA viral genome, the appropriate gene which encodes the protein of interest must be identified. With some DNA viruses, the cloned fragment containing the appropriate gene can be identified by recombinational mapping (Roizman et al., 1982). In other cases, the entire nucleotide sequence of large portions of the genome must by determined to locate the sequence of the desired gene(s). A third procedure involves hybridization selection of viral messenger RNA using cloned DNA fragments of the viral genome. Subsequent in vitro translation of the selected mRNA and immunoprecipitation of the translation products with antibodies known to react with the viral protein can be utilized to select the genomic fragment containing the gene. Once located, the gene can be subcloned and used to direct the expression of the viral protein.

1.2. Organisms used for expression of viral polypeptides

Currently three classes of organisms have been used to express proteins using rDNA techniques. The first are the procaryotic organisms, of which *E. coli* is genetically the best understood and most widely used. Since most of the cloning is done in *E. coli*, it is natural to attempt protein expression in the same organism. The advantages of using *E. coli* are the high level of expression attainable with several expression plasmids and the fact that *E. coli* can be grown to high cell densities at a relatively low cost. A major drawback of *E. coli* expression systems is the limited protein secretion. Other procaryotes are being developed which have a greater secretory ability (e.g. *Bacillus subtilis* and the pseudomonads). The knowledge of the genetics of these is not as advanced as for *E. coli*, but is rapidly increasing.

Among eucaryotes expression of heterologous genes has been achieved in mammalian cells and in the unicellular organism, yeast. Yeast cells can be grown economically on a large scale and have some capability of secretion and glycosylation (Novick et al., 1981). From a secretory viewpoint there are two major differences between yeast cells and higher eucaryotes. Yeast cells have a defined cell wall and oligosaccharide chains which are comprised of asparagine-linked core oligosaccharides like higher eucaryotes, but also large repeating oligosaccharide units called outer chains (Ballou, 1976). These two differences appear to have an effect on the secretion of at least one recombinant protein, the hepatitis B surface antigen, as will be discussed in greater detail below. The third alternative for an expression system is mammalian cells in culture. These provide a more 'natural' substrate for expression of viral antigens in vaccine production, since the cells can be engineered to produce viral proteins in a manner similar to their synthesis during a natural virus infection. This includes the proper folding, processing, glycosylation and localization within the cell. The drawback to using mammalian cells in culture is the high cost of vaccine production relative to procaryotes and yeast.

2. Expression of viral polypeptides in bacteria

E. coli is genetically the best defined organism and has been the primary organism for cloning procaryotic and eucaryotic genes. It was logical that it would be the organism of choice for initially expressing most cloned genes. This turned out to be a very successful approach for the production of large quantities of several relatively small proteins (about 20 kDa or less) such as insulin (Goeddel et al., 1979b; Chance et al., 1981), the interferons (Goeddel et al., 1980; Gray et al., 1982), somatostatin (Itakura et al., 1977) and human growth hormone (Goeddel et al., 1979a; Olson et al., 1981). When proteins are expressed at high levels (about 20% of total protein) within *E. coli* cells, however, they sometimes become sequestered into inclusion bodies (Williams et al., 1982). Recovery of proteins from inclusion bodies in some cases requires solubilization with strong denaturants such as guanidine HCl, urea or detergents (Goeddel et al., 1979b; Kleid et al., 1981). Once these agents are removed, however, many of the proteins become fully active, presumably because of proper refolding into a native conformation (Itakura et al., 1977; Goeddel et al., 1979b; Goeddel et al., 1980; Chance et al., 1981; Olson et al., 1981; Gray et al., 1982).

2.1. VP1 Protein of foot-and-mouth disease virus

Vaccines that would appear to have the greatest likelihood of successful expression in *E. coli* are those in which an isolated viral protein or protein fragment has been shown to retain some of its antigenic determinants after exposure to denaturing conditions. As early as 1973 it was demonstrated that the VP1 protein of foot-and-mouth disease virus (FMDV) by itself was capable of raising neutralizing antibodies (Laporte et al., 1973) and that vaccination of swine and cattle with VP1 protected the animals from infection with FMDV (Bachrach et al., 1975). Subsequently, it was shown that cyanogen bromide and proteolytic fragments of VP1 also retained some of their 'native' antigenic determinants and were able to protect animals from FMDV infection (Kaaden et al., 1977; Bachrach et al., 1979; Strohmaier et al., 1982).

Several laboratories have cloned the VP1 region of the FMDV genome in *E. coli* and determined its nucleotide sequence (Boothroyd et al., 1981; Kupper et al., 1981; Kleid et al., 1981). The cloned VP1 gene was expressed as a chimeric protein fused either to 99 amino acids of the MS2 replicase protein (Kupper et al., 1981) or to 190 amino acids from the tryptophan operon of *E. coli* termed LE' (Kleid et al., 1981). In the case of the LE'-VP1 fusion protein some of the antigenic determinants normally displayed on the surface of virions were retained, since animals immunized with LE'-VP1 were protected from FMDV infection (Kleid et al., 1981). Not all of the antigenic determinants on virion-associated VP1 are retained on the isolated protein (Meleon et al., 1983) and, likewise, they are not all duplicated on recombinant VP1 expressed in *E. coli*. It is clear, however, that the recombinant VP1 can serve as an effective vaccine against an FMDV infection (Kleid et al., 1981; Moore, 1983; Kleid et al., 1984).

The VP1 protein of FMDV is a good example of how several new technologies (rDNA and chemical synthesis of peptides) as well as existing techniques (proteolytic fragments of VP1) can complement one another when trying to elucidate important immunogenic regions on a polypeptide. Initially, a 13-kDa cyanogen bromide fragment of VP1 was demonstrated to protect against infection with FMDV (Kaaden et al., 1977; Bachrach et al., 1979). Using proteases to generate smaller fragments, two regions of VP1 were found to be exposed on the surface of virions. For the A12 serotype this region included amino acid residues 136–144 (Robertson et al., 1983), whereas with the 01 Kaufbeuren (01K) serotype two exposed regions spanning amino acids 138–154 and 200 to the carboxy terminus of VP1 were discovered (Strohmaier et al., 1982). Proteolytic fragments from these two regions of the 01K serotype induced neutralizing antibodies (Strohmaier et al., 1982). Additional data from amino acid sequence comparisons (Makoff et al., 1982) and from guinea pig protection experiments with chemically synthesized peptides (Bittle et al., 1982; Pfaff et al., 1982) confirmed that these two general regions of VP1 contained antigenic determinants which induced neutralizing antibodies, although these may not be the only sites displayed on the surface of virions (Meleon et al., 1983). Peptides spanning amino acids 131–167 of VP1 of the A12 serotype and amino acids 137–157 of VP1 of the A24 serotype were also expressed in *E. coli* as LE' fusion proteins which elicited neutralizing antibody titers that were just as high as those elicited by LE'-VP1 (full length) fusion proteins (Table 11.1).

When employing immunogens of reduced size for a vaccine, it is of utmost importance to establish that the immunogen can induce antibodies that protect the *target animal* against infection rather than a model animal system in which virus neutralization may not have identical requirements. Virus neutralization is a complex interaction between virions, antibodies and cells which is not well understood at the

TABLE 11.1

Mouse neutralization titers elicited in guinea pigs

Antigen	Mouse neutralization titers[a]	
	4 week	8 week
A24-LE'-VP1[b]	1.3	3.2
A24-LE'-VP1-fragment[c]	1.7	3.6
A12-LE'-VP1[d]	3.3	4.0
A12-LE'-VP1-fragment[e]	4.4	4.6

[a] Mouse neutralization titers were determined by the suckling mouse protection assay (Skinner, 1952).
[b] 50 μg of total protein in water was emulsified with Freund's incomplete adjuvant and injected on weeks 0, 4 and 6. The antigen contained the entire VP1 protein (A24 serotype) fused with 17 amino acids from the tryptophan (*trp*) operon of *E. coli*.
[c] 50 μg of total protein in 6 M urea, 0.05 M 2-mercaptoethanol, 0.014 M Tris (pH 8.6) was emulsified with Freund's incomplete adjuvant and injected on weeks 0, 4 and 6. The antigen contained amino acids 137–157 of the VP1 protein (A24 serotype) fused with 190 amino acids from the *trp* operon of *E. coli* (Kleid et al., 1981).
[d] 40 μg of total protein in urea buffer described in 'c' was emulsified in Freund's adjuvant and injected on weeks 0, 2 and 4. The antigen contained the entire VP1 protein (A12 serotype) fused with 17 amino acids of the *trp* operon of *E. coli*.
[e] 100 μg of total protein in urea buffer described in 'c' was emulsified with Freund's adjuvant and injected on weeks 0, 2 and 4. The antigen contained a tandem peptide of amino acids 131–167 of VP1 (A12 serotype) fused with 190 amino acids of the *trp* operon of *E. coli* (Kleid et al., 1981).

present time (Emini et al., 1983; Icenogle et al., 1983). In some cases, protection of mice or guinea pigs from FMDV infection may not correlate with protection in cattle. For example, although the chemically synthesized peptides (01 serotype) protect guinea pigs, they have so far been unable to protect the target animal, cattle (Wilson, 1984). On the other hand, the LE'-fusion peptides expressed in *E. coli* (A12 and A24 serotypes) not only protected mice (Table 11.1), but also protected cattle (Moore, 1983; Kleid et al., 1984). The differences observed in the cattle protection experiments between chemically synthesized and recombinant peptides could be due to several factors. The recombinant peptides contain a well-defined biosynthetically linked carrier in the fusion protein rather than a heterogeneous mixture of peptides chemically coupled to a carrier protein. In addition, the peptides studies were derived from different serotypes of FMDV and slightly different amino acid regions of VP1 were contained in each peptide. Whether these factors or others were responsible for the observed differences in cattle protection will have to await further testing.

2.2. Other viruses

Like FMDV, poliovirus is also a member of the Picornaviradae family with similar morphological features. A neutralizing site on the polio virion has been localized to an 8 amino acid region of VP1, which, like the FMD virus VP1, is located on the virion surface (Evans et al., 1983; Minor et al., 1983). A panel of monoclonal antibodies directed against this neutralizing site, however, did not react with isolated poliovirus VP1 (Ferguson et al., 1984). In addition, antibodies produced against isolated poliovirus VP1 do not neutralize the virus nor do they react with whole virions (Meleon et al., 1979). Similarly, a chimeric protein of poliovirus VP1 fused with the β-lactamase gene and expressed in *E. coli* could be immunoprecipitated with anti-VP1 antisera, but not with antivirion antisera (Van der Werf et al., 1983). Possibly the poliovirus VP1 requires refolding to regain important 'native' antigenic determinants. These data make it clear that poliovirus VP1 does not have the same structural and immunological characteristics as FMDV VP1.

Other medically important viruses which would be candidates for a rDNA subunit vaccine that have been expressed in *E. coli* are lipid enveloped viruses containing membrane glycoproteins. Many of these glycoproteins have been isolated from purified virions with nonionic detergents which solubilize the viral envelope, but do not denature the protein. These solubilized glycoproteins either in the presence of detergent or after detergent removal will in some cases immunize animals against virus infection (Dietzschold et al., 1978; Balcarova et al., 1981). Very little is known, however, about whether these envelope glycoproteins will retain any of their 'native' antigenic determinants after exposure to denaturing conditions or after cleavage into peptide fragments (Jackson et al., 1979).

In general, expression of most of these envelope glycoprotein genes in *E. coli* resulted in the synthesis of proteins which could not induce neutralizing antibodies in animals. Two groups have been successful in expressing the hemagglutinin (HA) of influenza virus in *E. coli*. In one instance the *E. coli*-derived HA antigen bound antibodies raised to the native HA protein (Emtage et al., 1980) while in the second case antibodies to LE'-HA fusion protein reacted with isolated detergent-solubilized authentic HA or HA on the surface of virions (Davis et al., 1983). In neither case were any animal protection data presented. Similar findings were reported for the rabies virus glycoprotein expressed in *E. coli*; however, in this case solubilization of a partially purified *E. coli* extract with guanidine hydrochloride resulted in a 30-fold increase in antigenicity, indicating that many of the native antigenic determinants could be restored by this solubilization procedure (Yelverton et al., 1983). It is not clear whether solubilization played any role in the activation of antigenic determinants displayed by herpes simplex virus type 1 (HSV-1) glycoprotein D expressed in *E. coli* as a chimeric protein fused to β-galactosidase. When this chimeric protein was solubilized with sodium dodecyl sulfate and purified by polyacrylamide gel electrophoresis, it elicited antibodies which neutralized the virus in vitro (Watson et al., 1982; Weis et al., 1983).

A great deal of effort has been directed at expressing hepatitis B surface antigen (HBsAg) in both procaryotes and eucaryotes, primarily because acceptable in vitro cell culture systems are not yet available for vaccine production. Current vaccines containing HBsAg purified from the plasma of hepatitis B carriers, have certain known and perceived risks and require a labor-intensive and costly purification procedure (Hilleman et al., 1981). Unlike the envelope glycoproteins discussed previously, HBsAg is a very hydrophobic integral membrane protein which assembles into 22 nm lipoprotein particles comprised of fully immunogenic disulfide-bonded dimers (Skelly et al., 1979; Mishiro et al., 1980). The monomers are 20 to 1000-fold less immunogenic than dimers (Mishiro et al., 1980). Several groups have expressed HBsAg in *E. coli* as a fusion protein or by direct expression (Pasek et al., 1979; Charnay et al., 1980; Edman et al., 1981; Fujisawa et al., 1983). In each case the expression levels were low, possibly due to the presence of signal peptides or the overall hydrophobicity of HBsAg; however, the expression levels were enhanced when the amino terminal 22 to 32 amino acids were deleted (Edman et al., 1981; Fujisawa et al., 1983). Although the HBsAg expressed in *E. coli* in some cases reacted with antibody to serum-derived HBsAg, it is doubtful at this time that this system would ever become a viable approach for HBsAg vaccine production since the expression levels are low and the immunogenicity is minimal, probably due to the lack of formation of dimers or 22 nm lipoprotein particles (Charnay et al., 1980; Edman et al., 1981; Fujisawa et al., 1983).

3. Expression of viral polypeptides in yeast

Expression of viral polypeptides in yeast would be desirable because, unlike higher eucaryotes, there has been a great deal of experience in fermenting yeast on an industrial scale. This could be done inexpensively and without any major technological innovations. Unlike higher eucaryotes, however, yeast cells have a rigid cell wall and large repeating oligosaccharide units known as 'outer chains' (Ballou, 1976). Even with these differences, there is evidence that some secretory proteins of higher eucaryotes, such as interferon and human growth hormone, can be secreted when expressed in yeast cells using either homologous or heterologous signal sequences (Hitzeman et al., 1983; 1984).

The only published report of a viral antigen that has been expressed in yeast cells is HBsAg. This has been reported by several groups and in each case all of the HBsAg was detected within the cells in an unglycosylated form (Valenzuela et al., 1982; Hitzeman et al., 1983; Miyanohara et al., 1983). One major difference between HBsAg and other secretory proteins is that HBsAg is an extremely hydrophobic integral membrane protein which forms 22 nm lipoprotein particles (Tiollais et al., 1981) that may not be able to be secreted across the yeast cell wall. Upon disruption of yeast cells the HBsAg can be recovered as a heterogeneous mixture of particles which have some of the biochemical, biophysical and immunologic properties of serum-derived particles (Valenzuela et al., 1982; Hitzeman et al., 1983; Miyanohara et al., 1983). The antigenicity of HBsAg either in a crude yeast cell extract or as purified particles was low (Hitzeman et al., 1983; McAleer et al., 1984). However, vaccination of chimpanzees with the yeast-derived particles did protect the animals from infection with hepatitis B virus.

4. Expression of viral polypeptides in mammalian cells

Viruses, as obligate parasites, depend entirely upon the host cell's biosynthetic machinery for functions such as protein synthesis, carbohydrate addition and processing, post-translational modifications and transport of membrane proteins to the cell surface. Most of the viruses which will be discussed are pathogenic to mammalian cells and can be propagated in several different cell lines with similar biosynthetic results. In certain cases there are differences in the sialic acid content of viral membrane glycoproteins depending on the host cell, but for the most part the oligosaccharide content and processing is determined by the glycoprotein itself (Patzer et al., 1979). On the basis of these observations, it would be expected that the expression of viral polypeptides in mammalian cells would result in the proper synthesis, processing and compartmentalization.

4.1. Expression of hepatitis B surface antigen

The synthesis and secretion of HBsAg from mammalian cells in culture has been accomplished by several groups using either the entire HB virus genome or fragments of it which contain the entire transcription unit for HBsAg (Hirschman et al., 1980; Dubois et al., 1980; Christman et al., 1982; Stenlund et al., 1983; Wang et al., 1983), or using the SV-40 early or late transcriptional control elements (Moriarty et al., 1981; Liu et al., 1982; Crowley et al., 1983). There appeared to be no difference in the HBsAg polypeptides synthesized, regardless of the transcriptional control elements or whether the HBV sequences were stably integrated into the cellular genome (Dubois et al., 1980; Christman et al., 1982) or replicated as extrachromosomal plasmids (Wang et al., 1983; Stenlund et al., 1983).

The assembly and secretion of HBsAg 22 nm lipoprotein particles is a complex process which is very slow relative to the transport of other viral envelope proteins to the cell surface, e.g. the vesicular stomatitis virus G protein (Fig. 11.1; Patzer et al., 1984). The morphogenesis of the 22 nm particle involves multiple post-translational events such as the glycosylation of only one of two identical polypeptides (Peterson, 1981), assembly of HBsAg into 22 nm lipoprotein filaments or spherical particles and secretion of the particles from the cell. All of the post-translational events occur even though the only HBV DNA sequences incorporated into the cells were

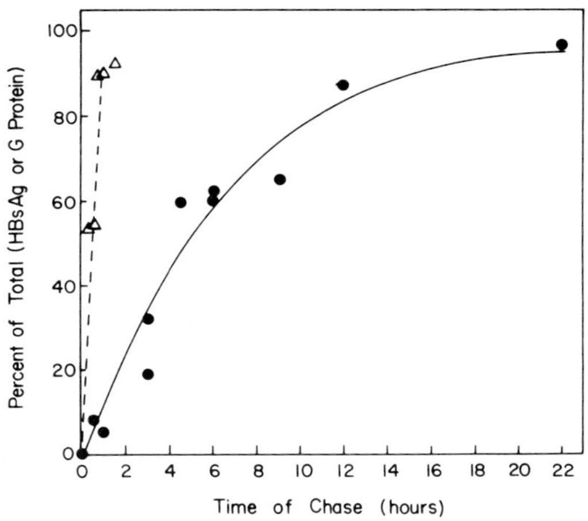

Fig. 11.1. The kinetics of secretion of HBsAg (•) and G protein of vesicular stomatitis virus (△). Cells were pulse-labeled with ^{35}S-methionine and chased in unlabeled medium containing 2 mM methionine. Radiolabeled HBsAg or G protein were immunoprecipitated and the precipitates were analyzed by sodium dodecyl sulfate polyacrylamide gel electrophoresis. The radiolabeled HBsAg or G polypeptide gel bands were quantified by scanning densitometry and the amount in the plasma membrane (G) or secreted into the media (HBsAg) was calculated as a percent of the total.

those encoding the mature fully processed form of HBsAg (Liu et al., 1982; Crowley et al., 1983), suggesting that all of the information necessary for these post-translational events is embodied within the structural sequences of HBsAg.

HBsAg particles secreted by mammalian cells in culture are biochemically, biophysically and morphologically indistinguishable from serum-derived HBsAg particles (Dubois et al., 1980; Moriarty et al., 1981; Christman et al., 1982; Liu et al., 1982; Stenlund et al., 1983; Wang et al., 1983). In addition, immunological analyses suggest that cell culture and serum-derived HBsAg are very similar antigenically. Cell culture-derived HBsAg is immunoprecipitated by antibodies raised against serum-derived HBsAg (Moriarty et al., 1981). More importantly, cell culture-derived HBsAg competes for all of the antigenic determinants on serum-derived HBsAg that bind to human antibodies to Heptavax B® (Fig. 11.2; Patzer et al., 1984a). The immunogenic potency of the cell culture-derived HBsAg has been established in mice (Table 11.2) and chimpanzees (Patzer et al., 1984a). The dose at which 50% of mice become seropositive is lower for cell culture than for serum-derived HBsAg (Table 11.2). In addition, each of six chimpanzees became seropositive for anti-HBs antibodies after only one 20 μg injection of cell culture-derived HBsAg and all had anti-HBs titers greater than 5000 (Ausab units, Abbott Laboratories) after two injections (Patzer et al., 1984a). These results establish that cell culture-derived HBsAg is very immunogenic and may provide a less expensive alternative to the serum-derived vaccine while lacking many of its perceived risks.

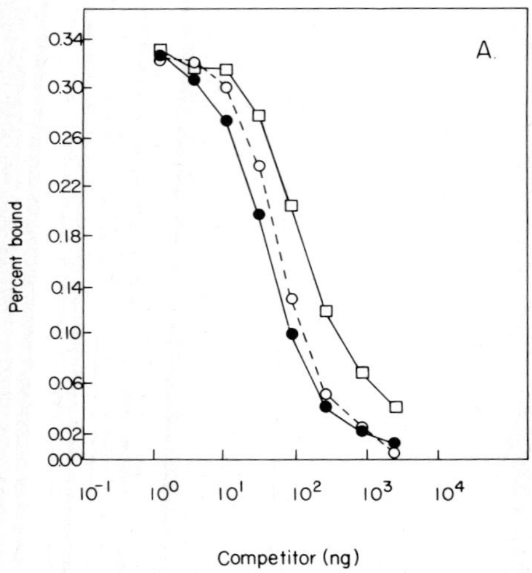

Fig. 11.2. Competition curves obtained using two human anti-Heptavax B antisera (panel A or B) and ^{125}I-HBsAg (serum-derived, ad subtype). The competitors were the ad subtype of cell culture-derived HBsAg (○), the ad subtype of serum-derived HBsAg (●) or the ay subtype of serum-derived HBsAg (□). The liquid phase competition assays contained ^{125}I-serum-derived HBsAg, unlabeled competitors and human anti-Heptavax B® antisera in the initial incubation (16–18 hours at 4° C). Antigen-antibody complexes were subsequently precipitated by incubating for 2 hours at room temperature with goat antihuman antibody. The percent of total ^{125}I-HBsAg precipitated was plotted as a function of the amount of competitor added.

TABLE 11.2

Mouse potency assay[a]

Dose (μg)	Heptavax B®	Cell culture vaccine
	Seroconversion[b]	
0.039	5	75
0.156	55	95
0.625	85	100
2.50	90	100
10.00	90	100
	ED_{50}(μg)	
	0.20	<0.039

[a] Mice were injected intraperitoneally with 1 ml of vaccine containing an aluminum gel adjuvant. 28 days after injection the mice were exsanguinated and the plasma fraction was tested for the presence of anti-HBs antibodies by the Ausab radioimmunoassay (Abbott Laboratories).
[b] % Seroconversion = (seropositive mice/total mice) × 100, with 20 mice per group.

4.2. Expression of viral envelope glycoproteins

Several viral envelope glycoproteins have been expressed in mammalian cells under the control of SV-40 promoters. These include the hemagglutinin (HA) and neuraminidase (NA) proteins of influenza (Gething and Sambrook, 1981; Sveda et al., 1982; Hartman et al., 1982; Markoff et al., 1984), the glycoprotein G of vesicular stomatitis virus (VSV) (Rose and Bergmann, 1982) and the glycoprotein D (gD) of herpes simplex virus type 1 (HSV-1) (Berman et al., 1983; Lasky et al., 1984). In each instance the mature form of the protein was expressed on the cell surface just as the authentic envelope protein during a virus infection (Gething and Sambrook, 1981; Sveda et al., 1982; Rose and Bergmann, 1982; Berman et al., 1983; Markoff et al., 1984). Based on their migration rate in SDS-polyacrylamide gels, the recombinant HA and NA of influenza and G protein of VSV appear to be glycosylated to the same extent as their authentic viral counterparts (Gething and Sambrook, 1981; Sveda et al., 1982; Rose and Bergmann, 1982). The recombinant gD of HSV-1 expressed in Chinese hamster ovary (CHO) cells appeared by SDS-PAGE analysis to be larger than gD synthesized during HSV infection of human embryonic lung or human lung carcinoma cells (Berman et al., 1983). Similar size differences in HSV glycoproteins have been ascribed to differences in glycosylation, which varies depending on the infected host cell (Pereira et al., 1981). The rates of glycosylation and transport to the cell surface for the recombinant HA, G and gD glycoproteins mimicked the rates for the authentic glycoproteins during the respective virus infections, indicating that these characteristics are independent of the other viral-encoded proteins (Gething and Sambrook, 1982; Rose and Bergmann, 1982; Berman et al., 1983).

The recombinant influenza HA and NA glycoproteins expressed on the cell surface displayed the same biological activities as the virion-associated glycoproteins (Gething and Sambrook, 1981; Sveda et al., 1982; Markoff et al., 1984). Immunologically, each of the full-length recombinant proteins expressed on the surface of mammalian cells retained their native antigenic determinants since they were immunoprecipitated by antibodies to their authentic counterparts. Further evidence for the native conformation of the HSV gD was demonstrated by its reactivity with a battery of rodent and human antisera to native proteins as well as mouse monoclonal antibodies to native gD, which included a neutralizing monoclonal antibody (Berman et al., 1983). The potential efficacy of the cell surface-expressed gD was provided by immunogenicity studies in mice that demonstrated protection against HSV infection (L. Lasky, personal communication).

A more feasible system for vaccine production from mammalian cells would be one in which the protein is secreted into the cell culture medium. Theoretically, this could be accomplished by removing the DNA sequences encoding the hydrophobic domain of the glycoprotein which spans the plasma membrane (transmembrane region) and the charged carboxy terminal region of the protein which presumably serves to anchor the glycoprotein in the membrane. Expression of the glycoprotein gene devoid of these sequences should result in secretion from the cell of a truncated protein that presumably would retain most or all of its native antigenic determinants. The conversion from membrane-bound to secreted form occurs naturally with immunoglobulins (Alt et al., 1980; Early et al., 1980). The secreted form of immunoglobulin retains all of the antigen binding characteristics of the membrane-bound form, but lacks the transmembrane hydrophobic region and the charged carboxy-terminal anchor. Purification of a secreted protein for a vaccine would not require lysis of the cells or separation of the immunogen from a large number of other cellular proteins. The only drawback to using a monomeric secreted protein is that it may exhibit reduced immunogenicity (Balcarova et al., 1981), although this probably can be overcome by using more antigen or through the use of adjuvants.

The HA of influenza, G protein of VSV and gD of HSV-1 have all been expressed in the secreted or truncated form lacking the hydrophobic transmembrane domain and the carboxy-terminal anchor. In each case the truncated form was secreted from the cell at a 4 to 8-fold slower rate than the full-length protein reached the cell surface (Gething and Sambrook, 1982; Rose and Bergmann, 1982; L. Lasky, personal communication). This slower rate of export does not appear to have a measurable impact on the total amount of protein produced, since an unlimited quantity of protein can be secreted into the media, whereas there is a finite amount of protein that can be incorporated into the plasma membrane.

Expression of truncated forms of membrane proteins does not always lead to secretion from the cell. A shift in reading frame upstream from the transmembrane region of the influenza virus HA resulted in a completely new set of amino acids at the carboxy terminus and termination of the protein nine amino acids before authentic HA (Sveda et al., 1984). This altered protein contained a putative hydrophobic transmembrane domain and was glycosylated normally, but accumulated intracellularly. Similarly, deletion of the nucleotide sequences encoding the amino-terminal 'signal sequence' of HA resulted in an intracellular localization and degradation of the altered protein (Gething and Sambrook, 1982; Sekikawa and Lai, 1983). The NA protein of influenza virus has been reported to have an inverted orientation relative to HA, and would therefore contain the hydrophobic transmembrane region at its amino terminus (Blok et al., 1982). Deletion of this region of NA, however, resulted in intracellular localization and lack of glycosylation, presumably because the amino-terminal signal sequence is also contained within this region (Markoff et al., 1984).

The secreted forms of HA of influenza, G protein of VSV and gD of HSV-1 all retained some of their native antigenic structure, as evidenced by their ability to be immunoprecipitated by antisera to the authentic proteins (Sveda et al., 1982; Gething et al., 1982; Lasky et al., 1984). Secreted gD also reacted with rabbit and human polyclonal anti-HSV antibodies and mouse monoclonal anti-gD antibodies (Lasky et al., 1984). Mice immunized with secreted gD developed antibodies which neutralized HSV-1 and HSV-2 in an in vitro plaque reduction assay, and protected the mice from a lethal challenge of HSV-1 or HSV-2 (Lasky et al., 1984). These efficacy experiments clearly demonstrate that mammalian cells in culture can be engineered to secrete a truncated viral envelope protein which can be used as a vaccine for protection against subsequent viral infections.

5. Conclusion

During the last century the medical community has successfully administered both live and killed vaccines to protect people from infectious disease. The track record of these vaccines has been extraordinarily good with respect to protection and duration of immunity. Unfortunately with some vaccines there is documented evidence for serious vaccine-associated complications and illness (Fulginiti, 1982). It is apparent that we now stand on the threshold of a new era in vaccine development where virtually any viral protein can be produced in vast quantities. These technological developments will make available new and more effective vaccines and may lead to the replacement of those that have associated risks. In addition, the new technology will provide an opportunity to study the structural and immunological details of proteins, which will help in the design of even more effective vaccines. It seems reasonable to expect that within this decade subunit vaccines will be available to fight our most devastating viral diseases.

6. Acknowledgments

The authors wish to thank P. Berman, D. Dowbenko, T.J. Harris, D. Kleid, L. Lasky and C. Simonsen for providing preprints of their manuscripts. We would also like to thank J. Arch for typing the manuscript.

7. References

Alt, F.W., Bothwell, A.L.M., Knapp, M., Siden, E., Mather, E., Koshland, M. and Baltimore, D. (1980) *Cell* 20, 293–301.

Bachrach, H.L., Moore, D.M., McKercher, P.D. and Polatnick, J. (1975) *J. Immunol.* 115, 1636–1641.

Bachrach, H.L., Morgan, D.O. and Moore, D.M. (1979) *Intervirology* 12, 65–72.

Balcarova, J., Helenius, A. and Simons, K. (1981) *J. Gen. Virol.* 53, 85–92.

Ballou, C. (1976) *Adv. Microbiol. Physiol.* 14, 93–158.

Berman, P.W., Dowbenko, D. and Lasky, L.A. (1983) *Science* 222, 524–527.

Bittle, J.L., Houghten, R.A., Alexander, H., Shinnick, T.M., Sutcliffe, J.G., Lerner, R.A., Rowlands, D.J. and Brown, F. (1982) *Nature* 298, 30–33.

Blok, J., Air, M., Laver, W.G., Ward, C.W., Lilley, G.G., Woods, E.F., Roxburgh, C.M. and Inglis, A.S. (1982) *Virology* 119, 109–121.

Bolivar, F., Rodriguez, R.L., Greene, P.J., Betlach, M.C., Heyneker, H.L., Boyer, H.W., Crosa, J.H. and Falkow, S. (1977) *Gene* 2, 95–102.

Boothroyd, J.C., Highfield, P.E., Cross. G.A.M., Rowlands, D.J., Lower, P.A., Brown, F. and Harris, T.J.R. (1981) *Nature* 290, 800–802.

Cashdollar, L.W., Esparza, J., Hudson, G.R., Chmelo, R., Lee, P.W.K. and Joklik, W.K. (1982) *Proc. Natl. Acad. Sci. USA* 79, 7644–7648.

Chance, R.E., Hoffman, J.A., Kroeff, E.P., Johnson, M.G., Schirmer, E.W., Bromer, W.W., Ross, M.J. and Wetzel, R. (1981) In *Proceedings of the Seventh American Peptide Symposium* (D.H. Rich and E. Gross, Eds.) Pierce Chemical Co., Rockville IL. pp. 721–738.

Charnay, P., Gervais, M., Louise, A., Falibert, F. and Tiollais, P. (1980) *Nature* 286, 893–895.

Christman, J.K., Gerber, M., Price, P.M., Flordellis, C., Edelman, J and Acs, G. (1982) *Proc. Natl. Acad. Sci. USA* 79, 1815–1819.

Crowley, C.W., Liu, C.C. and Levinson, A.D. (1983) *Mol. Cell. Biol.* 3, 44–55.

Davis, A.R., Bos, T., Ueda, M., Nayak, D.P., Dowbenko, D. and Compans, R.W., (1983) *Gene* 21, 273–284.

Dietzschold, B., Cox, J.H., Schneider, L.G., Tadeusz, J.W. and Koprowski, H. (1978) *J. Gen. Virol.* 40, 131–139.

Dubois, M.F., Pourcel, C., Rousset, S., Chany, C. and Tiollais, P. (1980) *Proc. Natl. Acad. Sci. USA* 77, 4549–4953.

Early, P., Rogers, J., Davis, M., Calame, K., Bond, M., Wall, R. and Hood, L. (1980) *Cell* 20, 313–319.

Edman, J.C., Hallewell, R.A., Valenzuela, P., Goodman, H.M. and Rutter, W.J. (1981) *Nature* 291, 503–506.

Emini, E.A., Kao, S.Y., Lewis, A.J., Crainic, R. and Wimmler, E. (1983) *J. Virol.* 46, 466–474.

Emtage, J.S., Tacon, W.C.A., Catlin, G.H., Jenkins, B., Porter, A.G. and Carey, N.H. (1980) *Nature* 283, 171–174.

Evans, D.M.A., Minor, P.D., Schild, G.S. and Almond, J.W. (1983) *Nature* 304, 459–462.

Ferguson, M., Minor, P.D., Magrath, D.L., Qui, Y.H., Spitz, M. and Schild, G.C. (1984) *J. Gen. Virol.* 65, 197–201.

Fujisawa, Y., Ito, Y., Sasada, R., Ono, Y., Igarashi, K., Marumoto, R., Kikuchi, M. and Sigino, Y. (1983) *Nucleic Acids Res.* 11, 3581–3591.

Fulginiti, V.A. (1982) *J. Pediatr.* 4, 487–494.

Gething, M.J. and Sambrook, J. (1981) *Nature* 293, 620–625.

Gething, M.J. and Sambrook, J. (1982) *Nature* 300, 598–603.

Goeddel, D.V., Heyneker, H.L., Hozumi, T., Arentzen, R., Itakura, K., Yansura, D.G., Ross, M.J., Miozzari, G., Crea, R. and Seeburg, P. (1979a) *Nature* 281, 544–548.

Goeddel, D.V., Kleid, D.G., Bolivar, F., Heyneker, H.L., Yansura, D.G., Crea, R., Hirose, T., Kraszewski, A., Itakura, K. and Riggs, A.D. (1979b) *Proc. Natl. Acad. Sci. USA* 76, 106–110.

Goeddel, D.V., Yelverton, E., Ullrich, A., Heyneker, H.L., Miozzari, G., Holmes, W., Seeburg, P.H., Dull, T., May, L., Stebbing, N., Crea, R., Maeda, S., McCandliss, R., Sloma, A., Tabor, J.M., Gross, M., Familletti, P.C. and Pestka, S. (1980) *Nature* 287, 411–416.

Goodman, H.M. and MacDonald, R.J. (1979) *Methods Enzymol.* 68, 75–90.

Gray, P.W., Leung, D.W., Pennica, D., Yelverton, E., Najarian, R., Simonsen, C.C., Derynck, R., Sherwood, P.J., Wallace, D.H., Berger, L.L., Levinson, A.D. and Goeddel, D.V. (1982) *Nature* 295, 503–508.

Hardy, K., Stahl, S. and Kupper, H. (1981) *Nature* 293, 481–483.

Hartman, J.R., Nayak, D.P. and Faree, G.C. (1982) *Proc. Natl. Acad. Sci. USA* 79, 233–237.

Hilleman, M.R., Buynak, E.B., McAleer, W.J., McLean, A.A., Provost, P.J. and Tytell, A.A. (1981) *Viral Hepatitis International Symposium* (W. Szmuness, H.J. Alter and J.P. Maynard, Eds.) The Franklin Press, Philadelphia.

Hirschman, S.Z., Price, P., Garfinkel, E., Christman, J. and Acs, G. (1980) *Proc. Natl. Acad. Sci. USA* 77, 5507–5511.

Hitzeman, R.A., Chen, C.Y., Hagie F.E., Patzer, E.J., Liu, C.C., Estell, D.A., Miller, J.V., Yaffe, A., Kleid, D.G., Levinson, A.D. and Opperman, H. (1983) *Nucleic Acids Res.* 11, 2745–2763.

Hitzeman, R.A., Leung, D.W., Perry, L.J., Kohr, W.J., Levine, H.L. and Goeddel, D.V. (1983) *Science* 219, 620–625.

Hitzeman, R.A., Chen, C.Y., Hagie, F.E., Lugovoy, J.M. and Singh, A. (1984) In *Recombinant DNA Products: Insulin-Interferon-Growth Hormone* (A.P. Bollon, Ed.) CRC Press, Boca Raton.

Icenogle, J., Hong, S.W., Duke, G., Gilbert, S., Rueckert, R. and Anderegg, J. (1983) *Virology* 127, 412–425.

Itakura, K., Hirose, T., Crea, R., Riggs, A., Heyneker, H.L., Bolivar, F. and Boyer, H. (1977) *Science* 198, 1056–1063.

Jackson, D.C., Brown, L.E., White, D.O., Dopheide, T.A.A. and Ward, C.W. (1979) *J. Immunol.* 123, 2610–2617.

Kaaden, O.R., Adam, K.H. and Strohmaier, K. (1977) *J. Gen. Virol.* 34, 397–400.

Kleid, D.G., Yansura, D., Small, B., Dowbenko, D., Moore, D.M., Grubman, J.J., McKercher, P.D., Morgan, D.O., Robertson, B.H. and Bachrach, H.L. (1981) *Science* 214, 1125–1129.

Kleid, D.G., Yansura, D.G., Dowbenko, D., Weddell, G.N., Hoatlin, M.E., Clayton, N., Shire, S.J., Bock, L.A., Ogez, J. Builder, S., Patzer, E.J., Moore, D.M., Robertson, B.H., Grubman, M.J. and Morgan, D.O. (1984) Journal In Press.

Kupper, H., Keller, W., Kurz, C., Forss, S., Schaller, H., Franze, R., Strohmaier, K., Marquardt, O., Zaslavsky, V.G. and Hofschneider, P.H. (1981) *Nature* 289, 555–559.

Laporte, J., Grosclaude, J., Wantyghem, S., Bernard, P. and Rouse, P. (1973) *C.R. Acad. Sci.* 276, 3399–3402.

Lasky, L.A., Dowbenko, D., Simonsen, C.C. and Berman, P.W. (1984) *Biotechnology* 2, 527–532.

Liu, C.C., Yansura, D. and Levinson, A.D. (1982) *DNA* 1, 213–221.

Makoff, A.J., Paynter, C.A., Rowlands, D.J. and Boothroyd, J.C. (1982) *Nucleic Acids Res.* 10, 8285–8295.

Markoff, L., Lin, B.C., Sveda, M.M. and Lai, C.J., (1984) *Mol. Cell Biol.* 4, 8–16.

McAleer, W.J., Buynak, E.B., Maigetter, R.Z., Wampler, D.E., Miller, W.J. and Hilleman, M.R. (1984) *Nature* 307, 178–181.

Meleon, R.H., Rowlands, D.J. and Brown, F. (1979) *J. Gen. Virol.* 45, 761–763.

Meleon, R.H., Briaire, J., Woortmeyer, R.J. and Van Zaane, D. (1983) *J. Gen. Virol.* 64, 1193–1198.

Minor, P.D., Schild, G.C., Bootman, J., Evans, D.M.A., Ferguson, M., Reeve, P., Spitz, M., Stanway, G., Cann, A.J., Hauptmann, R., Clarke, L.D., Mountford, R.C. and Almond, J.W. (1983) *Nature* 301, 674–679.

Mishiro, S., Imai, M., Takahashi, K., Machida, A., Gotanda, T., Miyakawa, Y. and Mayumi, M. (1980) *J. Immunol.* 124, 1589–1593.

Miyanohara, A., Toh-E, A., Nozaki, C., Hamada, F., Ohtomo, N. and Matsubara, K. (1983) *Proc. Natl. Acad. Sci. USA* 80, 1–5.

Moore, D.M. (1983) *Genetic Engineering: Applications to Agriculture.* Beltsville Symposium 7. (L.D. Owens, Ed.) Rowman and Allanheld, Totowa. pp. 89–106.

Moriarty, A.M., Hoyer, B.H., Shih, J.W.K., Gerin, J.L. and Hamer, D.H. (1981) *Proc. Natl. Acad. Sci. USA* 78, 2606–2610.

Novick, P., Ferro, S. and Schekman, R. (1981) *Cell* 25, 461–469.

Olson, K.C., Fenno, J., Lin, N., Harkins, R.N., Snider, C., Kohr, W.H., Ross, M.M., Fodge, D., Prender, G. and Stebbing, N. (1981) *Nature* 293, 408–411.

Pasek, M., Goto, T., Gilbert, W., Zink, B., Schaller, H., MacKay, P., Leadbetter, G. and Murray, K. (1979) *Nature* 282, 575–579.

Patzer, E.J., Wagner, R.R. and Dubovi, E.J. (1979) *CRC Crit. Rev. Biochem.* 165–217.

Patzer, E.J., Simonsen, C.C., Nakamura, G.R., Hershberg, R.D., Gregory, T.J. and Levinson, A.D. (1984a) *The Proceedings of the 1984 Viral Hepatitis Symposium.*

Patzer, E.J., Nakamura, G.R. and Yaffe, A. (1984b) *J. Virol.* 51, 346–353.

Pereira, L., Dondero, D., Norrild, B. and Roizman, B. (1981) *Proc. Natl. Acad. Sci. USA* 78, 5202–5206.

Peterson, D.L. (1981) *J. Biol. Chem.* 256, 6975–6983.

Pfaff, E., Mussgay, M., Bohm, H.O., Schulz, G.E. and Schaller, H. (1982) *EMBO J.* 7, 869–874.

Robertson, B.H., Moore, D.M., Grubman, M.J. and Kleid, D.G. (1983) *J. Virol.* 46, 311–316.

Roizman, B., Warran, J., Thuming, C., Fanshaw, M., Norrild, B. and Meignier, B. (1982) In *Herpes Virus of Man and Animal: Standardization of Immunological Procedure* (M. Bonneau, W. Hennessen and S. Karker, Eds.) pp. 287–309.

Rose, J.K. and Bergmann, J.E. (1982) *Cell* 30, 753–762.

Sekikawa, K. and Lai, C.J. (1983) *Proc. Natl. Acad. Sci. USA* 80, 3563–3567.

Skelly, J., Howard, C.R. and Zuckerman, A.J. (1979) *J. Gen. Virol.* 44, 679–689.

Stenlund, A., Lamy, D., Moreno-Lopez, J., Ahola, H., Pettersson, U. and Tiollais, P. (1983) *EMBO J.* 5, 669–673.

Strohmaier, K., Franze, R. and Adam, K.H. (1982) *J. Gen. Virol.* 59, 295–306.

Sveda, M.M., Markoff, L.J. and Lai, C.J. (1982) *Cell* 30, 649–656.

Sveda, M.M., Markoff, L.J. and Lai, C.J. (1984) *J. Virol.* 49, 223–228.

Tiollais, P., Charnay, P., and Vyas, G.N. (1981) *Science* 213, 406–411.

Valenzuela, P., Medina, A., Rutter, W.J., Ammerer, G. and Hall, B.D. (1982) *Nature* 298, 347–350.

Van der Werf, S., Dreano, M., Bruneau, P., Kopecka, H. and Girard, M. (1983) *Gene* 23, 85–93.

Wang, Y., Stratowa, C., Schaefer-Ridder, M., Doehmer, J. and Hofschneider, P.H. (1983) *Mol. Cell Biol.* 3, 1032–1039.

Watson, R.J., Weis, J.H., Salstrom, J.S. and Enquist, L.W. (1982) *Science* 218, 381–384.

Weis, J.H., Enquist, L.W., Salstrom, J.S. and Watson, R.J. (1983) *Nature* 302, 72–74.

Williams, D.C., Van Frank, R.M., Muth, W.L. and Burnett, J.P. (1982) *Science* 215, 687–688.

Wilson, T. (1984) *Biotechnology* 29–39.

Yelverton, E., Norton, S., Obijeski, J.F. and Goeddel, D.V. (1983) *Science* 219, 614–620.

Eds. M.H.V. van Regenmortel and A.R. Neurath
Immunochemistry of viruses. The basis for serodiagnosis and vaccines.

Inactivated and live vaccines 12

PHILIP A. BRUNELL

Department of Pediatrics, The University of Texas Health Science Center, 7703 Floyd Curl Drive, San Antonio, TX 78284, U.S.A.

The impact of virus diseases can be modified by prior exposure to the offending agent or one of its components. It is generally recognized that childhood diseases, e.g. measles or chickenpox, generally occur once and that the individual does not develop a similar illness following a subsequent exposure. Immunization with an appropriate vaccine also can modify or prevent illness. Vaccination, which is the term often used synonymously with immunization, refers to what is generally accepted to be the first vaccine, vaccinia. It is somewhat ironic that centuries later we return to vaccinia virus with a new strategy, namely that we can insert a DNA fragment of another virus into its genome and have vaccinia virus provide a vector for this inserted genetic material. The host cell infected with this chimeric virus then makes proteins coded by the DNA we choose to insert as well as vaccinia-directed proteins (Smith et al., 1983).

1. Vaccine strategy

The strategy for prevention of viral disease will vary for different agents. In general it will be dependent upon the pathogenesis and epidemiology of the virus and the options for production of vaccines. The concept of eradication by immunization of a sufficient proportion of the population to prevent person to person transmission is appropriate for certain diseases, e.g. smallpox. When an insect vector is involved in viral transmission, its elimination is paramount with immunization playing a secondary role, e.g. yellow fever. To prevent congenital rubella, some countries have elected to attempt to eradicate rubella by universal immunization of all children together with protecting women of childbearing age against infection, while others have chosen only the latter. Prevention of diseases with a relatively low risk of exposure, e.g. rabies, is often accomplished following a bite of an animal suspected of being rabid, whereas those at higher risk of exposure, e.g. veterinarians, may be immunized before exposure is likely to occur.

The pathogenesis of the infection may determine the vaccine strategy. Poliovirus produces an enteric infection that occasionally causes paralysis when the virus reaches the central nervous system via the blood. By parenteral injection of killed vaccine, one can raise antibody which will neutralize virus in the blood and prevent paralysis. In fact, killed antigen injected parenterally also might stimulate antibody-producing cells to produce secretory antibody in the intestinal tract. Orally administered live attenuated vaccine acts directly on those cells to produce intestinal immunity. Thus, the likelihood of infection of the intestinal tract by wild polio virus is diminished, and its transmission reduced. In addition, some of the locally produced antibody appears in the blood stream and can also prevent central nervous system involvement.

To effectively prevent disease, a vaccine must stimulate the relevant components of the host's immune response repertoire. The presence of antibody appears to correlate well with protection against certain viruses. Thus a vaccine designed to raise antibody against such a virus can be expected to prevent or modify an infection caused by this agent. Traditionally, serum antibody has been measured to evaluate vaccine efficacy. It is clear, however, that some diseases, e.g. zoster, may occur in the face of serum antibody (Brunell et al., 1975). It is recognized, moreover, that the depressed cellular immunity presages the onset of zoster and the frequency of this condition is increased in those with decreased cell-mediated responses (Arvin et al., 1983; Meyers et al., 1980). A strategy for the prevention of zoster, therefore, would probably require a vaccine that would stimulate the relevant cell-mediated response. It is notable that initial infection with varicella-zoster virus can be prevented by antibody (Brunell et al., 1969), whereas recrudescence, i.e. zoster, is prevented by another component of the host response repertoire.

An immunization plan must take into account not only the vaccine, but also the recipient. Although most vaccines are used in the normal host, it is often the immunocompromised, the very young or the aged who suffer most from certain viral infections. Successful immunization of the immunocompromised may be difficult or impossible because of their inability to respond to antigenic stimulation. What is more, live virus vaccine replication may be unrestrained and result in serious injury or even death of the vaccinee.

Early immunization is desirable to ascertain that infants are protected against infection prior to exposure. In many instances, this cannot be accomplished effectively because maternal antibody may neutralize vaccine virus and prevent successful immunization. Thus, immunization with parenterally administered live vaccines are usually deferred until passively acquired maternal antibody is catabolized.

Early exposure to some antigens has occasionally diminished responses to reimmunization (Murphy et al., 1980). In addition, some immunizations have tended to sensitize rather than protect against infection (Fulginity et al., 1967).

2. Types of vaccines

Vaccine preparations are either live or inactivated. Live attenuated viruses are designed to infect the vaccinee in a manner that will result in less morbidity than the natural infection, and yet produce immunity simulating that produced by the disease itself. This approach has obvious advantages as well as disadvantages. Conceptually, it is appealing in that the relevant components of host response are likely to be stimulated and the vaccine, as in the case of oral poliovaccine, can often be administered by the natural route. Most important, live vaccines usually produce permanent immunity. Vaccinia, however, required periodic reimmunization to assure protection (World Health Organization, 1972). Reinfection of those immunized with live vaccines does occur (Krugman, 1971), but is asymptomatic. Thus except for vaccinia, a single successful immunization with a live virus vaccine produces permanent protection against disease.

It must be emphasized that live vaccines produce *less* morbidity than the natural infection. Administration of any live vaccine is associated with some risk. Injury from immunization is most likely to occur when live vaccine virus inadvertently infects the immunocompromised. This may occur if they inadvertently receive live vaccines or if they are exposed to persons who have been given certain live vaccines. Poliovaccine virus is excreted in the stool. Immunocompromised contacts infected with poliovaccine virus from contacts have developed paralysis. Vaccinia virus has also been shown to spread from the vaccinee to contacts with eczema, producing eczema vaccination. There is always the concern that repeated passage of vaccine virus through the human host could result in reversion of an attenuated vaccine to greater virulence. The instability of oral poliovaccine in human passage, for instance, is well recognized (Nottay et al., 1981). In addition, the presence of antibody may neutralize live vaccine virus. Live vaccines should not be given within two months following the administration of immunoglobulin, and their administration must be deferred until such time as maternal antibody will not interfere with immunization, usually after the first birthday. Finally, the viability of live virus vaccines must be maintained. Problems have been encountered, particularly in tropical areas, because of lapses in the 'cold chain'.

There are a variety of killed vaccine preparations, the prototype of which is whole inactivated influenza vaccine. Perhaps the relative ineffectiveness of this vaccine, as compared to other live vaccines, illustrates some of the basic problems of killed vaccines. Whereas infection with a live vaccine presents the host with all of the relevant antigens, killed vaccines do not replicate and thus may provide the vaccinee with a restricted selection of antigens. Thus, the search for a better influenza vaccine has involved the search for relevant antigens such as hemagglutinins, neuraminidase and others.

In some instances the inactivation process had resulted in the selective destruction of specific viral antigens. In the case of measles vaccine, the destruction of the fusion protein in the process of producing a killed antigen may have been the cause of the recipients of this preparation developing 'atypical measles' when subsequently exposed to wild measles virus. Infection of these individuals produced an accelerated and exaggerated response to other viral components, yet failed to prevent virus replication, presumably because of the absence of antifusion protein antibody (Choppin et al., 1981).

Because they are not infectious, killed vaccines are ideal for use in the immunocompromised host and there is no concern about spread of vaccine to contacts. Since these vaccines do not replicate they can be given in the presence of passively acquired antibody. In contrast to live vaccines, however, immunity is generally not long lasting; repeated injections are generally required to maintain immunity as well as to evoke an initial immune response.

Live viral vaccines are expected to replicate and thus amplify the amount of material administered, whereas relatively large quantities of killed antigen must be inoculated. This generally increases their cost and their reactogenicity, particularly of first generation killed vaccines. There is a need to develop more purified and cheaper vaccines.

The first generation killed vaccines were produced as live vaccines, namely by propagating the agents in small animals, in eggs or in tissue culture and then inactivating them. Adjuvants were usually added to enhance the immune response. Subsequently, vaccines were concentrated to increase antigenicity and purified to remove host cell components. The inability to propagate hepatitis B virus in traditional host systems necessitated a new approach, namely the use of blood from human hepatitis B carriers as a source of antigen for the production of this vaccine.

The ability to clone DNA fragments into hosts capable of synthesizing relevant antigens and the production of monoclonal antibodies against epitopes on these antigens have provided the potential to produce large quantities of highly purified antigens against specific viral components. The cloning of DNA in vaccinia virus, which could be used as a vector to transmit the genetic information needed to create relevant antigens by the vaccinees own cells provide additional possibilities for vaccine production. Finally, the synthesis of chemically characterized antigens and their coupling to carriers provides yet another novel approach.

It is essential that vaccines be safe and of consistent potency. Live vaccines can be standardized using an agreed assay of infectious units. Each batch must then contain sufficient infectivity to immunize a vaccinee using limits of potency established in vaccine trials. The potency of killed antigens must be established by experimentally determining the amount of antigen required to evoke an immune response. Ideally, this should be expressed in weight of the chemically defined antigen. Recently, weight rather than hemagglutinin (HA) units has been used to define the potency of influenza vaccine. The older (HA) assay varied depending upon the dispersion of the antigen and other factors. In addition to the quantity of killed antigen, the characteristics of the antigen, e.g. the D antigen of polio or the hemagglutinin of influenza, may be specified.

For efficient use of vaccines, it is desirable to immunize against multiple agents at a single visit. These may be given in a single preparation, e.g. measles, mumps and rubella, or separate preparations given at the same visit, e.g. poliovaccine given at the same visit as diphtheria, tetanus and pertussis. In combining immunizing agents it must be ascertained that the morbidity is not increased and that there is no interference between the vaccines. It is recognized, for instance, that yellow fever and cholera vaccines may reduce the antibody response to the other component if given simultaneously (Felsenfield et al., 1973).

Safety of vaccines must be assured. This is ascertained during development of vaccines by rigid control of each lot that is used following licensure and by a surveillance of reactions to vaccines already in use. The latter is particularly important to detect adverse effects that are too infrequent to have been found in initial evaluations, e.g. Guillian-Barre syndrome following swine influenza vaccine. In addition, some adverse reactions may be recognized only years after the introduction of certain vaccines, e.g. atypical measles in killed measles virus recipients. It is recognized, for instance, that some early lots of poliovaccine were later found to contain SV40, a virus which was subsequently found to produce tumors in primates. A surveillance of recipients of these early lots has failed to reveal an increase in tumors (Mortimer et al., 1981) Now, each lot of vaccine must be tested for the presence of adventitious agents and for evidence of oncogenic potential. In addition, vaccines cannot be produced in cell substrates derived from tumors. Finally, each lot must be carefully checked for sterility and inactivated vaccines tested to ascertain that there are no residual live viruses.

The effectiveness of a vaccine is judged by whether vaccinees are protected when they have an exposure intimate enough to cause infection of an unimmunized individual. There are many indirect measurements of protection, e.g. declining rates of disease or antibody determinations. The former may in part reflect decreased opportunities for exposure. Despite the fact that the incidence of measles and rubella are at an all-time low, epidemics among college students are still being reported. Presumably, these individuals escaped infection during childhood because of lack of exposure. When an adequate number of these susceptibles come together and are exposed epidemics occur. The reasons for these individuals being unprotected are multiple. Even under ideal circumstances, there will be some vaccine failures as no immunizing agent protects all vaccinees. It has been claimed that under routine conditions of use, vaccines may not achieve the success claimed for them during initial trials under experimental conditions (Balfour et al., 1978). Indeed, improper storage of vaccine has been cited as a cause of vaccine failure (Lerman and Gold, 1971). There has been some concern about the duration of immunity conferred by measles vaccine, but studies have indicated that the failure rate does not increase with greater intervals since immunization (Shasby et al., 1977).

Antibody response following immunization is the putative measure to estimate vaccine efficacy. Unfortunately, the assays used have seldom been standardized for sensitivity and specificity. Instead, assays for each virus generally have been adopted by acclamation. The neutralization test is frequently used as the standard against which assays are compared. Apparently the reverence for this standard may be ill founded, as antibody which fails to neutralize in vitro may protect in vivo (Balachandran et al., 1982). Furthermore, the sensitivity and specificity of neutralization tests can be altered by simply changing the amount of virus used in the test (Shehab et al., 1983).

3. Specific Vaccines

3.1. Vaccinia

Almost 200 years ago, Dr. Edward Jenner inoculated a young boy, James Phipps, with pus from cowpox obtained from a lesion on the finger of a dairymaid Sarah Nelmes. He subsequently inoculated the boy with smallpox pus and showed he was protected. Thus it was demonstrated that a human could be protected by virus derived from a cow and the virus attenuated in the process. Routine immunization of infants in the United States lead to the eradication of the disease, the last case being recorded in 1949. Subsequently, a world-wide program was undertaken which appears to have eliminated smallpox from this planet. The campaign involved routine immunization and intensive surveillance of susceptible cases, combined with concentrated vaccination campaigns in areas where cases were recognized (Henderson, 1975). Smallpox vaccine has been banned in the United States, except for those working with the virus in the laboratory. Careful laboratory containment of smallpox is essential, as research laboratories are probably the last remaining reservoir of virus, and laboratory-centered outbreaks have occurred (Breman and Arita, 1980).

The smallpox vaccine used in the United States at the time of eradication was similar to that originally developed in that it was produced by scarification of the skin of calves. The major advances which facilitated the eradication program included the development of a lyophilized preparation which would retain infectivity even in tropical areas and a prong for application of the vaccine. The inoculation was done as in the earlier days by scarification of the skin.

Revaccination was recommended every seven years as it was recognized that immunity produced by a single vaccination was not lifelong. Attenuated cases had occurred in vaccinated individuals following exposure even within this interval (World Health Organization, 1972). The decision to discontinue routine vaccination against smallpox in the United States was based on an analysis which revealed that the morbidity from immunization was greater than one would have predicted from the occasional importation of a case of smallpox from an endemic area. This was based on the low probability of an importation causing secondary cases given the relatively minimal communicability of smallpox and the high rate of immunity in the United States at that time (Lane et al., 1971).

The side effects resulting from smallpox vaccination were due to inadvertent administration of the material to individuals at high risk, auto-inoculation, superinfection and effects which were intrinsic to the vaccine. The ease with which the vaccine virus could be auto-inoculated and spread to contacts was a major problem. Auto-inoculation into the eye, particularly by infants, caused corneal scarring. Bacterial infection of the vaccination site was a problem, particularly in areas of poor hygiene; infection with *Clostridium tetanus* occurred in tropical areas. Although vaccine was contraindicated in those with severe skin disease, e.g. eczema, or in those with immunodeficiencies, these individuals would occasionally be infected by coming in contact with vaccinees. Eczema vaccination was a serious complication of immunization. Of even greater gravity was vaccinia necrosis which occurred in individuals with cell-mediated defects. Occasionally, this would occur in individuals who were ill advisedly treated for recurrent herpes by smallpox vaccination. This practice, which is contraindicated, is one of the main reasons for removing the vaccine from general use. Other complications include erythema multiforme, encephalitis and generalized vaccinia (Goldstein et al., 1975).

There has been renewed interest in vaccinia virus as a vector for cloned DNA (Smith et al., 1983). The most immediate use for this would be the immunization of newborns to prevent perinatal transmission of hepatitis B. In many areas, e.g. the sub-Sahara region, where hepatitis is a major problem, there is extensive experience with the World Health Organization smallpox eradication program. This would appear to be an ideal site to utilize vaccinia as a vector for hepatitis virus DNA. The use of vaccinia as a vector does not require that immunity be produced against smallpox. Thus, one may be able to use strains of smallpox virus that were overattenuated or could be given parenterally. In addition, the smallpox DNA could be further manipulated to reduce virulence of the vaccinia virus. Tissue culture-propagated virus might conceivably reduce some side effects due to adventitious agents in the calf pulp preparation. The inadvertent use in individuals with underlying disease would still pose a hazard.

3.2. Poliovaccine

The introduction of inactivated vaccine in the United States in 1955 lead to a striking decline in the number of cases of paralytic poliomyelitis. There were more than

18 000 cases reported in 1954. Cases were still reported in individuals considered to be fully immunized, and despite the widespread use of the vaccine, localized epidemics still occurred. Inactivated vaccine was largely replaced by live attenuated poliovaccine (LPV) in early 1963. This resulted in the virtual disappearance of paralytic disease in the United States. An average of 13 cases per year have occurred during the past decade, approximately half of which are associated with the use of LPV. The risk of vaccine-associated cases of paralysis is estimated to be one case in a recipient for every 11.5 million persons vaccinated with oral poliovaccine (OPV), one case in a household contact for every 3.9 million vaccinated and one case in a community contact for every 22.9 million vaccinated (Nightingale, 1977).

In examining the cases of paralysis several categories are encountered (Immunizations Practices Advisory Committee, 1982a). There are cases occurring in immunocompromised individuals who received the vaccine inadvertently. Live vaccines are generally contraindicated for this group. Paralysis has occurred both in immunodeficient vaccine recipients and contacts of vaccine recipients. The cases in contacts again illustrate the fact that polio is an enteric infection; the virus, whether wild or attenuated, is excreted in the stool and may infect contacts. There are also sporadic cases and epidemics of polio in normal individuals. The epidemics have occurred in groups who, for religious reasons, rejected immunization. The occurrence of the epidemics indicates that wild polio virus is still present in the United States and reinforces the need for continued universal immunization. A large number of the endemic cases are due to type 3, which suggests that these too may be related to vaccine strains rather than wild types. The prevalent strains in unimmunized populations is more likely to be type 1, while types 2 and 3 LPV strain are the most common isolates from vaccine recipients. Finally, assignment of the etiology of these cases must await testing of the isolates by nucleic acid analysis and monoclonal antibody studies to determine whether they are wild or vaccine-derived. The use of these techniques has confirmed earlier studies as to the genetic lability of LPV fed to human subjects (Nottay et al., 1981).

There is renewed interest in the use of inactivated poliovaccine (IPV) as the number of cases of paralysis attributable to the use of LPV approaches or exceeds that resulting from natural infection. What is more, the newer IPV is far more antigenic than the preparations used in the United States in the 1950s. Indeed, only 2 doses of IPV given to children appears to produce antibody in almost all the recipients (Kew et al., 1981). Studies in the United States indicate comparable serologic response in children receiving OPV (McBean et al., 1984). Finally, OPV does not appear to be as effective in inducing serologic response in tropical countries as in more temperate areas (Horstman, 1982). In Scandinavia, however, the routine use of IPV appears to have resulted in the eradication of polio. A number of explanations have been proposed for the relatively poor performance of OPV in tropical areas. Some have suggested that problems with maintenance of the 'cold chain' has resulted in loss of potency of the vaccine. Polio virus is known to be a relatively stable virus, however, and the amount of virus in the vaccine preparation far exceeds the amount needed to immunize. Others have proposed that the prevalence of interfering enteroviruses are responsible for the poor responses. Finally, breast feeding by hyperimmune mothers has been suspected. It is recognized that breast feeding can interfere with immunization with LPV in the newborn period (Warren et al., 1964). LPV provides many advantages for use in developing countries, not the least of which are lower cost than IPV and the ease of administration.

OPV is generally administered as a 'trivalent' preparation. The quantities of each of the three strains is balanced to achieve the greatest likelihood of seroconversions with multiple strains. The strain most likely to interfere with the others, type 3, is present in the lowest titer. An interval of at least 6 weeks between doses is chosen to minimize interference between strains of poliovaccine virus. To assure that immunization against all of the strains is achieved, the vaccine is fed on multiple occasions. Two feedings of trivalent vaccine will usually produce immunity against all three strains in more than 90% of vaccine recipients (Cabasso et al., 1964). Ordinarily a third dose is given. For routine childhood immunization in the United States two or three doses are given at intervals of at least 6 weeks during the first 6 months of life. Additional doses are given during the second year of life and before entry to school.

3.3. Measles vaccine

Measles is a childhood disease that is highly contagious. Even uncomplicated measles is a severe illness. Complications include encephalitis, often with permanent brain damage, ear infections and pneumonia. Measles is a significant cause of childhood mortality in developing countries. The introduction of measles vaccine in the United States in 1963 was a welcome addition to the vaccine repertoire.

The initial measles vaccines were quite reactogenic, and for that reason were given at the same time in a separate site from measles immunoglobulin. The latter, containing a measured amount of measles antibody, was designed to attenuate the reactions caused by live measles vaccine (LMV) in a fashion similar to the use of immunoglobulin to attenuate natural measles prior to the advent of the LMV. The subsequent development of further attenuated LMV made the use of globulin unnecessary.

A killed measles vaccine (KMV), which was less reactogenic, was given as an alternative to LMV. It was found that when recipients of KMV were exposed to natural measles, they developed a severe illness called 'atypical measles' (Fulginiti et al., 1967). During the inactivation of measles virus it was found that the F (fusion) protein was destroyed by formalin. It was postulated that the absence of antibody to this protein permitted infection to occur with the overproduction of other measles virus components which participated in the formation of immune complexes (Choppin et al., 1981). Although the use of KMV was discontinued in 1967, cases of atypical measles continued to occur decades later. Administration of LMV to previous recipients of KMV, although recommended, did not always prevent the occurrence of atypical measles (Fulginiti and Helfer, 1980).

Initially, LMV was given at 9 months of age until it was later learned that maternal antibody was still present in a substantial number of infants during the first year of life and could interfere with successful immunization by neutralization of vaccine virus. The inability to successfully immunize young infants was a problem, particularly in endemic areas where measles often is contracted at an early age; morbidity in those under a year, moreover, is quite high. Reimmunization of those immunized early was found to produce a poorer antibody response than a single immunization with LMV after a year of age (Wilkins and Wehrle, 1979). The actual proportion of infants seroconverting, however, was essentially the same (Murphy et al., 1983).

At the present time, a single injection of LMV is recommended after the first birthday. In areas of low incidence it is often given at 15 months of age, as some believe it may be slightly more effective than at one year (Brunell, 1978a). In the United States, LMV is usually given in combination with rubella and mumps in a single preparation. Although it had been suggested that measles vaccine, as ordinarily used in the community may be less effective than was originally reported in experimental trials (Balfour and Amren, 1978), sensitive serological techniques have revealed that the LMV currently in use produces a seroconversion in more than 98% of recipients (Brunell et al., 1983). The marked decline in the number of cases of measles attests to the effectiveness of the vaccine and immunization programs. Indeed, one plans for the eradication of measles in the United States. Measles remains a serious problem in developing countries. Administration of vaccine intranasally has been found to be highly effective and may be particularly useful in these areas because of ease of administration (Sabin et al., 1983).

In addition to a striking reduction in cases of measles, there has been a marked decline in the number of cases of subacute sclerosing pan-encephalitis (SSPE). This disease is a degenerative disease of the nervous system caused by measles virus. Affected children have usually had measles at a very young age. The latent period for development of SSPE is usually about seven years. Measles virus can be isolated by cocultivation of brain tissue of the victims of the disease (Horta-Barbosa et al., 1969). Defective synthesis of antibody against the matrix protein of measles virus appears to be found in these patients (Wechsler et al., 1979).

3.4. Rubella vaccine

Rubella is a relatively mild illness. When it occurs early in pregnancy, however, it can produce serious malformations in the developing fetus. Following the pandemic of 1964, it was estimated that between 20 000 and 40 000 infants were born in the United States alone with congenital rubella. Since the introduction of rubella vaccine in the United States in 1969, there has been a striking reduction in cases of both rubella and congenital rubella.

The vaccine is relatively innocuous in children, but tends to produce occasional joint involvement in adults (Weibel et al., 1972). It has been suggested that this results from immune complexes produced by immunization of individuals who are already immune (Tingle et al., 1983). There was concern about the use of live attenuated rubella vaccine in adult women for fear that vaccine virus might produce teratogenic effects in the fetus. It was well recognized, for instance, that HPV 77 and Cendehill-derived vaccine viruses could be isolated from fetuses of women who had undergone elective abortion after receiving rubella vaccine. Fortunately, women who elect to carry their pregnancies to term have not given birth to infants with malformations attributable to the vaccine. The RA 23/7-derived vaccine currently in use in the United States appears to be even less likely to infect the fetus than earlier vaccines (Centers for Disease Control, 1983). It is recommended, however, that women who are pregnant not receive vaccine and those who are immunized with rubella vaccine not become pregnant for three months after immunization.

It is well recognized that rubella vaccine virus can be recovered from the throat of some vaccinees. Despite this, vaccinees are not likely to spread vaccine virus to contacts (Hilleman et al., 1968b). For this reason, pregnancy is not deemed to be a contraindication to immunization of children in the same household. Vaccine virus is present in lower titers in the throats of vaccinees than is wild virus in the throats of patients with natural rubella. In addition, experiments attempting to administer early strains of rubella vaccine by the respiratory route were generally unsuccessful. Greater success has been achieved with RA27/3, the strain currently used to prepare rubella vaccine (Plotkin et al., 1973).

Rubella vaccine is given as a single dose. It is anticipated that this will produce lifelong protection against the development of rubella embryopathy, although reinfection of vaccinees as evidenced by rises in antibody do occur. In the United States it is recommended that rubella vaccine combined with mumps and measles vaccine in the same preparation be given to all children after the first birthday. In this way, it is hoped both to eradicate rubella by universal immunization and to prevent the occurrence of rubella in pregnancy by immunizing girls before they start childbearing.

Immunization of women who are susceptible to rubella either combined with a program of childhood immunization as in the United States or as a sole measure is used to prevent the birth of infants with congenital rubella. It is desirable not to immunize pregnant women with rubella vaccine. For this reason, efforts are made to immunize prior to the onset of puberty. In addition, serologic screening had been recommended as a routine at one time to avoid the immunization of pregnant women who may have in fact been immune. By not immunizing these women, one avoids the problem of anxiety associated with the inadvertent immunization of the occasional women who may be or become pregnant. The use of vaccines, e.g. RA 27/3, which are less likely to affect the fetus and the realization that fetal malformations, even with other vaccines, are unlikely (Centers for Disease Control, 1983) has led to rescission of the recommendation for routine serologic screening. Some physicians still elect to obtain blood at the time of immunization and test it only if the vaccinated woman should become pregnant during the next three months. If so, the serum can be tested and if the woman is found to have been immune at the time of immunization, anxiety about the effect of vaccine on the fetus will have been allayed.

3.5. Mumps vaccine

Mumps is a childhood disease characterized by paratoid swelling. Complications, e.g. deafness or encephalitis, are rare. The disease is more severe in adults; orchitis in males is a complication producing high morbidity. The introduction of a live vaccine in 1967 has led to a striking reduction of reported cases (Immunization Practices Advisory Committee, 1982c). Because the disease is relatively benign, most of the vaccine administered is given in combination with rubella and measles. It is recommended for children after the first birthday. Reactions to immunization are rare and immunity appears to persist indefinitely (Hilleman et al., 1968a). Many cases of parotoid swelling in those who have received mumps vaccine previously are due to causes other than mumps (Brunell et al., 1972). A killed vaccine was used in the United States at one time and is now used in Sweden (Ukkonen and Penttinen, 1981).

3.6. Rabies vaccine

Rabies vaccine was developed by Pasteur about a hundred years ago. His original vaccine was prepared by serial passage of virus in rabbits by intracerebral inoculation until the virus was 'fixed'. This term was used to distinguish it from 'street' or wild virus. The original brain vaccine was replaced by a duck embryo (DE) propagated product. The vaccine currently in use in the United States is prepared in human diploid cells and is inactivated.

The current vaccine is far more immunogenic than DE vaccine (Anderson et al., 1980; Immunization Practices Advisory Committee, 1980). Fewer doses are required for post-exposure prophylaxis, yet much higher antibody titers are achieved. Furthermore, the diploid vaccine produces far less morbidity than the DE vaccine. Local reactions and fever associated with both vaccines were similar and relatively insignificant (Hafkin et al., 1978). Neurologic complications or anaphylaxis, particularly with brain-derived preparations, were a major concern, but are far less frequent with DE vaccine and rare with the diploid vaccine (Centers for Disease Control, 1980). The cost of the vaccine, however, is a major deterrent to more widespread use of the product prepared in diploid cells, particularly in developing countries.

Rabies vaccine is unique in that Pasteur used it initially for post-exposure prophylaxis. This is feasible as the incubation period of rabies is relatively long. The vaccine was used for many years without objective evidence that it was effective, despite the fact that the original vaccine was associated with severe morbidity. It was probably not appreciated, moreover, that bites of rabid animals do not always result in the victim developing rabies even in the absence of treatment (Gremliza, 1953).

Rabies virus spreads to the central nervous system from the site of the bite via peripheral nerves. An essential part of prevention of rabies is the inactivation of virus at the site of the bite as soon as possible. This is accomplished by thorough washing, preferably with alcohol solution or another agent that will inactivate rabies virus, and the infiltration of the wound with rabies immunoglobulin to neutralize as much rabies virus as possible. A human product has largely replaced equine material in the United States. Approximately half the dose is infiltrated around the wound and the remainder given intramuscularly. Vaccine-induced antibody will appear between 7 and 14 days following immunization. Failure to administer globulin has resulted in death despite the use of diploid vaccine (Devriendt et al., 1982). A dose is chosen that will not be excessive to avoid inhibiting the effect of vaccine, which is given at the same time at a different site. Doses of vaccine are then repeated 3, 7, 14 and 28 days later (Centers for Disease Control, 1980). Serologic testing is not required after administration of diploid vaccine because of the uniformly excellent response. It is recommended following duck embryo vaccine as approximately 20% of vaccine recipients may not have an adequate antibody response following completion of the recommended series of doses. Diploid vaccine is ordinarily given subcutaneously. Efforts have been made to give small amounts of the vaccine intracutaneously to decrease the cost of the vaccine (Ajjan et al., 1980).

Routine pre-exposure immunization of those considered to be at increased risk, e.g. veterinarians and dog catchers, is recommended. For this purpose only three doses are required. Boosters are recommended every two years or at the time of exposure. Serologic testing is recommended only for laboratory workers who are intensively exposed to rabies virus (Centers for Disease Control, 1980).

3.7. Influenza virus vaccine

Influenza produces severe respiratory disease, particularly in the very young and the elderly. It may be life threatening in the latter group as well as in those with underlying cardiovascular disease. Influenza virus spreads readily from person to person and epidemics are common. The virus is very likely to change antigenically and vaccines are most effective when they are closely related antigenically to the infecting strain. Pre-existing immunity from previous infection protects against reinfection by the same strain and to some extent by related strains. Changes in the hemagglutinin or neuraminidase of the virus usually presage the occurrence of epidemics or pandemics. Efforts are made to prepare vaccines containing the circulating virus types, matching them as closely as possible antigenically with the prevalent strains. Antigenic drift, i.e. more subtle changes within a given type, also occur for influenza A, e.g. from A/Brazil/11/78 to A England/333/80. Vaccines are reviewed yearly to match them as closely as possible to the strain most likely to be prevalent in the coming season.

In order to prepare vaccines that are current, a constant worldwide surveillance tracking the occurrence of various strains and types is necessary. Vaccines are prepared in embryonated eggs and sufficient lead time is needed to produce the appropriate vaccine for each season. Occasionally, as in the case of swine influenza in the 1970s, vaccine is prepared and an epidemic fails to materialize.

Although newer vaccines are far more purified, reactions to contaminating egg proteins occur occasionally in egg-sensitive individuals. Influenza vaccine also produces frequent but relatively mild reactions at the site of injection and low grade fever: significant reactions are rare. Guillian-Barre syndrome appeared to be associated with administration of swine influenza, but not other influenza vaccines (Centers for Disease Control, 1983).

Influenza vaccine is killed. It is produced as a whole virion or detergent-'split' vaccine. The former, although somewhat more immunogenic, is quite reactive in children and is therefore not recommended for those under the age of 13. High risk individuals above this age usually receive a single dose annually. Those 12 years and under receive two doses a month apart unless they received a similar vaccine the previous year, in which case they would receive a single dose. The quantity of antigen given is reduced in children under 3 years of age. There is little experience with influenza vaccine in infants under 6 months (Centers for Disease Control, 1983b).

Vaccine composition is updated annually. The current vaccines contain 15 μg each of influenza B, A(H3N2) and A (H1N1) types. The effectiveness of influenza vaccines is less than optimal. Efforts to improve influenza vaccines, including the use of live vaccines, are in progress (Kilbourne et al., 1974). Serologic response is arbitrarily judged by the proportion of vaccine recipients achieving a hemagglutination inhibition titer of 1:40 postimmunization. Actual protective effect has been difficult to assess precisely.

3.8. Hepatitis B virus vaccine

One of the causes of hepatitis is hepatitis B virus (HBV). Of the many etiologic agents, HBV is the most likely to cause chronic liver disease. This agent is transmitted by parenteral inoculation, very intimate contact, e.g. sexual, and from mother to infant at the time of delivery. On a global basis, perinatal transmission is probably the most important method of perpetuation of HBV. Infection, particularly early in life, may lead to a permanent carrier state.

An enormous amount of HBsAg, the surface component of the virus, can be demonstrated in the blood of carriers. HBsAg derived from the blood of carriers has been used to prepare HBV vaccine, as the virus has not been successfully propagated. The HBsAg is extracted from the blood by ultracentrifugation, treated with detergents and formalin and put through a variety of safety tests including inoculation of susceptible chimpanzees (Gerety et al., 1979). This method of production has been costly and alternative methods are being sought.

Because the original vaccine was derived from blood, there was concern about the possibility that adventitious agents, e.g. the organism causing AIDS, might be present in the vaccine. To date there is no evidence to support an increase in AIDS in HBV vaccine recipients. Indeed, the side effects following the use of HBV vaccine have been relatively few (Immunization Practices Advisory Committee, 1983a).

The vaccine has been used mainly in individuals at high risk of developing HBV infection. Because of the high cost, serologic screening is recommended when groups with a high probability of already being immune are to be immunized (Immunization Practices Advisory Committee, 1982b). A major need in underdeveloped areas which cannot be addressed because of the present cost of the vaccine is the protection of infants of carrier mothers. Mothers who are also HBsAg positive have a very high risk of transmitting the infection to their infants. Female infants will in turn eventually infect their infants. There is good evidence to suppose that early infection may also predispose the development of hepatocellular cancer.

Protection of infants is usually accomplished by a combination of passive immunization with hepatitis B immunoglobulin and active immunization with the vaccine. A number of regimens have been used quite successfully in preventing neonatal infection (Beasley et al., 1983). The vaccine has also been very effective in preventing infection in a variety of high risk groups. Patients on chronic hemodialysis are more difficult to immunize and are given double the usual dose of 40 μg. Immunization is accomplished by giving two doses one month apart and a third dose six months following the first (Krugman, 1983).

3.9. Miscellaneous vaccines

There are a variety of vaccines used under special circumstances, e.g. adenovirus vaccine for those entering the armed services and yellow fever vaccine for travels to endemic areas (Immunization Practices Advisory Committee, 1978). In addition, various countries have vaccines to address particular needs of their region, e.g. Japanase B encephalitis vaccine. Additional vaccines have been prepared for laboratory workers working with hazardous agents, e.g. certain arboviruses.

4. Experimental vaccines

A chickenpox vaccine propagated in human diploid cells was developed by Japanese workers and has been tested in both normal children (Takahashi et al., 1974) and children with leukemia (Brunell et al., 1982). It appears to be safe and fairly effective. Long-term observations are required to determine the duration of immunity conferred and the incidence of zoster following immunization (Brunell, 1978b).

Work has continued on the development of vaccines to other herpesviruses, e.g. cytomegalovirus and genital herpes. The cytomegalovirus vaccine has been tested in nursing personnel (Elek and Stern, 1974) and patients with renal transplants (Glazer et al., 1979). This vaccine is designed to prevent congenital infection in the former group and severe cytomegalovirus pneumonia in the latter. Herpes simplex virus vaccines have been constructed by genetic engineering for the prevention of genital herpes (Roizman et al., 1982). Formerly killed herpes vaccines were used to treat recurrent herpes without evidence of efficacy (Kernm and Schiff, 1964).

Vaccines against respiratory virus infections are under investigation. Although a variety of agents are being studied, there appears to be greatest need for a vaccine against respiratory syncytial virus, the cause of bronchiolitis in infancy. Although earlier killed vaccines appeared to sensitize rather than protect and earlier live vaccines had a tendency to revert to virulence, more recent vaccines show greater promise (Buynak, 1979).

Rotavirus is the major cause of diarrhea in infants. A live virus vaccine prepared from calf rotavirus has shown encouraging results (Vesikari et al., 1983). Additional efforts are being made to prepare a glycoprotein vaccine.

5. References

Ajjan, N., Soulebot, J.P., Trau, R. and Biron, G. (1980) *JAMA* 244, 2528–2531.

Anderson, L.J., Sikes, R.K., Langkop, C.W., Mann, J.M., Smith, J.S., Winler, W.G. and Deitch, M.W. (1980) *J. Infect. Dis.* 142, 133–138.

Arvin, A.M., Pollard, R.B., Rasmussen, L.E. et al. (1983) *J. Clin. Invest.* 65, 869–878.

Balachandran, N., Bacchetti, S. and Rawls, W.E. (1982) *Infect. Immun.* 37, 1132–1137.

Balfour, H.H. Jr. and Amren, D.P. (1978) *Am. J. Dis. Child.* 132, 573–577.

Beasley, R.P., Lee, G.C., Roan, C., Hwang, L., Lan, C., Huang, F. and Chen, C. (1983) *Lancet* ii, 1099–1102.

Breman, J.G. and Arita I. (1980) *N. Engl. J. Med.* 303, 1263–1272.

Brunell, P.A., Ross, A., Miller, L.H. and Kuo, B. (1969) *N. Engl. J. Med.* 280, 1191–1194.

Brunell, P.A., Brickman, A., Steinberg, S. and Allen, E. (1972) *Pediatrics* 50, 441–444.

Brunell, P.A., Gershon, A.A., Uduman, S.A. and Steinberg, S. (1975) *J. Infect. Dis.* 132, 49–54.

Brunell, P.A. (1978a) *Pediatrics* 62, 1038–1041.

Brunell, P.A. (1978b) *JAMA* 239, 1034–1035.

Brunell, P.A., Geiser, C., Shehab, Z. and Waugh, J.E. (1982) *Lancet* ii, 1069–1073.

Brunell, P.A., Weigle, K., Murphy, D., Shehab, Z. and Cobb, E. (1983) *JAMA* 250, 409–1412.

Buynak, E.B., Weibel, R.E., Carlson, A.L.J., McLean, A.A. and Hilleman, M.R. (1979) *Proc. Soc. Exp. Biol. Med.* 160, 272–277.

Cabasso, V.J., Nozell, H., Ruegsegger, J.M. and Cox, H.R. (1964) *N. Engl. J. Med.* 270, 443–446.
Centers for Disease Control (1980) *Morb. Mort. Weekly Rep.* 29, 609–610.
Centers for Disease Control (1983) *Morb. Mort. Weekly Rep.* 32, 429–436.
Choppin, P.W., Richardson, C.D., Merz, D.C. and Scheid, A. (1981) *Perspectives in Virology XI* (M. Pollard, Ed.) Alan R. Liss, New York. pp. 57–75.
Devriendt, J., Staroukine, M., Costy, F. and Vanderhaeghen, J.J. (1982) *JAMA* 248, 2304–2306.
Elek, S.D. and Stern, H. (1974) *Lancet* i, 1–5.
Felsenfeld, O., Wolf, R.H., Gyr, K., Grant, L.S., Dutta, N.K., Zarifi, A.Z. and Zafari, Y. (1973) *Lancet* i, 457–458.
Fulginiti, V.A. and Helfer, R.E. (1980) *JAMA* 244, 804–806.
Fulginiti, V.A., Eller, J.J., Donnie, A.W. et al. (1967) *JAMA* 202, 1075–1080.
Gerety, R.J., Tabor, E., Purcell, R.H. and Tyeryar, F.J. (1979) *J. Infect. Dis.* 140, 642–648.
Glazer, J.P., Friedman, H.M., Grossman, R.A. et al. (1979) *Ann. Intern. Med.* 91, 676–683.
Goldstein, J.A., Neff, J.M., Lane, J.M. and Koplan, J.P. (1975) *Pediatrics* 55, 342–347.
Gremliza, L. (1953) *Z. Tropenmed. Parasitol.* 4, 382–389.
Hafkin, B., Hattwick, M.A.W., Smith, J.S., Alls, M.E., Yager, P.A., Corey, L., Hoke, C.H. and Baer, G.M. (1978) *Am. J. Epidemiol.* 107, 439–443.
Henderson, D.A. (1975) *J. Clin. Pathol.* 28, 843–849.
Hilleman, M.R., Buynak, E.B., Weibel, R.E. and Stokes J. Jr. (1968a) *N. Engl. J. Med.* 278, 227–232.
Hilleman, M.R., Buynak, E.B., Weibel, R.E. and Stokes J. Jr. (1968b) *N. Engl. J. Med.* 279, 300–303.
Horstmann, D.M. (1982) *J. Infect. Dis.* 146, 540–551.
Horta-Barbosa, L., Fucillo, D.A., Zeman, W. and Sever, J.L. (1969) *Nature* 221, 974.
Immunization Practices Advisory Committee (1978) *Morb. Mort. Weekly Rep.* (Centers for Disease Control) 27, 268–270.
Immunization Practices Advisory Committee (1980) *Morb. Mort. Weekly Rep.* (Centers for Disease Control) 29, 265–280.
Immunization Practices Advisory Committee (1982a) *Morb. Mort. Weekly Rep.* (Centers for Disease Control) 31, 22–34.
Immunization Practices Advisory Committee (1982b) *Morb. Mort. Weekly Rep.* (Centers for Disease Control) 31, 417–328.
Immunization Practices Advisory Committee (1982c) *Morb. Mort. Weekly Rep.* (Centers for Disease Control) 31, 617–624.
Immunization Practices Advisory Committee (1983a) *Morb. Mort. Weekly Rep.* (Centers for Disease Control) 32, 135–143.
Immunization Practices Advisory Committee (1983b) *Morb. Mort. Weekly Rep.* (Centers for Disease Control) 32, 333–337.
Kernm, A.B. and Schiff, B.L. (1964) *Arch. Dermatol.* 89, 844–845.
Kew, O.M., Nottay, B.K., Hatch, M.H., Nakano, J.H. and Obijeski J.F. (1981) *J. Gen. Virol.* 56, 337–347.
Kilbourne, E.D., Chanock, R.M., Choppin, P.W., Davenport, F.M., Fox, J.P., Gregg, M.B., Jackson, G.G. and Parkman, P.D. (1974) *J. Infect. Dis.* 129, 750–771.
Krugman, S. (1971) *J. Pediatr.* 78, 1–16.
Krugman, S. (1983) *Lab. Med.* 14, 727–732.
Lane, J.M., Millar, J.D. and Neff, J.M. (1971) *Annu. Rev. Med.* 22, 251.
Lerman, S.J. and Gold, E. (1971) *JAMA* 216, 1311–1314.
McBean, A.M., Thoms, M.L., Johnson, R.H., et al. (1984) *Rev. Infect. Dis.* 6 Suppl. 2, 552–555.
Meyers, J.D., Flournoy, N. and Thomas, E.D. (1980) *J. Infect. Dis.* 141, 479–487.
Mortimer, E.A. Jr., Lepow, M.L., Gold, E., Robbins, F.C., Burton, G.J. and Fraumeni, J.F. Jr., (1981) *N. Engl. J. Med.* 305, 1517–1518.

Murphy, M.D., Brunell, P.A., Lievens, A.W. and Shehab, Z.M. (1984) *Pediatrics* 74, 90–93.
Nightingale, E.O. (1977) *N. Engl. J. Med.* 297, 249–253.
Nottay, B.K., Kew, O.M., Hatch, M.H., Heyward, J.T. and Obijeski, J.F. (1981) *Virology* 108, 405.
Plotkin, S.A., Farquhar, J.D. and Ogra, P.L. (1973) *JAMA* 225, 585–590.
Roizman, B., Warren, J., Thuning, C.A., Fanshaw, M.S., Norrild, B. and Meignier, B. (1982) *Dev. Biol. Standard* 52, 287–304.
Sabin, A.B. (1977) *JAMA* 238, 1731–1733.
Sabin, A.B. et al. (1983) *JAMA* 249, 2651.
Shasby, D.M., Shope, T.C., Downs, H., Herrmann, K.L. and Polkowski, J. (1977) *N. Engl. J. Med.* 296, 585–589.
Shehab, Z.M., Brunell, P.A. and Cobb., E. (1983) *J. Infect. Dis.* 148, 472–476.
Smith, G.L., Mackett, M. and Moss, B. (1983) *Nature* 302, 490–495.
Takahashi, M., Otsuka, T., Okuno, Y., Asano, Y., Yzaki, T. and Isomura, S. (1974) *Lancet* ii, 1288–1290.
Tingle, A.J., Yang, T., Allen, M., Kettyls, G.D., Bryce Larke, R.P. and Schulzer, M. (1983) *Infect. Immun.* 40, 22–28.
Ukkonen, P. and Penttinen, K. (1981) *J. Infect. Dis.* 144, 496–497.
Vesikari, T., Islauri, E., Delem, A., D'Hondt, E., André, F.E. and Zissis, G. (1983) *Lancet* ii, 807–811.
Warren, R.J., Lepow, M.L., Bartsch, G.E. and Robbins, F.C. (1964) *Pediatrics* 34, 4–13.
Wechsler, S.L., Weiner, H.L. and Fields, B.N. (1979) *J. Immunol.* 123, 884–889.
Weibel, R.E., Stokes J. Jr., Buynak, E.B. and Hilleman, M.R. (1972) *JAMA* 222, 805–807.
Wilkins, J. and Wehrle, P.F. (1979) *J. Pediatr.* 94, 865–869.
World Health Organization Expert Committee on Smallpox Eradication (1972) (2nd Report) pp. 5–64.

Eds. M.H.V. van Regenmortel and A.R. Neurath
Immunochemistry of viruses. The basis for serodiagnosis and vaccines.

The use of vaccines for controlling viral diseases of animals

13

LORNE A. BABIUK

Department of Veterinary Microbiology, Western College of Veterinary Medicine, University of Saskatchewan, Saskatoon, Saskatchewan S7N OWO, Canada

One of the major triumphs of modern medicine during the last two centuries has been the development and use of viral vaccines to control economically important diseases of animals. However, even though it is nearly 200 years since Jenner first showed that protection from smallpox in man was possible by vaccination, very few animal virus diseases have been eliminated from our environment. The reasons for this rather dismal success rate are multifold and may be related to the patterns of pathogenesis of many viral infections, the multiplicity of hosts and antigenic variability. Thus before eradication programs can be implemented more knowledge about these viruses and virus-host interactions including mechanisms of immunity is mandatory. However, even with our continuing advances in these areas, eradication will not be possible without a great deal of cooperation between producers, veterinarians and government agencies. Such cooperation was used during the smallpox eradication program in humans. Unfortunately, in veterinary medicine such programs will only be introduced if they are first proven to be cost effective and eradication is the ultimate goal. To date economics dictates that we reduce economic losses due to the many virus diseases rather than eliminate them. In countries where a virus does not exist, elaborate monitoring schemes have been developed to control the introduction of such 'exotic viruses' before the agents get established. In countries where the virus is endemic, the success in controlling the infection depends on the viral agent involved, the type of immune response that is involved in clearing the virus as well as the specific vaccine that has been developed and used to control the disease. The present review will attempt to discuss the various types of vaccines used for controlling the major viral diseases in veterinary medicine, some of the practical problems that are most relevant to veterinary medicine as well as the potential for improving present vaccines by modern biotechnological methods. In the field of veterinary medicine a considerable amount of fundamental information regarding the specific antigens involved in eliciting protective immunity is required before such approaches can be used for many viral diseases.

1. Practical problems of implementing effective vaccine programs in animals

Optimal immune responses will only occur if the animal is healthy and immunologically competent. In many instances it is not possible to determine whether an animal is already incubating either the specific virus for which protection is being sought or another virus. If a heterologous field strain virus is immunosuppressive then only a limited response to the vaccine virus may occur. An even more serious situation may occur if the attenuated vaccine virus can replicate to a greater extent because of the reduced immunocompetence of the host. In these situations the vaccine virus may in itself actually produce a disease. Such occurrences have been reported for canine distemper-canine parvovirus combinations and may occur in cats infected with feline leukemia virus (Schultz, 1982). Even if overt immunosuppression does not occur the induction of interferon by one virus may limit the replication of the attenuated vaccine virus and thus reduce the antigenic mass below a threshold level required for adequate immunity to develop.

It has become a common practice to vaccinate animals after a certain percentage have begun to exhibit clinical signs. Whether such a practice is of value must be questioned on scientific grounds even though experience in the field suggests that vaccination of animals incubating the disease may in some cases reduce or prevent infection (Schultz, 1982). The basis of this protection, if it occurs at all, is not scientifically evident and it may only be effective in those animals that have not yet been exposed.

The predominance of live attenuated vaccines in veterinary medicine requires the understanding of viral inactivation kinetics upon storage prior to as well as after reconstitution. In many instances situations arise where the vaccine is not or cannot be stored properly. This leads to virus inactivation and 'vaccine breaks' can occur. The development of more thermostable attenuated vaccines should greatly alleviate this problem (Heymann et al., 1982).

The age of the animal at the time of immunization is very important in many viral diseases. In most cases, the animal's immune system is sufficiently mature at birth to respond to immunization but the presence of passive antibody acquired from the dam may significantly alter responses. This antibody protects the neonate from infection during the first few weeks of life and also interferes with immunization with attenuated vaccines. Thus, immunization must occur as soon as possible after passive antibody decays. Unfortunately, unless serological testing is conducted on each individual animal, it is not possible to predict precisely when this time occurs. If immunization is conducted too early no response will occur and if delayed the animal could be exposed to virulent field strains while unprotected.

Another consideration is the route of vaccine administration. Differences in immune responses do occur between vaccine administered intramuscularly, subcutaneously and intradermally. Different vaccines produce different responses depending on the route of administration and therefore the route should not be changed from what is recommended by the manufacturer, even though an alternate route may be more convenient. In the development of vaccines, knowledge about the pathogenesis of the disease and immune responses must be considered for judicious designing of vaccine protocols. Thus, if the virus is strictly localized in mucosal surfaces, systemic immunity may be of limited value. However, if the virus is both local and systemic then parenteral immunization will be effective in preventing systemic spread. It must be emphasized that the locally administered vaccines must be much more highly attenuated than parenterally administered vaccines. Concomittant with increased attenuation, the rate of replication in the animal and also the level of immunity induced is often reduced. Thus in these situations one must consider a compromise between the level of immunity achievable and the type of protection desired.

The final consideration is delivery of the vaccine. In many cases wild animals may be reservoirs of the virus and effective control must be directed at controlling infection of these reservoir populations. An example is rabies virus, with which a variety of animals can be infected in nature. These wild animals cannot be immunized by conventional methods. Therefore, interest has been generated in oral administration of attenuated or killed antigen in bait (Hafliger et al., 1982; Steck et al., 1982; Schneider and Cox, 1983). The oral administration of killed antigen is questioned at present since there is very little evidence that adequate protection is provided by such a route (Lawson et al., 1982). However, the use of live rabies virus in bait must insure that species other than the target species which may eat the bait are not severely affected by the vaccine virus (Schneider and Cox, 1983). Unfortunately at present we do not know what effect these live attenuated virus vaccines may have on all the different wild species.

2. Immune responses to viral infection

Implementation of effective immunization regimes for viral infections requires information regarding the specific immune response that is involved in providing protection as well as the antigens that elicit the appropriate response. In many cases animal vaccines only prevent development of clinical disease without affecting infection, and many manufacturers try to correlate protection with the vaccine's ability to induce a systemic antibody response. Unfortunately, in many local infections, such as in the respiratory or gastrointestinal tract, parenterally administered vaccines which induce systemic humoral immunity will not protect the mucosal surface against viral replication. Thus a humoral antibody response should not be used as proof of efficacy. In fact the presence of antibody may enhance the infectivity of some viruses (Hawks and Lafferty, 1967; Peiris and Porterfield, 1981). Possibly the most striking effect of how an antibody response to an inappropriate antigen can even be detrimental has been demonstrated with measles virus (Isacson, 1968; Norrby, 1983) and respiratory syncytial virus (RSV) (Kim et al., 1969). In these instances the use of an inactivated vaccine actually produced a more severe disease upon exposure to virulent virus (Chanock et al., 1968; Kim et al., 1969). Thus, as with RSV and other paramyxoviruses, many viruses have more than one protein that is involved in inducing protection. Furthermore one antigen may induce one type of immune response and a second antigen on the same virus may induce a completely different response. For example, an antibody response to the hemagglutinin of influenza virus can neutralize that specific virus. This response is restricted to the specific hemagglutinin type. However, cytotoxic T cells also appear to recognize the matrix protein, whose antigenic structure appears to be conserved between all A strains of influenza virus (Biddison et al., 1977; Bennink et al., 1978; Braciale et al., 1981), demonstrating that neutralizing antibody can be directed at one virion protein but others may be important in other mechanisms of defense. Thus a considerable amount of basic virological and immunological knowledge is required concerning the actual antigens and types of immune responses involved in induction of immunity if we hope to design a vaccine that will be most effective. Unfortunately many virus infections encountered in veterinary medicine have not been adequately characterized at the molecular level nor have the host immune responses to these viruses been characterized. If this were achieved then it may be possible to engineer better vaccines to control diseases for which we already have at least partially effective vaccines as well as for those where no effective immunization regimes have been implemented. For example, in African swine fever, hog cholera and bovine viral diarrhea only limited information is available concerning which immune responses are important in preventing infection and in assisting recovery from infection.

3. Types of vaccines

The vaccines used in veterinary medicine can be classified into live and killed vaccines (Table 13.1). To date the majority of viral vaccines used in veterinary medicine are live attenuated strains. However, in certain cases where it has been difficult to attenuate specific viruses, attenuated or virulent heterologous cross-reacting viruses have been used to help elicit an immune response against the pathogenic viruses. In addition, homologous virulent virus has also been used in certain isolated situations. In these cases the virulent virus is either given at an age when the animal

TABLE 13.1

Vaccines for control of veterinary viruses

Virus type		Example
Live	Homologous – Virulent	Infectious larygotracheitis, Orf
	Homologous – Attenuated	Canine distemper
	Heterologous – Virulent	Bovine viral diarrhea for hog cholera
	Heterologous – Attenuated	Measles for canine distemper
	Temperature sensitive	Bovine herpes-1
Killed	Inactivated whole virus	Equine influenza
	Polypeptide subunit	Experimental foot-and-mouth disease
	Synthetic peptide	Experimental foot-and-mouth disease
Autogenous	Formalized warts	Papilloma virus

is not very susceptible to the specific disease or the virus is introduced into an unnatural site. In the latter case replication occurs and immunity is produced before the virus reaches the normal site of replication where it can induce disease. An example of such a vaccine is infectious laryngotracheitis virus, a herpes virus infection of the respiratory tract of chickens (Beaudette and Hudson, 1933). This virus vaccine is instilled on the cloaca. Another example is Orf, a poxvirus of sheep where the virus is instilled on the unwooled area of the flank. The major disadvantages of these types of vaccines is the high probability of transfer of the virulent virus to the natural site of infection before immunity is developed. A means of avoiding such problems is the use of heterologous virus which can replicate in the host but does not induce clinical disease. If these heterologous vaccines are not attenuated there is a risk of transfering the virus from the vaccinated animals to the susceptible host and subsequent development of disease. For example, virulent bovine viral diarrhea virus (BVD), can protect pigs against hog cholera, but if cattle are exposed to the pigs shedding BVD they may suffer from the disease.

The main advantage of live attenuated vaccines is that they replicate in the host and produce a wide range of immunological responses very similar to those induced by the natural infection. Furthermore duration of immunity following vaccination with live attenuated viruses is generally longer than with killed virus vaccines. However, with the use of these vaccines it is imperative that they be sufficiently attenuated so as not to cause disease. This is especially important in those vaccines that are aimed at eliciting local immunity by intranasal inoculation and are given by the natural route of infection, for instance feline rhinotracheitis and calici viruses. Unfortunately in some cases the degree of attenuation required to prevent disease is so great that the virus replicates poorly in vivo and thus does not induce a satisfactory immune response.

One of the problems with attenuated vaccines is that very little knowledge is available concerning the mechanism of attenuation. Therefore it is not possible to insure that they will not revert to virulence or recombine with virulent viruses in vivo. This is especially important in vaccine viruses which can establish a persistent or latent infection. Another major disadvantage is that of interference between viruses (Schultz, 1982). This may be a greater problem in veterinary medicine than it is in human medicine. Interference can occur in two ways: different viruses in combination vaccines can interfere with each other or there may be interference from a field virus present in the animal at the time of vaccination. The trend presently being adopted by most biological companies is to combine many viral vaccines in a single dose. This approach may not provide the best immunity to all or any one of the agents. Unfortunately this practice is encouraged because of the risk of exposure to many of the infectious agents at a very early age. A second reason for the increased interest in combining multiple viruses in one vaccine is that it reduces the number of trips that the client must make to the veterinarian. Practical and economical considerations thus often dictate what is done even though such procedures may not be scientifically ideal. However, one must discourage the combination of vaccines by the veterinarian, especially if one of them is a killed virus vaccine, since the killed vaccine may contain preservatives that are detrimental to the live virus vaccine.

Killed virus vaccines are beginning to return in popularity. One major reason for increased interest in killed rabies virus vaccine is the psychological assurance that accidental injection of the vaccine will not lead to infection. Although the risk of developing rabies from an attenuated vaccine is slim, the fact that it is a live vaccine does carry this risk. Another more important impetus for the resurgence of interest in killed vaccines has been the advances in recombinant DNA technology and an emerging understanding of the individual epitopes and amino acid sequences or neutralizing sites on viral polypeptides. This knowledge provides an opportunity to economically produce large quantities of efficacious killed vaccines for use in veterinary medicine. However, as previously mentioned a considerable amount of information regarding the antigenic structures of animal viruses still remains to be obtained before such vaccines can replace the conventionally produced vaccines.

Autogeneous vaccines produced by formalin inactivation of homogenizing warts are often used to control warts caused by papilloma virus. These vaccines are given intradermally or subcutaneously. The value of such a vaccine may be questioned since many warts often regress spontaneously even without such treatment.

3.1. Small animal vaccines

The most commonly used vaccines for small animals are listed in Table 13.2. The vaccines for small animals are among the best in veterinary medicine with respect to preventing the disease syndrome for which they are designed. These vaccines have

TABLE 13.2.

Common vaccines used in small animal veterinary medicine

Species	Disease	Vaccine	Comments
Canine	Distemper	Modified live distemper virus Modified live measles virus	Very effective Only for primary immunization
	Infectious hepatitis	Modified live canine adenovirus serotype 1 and 2	CAV-1 may cause blue eye upon repeated exposure; CAV-2 can be used to boost immunity to CAV-1
	Respiratory	Modified live parainfluenze Modified live adeno-2	
	Parvovirus	Modified live and killed canine parvovirus and feline panleukopenia	
	Rabies	Modified live or killed rabies	Very effective if used annually, killed vaccine gaining popularity
Feline	Panleukopenia	Modified live or killed panleukopenia	Very effective
	Rhinotracheitis	Modified live feline rhinotracheitis	May cause mild clinical disease; does not eliminate carrier states
	Calicivirus	Modified live or killed feline calicivirus	As for rhinotracheitis
	Rabies	Modified live or killed rabies	Killed vaccine may be preferred

greatly reduced losses in dogs and cats. In most cases the viruses are ubiquitous, with the consequence that most animals must be vaccinated annually against the multitude of viruses that are present. Most of these vaccines are given in combination in an attempt to induce immunity to all of the agents as early as possible and also reduce the number of visits to a veterinary clinic. As a result of vaccination, adults transfer antibody to the young in the colostrum. This antibody (lgG) is absorbed from the intestine following feeding and protects the animals during the first weeks of life. This protection may range anywhere from 6 to 12 weeks depending on the quantity of specific antibody ingested. However, as the antibody decays, the animal's susceptibility increases. The time that it takes for passive immunity to drop below protective levels varies depending on the initial level of antibody acquired. The amount of antibody ingested and absorbed varies from litter to litter as well as within animals from the same litter. Since it is not practical to test each animal to predict the time the antibody to a specific virus drops below a threshold protection level it is recommended that the first vaccination against all the viral agents be given at approximately 6–9 weeks of age; at which time some animals will respond. A second dose is generally given at 12–16 weeks of age. This second dose serves as a booster for those animals that responded to the first immunization and provides a primary immunization to those animals that still had antibody at the first injection. Annual revaccinations are recommended. Although it is probably not essential to immunize annually, since in most cases immunity will last longer than one year, if such protocols are followed minimal disease will occur.

Measles virus for primary immunization against canine distemper can be given to puppies at a slightly younger age than canine distemper virus since measles virus is not sensitive to neutralization with endogenous antibody to canine distemper virus and can therefore be used as the initial vaccine in younger animals if there is a concern of exposure to virulent canine distemper virus. Measles virus vaccine should only be used as a temporary measure to provide early protection and should not be used for boosters. It is especially important not to use it in pregnant bitches since this would result in passive transfer of measles antibody to the puppies and would nullify the advantage of measles vaccination of the puppies.

Annual revaccination with attenuated canine adenovirus-1 (CAV-1) may result in uveitis or corneal edema (blue eye), and annual revaccination may in fact be detrimental in these cases (Schultz, 1982). To overcome these problems, the trend is to revaccinate with canine adenovirus-2 (CAV-2), which provides cross-protection to CAV-1 as well as to CAV-2 but eliminates the adverse reactions of repeated CAV-1 immunization.

The trend in rabies vaccination of cats and dogs is to use killed virus vaccines. The reason for this is two fold. As mentioned previously some practitioners are concerned with accidental inoculation of themselves or their assistants with live virus. Secondly, some live rabies virus vaccines may cause problems in cats (Schultz, 1982).

The severe clinical signs caused by feline respiratory viruses (FRV and FCV) can be controlled by local immunization with attenuated vaccines, but these do not prevent infection with virulent virus (Gaskell and Wardley, 1977). Furthermore, they do not prevent virus shedding and may themselves induce some mild clinical disease for 1–2 days following vaccination (Povey, 1976). However, improvement of these vaccines is continuing and it is hoped that all adverse reactions can be overcome. Due to the nature of these diseases it will probably be impossible to eliminate carriers by immunization. There is no evidence that these virus vaccines interfere with each other even if included with the attenuated feline panleukopenia virus.

3.2. Vaccines for large animals

The majority of vaccines presently used in large animal medicine are listed in Table 13.3. The efficacy of viral vaccines for large domestic animals varies greatly from those which are very effective to ones of questionable value. The reasons for this variability are dependent not only on the viral vaccine per se but also on the pathogenesis of the specific viral disease, as well as the management conditions under which the animals are raised. These factors must all be considered in any immunization regimes designed for large animals. Unfortunately this is not always the case and consequently 'vaccine breaks' occur.

3.2.1. Bovine viral vaccines

In North America the two disease complexes that cause the most economic losses in cattle are bovine respiratory disease and neonatal diarrhea. With respect to enteric infections, vaccines have been tried and are licenced for immunization of calves against rotavirus and coronavirus. These vaccines appear to have limited effectiveness (Saif and Smith, 1983; Waltner-Toews et al., 1983). The possible reason for this lack of activity in the field is that maternal colostral antibody neutralizes the vaccine virus before it can infect the calf enterocytes and initiate an immune response. Alternatively, in rotavirus infections it is highly likely that infection in the field occurs with different serotypes than the one used for immunization. Immunization of calves with one serotype does not protect against challenge with the heterologous serotype (Woode et al., 1983). Sequential infections with different serotypes has also been shown to occur in humans, further supporting the latter possibility (Fonteyne et al., 1978). The lack of effectiveness of active immunization of the calf has led to interest in parenteral immunization of pregnant cattle at mid-gestation and again in late gestation (Dauvergne et al., 1983; Eichhorn et al., 1983). This practice is based on the hope that hyperimmunization of the dam will increase antibody levels in milk and thus provide passive protection while the calf is feeding on antibody-containing milk. Such an approach can significantly reduce disease incidence under experimental conditions (Snodgrass et al., 1982; Saif et al., 1983). However, the efficacy of the rota-corona attenuated vaccine presently licenced for

TABLE 13.3

Common vaccines used in large animal veterinary medicine

Species	Disease	Vaccine type	Comments
Equine	Rhinopneumonitis	Attenuated or killed	Require frequent revaccination (every 3–4 months) for effective immunity
	Influenza	Killed	Should contain both influenza serotypes; vaccinate every 3–4 months
	Encephalomyelitis	Killed trivalent WEE, EEE, VEE (attenuated VEE)	Effective if given annually; can be bi- or trivalent
	African horse sickness	Attenuated or killed	Contains 9 different serotypes, both polyvalent and monovalent
	Equine viral arteritis	Attenuated	Vaccination every year effective experimentally
Swine	Transmissible gastroenteritis	Attenuated	Not as effective as virulent virus
		Virulent	Pregnant sows fed infected intestines containing virulent virus 1 month before farrowing; maintains virulent virus in environment
	Rotavirus	Attenuated	Efficacy questionable in piglets due to presence of maternal antibody
	Pseudorabies	Attenuated and killed	Problems in differentiating vaccinated animals from carriers of virulent virus
	Hog cholera	Attenuated or killed heterologous bovine viral diarrhea, virulent virus and antisera	Animals may be carriers

	Porcine parvovirus	Killed	Reduce reproductive failures
	African swine fever	Experimental killed and live	No effective immunization procedure; immunity incomplete; control by eradication best method.
	Porcine enterovirus	Attenuated	Only against some serotypes
Bovine	Bovine herpes virus-1 Parinfluenza Bovine viral diarrhea	Attenuated and inactivated	Effective if used properly: outbreaks occur in vaccinated feedlot cattle before immunity is developed; live bovine viral diarrhea may be immunosuppressive
	Enteric complex, rota, corona	Attenuated	Efficacy questioned: only one serotype of rotavirus used in vaccine; multiple rotavirus serotypes occur in nature; local immunity important; oral administration of newborns or hyperimmunization of dam before parturition
	Rinderpest	Attenuated goat-propagated live virus	Most widely used vaccines have reduced but not eliminated epidemics
	Foot-and-mouth disease	Inactivated and attenuated	Must contain type and subtype present in the region; immunity is of short duration; modified live vaccines have very restricted use
Sheep and goat	Blue tongue	Attenuated	Should not be administered to pregnant ewes; must contain serotypes present in specific locations
	Poxviruses (sheep pox, goat pox, Orf)	Virulent and attenuated	Scarify unwooled area of thigh; second dose in 2 to 3 months is recommended
	Louping ill	Inactivated	Excellent immunity; immunize lambs
	Nairobi sheep disease	Inactivated mouse brain-adapted virus	Most animals in enzootic areas have antibody

hyperimmunizing the dam and thus providing protection to the offspring has been questioned (Saif and Smith, 1983). The apparent reason for the ineffectiveness of the vaccine could be related to the low levels of virus in the vaccine. The use of higher levels of virus in the vaccine would appear to make this approach to immunization highly effective, especially if all rotavirus serotypes are included.

The value of vaccines for controlling respiratory diseases in cattle has also been questioned (Martin, 1983). The possible reason for the lack of efficacy is that animals are often weaned and transported to sales barns, where they are mixed with other animals and their pathogens before being transported to feedlots where they are immunized. Thus a number of animals are probably infected prior to vaccination and before adequate immunity develops. If immunization occurs prior to weaning and transportation then these vaccines may have a much better performance record. A second possible reason why these vaccines do not significantly reduce bovine respiratory disease is that this disease is a complex, and that etiological agents other than the ones present in the vaccine could be inducing the disease (Rosenquist, 1984). The development of vaccines for other agents as well as immunizing prior to movement and mixing of cattle should greatly reduce disease incidence. Thus the potential for reducing disease incidence due to respiratory and enteric infections in cattle is feasible if both the vaccines and the management regimes are considered in the overall control programs. The fact that good immunity can be induced under experimental conditions further supports this contention (Jericho et al., 1982).

The most dreaded bovine disease in many countries is foot-and-mouth disease. The reason for such concern is based on the multiplicity of susceptible hosts that can be infected with this virus, its highly contagious nature and the large number of antigenic variants. The presence of these variants makes it difficult to insure adequate immunity even in vaccinated animals. In countries where this virus is not endemic, the disease is controlled by strict quarantine. If the virus is accidentally introduced into the country a slaughter policy is the first consideration. In those countries where the virus is endemic, losses are controlled by immunizing with inactivated vaccines prepared from infected bovine tissue or tissue culture-grown virus. An attenuated vaccine has also been developed but its use is limited to very few countries. The importance of this disease has stimulated an interest in preparing and testing synthetic or subunit vaccines prepared by recombinant DNA technology.

Rinderpest vaccination programs vary from region to region and often use locally produced vaccine from local virus strains. These include avian, goat, rabbit and tissue culture-modified virus vaccines. Since this disease can occur in remote areas, consideration must be given to maintaining the viability of the live virus in the vaccine prior to its administration.

3.2.2. Equine viral vaccines

Four common viral-induced equine diseases transmitted by insect vectors include African horse sickness and Western, Eastern and Venezuelan equine encephalomyelitis. Although control may be achieved by controlling vectors, active immunization is the method of choice for all of these diseases. The multiple serotypes found in African horse sickness dictates that the vaccine include all 9 serotypes in order to provide adequate protection. The three encephalomyelitis viruses are often combined in one inactivated vaccine and given annually before peak vector activity and transmission occurs.

Equine influenza and equine rhinopneumonitis (ERP) are the most common causes of respiratory tract infection at race tracks and horse shows, where horses from various sources are mixed. Immunity to these viruses following vaccination appears to be very short lived and it is recommended that horses be immunized every 3–4 months (Ardans, 1982). For best protection they should be immunized 2–3 weeks before going to a rack track or sales barn. Even with such immunization regimes, horses can still get infected, but the resulting disease is often milder. Also, although immunization with ERP has reduced abortions, it has not eliminated them. Thus it is obvious that considerable knowledge about these viruses and the immune responses which they elicit is required before adequate control can be achieved.

3.2.3. Swine viral vaccines

Neonatal diseases of pigs (transmissible gastroenteritis, porcine coronavirus and rotavirus) can often be controlled by hyperimmunizing the sows prior to parturition. This can be done with virulent or attenuated viral vaccines. If immunity is developed in sows they can transfer passive antibody (SIgA) to the piglets. This antibody will protect the young piglets by neutralizing any virus that they may ingest for as long as the antibody remains in the lumen. The practice of feeding virulent transmissible gastroenteritis virus to immunize pregnant sows or intramammary injection is very effective in hyperimmunizing the sow (Djurickovic and Thorsen, 1970; Bohl et al., 1972). However, this practice insures the continued presence of virus in the environment and attenuated vaccines have therefore been developed for transmissible gastroenteritis. Unfortunately attenuated vaccines do not appear to be as effective as the virulent strains (Bachmann and Hess, 1982). The use of these attenuated vaccines in neonates also does not appear to be effective since they do not appear to be able to overcome the passive immunity obtained from the dam. To circumvent these deficiencies some producers hyperimmunize the sows to provide early protection. A few days prior to weaning the piglets are removed from the sow for approximately 12 hours to allow the antibody in the lumen to decrease and hopefully allow infection with field virus. They are then returned to the sow to once again ingest IgA which limits the replication of the field virus but does not interfere with induction of active immunity. In these cases there is no guarantee that animals will get infected with field strains during the 12-hour period and immunity may not develop. However, such a practice appears to be at least partially successful.

Hog cholera can be partially controlled by vaccination with heterologous bovine viral diarrhea virus (BVDV), inactivated hog cholera vaccines, attenuated vaccines or a combination of virulent hog cholera and antisera (Kamijio et al., 1977; Gillespie and Timoney, 1981). A major question that must be addressed when using vaccines to control hog cholera and another porcine virus, pseudorabies, is whether it is desirable to eradicate these diseases or just control them. Many countries are now trying to eradicate these diseases by serologically testing all animals and disposing of those that have antibody titres. If animals are vaccinated with the particular vaccine it is impossible to differentiate between animals that were naturally exposed and may be carriers and those who have developed serological responses due to vaccination programs. Introduction of mass immunization programs may prolong the problem.

The intranasal vaccines against pseudorabies appear to give better protection than do the intramuscular vaccines; however, they may allow shedding of vaccine virus and maintain the virus in the environment. Unfortunately vaccination must be repeated at 6 month intervals (Schoneweis and Henry, 1982), a fact that further supports eradication programs.

3.2.4. Sheep and goat viral vaccines

It is possible to protect sheep against sheep pox by immunizing with goat pox but not vice versa. The best method of controlling the spread of sheep and goat pox is to reduce the import of animals from endemic areas. In the case of Orf, although this virus occurs worldwide, immunization is only recommended in endemic areas.

Bluetongue virus vaccine should contain the serotypes that are present in the specific locations for most effective protection to be obtained. These vaccines are generally not very well attenuated and pregnant ewes should therefore not be immunized since the vaccine virus can cause congenital infections. Furthermore immunization should occur after shearing to reduce losses in wool quality.

Inactivated louping ill vaccines produced from nerve tissue are very effective if administered to newborn lambs shortly after weaning when colostral antibody declines. This vaccine provides good humoral immunity which prevents the viremic spread of virus to the central nervous system. In this disease, as well as in Nairobi sheep disease, control can also be achieved by controlling the vectors responsible for transmission of these diseases.

3.3. Poultry diseases

Intensive avian management practices provide both challenges and unique opportunities for developing innovative methods of vaccine administration to large numbers of animals in a very short time period. Intensive rearing ensures the rapid spread of viral agents, but also allows mass administration of vaccines by aerosols or in water (MacDonald et al., 1981; Meszaros, 1983). Marek's disease virus (MDV) has been essentially controlled by immunizing with a related herpes virus of turkeys (HVT). This vaccine provides protection from transformation caused by MDV, the level of viremia and the degree of virus spread but does not prevent infection by the virus (Table 13.4). In some cases infection may occur in the first few days of life and thus vaccination at hatching may not induce immunity soon enough. To overcome this deficiency attempts have been made at vaccinating embryos with HVT on approximately the 18the day of incubation, when the eggs are being moved from the incubator to hatching trays. This approach does not add significantly to the additional handling of eggs, does not alter hatchability but does provide protection if birds are exposed to field strain of MDV soon after hatching (Sharma and Graham, 1982). Although embryo vaccination or vaccination at birth with HVT has significantly reduced losses to MDV, such losses have not been eliminated. Some highly virulent MDV or variable serotypes can overwhelm immunity produced by HVT. To overcome this problem bivalent or trivalent vaccines combining HVT with different serotypes of nononcogenic MDV are being employed (Witter, 1982; Sharma and Witter, 1983). The presence of maternal antibody has also been incriminated in being responsible for vaccine failures. Since this effect is more pronounced with homologous systems it may require alternate immunization of one generation with HVT and the next generation with nononcogenic MDV. Cell-associated antigen is less likely to be neutralized by maternal antibody and this may warrant a return to cell-associated viral vaccines (Sharma and Graham, 1982).

Many infections that are serious only in baby chicks may be controlled by hyperimmunizing the breeder hens. In these cases sufficient antibody will be transmitted to protect the newborn chicks during the most critical period (van der Heide et al., 1983). Such an approach is used in infectious bursal disease (IBDV) in which the virus replicates in the bursa of young chicks and causes severe immunosuppression (Nagi et al., 1983). Immunization of breeder flocks provides protection at the early stages when the bursa is most susceptible. To ensure continued protection beyond 3–4 weeks of age, when antibody declines, the chicks are reimmunized with various attenuated or inactivated IBDV vaccines in oil adjuvant or in drinking water. Depending on the degree of attenuation some may induce good active immunity even in the presence of passive antibody; however, they may also induce some damage if not attenuated sufficiently. The degree of attenuation is important with regard not only to overcoming passive antibody but also to making animals more susceptible to other infections. For example, IBDV and reovirus, if not attenuated sufficiently, may cause immunosuppression and make animals more susceptible to other virus diseases. In contrast, too high a level of attenuation may result in low immunogenicity. Since the level of maternal immunity varies between flocks and between individual birds in a flock, the timing and choice of vaccine must be selected with consideration towards the severity of the suspected field infection.

TABLE 13.4

Common avian viral vaccines

Disease	Vaccine type	Comments
Marek's disease	Herpes virus of turkeys Attenuated nononcogenic MD virus Cell-associated and cell-free vaccines	Does not prevent infection Reduces viremia and horizontal spread
Infectious laryngotracheitis	Attenuated and virulent	Conjunctival, aerosol administration, or in drinking water; virulent strains administered on cloaca may result in carriers
Fowlpox	Attenuated	Birds may be carriers; use only in regions where disease is a problem; booster at 12—20 weeks recommended subcutaneously or brushing feather follicles
Newcastle disease	Killed and attenuated vaccines	Administered by aerosol, drinking water or feather follicle route; injection of killed virus in adjuvants for laying hens
Infectious bronchitis	Attenuated vaccines	Drinking water or aerosols; multiple serotypes; 7—10 days of age
Fowl plaque	Attenuated and killed	Slaughter is best method of control
Avian encephalomyelitis	Attenuated and killed	Given in drinking water prior to laying; killed for laying hens
Infectious bursal disease	Attenuated and virulent	Given in drinking water or occular instillation; virulent virus produces bursal lesions
Avian arthritis	Attenuated	Subcutaneous or in drinking water

Although there are attenuated vaccines against infectious laryngotracheitis virus which may be added to the drinking water for mass immunization of chicks (Samberg et al., 1971), control has also been achieved by the use of fully virulent virus. The virus mixture is applied with a stiff brush on the mucus membrane of the cloaca. In these types of vaccination programs care must be taken not to aerosolize the virus or expose the upper respiratory tract of the chickens to the virus since introduction of this highly virulent virus into the respiratory tract may create problems.

Infectious bronchitis virus contains several serotypes. To insure protection, attempts have been made at using polyvalent vaccines (Andrade et al., 1983). Unfortunately some of these polyvalent vaccines are less immunogenic than monovalent ones (Meszaros, 1983). Thus the selection of suitable vaccines must take into account the specific serotypes circulating in a particular location.

In general, immunization of laying hens with attenuated vaccines should be avoided because this may cause a temporary drop in egg production. To obtain maternal antibody it is therefore recommended that immunization be performed prior to the laying period. This is especially important with virus diseases such as egg drop syndrome, in which vertical transmission can occur by viremic hens (Darbyshire and Peters, 1980). If immunization must be performed during laying it is best to do so with killed virus vaccines.

4. General assessment of present licensed vaccines

In general, the vaccines licensed for use in small animal medicine are effective because of the individual patient care received. Similarly poultry vaccines, considering the large numbers of animals immunized in concentrated periods, are relatively effective. In contrast, some of the vaccines presently used for large animals could be improved dramatically. However, in many cases improving the vaccine itself will not be sufficient to reduce economic losses because even the best vaccines cannot offer complete protection by themselves. Thus the best that can be hoped for is to reduce losses to an economically acceptable level. These losses can be further reduced by educational programs which emphasize overall management of animals. A primary example where changes in management are required is in the control of bovine respiratory diseases. Unfortunately many animals are already incubating the disease or will get infected shortly after vaccination. If this is the case, even the best vaccines would not be expected to give instant solid immunity. Thus the greatest possible benefit from the use of this vaccine is not gained.

In many countries it is not possible to monitor the quality of all the vaccines that are licensed. Problems with handling the vaccines includes storage, transportation and administration. If these factors are not considered they may negate the efficacy of some vaccines. This is especially important for live attenuated vaccines. Thus although vaccines can and do play an important role in reducing economic losses from viral diseases they should not be expected to replace proper hygienic and management regimes.

5. Potential for improving immunization regimes

At present many attenuated vaccines are selected from a pool of viruses that have undergone random mutation followed by selection for isolates with reduced virulence. Such an approach not only limits the ability to control attenuation but, more important, does not insure that back mutation or reversion will not occur. The recent construction of herpes virus mutants with specific deletions of nonessential genes appears to provide a better approach to directed mutagenesis. Thus neither the TK or the alpha 22 gene of herpes simplex appear to be required for replication in at least some cell populations (Roizman et al., 1981; Kit and Kit, 1983). Deletion of the alpha 22 gene resulted in attenuation whereas deletion of the TK gene generally results in reduced reactivation following infection (Tenser et al., 1981). Similarly adenovirus deletion mutants have been made that lack the transforming gene (Jones and Shenk, 1979). In addition to the work with herpes viruses cited above similar studies with reovirus, influenza and poliovirus demonstrate that identification and alteration of virulent genes are feasible in many virus families (Kew et al., 1981; Fields and Greene, 1982; Murphy et al., 1982). Thus the continued understanding of virulence and identification of genes affecting virulence should allow the construction of live viral vaccines for most animal viruses by deletion or replacement of these genes. In some cases pathogenesis and immunity may be influenced by more than one gene (Campbell et al., 1982). Whether alterations in all these genes will be required remains to be determined. However, it is conceived that although it may be more tedious to construct double deletions or alterations this would further reduce the probability of reversion.

The recent construction of hybrid viruses provides a unique opportunity to produce vaccines which will induce the desired immune response as well as one of longer duration (Gething and Sambrook, 1981; Smith et al., 1983). For example, in diseases where cellular immunity is essential, killed vaccines are of limited value. However these live hybrid vaccines may prove to be extremely effective in inducing immunity of long duration. Although one could theoretically use a number of different viruses in these hybrid vaccines, poxviruses may be the one of choice for a number of reasons. (*a*) Most animal species can be infected with at least one pox virus. (*b*) Apathogenic poxviruses can be identified for most species. (*c*) The size of the poxvirus genome could possibly accommodate a number of different genes from a number of different viruses. Since most species are immunized simultaneously with multiple virus vaccines, a single construction could be made which would accommodate all the necessary agents. This approach would reduce some of the problems of using multiple vaccines. In many cases, such as poultry, a single immunization would be sufficient for the life of the animal. In other cases it may necessitate periodic boosters. If secondary immunization with the poxvirus-hybrid vaccines is not possible due to solid immunity induced to the poxvirus, immunity to other viral antigens could be boosted by virus-specific vaccines presently in use. Alternatively if the animal is already primed with the pox-hybrid vaccine then individual polypeptides produced by recombinant DNA technology, which at present are not very immunogenic as primary immunogens, may adequately boost immunity after the animal has been primed. These types of approaches should be tried in animals before they are developed for use in humans. Furthermore the possibility exists of developing good immunogens to viruses for which no vaccines are presently available either due to the inability to attenuate these viruses or to grow them in vitro to sufficient levels to make killed vaccines economical.

Subunit vaccines can be developed by a variety of methods. The easiest approach is to isolate individual viral capsid proteins or envelope glycoproteins. This approach has been tested for a number of viruses and immunity can be induced with these 'split product' vaccines (Klein et al., 1981; Cappel et al., 1982). However, immunity does not exceed that produced by whole virus. The extra work involved in producing these types of subunit vaccines will in most instances prevent their use in veterinary medicine strictly on economical grounds. Places where they may be considered is in equine medicine where the cost of the individual animal is high. The second approach is to use synthetic peptides (Bittle et al., 1982; Green et al., 1982; Lerner, 1983). This requires identification of individual epitopes on the viruses. Identification of antigenic sites by computer modeling and monoclonal antibodies are available, thus making it feasible to use this approach for almost any virus. The viruses that have received the most attention recently include influenza (Green et al., 1982), hepatitis B (Lerner et al., 1981), polio (Emini et al., 1983) and foot-and-mouth disease (Bittle et al., 1982), for which oligopeptides have been shown to induce immune responses and could induce at least partial protection upon challenge. In other instances the oligopeptides could prime the immune response so that exposure to intact virus resulted in an anamnestic response (Emini et al., 1983). If nonvariable regions could be identified for each virus such an approach would be very attractive. Unfortunately, considerable knowledge is still required concerning carriers and the form in which the peptide must be presented before such approaches can be adopted as routine. At present it is not possible to predict whether a specific carrier or adjuvant will improve immunogenicity. Furthermore some adjuvants may prove to be excellent in experimental animals but unacceptable in food-producing or pet animals.

Production of subunit vaccines by recombinant DNA technology is well under way in veterinary medicine with rabies and foot-and-mouth disease virus being prime examples (Kleid et al., 1981; Yelverton et al., 1983). The observation that neutralizing antibody and protection can be elicited with such vaccines is very encouraging and opens the door for this method of producing vaccines for other viruses either in bacteria, yeast or eukaryotic cells (Davis et al., 1981; Watson et al., 1982; Berman et al., 1983; Hitzeman et al., 1983). This approach may be especially useful for those viruses for which no vaccines are presently available, such as feline and bovine leukemia virus. However, once again we must identify the protective antigens and the genes coding for these antigens before such approaches can be used for many viruses of veterinary importance.

An interest has been expressed in vaccination with anti-idiotypes. Thus immunization with an antibody that recognizes the idiotype of a neutralizing antibody should theoretically induce an immune response to the virus. Such an approach has been used to treat B cell lymphoma patients (Miller et al., 1982) and the recent report that anti-idiotype antibodies can induce neutralizing antibodies to rabies virus provides yet another approach to immunization (Reagan et al., 1983). This approach may prove to be as beneficial and result in products that are even more immunogenic than subunit vaccines.

According to present experience, subunit vaccines produced by recombinant DNA techniques or by chemical synthesis are not as immunogenic as the naturally produced antigens. As more knowledge is gained concerning how antigens are recognized and processed it should be possible to present these antigens in such a way that they are highly immunogenic and induce the appropriate immune response. In many cases this may require conformational alterations in the engineered molecule, incorporation of these antigens in liposomes (Six et al., 1980) or other antigen carriers. Alternatively it may require modification of the proteins to suppress one immunological response and enhance another (Adorini et al., 1979). Another possibility is to alter immunological responses by including various immunopotentiators with vaccine regimes.

Although this review concentrates on active immunization against viral diseases it must be emphasized that control by passive immunization may be appropriate in some instances. In the case of enteric infection in nenonatal animals local antibody present in the lumen appears to be very effective in providing protection. In cases where animals are weaned at an early age it may be appropriate to feed colostral antibody (Saif et al., 1983). Alternatively it may prove to be economical to feed monoclonal antibody in the diet during the critical susceptible period. Such an approach has been used to protect calves against enterotoxigenic *Escherichia coli* (Sherman et al., 1983). A similar approach might also be used in rotavirus and coronavirus infections.

The recent technology available for identifying important antigens involved in immunity combined with methodology to characterize immunological responses involved in recovery from virus infections provides a great potential for improving vaccines used in veterinary medicine over those presently licensed. Other factors that must be considered in immunization strategy include delivery systems, immunomodulators and adjuvants. These are presently being developed to improve the immunogenicity of vaccines. However, regardless of all the scientific knowledge that is available, vaccine strategies must also consider the practical problems of animal husbandry practices and politics. In some instances the best scientific approach is not always practical under field conditions and a compromise often has to be reached. Unfortunately, such compromises result in reduced efficacy of the vaccines and economic losses continue. Finally better regulations for standardization of vaccines and quality control at government levels must be implemented. At present these vary from country to country and even between regions of some countries. This cannot be allowed to continue if we hope to achieve global control of viral infections.

6. References

Adorini, L., Harvey, M.A., Miller, A. and Sercarz (1979) *J. Exp. Med.* 150, 293–306.

Andrade, L.F., Villegas, P. and Fletcher, D.J. (1983) *Avian Dis.* 27, 178–187.

Ardans, A.A. (1982) *J. Am. Vet. Med. Assoc.* 181, 1150–1153.

Bachmann, P.A. and Hess, R.G. (1982) In *Viral Infections of the Gastrointestinal tract* (D.A.J. Tyrell and A.Z. Kapikian, Eds.) Marcel Dekker, New York. pp. 361–397.

Beaudette, F.R. and Hudson, C.B. (1933) *J. Am. Vet. Med. Assoc.* 82, 460–476.

Bennink, J., Effros, R.B. and Doherty, P.C. (1978) *Immunology* 35, 503–509.

Berman, P.W., Dowbenkco, D., Lasky, L.A. and Simonsen, C.C. (1983) *Science* 22, 524–527.

Biddison, W.E., Doherty, P.C. and Webster, R.G. (1977) *J. Exp. Med.* 146, 690–696.

Bittle, J.L., Houghton, R.A., Alexander, H., Shinnick, T.M., Sutcliffe, J.G., Lerner, R.A., Rowlands, D.J. and Brown, F. (1982) *Nature* 289, 30–33.

Bohl, E.H., Gupta, R.K.P., Olquin, F.M.W. and Saif, L.J. (1972) *Infect. Immun.* 6, 289.

Braciale, T.J., Andrew, M.E. and Braciale, V.L. (1981) *J. Exp. Med.* 153, 910–923.

Campbell, D., Sweet, C., Hay, A.J., Douglas, A., Skekel, J.J., Mason, T.L. and Smith, H. (1982) *J. Gen. Virol.* 58, 387–398.

Cappel, R., Sprecher, S., Richaert, F. and De Cuyper, F. (1982) *Arch. Virol.* 73, 61–67.

Chanock, R.M., Parrott, R.H., Kapikian, A.Z., Kim, H.W. and Brandt, C.D. (1968) In *Perspectives in Virology* (M. Pollard, Ed.) Academic Press, New York. pp. 125–135.

Darbyshire, J.H. and Peters, R.W. (1980) *Avian Pathol.* 9, 277–290.

Dauvergne, M., Brun, A. and Soulebot, J.P. (1983) *Dev. Biol. Stand.* 53, 245–255.

Davis, A.R., Nayak, D.P., Veda, M., Hiti, A.L., Dowbenko, D. and Kleid, D.G. (1981) *Natl. Acad. Sci. U.S.A.* 78, 5376–5381.

Djurickovic, S. and Thorsen, J. (1970) *Vet. Rec.* 87, 62.

Eichhorn, W.P., Backmann, P.A., Baljer, G., Plank, P. and Schneider, P. (1983) *Dev. Biol. Stand.* 53, 237–243.

Emini, E.A., Jameson, B.A. and Wimmer, E. (1983) *Nature* 304, 699–703.

Fields, B.N. and Greene, M.I. (1982) *Nature* 300, 19–23.

Fonteyne, J., Zissis, G. and Lambert, J.P. (1978) *Lancet* 1, 983.

Gaskell, R.M. and Wardley, R.C. (1977) *J. Small Anim. Pract.* 19, 1–16.

Gething, M.J. and Sambrook (1981) *Nature* 293, 620–625.

Gillespie, J.H. and Timoney, J.F. (1981) *Hagan and Bruner's Infectious Diseases of Domestic Animals.* Cornell University Press, Ithaca. pp. 665–676.

Green, N., Alexander, H., Olson, A., Alexander, S., Shinnick, T.M., Sutcliffe, J.G. and Lerner, R.A. (1982) *Cell* 28, 477–487.

Sherman, D.M., Acres, S.D., Sadowski, P.L., Springer, J.A., [illegible]
plat, C.C. (1983) *Infect. Immun.* 42, 653–658.
Six, H.R., Kramp, W.J. and Kasel, J.A. (1980) In *Liposomes* [illegible]
Six, Eds.) Elsevier/North-Holland, Amsterdam, pp. 119–131.
Smith, G.L., Mackett, M. and Moss, B. (1983) *Nature* 302, 490–[illegible]
Snodgrass, D.R., Stewart, J., Taylor, J., Krautil, F.L. and Smith, [illegible] 32,
70–73.
Steck, F., Wandeler, A., Bichsel, P., Capt, S., Hafliger, U. and [illegible]
Microbiol. Infect. Dis. 5, 165–[illegible]
Tenser, R.B., Ressel, S. and Dunstan, M.E. (1981) *Virology* 112, 328–[illegible]
Van der Heide, L., Kalbac, M. and Brustolon, M. (1983) *Avian Dis.* [illegible]
Waltner-Toews, D., Martin, S.W. and Meek, A.H. (1983) In [illegible]
rhea. (S.D. Acres, Ed.) VIDO, Saskatoon. In press.
Watson, R.J., Weiss, J.H., Salstrom, J.S. and Enquist, L. [illegible]
Winter, R.L. (1982) *Avian Pathol.* 11, 49–62.
Woode, G.N., Kelso, N.E., Simpson, T.F., Gaul, S.K., Evans, L.E. and [illegible]
Microbiol. 18, 358–364.
Yelverton, E., Norton, S., Obijeski, J.F. and Goeddel, D.V. [illegible]

immunogenicity of vaccines. However, regardless of all the scientific knowledge that is available, vaccine strategies must also consider the practical problems of animal husbandry practices and politics. In some instances the best scientific approach is not always practical under field conditions and a compromise often has to be reached. Unfortunately, such compromises result in reduced efficacy of the vaccines and economic losses continue. Finally better regulations for standardization of vaccines and quality control at government levels must be implemented. At present these vary from country to country and even between regions of some countries. This cannot be allowed to continue if we hope to achieve global control of viral infections.

6. References

Adorini, L., Harvey, M.A., Miller, A. and Sercarz (1979) *J. Exp. Med.* 150, 293–306.

Andrade, L.F., Villegas, P. and Fletcher, D.J. (1983) *Avian Dis.* 27, 178–187.

Ardans, A.A. (1982) *J. Am. Vet. Med. Assoc.* 181, 1150–1153.

Bachmann, P.A. and Hess, R.G. (1982) In *Viral Infections of the Gastrointestinal tract* (D.A.J. Tyrell and A.Z. Kapikian, Eds.) Marcel Dekker, New York. pp. 361–397.

Beaudette, F.R. and Hudson, C.B. (1933) *J. Am. Vet. Med. Assoc.* 82, 460–476.

Bennink, J., Effros, R.B. and Doherty, P.C. (1978) *Immunology* 35, 503–509.

Berman, P.W., Dowbenkco, D., Lasky, L.A. and Simonsen, C.C. (1983) *Science* 22, 524–527.

Biddison, W.E., Doherty, P.C. and Webster, R.G. (1977) *J. Exp. Med.* 146, 690–696.

Bittle, J.L., Houghton, R.A., Alexander, H., Shinnick, T.M., Sutcliffe, J.G., Lerner, R.A., Rowlands, D.J. and Brown, F. (1982) *Nature* 289, 30–33.

Bohl, E.H., Gupta, R.K.P., Olquin, F.M.W. and Saif, L.J. (1972) *Infect. Immun.* 6, 289.

Braciale, T.J., Andrew, M.E. and Braciale, V.L. (1981) *J. Exp. Med.* 153, 910–923.

Campbell, D., Sweet, C., Hay, A.J., Douglas, A., Skekel, J.J., Mason, T.L. and Smith, H. (1982) *J. Gen. Virol.* 58, 387–398.

Cappel, R., Sprecher, S., Richaert, F. and De Cuyper, F. (1982) *Arch. Virol.* 73, 61–67.

Chanock, R.M., Parrott, R.H., Kapikian, A.Z., Kim, H.W. and Brandt, C.D. (1968) In *Perspectives in Virology* (M. Pollard, Ed.) Academic Press, New York. pp. 125–135.

Darbyshire, J.H. and Peters, R.W. (1980) *Avian Pathol.* 9, 277–290.

Dauvergne, M., Brun, A. and Soulebot, J.P. (1983) *Dev. Biol. Stand.* 53, 245–255.

Davis, A.R., Nayak, D.P., Veda, M., Hiti, A.L., Dowbenko, D. and Kleid, D.G. (1981) *Natl. Acad. Sci. U.S.A.* 78, 5376–5381.

Djurickovic, S. and Thorsen, J. (1970) *Vet. Rec.* 87, 62.

Eichhorn, W.P., Backmann, P.A., Baljer, G., Plank, P. and Schneider, P. (1983) *Dev. Biol. Stand.* 53, 237–243.

Emini, E.A., Jameson, B.A. and Wimmer, E. (1983) *Nature* 304, 699–703.

Fields, B.N. and Greene, M.I. (1982) *Nature* 300, 19–23.

Fonteyne, J., Zissis, G. and Lambert, J.P. (1978) *Lancet* 1, 983.

Gaskell, R.M. and Wardley, R.C. (1977) *J. Small Anim. Pract.* 19, 1–16.

Gething, M.J. and Sambrook (1981) *Nature* 293, 620–625.

Gillespie, J.H. and Timoney, J.F. (1981) *Hagan and Bruner's Infectious Diseases of Domestic Animals.* Cornell University Press, Ithaca. pp. 665–676.

Green, N., Alexander, H., Olson, A., Alexander, S., Shinnick, T.M., Sutcliffe, J.G. and Lerner, R.A. (1982) *Cell* 28, 477–487.

Haflinger, U., Bichsel, P., Wandeler, A. and Steck, F. (1982) *Zentralbl. Veterinaer Med. Reihe* B. 29, 604–618.

Hawkes, R.A. and Lafferty, K.J. (1967) *Virology* 33, 250–261.

Heymann, D.L., Smith, E.L., Nakano, J.H., Jato, J.G., Martin, G.E. and Maben, G.K. (1982) *Br. Med. J.* 285, 531–534.

Hitzeman, R.A., Chen, C.Y., Hagie, F.E., Patzer, E.G., Liv, C.C., Estell, D.A., Miller, J.V., Yaffe, A., Kleid, D.G., Levinson, A.D. and Oppermann, H. (1983) *Nucleic Acid Res.* 11, 2745–2753.

Isacson, P. (1968) In *Perspectives in Virology* (M. Pollard, Ed.) Academic Press, New York. pp. 141–149.

Jericho, K.W.F., Yates, W.D.G. and Babiuk, L.A. (1982) *Am. J. Vet. Res.* 43, 1776–1780.

Jones, N. and Shenk, T. (1979) *Cell.* 17, 683–689.

Kamijio, Y., Ohkuma, S.I., Shimizu, M. and Shimizu, Y. (1977) *Natl. Inst. Anim. Health Tokyo* 17, 133–140.

Kew, O.M., Nottay, B.K. and Hatch, M.H. (1981) *J. Gen. Virol.* 56, 337–347.

Kim, H.W., Canchola, J.G., Brandt, C.D., Pyles, G., Chanock, R.M., Jensen, K. and Parrott, R.H. (1969) *Am. J. Epidemiol.* 98, 422–434.

Kit, M. and Kit., S. (1983) *Med. Clin. North Am.* 65, 1129–1145.

Kleid, D.G., Yansura, D., Small, B., Dowbenko, D., Moore, D.M., Grubman, M.J., McKercher, P.D., Morgan, D.O., Robertson, B.H. and Bachrach, H.L. (1981) *Science* 214, 1125–1129.

Klein, R., Buivovici-Klein, E., Moser, A., Monka, R. and Hilfenhaus, J. (1981) *Arch. Virol.* 68, 73–80.

Lawson, K.F., Johnston, D.H., Patterson, J.M., Black, J.G., Rhodes, A.J. and Zalan, E. (1982) *Can. J. Comp. Med.* 46, 382–385.

Lerner, R.A. (1983) *Sci. Am.* 248, 66–84.

Lerner, R.A., Cores, N., Alexander, H., Lin, F.T., Sutcliffe, J.G. and Shinnick, T.M. (1981) *Proc. Natl. Acad. Sci. U.S.A.* 78, 3403–3408.

Martin, S.W. (1983) *Can. J. Comp. Med.* 24, 10–19.

McDonald, J.W., Randall, C.J., McMartin, D.A. and Dagless, M.D. (1981) *Avian Pathol.* 10, 295–301.

Meszaros, J.F. (1983) *World Poultry Sci.* 39, 132–142.

Miller, J.F.A.P., Maloney, D.G., Warnke, R. and Levy, R. (1982) *N. Engl. J. Med.* 306, 517–522.

Murphy, B.R., Sly, M.D. and Tierney, L.L. (1982) *Science* 218, 1330–1332.

Nagi, S.A., Marquez, B. and Sahin, N. (1983) *Avian Dis.* 27, 623–631.

Norrby, E. (1983) *Arch. Virol.* 76, 163–177.

Peiris, J.S.M. and Porterfield, J.S. (1981) *J. Gen. Virol.* 57, 119–125.

Povey, R.C. (1976) *Can. Vet. J.* 17, 163–200.

Reagan, K.J., Wunner, W.H., Wiktor, T.J. and Koprowski, H. (1983) *J. Virol.* 48, 660–666.

Rosenquist, B.D. (1984) *N.A. Symp. Bovine Resp. Dis.* (R. Loan, Ed.) pp. 363–376.

Roizman, B., Warrne, J. and Thuning, C.A. (1981). *Dev. Biol. Stand.* 52, 287–304.

Roth, J.A. (1984) *N.A. Symp. Bovine Resp. Dis.* (R. Loan, Ed.) pp. 143–192.

Roth, J.A. and Kaeberle, M.L. (1982) *J. Am. Vet. Med. Assoc.* 180, 894–901.

Saif, L.J. and Smith, K.L. (1983). In *4th Int. Symposium on Neonatal Diarrhea.* (S.D. Acres, Ed.) VIDO, Saskatoon. pp. 394–423.

Saif, L.J., Redman, D.R., Smith, K.L. and Theil, K.W. (1983) *Infect. Immun.* 41, 1118–1131.

Samberg, V.E., Cuperstein, E., Bendheim, W. and Aronovici, I. (1971) *Avian Dis.* 15, 413–417.

Schneider, L.G. and Cox, J.H. (1983) *Tierarztl. Umschau.* 38, 315–324.

Schultz, R.D. (1982) *J. Am. Vet. Med. Assoc.* 181, 1142–1149.

Schoneweis, D.A. and Henry, S.C, (1982) *J. Am. Vet. Med. Assoc.* 181, 1154–1157.

Sharma, J.M. and Graham, C.K. (1982) *Avian Dis.* 26, 800–870.

Sharma, J.M. and Witter, R.L. (1983) *Avian Dis.* 27, 453–463.

Sherman, D.M., Acres, S.D., Sadowski, P.L., Springer, J.A., Bray, B., Raybould, T.J.G. and Muscoplat, C.C. (1983) *Infect. Immun.* 423, 653–658.
Six, H.R., Kramp, W.J. and Kasel, J.A. (1980) In *Liposomes and Immunobiology* (B.H. Tom and H.R. Six, Eds.) Elsevier/North-Holland, Amsterdam. pp. 119–131.
Smith, G.L., Machett, M. and Moss, B, (1983) *Nature* 302, 490–495.
Snodgrass, D.R., Stewart, J., Taylor, S., Krautil, F.L. and Smith, M.L. (1982). *Res. Vet. Sci.* 32, 70–73.
Steck, F., Wandeler, A., Bichsel, P., Capt., S., Hafliger, W. and Schneider, L. (1982) *Comp. Immun. Microbiol. Infect. Dis.* 5, 165–171.
Tenser, R.B., Ressel, S. and Dunstan, M.E. (1981) *Virology* 112, 328–341.
Van der Heide, L., Kalbac, M. and Brustolon, M. (1983) *Avian Dis.* 27, 698–706.
Waltner-Toews, D., Martin, F.W. and Meek, E.H. (1983) In *4th Int. Symposium on Neonatal Diarrhea.* (S.D. Acres, Ed.) VIDO, Saskatoon. In press.
Watson, R.J., Weiss, J.H., Salstrom, J.S. and Enquist, L. (1982) *Science* 218, 381–384.
Witter, R.L. (1982) *Avian Pathol.* 11, 49–62.
Woode, G.N., Kelso, N.E., Simpson, T.F., Gaul, S.K., Evans, L.E. and Babiuk, L.A. (1983) *J. Clin. Microbiol.* 18, 358–364.
Yelverton, E., Norton, S., Obijeski, J.F. and Goeddel, D.V. (1983) *Science* 219, 614–620.

Eds. M.H.V. van Regenmortel and A.R. Neurath
Immunochemistry of viruses. The basis for serodiagnosis and vaccines.

Antigenic structure of influenza viruses

14

G.M. AIR[1] and W.G. LAVER[2]

[1]Department of Microbiology, University of Alabama at Birmingham, Birmingham, AL 35294, U.S.A. and [2]Influenza Research Unit, John Curtin School of Medical Research, Australian National University, P.O. Box 334, Canberra, ACT 2601, Australia

Influenza viruses are classified into three types – A, B and C – on the basis of their type-specific nucleoprotein and matrix protein antigens. Type A influenza viruses are further classified into subtypes based on the antigenic characteristics of their surface antigens, hemagglutinin (HA) and neuraminidase (NA). Thirteen distinct HA subtypes and nine NA subtypes are now recognized in the nomenclature system for influenza A viruses recommended by the World Health Organization (WHO, 1980).

1. Properties of influenza A, B and C viruses

The highly pleomorphic particles of type A influenza virus (Fig. 14.1) are enclosed by a lipid envelope, derived from the plasma membrane of the host cell, to which the HA and NA antigens are attached by sequences of hydrophobic amino acids at one end of the molecules. Both surface antigens are glycosylated and some of the carbohydrate side chains possess antigenic activity characteristic of the host cell in which the virus grew.

Within the lipid envelope lies the matrix protein, which is believed to be structural in function. Within the matrix shell are eight single-stranded RNA molecules of negative sense (that is, the virion RNA is complementary to the messenger RNA) associated with a nucleocapsid protein and three large proteins, PA, PB1 and PB2, which are responsible for RNA replication and transcription (Krug, 1983). At least three virus-encoded nonstructural proteins are also found in infected cells. Table 14.1 summarizes current information on the eight genes of the influenza virus and the molecular weights of the proteins encoded by them.

TABLE 14.1

Influenza virus genome RNA segments and coding assignments

Segment	Length[a] (nucleotides)	Encoded polypeptide[b]	Nascent polypeptide length[c] (amino acids)	Mature polypeptides	Approx. no. molecules per virion[d]	Remarks
1	2341	PB2	759		30–60	Host cell RNA cap binding; component of RNA transcriptase[e]
2	2341	PB1	757		30–60	Initiation of transcription; possible endonuclease activity; component of RNA transcriptase[e]
3	2233	PA	716		30–60	Elongation of mRNA chains?; component of RNA transcriptase[e]
4	1778	HA	566	HA1 326 HA2 222	500	Surface glycoprotein; trimer; major antigenic determinant
5	1565	NP	498		1000	Associated with RNA segments to form ribonucleoprotein; structural component of RNA transcriptase; type-specific antigen

6	1413	NA	454	454	100	Surface glycoprotein; neuraminidase activity; tetramer; surface antigen
7	1027	M_1	252	252	3000	Major protein component of virus; underlies lipid bilayer; type-specific antigen
		M_2	96			Spliced mRNA; non-structural protein; function unknown[f]
		?	?9			Spliced mRNA; peptide predicted by nucleotide sequence only[f]
8	890	NS_1	230			Nonstructural protein; function unknown
		NS_2	121			Spliced mRNA; nonstructural protein; function unknown.[f]
Total	13588					

[a] For A/PR/8/34 strain: Winter et al.; see Lamb (1983) for full references.
[b] Determined by biochemical and genetic approaches.
[c] Determined by nucleotide sequence analysis and protein sequencing.
[d] Adapted from Compans and Choppin (1975).
[e] From Krug (1983).
[f] From Lamb (1983).

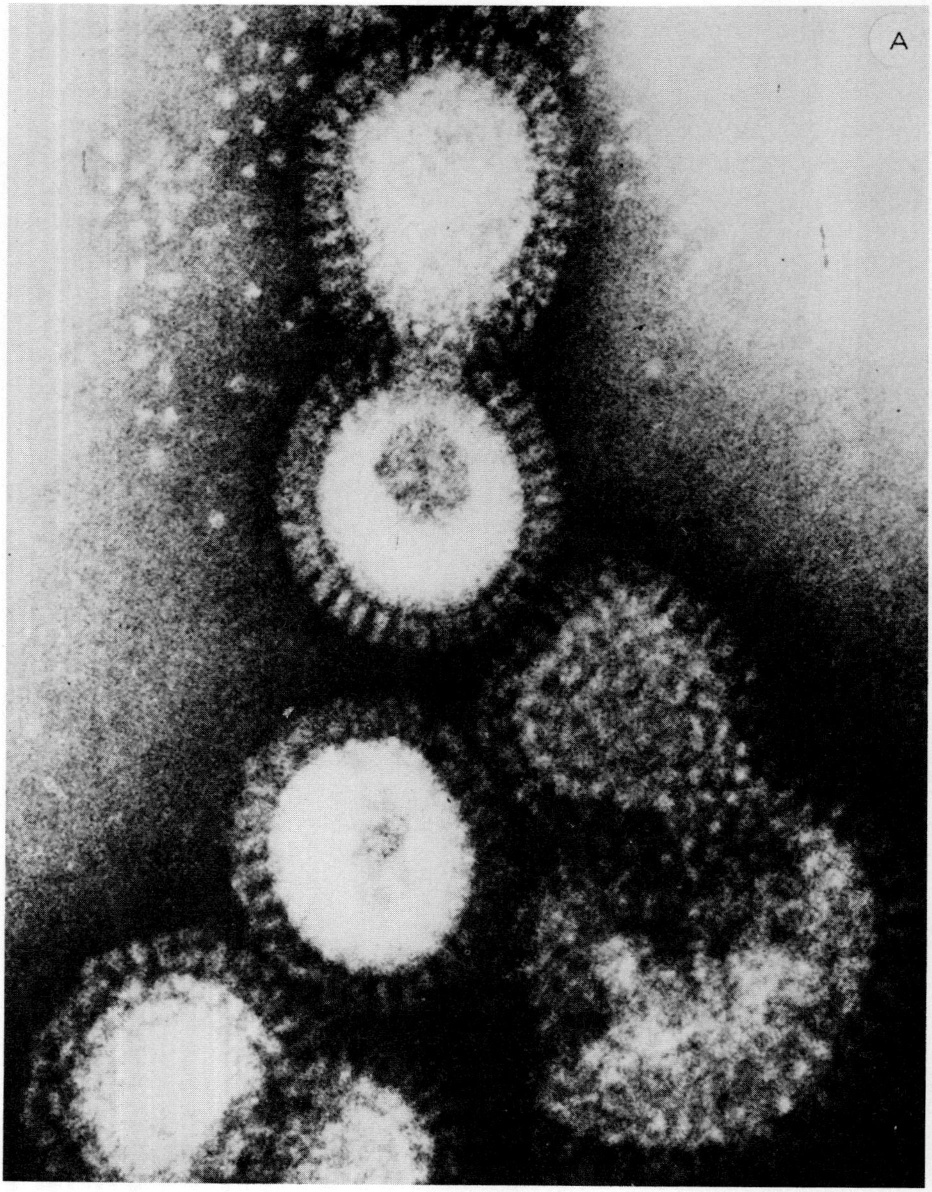

Fig. 14.1 Electron micrograph and diagram showing the surface structure of particles of influenza type A virus. There are eight RNA segments in the virus, each of which codes for one (in some cases for two, using overlapping reading frames) of the virus proteins.

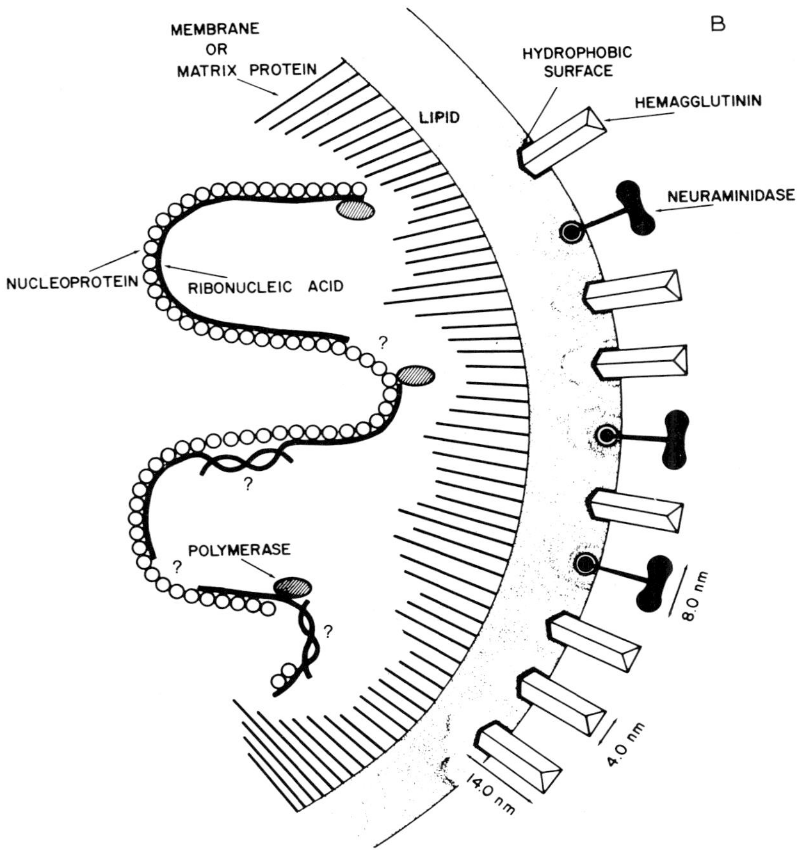

The properties of type B influenza virus particles are similiar to those of type A (except that an additional nonstructural protein, NB, is present; Lamb, 1983); but type C influenza virus differs considerably from the other two types.

Although the type C virus is much less well characterized than influenza A and B viruses, it differs most significantly in possessing a unique surface glycoprotein which resembles the HA of type A and B viruses in many of its biochemical properties.

The glycoproteins of the type C virus form characteristic hexagonal arrays on the surface of both spherical and filamentous forms of the virus (Fig. 14.2). Such arrays have not been observed on influenza A or B viruses, and it has been suggested (Herrler et al., 1981) that the presence of morphologically distinct trimeric HA and tetrameric NA of different symmetry may prevent lattice formation in these strains. Herrler et al. (1981) reported that the influenza C glycoprotein spikes viewed end-on in preparations negatively stained with sodium phosphotungstate (pH 7.2) have

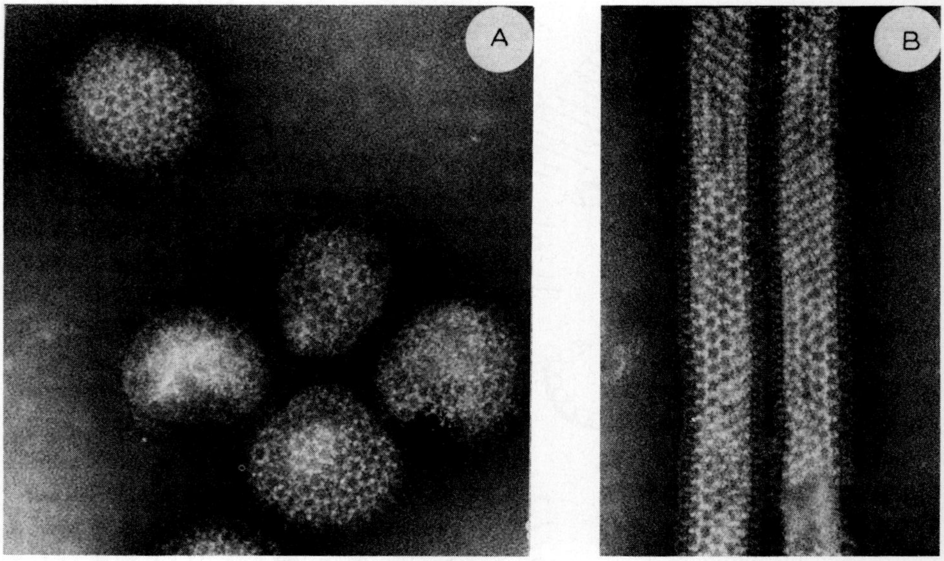

Fig. 14.2 Electron micrographs of influenza C virions. A, C/JHB/-1/66 virions grown in embryonated eggs. B, Filamentous influenza C/JHB/1/66 particles grown in chick embryo fibroblast cells. Negative staining with sodium phosphotungstate, pH 7.0. Micrographs courtesy of Dr. R.W. Compans.

a rounded end, in contrast with the triangular shape observed for influenza A HA (Wrigley, 1979).

However, the influenza C glycoprotein is clearly seen to be a trimer in specimens prepared with uranyl stains. Three-dimensional reconstructions from naturally occurring hexagonal arrays show that at low resolution (3 nm) the influenza C glycoprotein exhibits similar features to the HA glycoprotein of influenza A (Hewat et al., 1984). Both have a triangular stalk near the membrane. Further from the membrane, the stalk becomes broader and the monomers more separated, leaving an open centre. The molecule narrows at the top. The regions of greatest contact between adjacent trimers in the arrays are situated nearer the distal end of the molecule. These contact zones can be related to equivalent zones on the influenza A HA. Differences between the structure of the influenza A HA glycoprotein determined by x-ray analysis and reconstructions of the influenza C glycoprotein are greatest at either end of the molecule where the reconstructions are least reliable. Ordered glycoprotein arrays have not been observed on influenza C virions incubated at low pH. The staining patterns of glycoproteins on intact virions are essentially determined by the pH at which the virus is incubated and the stain type, but not the pH of the stain.

2. Antigenic variation

Two distinct kinds of antigenic variation have been demonstrated in influenza viruses, antigenic drift and major antigenic shifts. The first consists of relatively minor changes in the surface glycoproteins that occur gradually within a family of strains, all of which are clearly related to each other with respect to both internal and surface antigens. Among influenza A strains infecting man, each successive variant replaces pre-existing ones. This may be due to a selective advantage possessed by antigenic variants in overcoming immunological host resistance. Although antigenic variation occurs in influenza B viruses, it does not follow a sequential pattern as in influenza A viruses (Lu et al., 1983).

No evidence has been obtained for any antigenic changes in influenza C viruses since the first isolation of these agents in 1947. This antigenic stability resembles that of other respiratory agents, such as the parainfluenza viruses and respiratory syncytial virus, rather than influenza A or B viruses. Nucleotide sequences of the glycoprotein genes of two strains of influenza C (C/Johannesberg/66; Pfeifer and Compans, 1984 and C/California/78; Nakada et al., 1984) have been determined, and show 25 differences in amino acid sequence of the proteins. This finding is consistent with the observation that influenza C viruses have not undergone significant antigenic variation. The reason for this difference in stability of the surface antigen is not understood, particularly since influenza C virus appears to share with the influenza A and B viruses most features of virus structure as well as virus replication.

The second kind of antigenic variation, which has been described only for influenza type A, involves much more dramatic antigenic changes. These are referred to as major antigenic shifts. These major shifts, which occur at intervals of 10–15 years, are marked by the sudden appearance of 'new' subtypes in the population which cause the major pandemics of influenza.

3. Structure of the hemagglutinin

In type A influenza virus, the HA accounts for about 25% of the virus protein. It is responsible for the attachment of virus to cells and for the penetration of the virus into the cell during the initial stages of infection. Antibodies to the HA neutralize the infectivity of the virus. Variation in the HA glycoprotein is mainly responsible for the continually occurring outbreaks of influenza and for our inability to control these by vaccination.

The HA monomer is coded by the fourth largest RNA segment and is synthesized as a single polypeptide chain which undergoes post-translational cleavage in at least three places (Fig. 14.3). An N-terminal signal sequence is removed and, depending on the host cell and virus strain, the molecule is cleaved, with the removal of one or

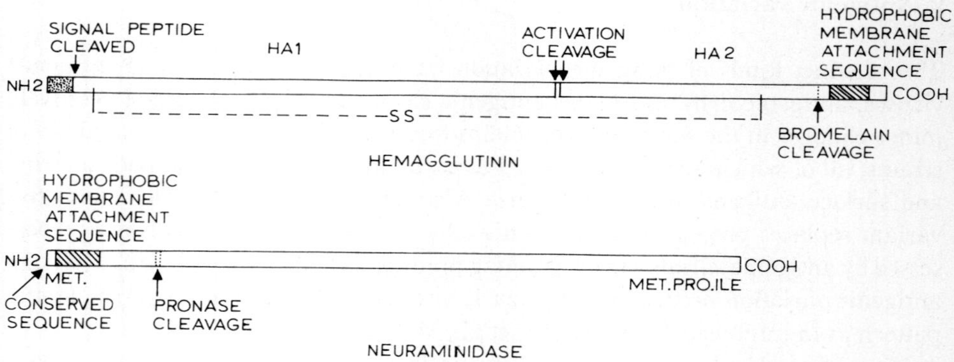

Fig. 14.3 Diagrammatic representation of certain features of the hemagglutinin and neuraminidase polypeptides. *Hemagglutinin:* The HA is synthesized as a single polypeptide. Following its synthesis, an N-terminal signal peptide is cleaved off and the molecule is cleaved further into HA1 and HA2 with the removal of one or more intervening amino acids. The latter cleavage is necessary for the virus to be infectious. HA1 and HA2 remain linked by a single disulfide bond and each HA spike contains three of these dimers. A sequence of hydrophobic amino acids near the C terminus of HA2 serves to anchor the HA in the lipid of the virus membrane. Treatment with bromelain removes this hydrophobic region without damaging the rest of the molecule, which, in some cases, can be crystallized. *Neuraminidase:* The neuraminidase is oriented in the virus membrane in the opposite way to the hemagglutinin. No post-translational cleavage of the NA polypeptide occurs, no signal peptide is split off, and even the initiating methionine is retained. No processing at the C terminus takes place; the C-terminal sequence Met-Pro-lle predicted from the gene sequence is found in intact NA molecules of the N2 subtype isolated from virus and in the pronase-released NA heads. A sequence of six polar amino acids at the N terminus of the NA polypeptide, which are totally conserved in the nine different NA subtypes, is followed by a sequence of hydrophobic amino acids which must represent the transmembrane region of the NA stalk. This sequence is not conserved at all between subtypes (apart from conservation of hydrophobicity). Pronase cleaves the polypeptide in the positions shown, removing the stalk and releasing the enzymically and antigenically active head of the NA, which, in some cases, can be crystallized.

more intervening residues, to give two polypeptide chains called HA1 and HA2, with molecular masses of 36 and 27 kDa respectively. HA1 and HA2 remain joined by a single disulfide bond and each HA 'spike' contains three of these HA1 and HA2 chains. A sequence of 25–32 hydrophobic amino acids near the C-terminus of HA2 anchors the HA in the virus membrane. Cleavage of the HA polypeptide into HA1 and HA2 is necessary for the virus particles to be infectious. Intact HA molecules can be isolated from detergent-disrupted virus particles. Soluble HA, lacking the hydrophobic membrane attachment sequence can be obtained by digesting virus particles with bromelain. This HA (BHA) can be crystallized and the three-dimensional structure of Hong Kong (H3N2) BHA has been determined (Wilson et al., 1981). It seems likely that HA molecules from other subtypes have similar structures. The HA glycoprotein of influenza virus is a trimer built of two structurally distinct regions: a triple-stranded coiled coil of α-helices extending 7.6 nm from the membrane, and a globular region of antiparallel β-sheet that contains the receptor binding site; the variable antigenic determinants are located on top of this stem (Wiley et al., 1981). Each subunit has an unusual loop-like topology; it begins at the membrane, extends 13.5 nm distally and folds back to enter the membrane.

Examination of the amino acids of the N-terminal sequence of HA2 of several influenza A viruses of human and avian origin has revealed a highly conserved region of 10 residues. The sequence is homologous to the N-terminal region of the fusion glycoprotein (F) of the paramyxovirus, Sendai virus (Gething et al., 1978). The F protein mediates fusion of the Sendai virus envelope with the plasma membrane of host cells, a function which is considered essential for virus infectivity. It has therefore been postulated that the infectivity of influenza virus is dependent on a fusion function of HA analogous to that of the Sendai fusion protein. Virus with uncleaved HA is not infectious, presumably because the N-terminus of HA2 is not accessible in uncleaved HA. Although the three-dimensional structure of the influenza HA shows that after cleavage the N-terminal HA2 peptide is buried in the molecule, it has recently been shown that, after incubation in conditions required for membrane fusion (pH 5.0; Maeda and Ohnishi, 1980) a conformational change occurs in the influenza HA glycoprotein. As a result the N-terminus of HA2 becomes exposed and the molecule acquires the ability to hemolyse erythrocytes while the bromelain-released HA is able to bind to lipid vesicles (Shekel et al., 1982).

4. Antigenic shift

An antigenic shift in type A influenza virus occurs when a virus suddenly appears in the human population with a 'new' HA and sometimes also a 'new' NA. Since the first human influenza virus was isolated in 1933, antigenic shifts have occurred in 1957 when the H2N2 subtype (Asian influenza) replaced the H1N1 subtype, in 1968 when the Hong Kong (H3N2) virus appeared and in 1977 when the H1N1 virus reappeared. All these major antigenic shifts in the virus occurred in China and anecdotal records suggest that previous epidemics (probably caused by the sudden emergence of 'new' viruses) also had their origin in China. Although we are still uncertain how these new subtypes appear or reappear in the human population, some clues are emerging from the plethora of recent sequence data.

Sequences up to 350 nucleotides from the 3′ end of the HA gene and the predicted amino acid sequences from 32 type A influenza viruses, including representatives of each of the 13 known HA subtypes, have been determined (Air, 1981; Hinshaw et al., 1982). Cysteine residues and certain other amino acids are conserved in all sequences, indicating that the 13 HA subtypes evolved from a common ancestor and share a common basic structure. When the partial amino acid sequences are compared pairwise, the most distant subtypes are H1 and H3 (25% homology). Other

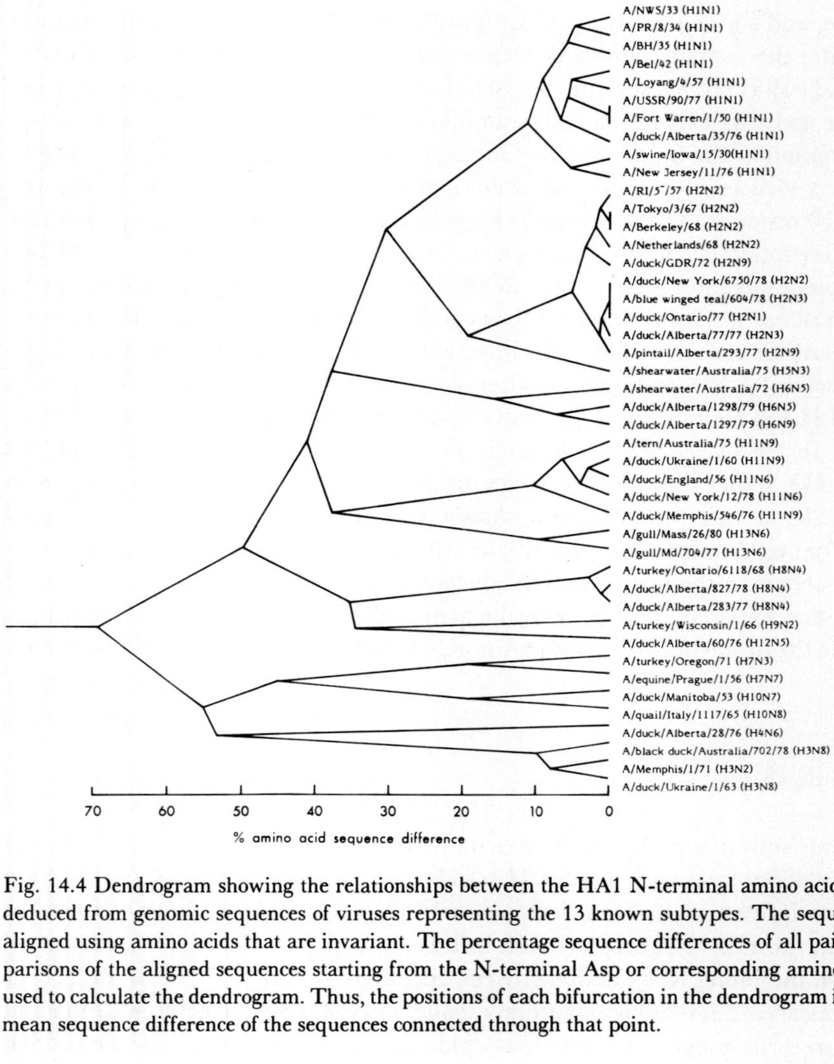

Fig. 14.4 Dendrogram showing the relationships between the HA1 N-terminal amino acid sequences deduced from genomic sequences of viruses representing the 13 known subtypes. The sequences were aligned using amino acids that are invariant. The percentage sequence differences of all pairwise comparisons of the aligned sequences starting from the N-terminal Asp or corresponding amino acid were used to calculate the dendrogram. Thus, the positions of each bifurcation in the dendrogram indicate the mean sequence difference of the sequences connected through that point.

subtypes are more similar to each other, the highest homology (80%) being between H2 and H5. Relationships between the partial sequences can be illustrated as a dendrogram (Fig. 14.4).

Within strains of a single subtype, the differences in deduced amino acid sequence are generally less than 10%, but differences in the N-terminal region of HA1 from two strains each of H7 and H10 viruses are close to 20%, almost equal to the difference between the amino acid sequences of the two closest subtypes (H2 and H5).

Whereas it is possible to imagine that the H2 viruses could eventually mutate into H5, the change to H3 must involve a more drastic mechanism. How, then, could the H3 hemagglutinin have replaced the H2 strain in the human population in 1968? First, the 'new' virus may have caused an epidemic in man many years previously and have remained hidden and unchanged in some unknown place ever since. Evidence for this kind of event has been obtained. The strain of 'Russian influenza' (H1N1) which reappeared in Anshan in northern China in May 1977 and subsequently spread to the rest of the world, seems to be identical, in all genes, to the virus which caused an influenza epidemic in 1950 (Nakajima et al., 1978; Scholtissek et al., 1978). Where was this virus for 27 years? As yet we have no answer.

Second, some of the 'new' viruses may be derived from animal or avian viruses. One strain of human influenza has been shown to be such a recombinant (reassortant). The Hong Kong (H3N2) virus contains the NA (and other) genes from an Asian (H2N2) strain of human influenza and a HA which is antigenically related to that of A/duck/Ukraine/63 (H3N8) and A/equine/2/Miami/63 (H3N8) viruses. The amino acid sequence homology between the HAs of A/duck/-Ukraine/63 and A/Aichi/2/68 viruses (both of subtype H3) is 96% (Fang et al., 1981; Ward and Dopheide, 1982). We do not know if it was an animal or bird virus which donated the HA gene during the recombination event that led to the formation of the Hong Kong strain. The donating virus could have been one which contained Hong Kong HA that had persisted from a much earlier human influenza epidemic and maintained unchanged in the same way as the Russian influenza virus. Antibodies in the sera of people who were born around 1900 suggest that a virus with a HA similar to that of the Hong Kong virus caused influenza at the time, but the NA of this virus was N8, not N2. It is possible that viruses such as equine (H3N8) and duck/Ukraine have evolved from the 1900 epidemic strain.

The third way in which new viruses could appear in the human population is by an animal or bird virus acquiring infectivity for man. This could be brought about by mutation, but the number of mutations required and the genes in which these would need to occur are as yet unknown.

It is possible that each of these mechanisms has operated at one time or another to produce a 'major shift' in influenza viruses infecting man. We do not know how the different subtypes of type A influenza arose, but it is reasonable to suppose that 'shuffling' of genes between influenza A strains is occurring all the time, most frequently in birds, but also in mammals, including man.

5. Antigenic drift

Antigenic drift in the HA of influenza A virus occurs by mutations in the gene, leading to an accumulation of amino acid sequence changes that alter the antigenic sites in such a way that they are no longer recognized by the host's immune system.

The HA from a large number of strains of influenza virus of types A and B have been completely sequenced (see Ward, 1981) but it is not possible to tell from this data precisely which of the sequence changes were responsible for changes in the antigenic properties of the HA.

6. Variants selected with monoclonal antibodies

Antigenic variants of several strains of HA have been selected by growing virus in the presence of monoclonal antibody to the HA (Laver et al., 1979; Gerhard et al., 1981; Caton et al., 1982; Daniels et al., 1983; Nakajima et al., 1983). The variants occurred at a frequency of 1 in 10^{-5} in virus stocks and did not bind at all to the antibody used for their selection. This dramatic change in antigenicity was found, in most cases, to be associated with single changes in the amino acid sequence of HA1 (Table 14.2) and these sequence changes were certainly related to antigenic changes.

The changes are scattered along the primary sequence of HA1 (Fig. 14.5) but in

```
Q D L P G N D N S T A T L C L G H H A V   20
P N G T L V K T I T D D Q I E V T N A T   40
E L V Q S S S T G K I C N N P H R I L D   60
G I D C T L I D A L L G D P H C D V F Q   80
N E T W D L F V E R S K A F S N C Y P Y  100
D V P D Y A S L R S L V A S S G T L E F  120
I T E G F T W T G V T Q N G G S N A C K  140
R G P G S G F F S R L N W L T K S G S T  160
Y P V L N V T M P N N D N F D K L Y I W  180
G I H H P S T N Q E Q T S L Y V Q A S G  200
R V T V S T R R S Q Q T I I P N I G S R  220
P W V R G L S S R I S I Y W T I V K P G  240
D V L V I N S N G N L I A P R G Y F K M  260
R T G K S S I M R S D A P I D T C I S E  280
C I T P N G S I P N D K P F Q N V N K I  300
T Y G A C P K Y V K Q N T L K L A T G M  320
R N V P E K Q T 328            HA2
```

Fig. 14.5 Amino-acid sequence of the HA1 polypeptide of A/Hong Kong/68 hemagglutinin (Verhoeyan et al., 1983). Residues which changed in variants selected with monoclonal antibodies are shown (Laver et al., 1981; Newton et al., 1983; Daniels et al., 1983). The sequence changes which occurred in monoclonal variants of other influenza viruses are shown in Table 14.2.

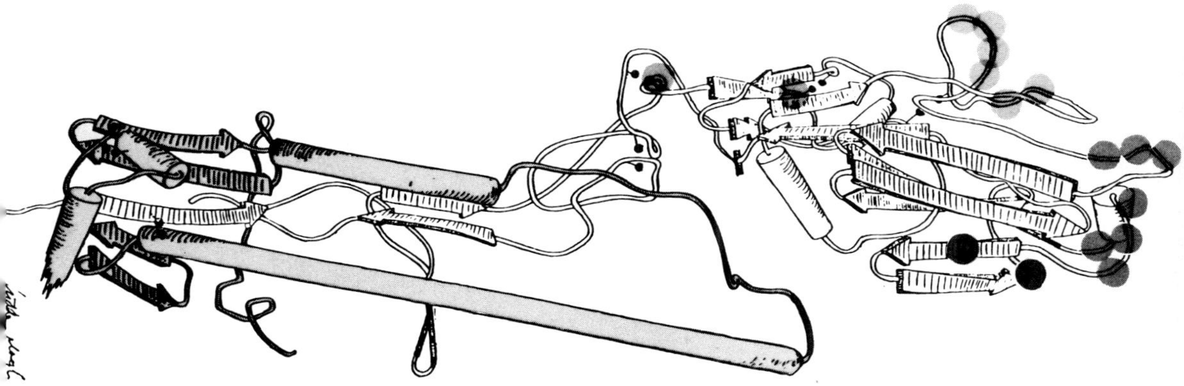

Fig. 14.6 This illustration (by Hidde Pleogh) of the Hong Kong/68 (H3) hemagglutinin monomer shows folding of the HA1 and HA2 polypeptides (Wilson et al., 1981) and the location of amino-acid sequence changes in variants selected with monoclonal antibodies which delineate four distinct antigenic sites (Wiley et al., 1981). These variants, which occur at a frequency of 1 in 10^5 of wild-type virus, have totally lost the ability to bind the particular monoclonal antibody used in the selection procedure.

the case of H3 HA cluster into four distinct sites (A—D) when the HA1 polypeptide folds into its three-dimensional structure (Fig. 14.6). Since the sequence changes affect antibody binding, it is clear that the sites must be involved in the antigenicity of the HA. Many of the sequence changes found in natural variants of Hong Kong influenza virus (field strains) also are within these sites (Both and Sleigh, 1981).

7. Antigenic variation in influenza B virus

Sequences of the HA gene of several field isolates of influenza B virus have been determined (Krystal et al., 1983; Verhoeyen et al., 1983; Berton et al., 1984) and it appears that variable areas in the influenza B virus HA structure correspond to antigenic regions A, B and D of influenza A viruses. No changes were found in the influenza B virus HAs in areas corresponding to antigenic region C of influenza A viruses.

The available sequence data for HA1 of influenza B HAs is shown in Fig. 14.7. Of 38 amino acid positions where variation is seen, 17 are conserved in all strains except B/Lee/40, and it has previously been noted that residue 90 of B/Lee/40 from another laboratory is valine, as in all other strains (Air and Compans, 1983). Otherwise substitutions tend to come and go over time, in contrast to the steady accumulation of amino acid substitutions with time found in HA1 of influenza A viruses of the H3 subtype (Both and Sleigh, 1981). Neither antigenic analyses (Lu et al., 1983) nor the sequence analyses indicate that antigenic drift occurs sequentially over time as in influenza A viruses.

TABLE 14.2

Amino acid sequence changes in the HA of variants selected with monoclonal antibodies

H1[a]			H3[b]			B[c]		
Position in HA1[d]	Sequence change	Antigenic region	Position in HA1	Sequence change	Antigenic region	Position in HA1[d]	Sequence change	Antibody used for selection
70 (78)	Leu to Pro	Cb	53	Asn to Lys	C	141 (137A)	Gly to Val	HK 21/6
71 (79)	L*qd* Pro	Cb	63	Asp to Asn	E	147 (142)	Thr to Ala	HK 74/1
73 (81)	Val to Glu	Cb	133	Asn to Lys	A	150 (145)	Asn to Ser	HK 21/6
74 (82)	Arg to Gly	Cb	135	Gly to Arg	A	162 (157)	Lys to Thr	HK 282/1
75 (83)	Ser to Pro	Cb	143	Pro to Leu	A	163 (158)	insert Asn	HK 16/1, 313/2, 430/1
115 (122)	Glu to Lys	Cb	143	Pro to Ser	A	166 (160A)	Lys to Ile	Ore 680/1, 710/1
124 (128)	Pro to Ser	Sa	143	Pro to Thr	A	166 (160A)	Lys to Asn	HK 74/1
125 (129)	Asn to Lys	Sa				196 (187)	Asn to Asp	Ore 401/2
125 (129)	Asn to Tyr	Sa	143	Pro to His	A	196 (187)	Asn to Lys	Ore 146/1
136 (140)	Ser to Pro	Ca2	144	Gly to Asp	A	197 (188)	Glu to Lys	Ore 280/2
139 (143)	Gly to Arg	Ca2	144	Gly to Ser	A	199 (190)	Gln to Lys	Ore 21/2
141 (145)	Ser to Ala	Ca2	145	Ser to Lys	A	201 (192)	Val to Leu	Ore 206/2
152 (156)	Glu to Gly	Sb	145	Ser to Asn	A	201 (192)	Val to Glu	Ore 113/2
154 (158)	Glu to Val	Sa	146	Gly to Ser	A	201 (192)	Val to Ala	Ore 152/2
155 (159)	Gly to Asp	Sb	156	Lys to Glu	B	202 (193)	Lys to Thr	Ore 238/4
156 (160)	Ser to Leu	Sa	157	Ser to Leu	B	202 (193)	Lys to Asn	Ore 10/4 HK 74/1
158 (162)	Pro to Leu	Sa	158	Gly to Glu	B	202 (193)	Lys to Arg	HK 365/1

159 (163)	Lys to Thr	Sa	188	Asn to Asp	B	202 (193)	Lys to Gly	HK 74/1
161 (165)	Lys to Glu	Sa	189	Gln to His	B	207 (196B)	Ser to Pro	Ore 128/2
161 (165)	insert Lys	Sa				207 (196B)	Ser to Leu	Ore 391/1
162 (166)	Asn to Lys	Sa	193	Ser to Arg	B	240 (225)	Pro to Thr	HK 74/1
163 (167)	Ser to Tyr	Sa	198	Ala to Glu	B	240 (225)	Pro to Gln	Ore 114/4
163 (167)	Ser to Phe	Sa	198	Ala to Val	B			
165 (169)	Val to Ala	Cal	205	Ser to Tyr	D			
169 (173)	Gly to Val	Cal	218	Gly to Trp	D			
169 (173)	Gly to Arg	Cal						
178 (182)	Ile to Met	Cal						
188 (192)	Gln to Leu	Sb						
188 (192)	Gln to Arg	Sb						
189 (193)	Asn to Ser	Sb						
189 (193)	Asn to Lys	Sb						
192 (196)	Gln to Arg	Sb						
194 (198)	Glu to Gly	Sb						
203 (207)	Ser to Leu	Cal						
220 (224)	Arg to Ile	Ca2						
221 (225)	Asp to Gly	Ca2						
236 (240)	Gly to Arg	Cal						
236 (240)	Gly to Glu	Cal						
270 (273)	Ser to Thr	Cal						

[a] Data from Caton et al. (1982).

[b] Data from Daniels et al. (1983). Laver et al. (1979; 1981) and Newton et al. (1983).

[c] Data from Hovanec and Air (1984) and Berton et al. (1984).

[d] Equivalent position in H3HA1 is given in parentheses.

```
                             10                  20                  30                  40                  50                  60
B/LEE/40    D R I C T G I T S S N S P H V V K T A T Q G E V N V T G V I P L T T T P T K S H F A N L K G T Q T R G K L C P N C F N C
B/MD/59                                                                                                                       L
B/HK/73                                                                                                                       L
B/SING/79                                                                                             K                   L
B/ORE/80                                                                                              K                   L

                             70                  80                  90                 100                 110                 120
B/LEE/40    T D L D V A L G R P K C M G N T P S A K V S I L H E V K P A T S G C F P I M H D R T K I R Q L P N L L R G Y E N I R L S
B/MD/59         M                       T I                           V
B/HK/73                                 T I         A                 V
B/SING/79                               T I         A                 V
B/ORE/80                                T I         A                 V

                            130                 140 *           *   150#                160   * *     #     170                 180
B/LEE/40    T S N V I N T E T A P G G P Y K V G T S G S C P N V A N G N G F F N T M A W V I P K D N N K T A I N P V T V E V P Y I C
B/MD/59       R         A                 T                     T     K       E         A V     N K         T     L
B/HK/73     A R     T   A                 I                     T             E         A V     - -         T     L
B/SING/79     R         A   R             I I                   T             A         A V     - -         T     L
B/ORE/80      R         A   R             I I                   T             A         A V       -         T     L

                                                      #
                            190           # #  #200 # *         #   210                 220                 230                 240#
B/LEE/40    S E G E D Q I T V W G F H S D D K T Q M E R L Y G D S N P Q K F T S S A N G V T T H Y V S Q I G G F P N Q T E D E G L K
B/MD/59     T K                             E   L   V I           K
B/HK/73     T K                             E       V K           K                                                   A           P
B/SING/79   T K                           T E       V K           K                                                         G     P
B/ORE/80    T K                           N E A     V K           K                                                         G     P

                            250                 260                 270                 280                 290                 300
B/LEE/40    Q S G R I V V D Y M V Q K P G K T G T I V Y Q R G I L L P Q K V W C A S G R S K V I K G S L P L I G E A D C L H E K Y G
B/MD/59                       I                               V
B/HK/73                                             A         V                         R       E
B/SING/79                                                     V
B/ORE/80                                                      V

                            310                 320                 330                 340                 350
B/LEE/40    G L N K S K P Y Y T G E H A K A I G N C P I W V K T P L K L A N G T K Y R P P A K L L K E R
B/MD/59
B/HK/73
B/SING/79
B/ORE/80
```

Fig. 14.7 Sequences of the HA1 polypeptide of influenza B strains. The sequence of B/Lee/40 (Krystal et al., 1982) is shown, together with differences in B/Maryland/59 (Krystal et al., 1983), B/Hong Kong/8/73 (Hovanec and Air, 1984), B/Singapore/222/79 (Verhoeyen et al., 1983) and B/Oregon/5/80 (Berton et al., 1984). Amino acids which change in variants of B/Oregon/5/80 (#) and B/Hong Kong/8/73 (★) are indicated.

The three-dimensional structure of the influenza B HA is not known but comparison of sequence data and computer-assisted model building suggested a structural similarity between influenza A and B HAs (Krystal et al., 1982).

To further investigate antigenic variation in influenza B viruses, variants have been selected with monoclonal antibodies and the sequence changes determined. Two strains have been used: B/Hong Kong/8/73 (Webster and Berton 1981; Hovanec and Air, 1984) and B/Oregon/5/80 (Berton et al., 1984). For comparison, amino acid positions in the HA1 peptide can be described using the corresponding A/Aichi/2/68 residue numbers (Wiley et al., 1981 as aligned by Krystal et al., 1983).

In the B/Hong Kong/8/73 HA1 polypeptide there are 2 areas of antigenic variability (Fig. 14.7). These areas would correspond to antigenic regions A and B of the A/Aichi/2/68 HA molecule (Fig. 14.6). Amino acid substitutions found in these areas include residues 137 and 142 (region A) and residues 157, 158 and 193 (region B). However, two variants have a further nucleotide substitution at position 1742 which changes amino acid 211 of the HA2 peptide. This residue is located within the transmembrane region of the B/Hong Kong/8/73 HA and is therefore not even present in the bromelain-released HA structure of A/Aichi/2/68 (Wilson et al., 1981), and is unlikely to be involved in antigenicity.

A noticeable feature of the sequences of HA genes of 'monoclonal' variants of B/Hong Kong/8/73 is the relatively high incidence of multiple nucleotide changes; 4 of the 8 'monoclonal' variants sequenced had more than one nucleotide change.

The nucleotide sequences of the HA1 portions of the HA genes of the parent strain and of 18 laboratory-selected antigenic variants of B/Oregon/5/80 have been determined (Berton et al., 1984). Comparison of the deduced amino acid sequences of the parental and variant HA1 polypeptides revealed 16 different amino acid substitutions at 9 positions. All amino acid substitutions resulted from single point mutations and no double mutants were detected. The amino acid changes cluster in 4 regions in the primary structure of the influenza B virus HA and appear to identify at least 2 immunodominant sites which correspond to proposed antigenic sites A and B on the influenza A virus H3 HA. The positions of change are indicated in Fig. 14.7. In studies of influenza A virus only single amino acid changes were found and monoclonal antibodies which recognized one region did not recognize any of the other antigenic sites. In both the influenza B virus studies monoclonal antibodies which belong to the same antigenic group recognize amino acid changes in regions corresponding to both antigenic sites A and B of the H3 HA (Berton et al., 1984; Hovanec and Air, 1984).

One interpretation of these results is that the loop regions of antigenic sites A and B are in closer proximity to each other in the influenza B than in the influenza A structure, although the basic three-dimensional structure of the influenza B HA may be similar to that of A/Aichi/2/68. The distance between antigenic sites A and B of the Aichi HA (Wilson et al., 1981) could possibly be spanned by an antibody-combining region, but results with H3 and H1 variants have not indicated that it is.

The sequence data of the variants suggest that there is a considerable overlap between the antigenic regions of the influenza B HA molecule. Distinct antigenic sites, as found in the HA of A/Aichi/2/68, may not exist as such in the HA of influenza B viruses. The HA1 peptide in type B viruses may be folded into a more compact structure, so that a relatively larger proportion of the sequence is recognized by a single monoclonal antibody.

8. Structure of the neuraminidase

The NA accounts for about 5–10% of the virus protein and exists as a mushroom-shaped spike on the surface of the virion. It is a tetramer with a box-shaped head, 10.0 × 10.0 × 6.0 nm, made out of four coplanar and roughly spherical subunits and a centrally attached stalk containing a hydrophobic region by which it is embedded in the viral membrane (Fig. 14.8). Several roles have been suggested for NA (acylneuraminyl hydrolase, EC 3.2.1.18). The enzyme catalyses cleavage of the α-ketosidic linkage between terminal sialic acid and an adjacent sugar residue. This reaction permits transport of the virus through mucin and destroys the HA receptor on the host cell, thus allowing elution of progeny virus particles from infected cells (Palese et al., 1974). The removal of sialic acid from the carbohydrate moiety of newly synthesized HA and NA is also necessary to prevent self-aggregation of the virus. In general, then, the role of NA may be to facilitate mobility of the virus both to and from the site of infection.

The NA molecule is composed of a single polypeptide chain, coded by RNA segment 6, and it is oriented in the virus membrane in the opposite way to HA (Fig. 14.3). No post-translational cleavage of the NA polypeptide occurs, no signal peptide is split off and even the initiating methionine is retained (Blok et al., 1982). No processing at the C-terminus takes place – the C-terminal sequence (-Met-Pro-Ile) predicted from the gene sequence for N2 NA is found in intact NA molecules isolated from virus, and in the pronase-released NA heads. A sequence of six polar amino acids at the N-terminus of the NA polypeptide, which is totally conserved in each of the 9 different NA subtypes, is followed by a sequence of hydrophobic amino acids which must represent the transmembrane region of the NA stalk. This sequence is not conserved at all between subtypes (apart from conservation of hydrophobicity). Pronase cleaves the polypeptide in the positions shown (Fig. 14.3), removing the stalk and releasing the enzymically and antigenically active head of the NA which, in some cases, can be crystallized (Fig. 14.9).

The three-dimensional structure of N2 neuraminidase heads determined from an electron density map at 0.29 nm resolution shows that each monomer is composed of six topologically identical β-sheets arranged in a propeller formation (Varghese et

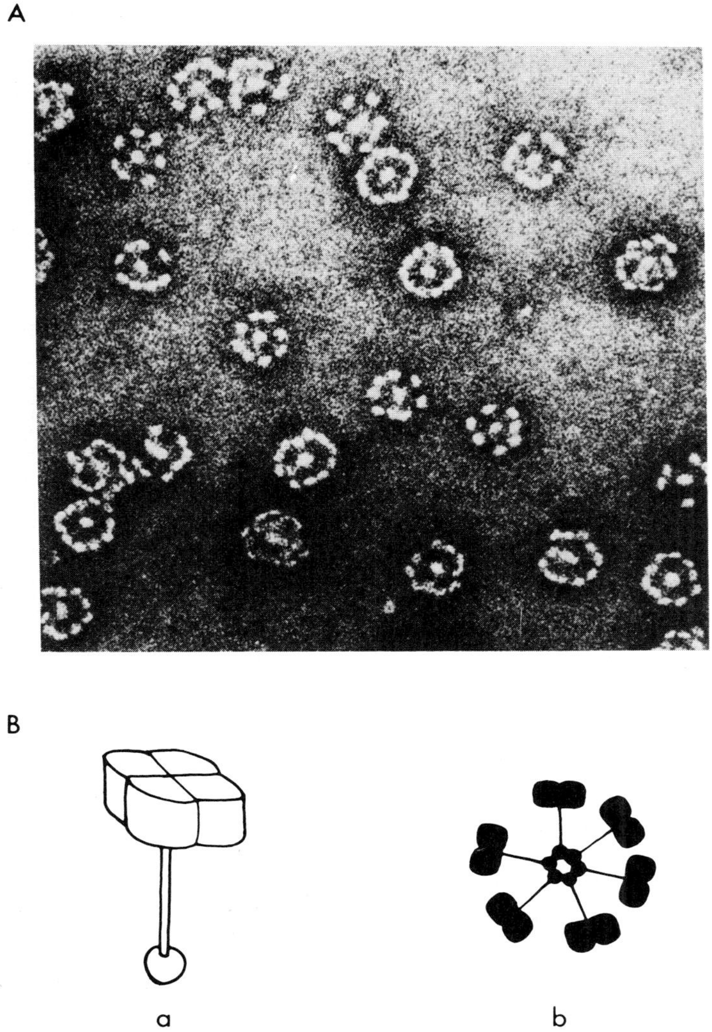

Fig. 14.8 Electron micrograph (A) and diagram (B) showing detergent-released neuraminidase molecules from which the detergent has been removed. The NA molecules have aggregated by the hydrophobic region near the end of the stalk (b), which serves to attach the NA to the lipid of the virus envelope. Treatment of virus particles with pronase cleaves the stalk and releases the head of the NA, which carries the enzymic and antigenic activities of the molecule and, in some cases, can be crystallized. (Electron micrograph courtesy of Nick Wrigley).

al., 1983). The tetrameric enzyme has circular 4-fold symmetry and is stabilized in part by metal ions bound on the symmetry axis. The catalytic sites are located in deep clefts which occur on the upper corners of the box-shaped tetramer. Sugar

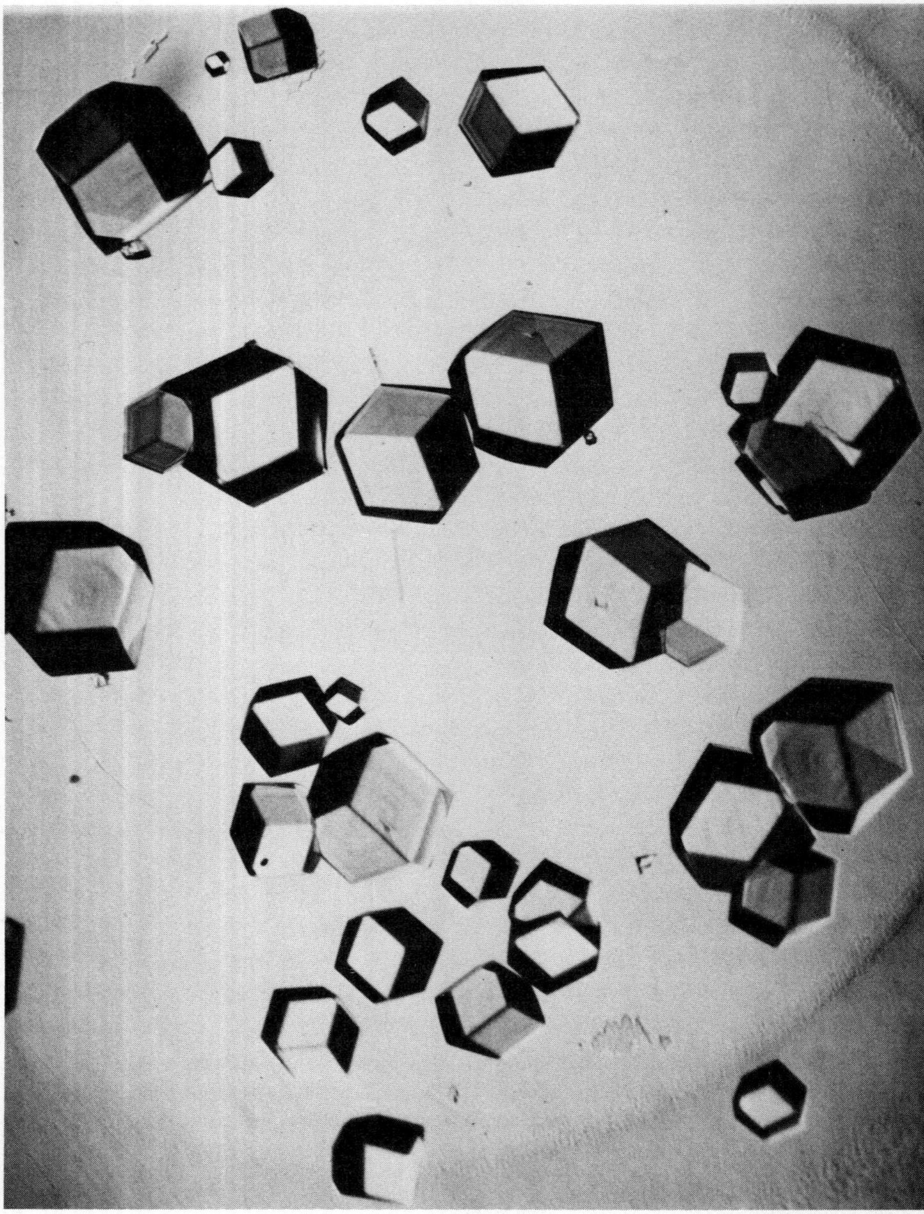

Fig. 14.9 Crystals of pronase-released neuraminidase heads of the N9 neuraminidase subtype suitable for x-ray diffraction studies. The crystals were grown from phosphate buffer solution by vapour diffusion.

residues are attached to four of the five potential glycosylation sequences, and in one case contribute to the interaction between subunits in the tetramer (Fig. 14.11).

Influenza B virus RNA segment 6 codes two distinct glycoproteins, NA and NB, using overlapping reading frames. The amino acid sequence of polypeptide NB deduced from the nucleotide sequence of the B/Lee/40 strain consists of 100 amino acids with a molecular weight of 11 242 (Shaw et al., 1982). The sequence contains four potential glycosylation sites, and the protein has been found to be glycosylated in infected cells. NB has not been found in virions. A protein analogous to NB has not been found in influenza A virus and this represents a major difference between the two virus types.

8.1. Antigenic variants of neuraminidase

A number of monoclonal antibodies to neuraminidase of the N2 subtype have been obtained. Competitive radioimmunoassays with these antibodies showed that the NA possessed at least 4 overlapping antigenic regions. Some of the antibodies effectively neutralized the infectivity of the virus, but careful experiments showed that these antibodies to the NA did *not* prevent the virus from infecting cells. However, low concentration of the antibodies in the fluid surrounding the cells prevented the release of virus from cells. These monoclonal antibodies could be used to select antigenic variants of the NA (Webster et al., 1982; 1984). These variants were found to have single amino acid sequence changes in the NA polypeptide (Fig. 14.10). The regions in which these changes occurred were certainly involved in the antigenicity of the NA. The location of the sequence changes on the three-dimensional structure of the NA (Colman et al., 1983) is shown in Figs. 14.11 and 14.12.

It is not known, however, if the amino acids which changed formed part of the antigenic determinants recognized by the monoclonal antibodies used to select the variants, or whether the sequence alterations caused conformational changes in the NA which in turn altered an antigenic site some distance away. Attempts have been made to answer this question by examining the effects of chemical modification of certain amino acid residues on the antigenic properties of the NA. N2 NA was reacted with succinic anhydride until 70% of the lysine was modified. Despite the drastic alteration in charge, the succinylated NA reacted with polyclonal antisera in immunodiffusion tests to almost the same extent as untreated NA (Fig. 14.13). This surprising finding suggests that lysines do not play a role in the antigenic sites of the NA (Laver et al., 1983). When the succinylated NA was tested with a panel of monoclonal antibodies some failed to bind, indicating that a small number of antigenic sites were destroyed, but why the majority were unaffected by succinylation remains a mystery. Similar results were obtained when other reagents were used, and when similar experiments were done using HA (Laver et al., 1981).

Despite the extensive studies using monoclonal antibodies and sequence analysis

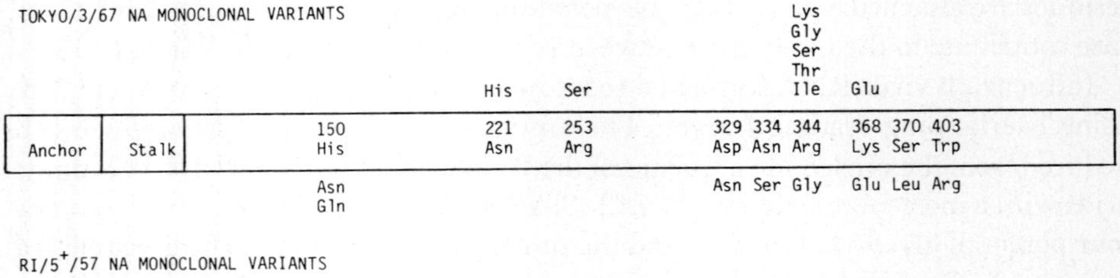

Fig. 14.10 Amino acid substitutions found in variants of neuraminidase selected with monoclonal antibodies of Tokyo/3/67 (Lentz et al., 1984) and RI/5$^+$/57 (Air and Els, unpublished results).

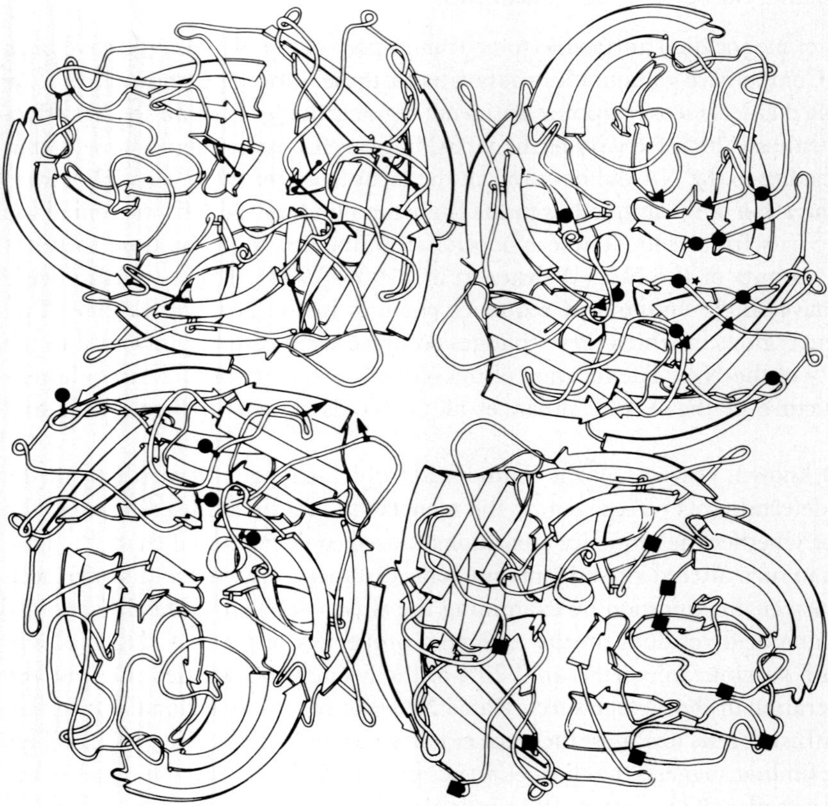

Fig. 14.11 Schematic diagram of the neuraminidase tetramer viewed from above down the symmetry axis. The four subunits highlight different features of the structure (Colman et al., 1983).
Top left: disulfide bonds (joined points). Bottom left: carbohydrate attachment sites at positions 86, 146, 200 and 234 (•) and metal ligands ASP 113 and ASP 114 (arrows). Bottom right: Residues which changed in variants selected with monoclonal antibodies (■). Top right: conserved acidic (●) and basic (▲) residues in influenza A and B neuraminidase and the sialic acid binding site (★).

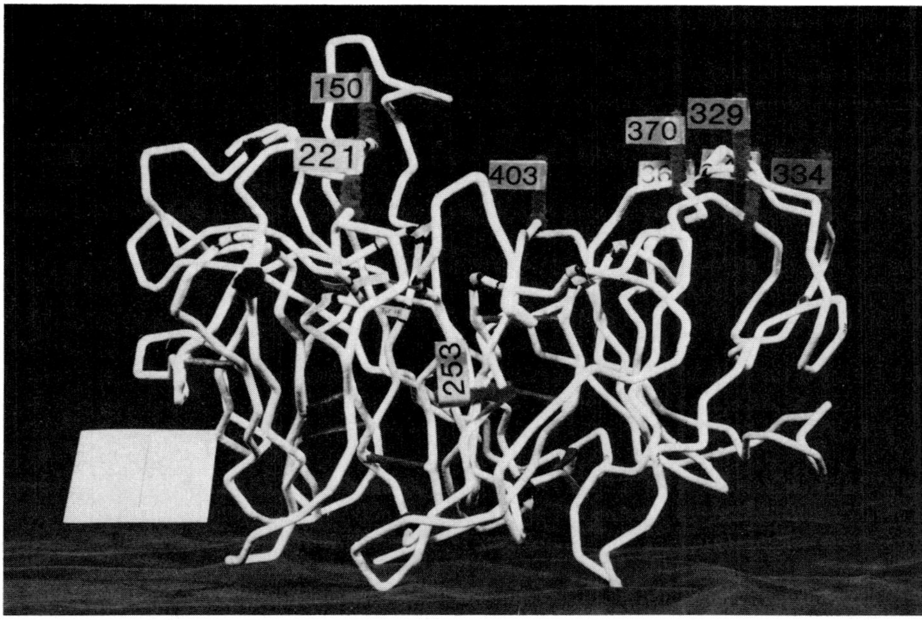

Fig. 14.12 Photograph of a wire model of N2 neuraminidase showing the amino acid positions of monoclonal variant changes. One subunit of the pronase-released head is shown and the bent wire represents the alpha carbon chain tracing. Amino acid residues which have changed in the monoclonal variants are flagged. Residues 329, 334, 344, 368 and 370 are located on the upper surface near one side which is on the outside of the tetramer. Residues 150 and 221 are similarly on the upper surface of the molecule. Residue 253 is located in a pocket on the underside of the protein. The cross approximately represents the 4-fold axis of symmetry in the mature tetramer and 'banded' residues are glycosylated. Data for constructing the model was from Varghese et al. (1983). It should be stressed that these are not the only antigenic regions on neuraminidase; variants can only be selected by those antibodies which prevent virus growth (Webster et al., 1984).

of variants, the nature and extent of the interaction between influenza antigens and antibody molecules is not known. Precise and unambiguous information about the structure of individual antigenic sites on HA and NA, and the way in which these change during antigenic drift, may come only from x-ray diffraction analyses of crystals of complexes between the Fab fragments of monoclonal antibodies and the antigens (Colman et al., 1982). This work is currently being undertaken.

9. Antigenic structure of internal proteins

The matrix protein and nucleoprotein are used to determine the type specificity of influenza viruses. It is known that antigenic variation in these proteins occurs inde-

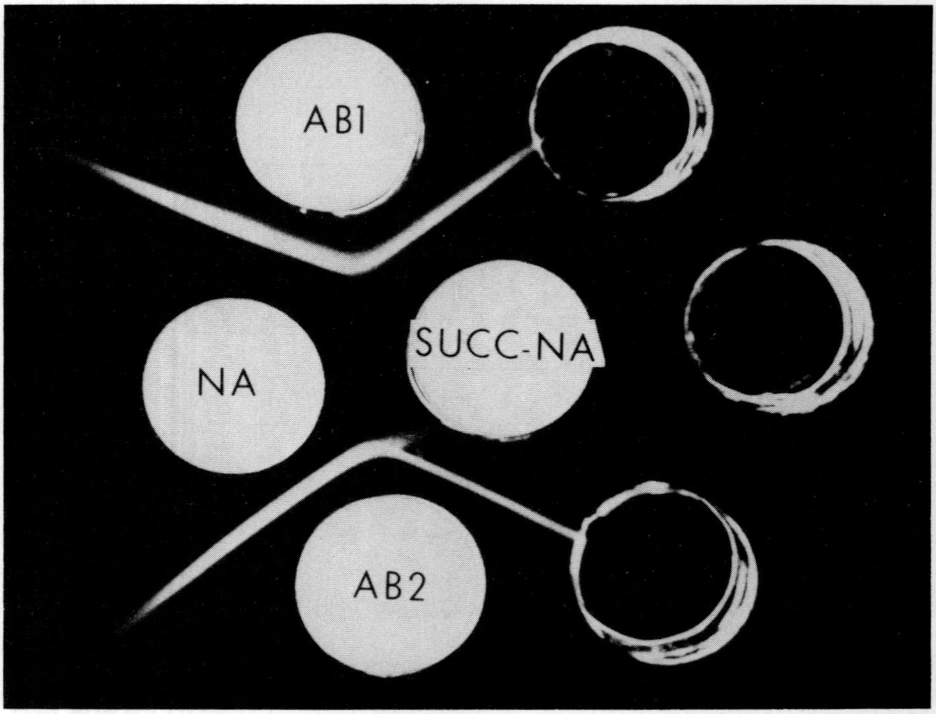

Fig. 14.13 Double immunodiffusion tests showing at the top a line of identity, and at the bottom a small spur between Tokyo NA and succinylated Tokyo NA when tested with polyclonal antisera from two individual rabbits.

pendently of the surface antigen subtype grouping, but molecular studies are so far lacking (see Webster et al., 1983).

10. Conclusions

The three-dimensional structures of both surface antigens (hemagglutinin and neuraminidase) of influenza virus are known, and studies with monoclonal antibodies and with variants selected by these antibodies have identified some of the sites which change during antigenic drift.

However, there are many questions still remaining. We do not know if the antigenic 'sites' identified by these analyses physically represent the contact sites of antibody interaction; if they do not, as appears to be the case with at least some of these sites, then we have no information on the nature or extent of the antigen-antibody interaction.

We do not know how single amino acid changes in the hemagglutinin and neuraminidase totally abolish binding of monoclonal antibodies, especially when the single changes are those generally considered conservative, such as arginine to lysine. We do not know why drift occurs in influenza viruses when other RNA viruses, such as measles, which yield variants with the same frequency as influenza when grown in the presence of monoclonal antibodies, do not vary antigenically in nature.

Drift in influenza A viruses follows a set path with changes accumulating over time and with older strains disappearing as new ones take over. Antigenic variants of influenza B virus appear to cocirculate, disappearing then and reappearing. Influenza C virus is reported to hardly vary at all. Why are these groups of orthomyxoviruses so different in their antigenic properties?

Although the experiments are far from complete, it appears that some areas of the surfaces of the antigens are more immunogenic than others, and it is possible that drift involves changes in immunogenicity as well as changes in specificity. It is not known if drift can continue indefinitely, or if it can reverse. We do not know how antigenic shift occurs (several mechanisms may operate), and although we may in due course understand, and hence predict, the course of antigenic drift, there are no indications as yet that we may foresee antigenic shifts.

The understanding of antigenic drift requires an understanding at the molecular level of the antigenic sites and the interaction of these sites with antibodies. At the present time, crystallographic analysis in conjunction with further antigenic and sequence analyses is the most likely way currently available by which these questions will be answered.

11. Acknowledgements

The authors were supported in part by grant A1-18203 and A1-19084 from the National Institute of Allergy and Infectious Diseases, and by grant CA-13148 from the National Cancer Institute. The writing of this joint review was greatly assisted by international telephone facilities donated by the Australian Overseas Telecommunications Commission.

12. References

Air, G.M. (1981) *Proc. Natl, Acad. Sci. U.S.A.* 78, 7639–7643.

Air, G.M. and Compans, R.W. (1983) In *Genetics of Influenza Viruses* (P. Palese and D.W. Kingsbury, Eds.) Springer-Verlag, New York. pp. 280–304.

Berton, M.T., Naeve, C.W. and Webster, R.G. (1984) *J. Virol.* 52, 919–927.

Blok, J., Air, G.M., Laver, W.G., Ward, C.W., Lilley, G.G., Woods, E.F., Roxburgh, C.M. and Inglis, A.S. (1982) *Virology* 119, 109–121.

Both, G.W. and Sleigh, M.J. (1981) *J. Virol.* 39, 663–672.

Caton, A.J., Brownlee, G.G., Yewdell, J.W. and Gerhard, W. (1982). *Cell* 31, 417–427.
Colman, P.M., Varghese, J.N. and Laver W.G. (1983) *Nature* 303, 41–44.
Colman, P.M., Gough, K.H., Lilley, G.G., Blagrove, R.J., Webster, R.G. and Laver, W.G. (1982) *J. Mol. Biol.* 152, 609–614.
Compans, R.W. and Choppin, P.W. (1975) In *Comprehensive Virology* (H. Fraenkel-Conrat and R.R. Wagner, Eds.) Vol IV. Plenum, New York. pp. 179–252.
Daniels, R.S., Douglas, A.R., Gonsalves-Scarano, F., Palu, G., Skehel, J.J., Brown, E., Knossow, M., Wilson, I.A. and Wiley, D.C. (1983) In *The Origin of Pandemic Influenza Viruses* (W.G. Laver, Ed.), Elseviers, Amsterdam. pp. 9–18.
Fang, R., Min Jou, W., Huylebroeck, D., Devos, R. and Fiers, W. (1981) *Cell* 25, 315–323.
Gerhard, W., Yewdell, J., Frankel, M. and Webster, R.G. (1981) *Nature* 290, 713–717.
Gething, M.J., White, J.M. and Waterfield, M.D. (1978) *Proc. Natl. Acad. Sci. U.S.A.* 75, 2737–2740.
Herrler, G., Nagele, A., Meier-Ewert, H., Bhown, A.S. and Compans, R.W. (1981) *Virology* 113, 439–451,
Hewat, E.A., Cusack, S., Ruigrok, W.R.H. and Verwey, C. (1984) *J. Mol. Biol.* 175, 175–193.
Hinshaw, V.S., Air, G.M., Gibbs, A.J., Graves, L., Prescott, B. and Karunakaran, D. (1982). *J. Virol.* 42, 865–872.
Hovanec, D.L. and Air, G.M. (1984) *Virology* In press.
Krug, R.M. (1983) In *Genetics of Influenza Viruses.* (P. Palese and D.W. Kingsbury, Eds.) Springer-Verlag, New York. pp. 70–98.
Krystal, M., Elliott, R.M., Benz, E.W., Young, J.F. and Palese, P. (1982) *Proc. Natl. Acad. Sci. U.S.A.* 79, 7800–7804.
Krystal, M., Young, J.F., Palese, P., Wilson, I.A., Skehel, J.J. and Wiley, D.C. (1983) *Proc. Natl. Acad. Sci. U.S.A.* 80, 4527–4531. Correction. *Proc. Natl. Acad. Sci. U.S.A.* 81, 1261.
Lamb, R.A. (1983). In *Genetics of Influenza Viruses* (P. Palese and D.W. Kingsbury, Eds.) Springer-Verlag, New York. pp. 21–94.
Laver, W.G., Air, G.M. and Webster R.G. (1981) *J. Mol. Biol.* 145, 339–361.
Laver, W.G., Webster, R.G. and Air, G.M. (1983). In *The Origin of Pandemic Influenza Viruses* (W.G. Laver, Ed.) Elsevier, Amsterdam. pp. 113–120.
Laver, W.G., Air, G.M. and Webster, R.G. (1981) *Virology* 111, 538–548.
Laver, W.G., Air, G.M., Webster, R.G., Gerhard, W., Ward, C.W. and Dopheide, T.A.A. (1979) *Virology* 98, 226–237.
Lentz, M.R., Air, G.M., Laver, W.G. and Webster, R.G. (1984) *Virology* 135, 257–265.
Lu, B.L., Webster, R.G., Brown, L.E. and Nerome, K. (1983) *Bull. WHO* 61, 681–687.
Maeda, T. and Ohnishi, S. (1980) *FEBS Lett.* 122, 283–287.
Nakada, S., Creager, R.S., Krystal, M., Aaronson, R.P. and Palese, P. (1984) *J. Virol.* 50, 118–124.
Nakajima, K., Desselberger, U. and Palese, P. (1978) *Nature* 274, 334–339.
Nakajima, S., Nakajima, K. and Kendal, A.P. (1983) *Virology* 131, 116–127.
Newton, S.E., Air, G.M., Webster, R.G. and Laver, W.G. (1983) *Virology* 128, 495–501.
Palese, P., Tobita, K., Ueda, M. and Compans, R.W. (1974) *Virology* 61, 397–410.
Pfeifer, J.B. and Compans, R.W. (1984) *Virus Res* 1, 281–296.
Scholtissek, C., van Hoyningen, V. and Rott, R. (1978) *Virology* 89, 613–617.
Shaw, M.V., Lamb, R.A., Erickson, B.W., Briedis, D.J. and Choppin, P.W. (1982) *Proc. Natl. Acad. Sci. U.S.A.* 79, 6817–6821.
Skehel, J.J., Bayley, P.M., Brown, E.B., Martin, S.R., Waterfield, M.D., White, J.M., Wilson, I.A. and Wiley, D.C. (1982) *Proc. Natl. Acad. Sci. U.S.A.* 79, 968–972.
Varghese, J.M., Laver, W.G. and Colman, P.M. (1983) *Nature* 303, 35–40.
Verhoeyen, M., Van Rompuy, L., Min Jou, W., Huylebroeck, D. and Fiers, W. (1983). *Nucleic Acids Res.* 11, 4703–4712.

Ward, C.W. (1981). *Curr. Topics Microbiol. Immunol.* 94/95, 1–74.
Ward, C. and Dopheide, T. (1982) In *Genetic Variation Among Influenza Viruses* (D.P. Nayak and C.F. Fox, Eds.) ICN-UCLA Symposia on Molecular and Cellular Biology, Vol XXI. Academic Press, New York. pp. 323–340.
Webster, R.G. and Berton, M.T. (1981) *J. Gen. Virol.* 54, 243–251.
Webster, R.G., Brown, L.E. and Laver, W.G. (1984) *Virology* 135, 30–42.
Webster, R.G., Laver, W.G. and Air, G.M. (1983). In *Genetics of Influenza Viruses* (P. Palese and D.W. Kingsbury, Eds.) Springer-Verlag, New York. pp. 127–168.
Webster, R.G., Hinshaw, V.S. and Laver, W.G. (1982) *Virology* 117, 93–104.
WHO (1980) *Bull. WHO* 58, 585–591.
Wiley, D.C., Wilson, I.A. and Skehel, J.J. (1981) *Nature* 289, 373–378.
Wilson, I.A., Skehel, J.J. and Wiley, D.C. (1981) *Nature* 289, 366–373.
Wrigley, N.G. (1979) *Br. Med. Bull.* 35, 35–38.

Eds. M.H.V. van Regenmortel and A.R. Neurath
Immunochemistry of viruses. The basis for serodiagnosis and vaccines.

Antigenic structure of paramyxoviruses

15

CLAES ÖRVELL and ERLING NORRBY

Department of Virology, Karolinska Institute, School of Medicine, and Department of Virology, National Bacteriological Laboratory, SBL S-10521 Stockholm 1, Sweden

The family Paramyxoviridae encompasses three genera – paramyxoviruses, morbilliviruses and pneumoviruses (Kingsbury et al., 1978). Each genus includes different virus types and most of these types have importance due to the diseases they cause in man and animals. During the last years there has been an impressive accumulation of data on the molecular and immunobiological properties of these viruses. This development is not due to improvement of techniques for handling the viruses in the laboratory, but instead to the availability of new technology, in particular recombinant DNA techniques and hybridoma methodology. The use of monoclonal antibodies has allowed a detailed characterization of structural-functional relationships in individual virus components. Furthermore, access to such antibodies has made it possible to identify a high frequency of occurrence of spontaneous and induced antigenic variations in these monotypic viruses. This chapter will describe the current state of knowledge of immunobiological characteristics of different members of the paramyxovirus family and the relevance of this knowledge to serodiagnosis and immune prophylaxis.

In this review we have adopted the following terminology for description of antigenic properties (Yewdell and Gerhard, 1981). An *epitope* represents a defined combination of amino acid residues on the surface of a protein antigen. By operational analysis, e.g. competition experiments or induction of mutational changes, individual or groups of epitopes can be defined to represent an *antigenic site*. These sites are interpreted to be topographically nonoverlapping. However, there is one caveat in this definition. This is that an antibody may interfere with a topographically nonoverlapping antigenic site via allosteric effects. For pragmatic reasons this limitation of the analysis is not further considered in this review.

1. The structural and nonstructural components of paramyxoviruses and their biological activities

The paramyxovirus genus includes mumps virus, four types of human parainfluenza virus, Sendai virus of mice (also known as hemagglutinating virus of Japan), Simian virus 5 (SV5) and Newcastle disease virus (NDV) of birds and a number of other less well-characterized types isolated from birds and mammals. Certain immunologic cross-reactions between types have been demonstrated over the years (Compans and Nakamura, 1978).

The morbillivirus genus includes measles virus, canine distemper virus (CDV), rinderpest virus (RV) of cattle, bovine paramyxovirus 107 and peste des petits ruminants virus (PPRV) of sheep and goats. There are also pronounced immunological relationships between these viruses (Stephenson and ter Meulen, 1979; Hall et al., 1980; Örvell and Norrby, 1980).

Respiratory syncytial (RS) virus, of both human and bovine type, and pneumonia virus of mice (PVM) are members of the third genus of paramyxoviruses, the pneumoviruses.

Members of the three genera of paramyxoviruses have been classified on the basis of distinguishing features of biological activities and morphology. Paramyxoviruses are enveloped viruses of usually spherical shape, their size varying from 150 to 400 nm in diameter, although they may exist in filamentous forms with a length of several micrometers (Choppin and Compans, 1975; Kingsbury et al., 1978; Compans and Klenk, 1979; Matsumoto, 1982). Within the viral envelope the linear RNA molecule is contained in a helical nucleocapsid. The nucleocapsid has a diameter of 13 (pneumoviruses) to 18 nm and a length of 1 μm. The virion envelope is composed of a lipid bilayer with a matrix (M) protein on the inside and two forms of transmembraneous spikes projecting 8–12 nm.

The two forms of spikes are made up of separate glycoprotein molecules. The larger glycoprotein has been shown to carry both hemagglutinating (HA) and neuraminidase (NA) activities in members of the paramyxovirus genus (Scheid et al., 1972; Scheid and Choppin, 1973; 1974; Seto et al., 1973; Tozawa et al., 1973; Jensik and Silver, 1976) and is therefore designated the hemagglutinating-neuraminidase (HN) protein in members of this genus. The large glycoprotein of measles virus of the morbillivirus genus carries HA but not NA activity and HA activity has been demonstrated in preparations of mouse pneumonia virus (Kingsbury et al., 1978). However, in remaining virus types of both the morbilli and pneumovirus genera neither HA nor NA activity has been demonstrated. Because of the lack of NA activity in these cases the large glycoprotein has been designated H or HA. In this review we use the term H. The fusion (F) protein is the smaller of the two glycoprotein molecules forming the second surface spike of the envelope. The F protein is involved in cell fusion, hemolysis and virus penetration into cells. In Sendai virus, the relative proportions of the two envelope peplomers have been estimated by immunological methods (Örvell and Norrby, 1977). HN to F occurred in a proportion of about 2 to 1.

Most knowledge on the structural and functional characteristics of paramyxovirus proteins has been derived from studies on members of the paramyxovirus genus, whereas comparatively less is known about members of the pneumovirus genus. However, the number of polypeptides and their structural allocation appears similar in all family members. Six structural proteins coded for by the polycistronic viral genome have been identified. These are, in addition to the three envelope proteins, the nucleo (N), polymerase (P) and large (L) proteins. These three proteins are located in the central part of the virus and are associated with the transcriptive complex. The size as well as structural and functional properties of equivalent proteins of the three paramyxovirus genera are summarized in Table 15.1.

The nucleoprotein is the most abundant protein of nucleocapsids and in infected cells. It surrounds the virion RNA protecting it from nucleases. The nucleoprotein of paramyxoviruses has been shown to exist in both a phosphorylated (Lamb and Choppin, 1977; Robbins and Bussell, 1979; Hall et al., 1980; Rima et al., 1980), and a nonphosphorylated form in Sendai virus (Lamb and Choppin, 1977). There is a considerable variation in the molecular mass of the nucleoprotein of different paramyxoviruses, which varies from 70 kDa in mumps virus (Örvell, 1978b; McCarthy and Johnson, 1980) to 44 kDa in RS virus (Cash et al., 1977; Bernstein and Hruska, 1981; Walsh and Hruska, 1983).

The polymerase protein is a phosphorylated protein (Lamb and Choppin, 1977; Robbins and Bussell, 1979; Hall et al., 1980). It is associated with nucleocapsids in infected cells (Portner and Kingsbury; 1976; Kristensson and Örvell, 1983) and in virions as part of the transcriptive complex (Stone et al., 1972; Marx et al., 1974; Buetti and Choppin, 1977). It is relatively more abundant in cells than in virions (Lamb et al., 1976; Portner and Kingsbury, 1976). RNA-dependent RNA polymerase activity has been described in detergent-disrupted virions and nucleocapsid preparations of paramyxoviruses (Robinson, 1971; Huang et al., 1971; Stone et al., 1972; Marx et al., 1974; Bernard and Northrop, 1974; Buetti and Choppin, 1977; Seifried et al., 1981), but no enzymatic activity has been demonstrated in isolated polymerase protein. There is a considerable size variation of the polymerase protein in different paramyxovirus types (Table 15.1).

A large protein has been identified in all paramyxoviruses studied. The large protein only represents a minor structural component of the virus and has been found in association with nucleocapsids (Buetti and Choppin, 1977; Stallcup et al., 1979). The function of the protein is not known, but it is assumed to be a part of the transcriptive complex.

The matrix protein is located inside virions as a structural component of the viral envelope. It has hydrophobic properties (Scheid and Choppin, 1973; 1974) and a high affinity for lipids and viral nucleocapsids (Shimizu and Ishida, 1975; Yoshida et al., 1976; Tyrrell et al., 1980). The protein has an important function in virus

TABLE 15.1

Structural proteins of paramyxoviruses

Designation	Relative molecular mass (kDa)			Location	Function
	Paramyxovirus	Morbillivirus	Pneumovirus		
N	55–70	58–60	44	Nucleocapsid	Protecting RNA
P	45–79	72–78	NI[a]	Nucleocapsid	In transcriptive complex
L	~200	~200	~200	Nucleocapsid	Not known. Possibly in transcriptive complex
Actin	43	43	43	Not known (not at virion surface)	Not known
M	34–40	34–37	28	Inside virion envelope	Virion assembly
HN/H	70–80	76–85	90	Spike	Adsorption to cells, NA activity in paramyxovirus
F_0	60–70	60	70	Spike	Fusion of cells, hemolysis, virus entry
F_1	48–61	40–41	48		
F_2	10–16	15–20	20		

[a] Not identified

assembly and it stabilizes the viral envelope (Matsumoto, 1982). The matrix protein of Sendai virus has been shown to exist both in a phosphorylated and nonphosphorylated form (Lamb and Choppin, 1977; Hsu and Kingsbury, 1982). The phosphorylated form of the protein was found predominantly in infected cells, whereas the nonphosphorylated form dominated in virions (Lamb and Choppin, 1977). The isolated matrix protein has a morphology varying with different conditions. The matrix component of measles virus at pH 3.0 appeared as a rounded particle with a diameter of 8 nm with a central accumulation of contrast of 1.5 nm. (Varsanyi et al., 1984). This structure was suggested to include 4–6 matrix protein polypeptides. Similar morphological characteristics were previously described for the matrix protein of Sendai virus under high salt conditions (Hewitt and Nermut, 1977; Heggeness et al., 1982). There are pronounced variations in the molecular mass of the matrix protein in different paramyxoviruses, from 28 kDa in RS virus (Cash et al., 1977; Bernstein and Hruska, 1981) to 40 kDa in mumps virus (Örvell, 1978b; McCarthy and Johnson, 1980). In the case of measles virus small intratypic strain variations in the molecular weight of matrix protein have been well documented (Wechsler and Fields, 1978; Wechsler et al., 1979; Hall et al., 1979).

The HN or H protein contains carbohydrate derived from the host cell (Compans and Klenk, 1979) and is the larger of the two glycoprotein molecules in the envelope. It is responsible for adsorption of virions to infected cells (Hsu et al., 1979; Merz et al., 1981). The protein is transmembraneous (Lyles, 1979; Bowen and Lyles, 1981; Bellini et al., 1983). The H spike of morbilliviruses has been reported to consist of a dimer of two glycoprotein molecules connected by disulfide bonds (Hardwick and Bussell, 1978; Lund and Salmi, 1981; Shapshak et al., 1982) and the HN spike of Sendai virus and NDV has been suggested to consist of disulfide-bonded dimers and tetramers (Ozawa et al., 1976; Hardwick and Bussell, 1979; Markwell and Fox, 1980). The isolated H peplomer of measles virus has a truncated conical shape (Varsanyi et al., 1984) similar in appearance to the HN peplomer of Sendai virus (Shimizu et al., 1974). The large glycoprotein spike can be selectively removed from the envelope by treating virions with various proteolytic enzymes (Shimizu and Ishida, 1975; Norrby and Gollmar, 1975; Örvell, 1976a; 1976b; 1978b; Tyrrell and Norrby, 1978; Örvell and Norrby, 1980; Merz and Wolinsky, 1983). The molecular weight of the large glycoprotein monomer appears not to vary to a great extent in different paramyxoviruses. Molecular weight differences between the large glycoprotein of different strains of measles, mumps and CDV have been reported (Rima et al., 1979; McCarthy and Johnson, 1980; Örvell, 1980).

The fusion (F) protein is the smaller of the two transmembraneous glycoprotein molecules forming the second surface spike of the envelope. A precursor protein, designated F_0, has been identified in paramyxoviruses (Samson and Fox, 1973; Nagai et al., 1976; Scheid and Choppin, 1977). This precursor molecule is biologically inactive and must be processed in order to exert its biological activity. The processing or cleavage of the precursor protein to the active fusion protein composed of two disulfide-linked subunits, F_1 and F_2, is dependent on the presence in the cells of certain proteolytic enzymes as demonstrated in the case of NDV, SV5 and Sendai virus (Nagai et al., 1976; Nagai and Klenk, 1977; Scheid and Choppin, 1976; 1977). F_1 and F_2 are separable under reducing conditions in polyacrylamide gels. F_2 is smaller than F_1 and has a higher concentration of carbohydrates (Scheid and Choppin, 1977). In members of the morbillivirus genus F_1 does not appear to contain detectable carbohydrates (Mountcastle and Choppin, 1977; Hardwick and Bussell, 1978; Tyrrell and Norrby, 1978; Örvell, 1980; Campbell et al., 1980). There is no major variation in the size of the F protein in different paramyxoviruses. The amino terminus of F_1 contains a highly conserved region (Scheid et al., 1978; Gething et al., 1978). Synthetic peptides with similar amino acid sequences can inhibit virus entry into cells and viral hemolysis (Norrby, 1971; Norrby and Sievertsson, 1974; Richardson et al., 1980). The isolated F peplomer has a club-like shape with a terminal knob (Shimizu et al., 1974; Varsanyi et al., 1984).

At least one nonstructural protein has been identified in paramyxovirus infected cells, but not in virions (Lamb et al., 1976; Peluso et al., 1977; Lamb and Choppin, 1978; Rima and Martin, 1979; Herrler and Compans, 1982). The nonstructural protein(s) in different paramyxoviruses have been designated C (Lamb et al., 1976), but are sometimes called V (Peluso et al., 1977) or S (small) proteins (Rima and Martin, 1979). No data are available on the functional or immunological properties of the nonstructural protein(s).

The host cell protein actin has been identified in purified preparations of paramyxoviruses (Lamb et al., 1976; Wang et al., 1976; Lamb and Choppin, 1978; Tyrrell and Norrby, 1978; Örvell, 1978b; 1980; Naruse et al., 1981). The protein is probably located inside the viral envelope since it is not degraded by proteolytic enzymes which remove the surface spikes of the virus. Tyrrell and Ehrnst (1979) have proposed that actin may mediate the communication between viral nucleocapsids and the cell membrane in virus budding. Actin may interact with hydrophobic proteins, and in NDV and Sendai virus, actin has been shown to interact with purified M component (Giuffre et al., 1982).

2. Structural-functional characterization of peplomer components

In paramyxoviruses both glycoprotein spikes are important to allow the virus to initiate an infection in susceptible cells. The HN protein anchors the virus to the cell surface and the F protein brings about a fusion between the viral envelope and the cell membrane, thereby releasing the viral nucleocapsid into cells (Choppin and Scheid, 1980). The HN protein may exist as a precursor glycoprotein, designated HN_0 in the case of NDV (Nagai et al., 1976). Cleavage of the precursor molecule of HN of NDV was a precondition for the biological activity of the protein (Nagai and Klenk, 1977). The function of the enzyme NA located on the HN protein of paramyxoviruses has not been established. Merz and Wolinsky (1981; 1983) have found that strains of mumps virus which caused more cell fusion and were more cytopathogenic contained less active NA. Virions adsorbed to cells eluted more rapidly in strains with more active NA, thereby allowing less time for the F protein to act. Reduction of NA activity by treatment of virions with chymotrypsin, which did not affect the F protein, resulted in conversion of a nonfusing mumps virus infection to one inducing fusion.

It has been known for some time that the HA and NA activities are associated with the same polypeptide, but the question remained whether they were located on the same or separate active sites on the molecule. Polyclonal sera prepared against the isolated HN protein contained hemagglutination-inhibition (HI) and neuraminidase-inhibiting (NI) as well as neutralization test (NT) antibodies (Seto et al., 1974; Urata and Seto, 1976; Örvell and Norrby, 1977; Dyke et al., 1978; Örvell, 1978a; Avery and Niven, 1979; Merz et al., 1981). Antibodies against the large glycoprotein of paramyxoviruses also inhibit hemolysis and cell fusion in an indirect way by preventing adsorption of the virus to cells (Seto et al., 1974; Norrby and Gollmar, 1975; Örvell and Norrby, 1977; Örvell, 1978a; Merz et al., 1981). It has not been possible to identify antibodies against only one of the biological activities of HA or NA by adsorption of polyclonal antisera against HN with different antigenic preparations. Portner (1981) suggested that different active sites were involved in the two biologic activities. He found, firstly, that a Sendai virus mutant which was temperature sensitive at 38°C for hemagglutination and attachment to host cells was not temperature sensitive for neuraminidase activity. Secondly monoclonal antibodies directed against HN were used to select antigenic variants. In some of the variants the HA but not the NA activity was inhibited. Finally an analogue of *N*-acetylneuraminic acid inhibited NA activity but not HA activity of Sendai virus and NDV. With the advent of hybridoma technology, epitope characterization of the HN molecule became possible and the question of whether one or two active sites were present in HA and NA became amenable to detailed analysis.

Four distinct antigenic sites were identified in parainfluenza 1 6/94 and NDV (Yewdell and Gerhard, 1982; Iorio and Bratt, 1983). Monoclonal antibodies against two sites of the HN glycoprotein inhibited HA but not NA activity, whereas a monoclonal antibody against a third site inhibited both activities (Yewdell and Gerhard, 1982). Örvell and Grandien (1982) studied 52 monoclonal antibodies against the HN protein of Sendai virus. The clones could be divided into four groups on the basis of their serologic reactivities. The first group of clones could not inhibit any biological activity (HA, NA, hemolysis or infectivity of the virus) in the absence of anti-γ-globulin. The second group blocked hemolysis, but did not block HA or NA activity. Monoclonal antibodies directed against the large glycoprotein which were able to block hemolysis but not HA and NA activity have been described in other studies on measles and Sendai virus (Togashi et al., 1981; Miura et al., 1982). The third group, which included six clones, blocked all biologic activities. These clones were directed against two distinct antigenic sites. The fourth group of seven monoclonal antibodies directed against one antigenic site could only block the NA activity of the virus. In a recent study 24 monoclonal antibodies directed against the HN glycoprotein of mumps virus were identified (Örvell, 1984). The clones, which were directed against a minimum of 10 distinct antigenic sites, could be divided into four groups similar to those defined in studies of the interaction of monoclonal antibodies with the HN protein of Sendai virus. One group of five monoclonal antibodies directed against one antigenic site could block NA but not HA activity. In contrast to the corresponding Sendai virus clones these monoclonal antibodies could also block hemolytic activity and infectivity of the virus. In these two studies no antibodies were found that could block HA but not NA of the virus when a large substrate, fetuin, was used in the neuraminidase inhibition (NI) test. In contrast, several groups described monoclonal antibodies that could block HA, but not NA activity when a small substrate, neuraminlactose, was used in NI tests (Miura et al., 1982; Server et al., 1982; Russell et al., 1983; Örvell, 1984). Hoshi et al. (1983), in studies on variants of NDV found that one monoclonal antibody which reacted to similar titers in HI tests with the homologous and one heterologous strain reacted in NI tests significantly higher with the homologous strain. All these studies are consistent with the view that HA and NA are located on distinct antigenically different parts of the HN molecule.

Although most studies have revealed four or less antigenic sites on the HN glycoprotein (Miura et al., 1982; Yewdell and Gerhard, 1982; Goswami and Russell, 1983; Iorio and Bratt, 1983; Hoshi et al., 1983; Nishikawa et al., 1983; Russell et al., 1983), recent studies on measles, mumps and canine distemper virus have identified a minimum of 9, 10 and 7 antigenic determinants, respectively (Sheshberadaran et al., 1983; Örvell, 1984; Örvell et al., 1984). No epitope characterization of the H protein of pneumoviruses has been reported as yet.

Polyclonal sera against the F component have been characterized in a number of studies (Seto et al., 1974; Norrby and Gollmar, 1975; Örvell, 1976a; b; 1978a; Örvell and Norrby, 1977; 1980; Avery and Niven, 1979; Merz et al., 1981). All studies have shown that polyclonal antibodies of this specificity are capable of blocking hemolysis but not HA or NA activity of the virus. Apart from the studies by Avery and Niven (1979) and Merz et al. (1981), these antibodies have shown little or no capacity to neutralize the infectivity of the virus. These two groups used a plaque reduction method to demonstrate neutralizing antibodies, whereas most other workers used end point neutralization in eggs or tissue culture tubes. It was shown in a recent study that when the plaque reduction method was used to determine neutralizing activity of antibodies against the F component, high titers were obtained which could not be reproduced when the same antibodies were characterized by end point neutralization (Örvell and Grandien, 1982). However, in a recent study Varsanyi et al. (1984) have shown that in the case of measles virus, polyclonal serum against the F protein showed a certain capacity to neutralize the infectivity of the virus when end point neutralization in tissue culture tubes was used as test system. In indirect neutralization, after addition of anti-γ-globulin to the test, more pronounced increases of titers have been observed in the case of antibodies against the F component than with antibodies against HN or H (Norrby and Gollmar, 1975; Örvell, 1976a; 1976b; Örvell and Norrby, 1977; Sheshberadaran and Varsanyi, unpublished results).

Only a few reports have appeared in the literature on monoclonal antibodies against the F component. In the case of measles, Sendai virus and mumps virus, 6, 7 and 12 monoclonal antibodies, respectively, have been described (Norrby et al., 1982; Örvell and Grandien, 1982; Örvell, 1984). Epitope characterization of the F protein of measles (Sheshberadaran et al., 1983), Sendai (Örvell and Grandien, 1982) and mumps virus (Örvell, 1984) revealed a minimum of 6,4 and 8 antigenic sites, respectively. In the case of Sendai virus, monoclonal antibodies directed against 3 distinct antigenic determinants could inhibit hemolysis, whereas antibodies against a fourth site could not block hemolysis of the virus (Örvell and Grandien, 1982). Five of the twelve monoclonal antibodies against mumps virus F protein contained demonstrable hemolysis-inhibiting titers (Örvell, 1984). As the titers of hemolysis-inhibiting antibodies were low, no clear distinction could be made between monoclonal antibodies with or without capacity to block hemolysis. None of the monoclonal antibodies against F protein could neutralize the infectivity of the three viruses. After addition of anti-γ-globulin to the test in indirect neutralization, monoclonal antibodies against 3 distinct antigenic determinants blocking hemolysis of Sendai virus were shown to neutralize the infectivity of the virus (Örvell and Grandien, 1982). After addition of anti-γ-globulin to the test no neutralization was observed with monoclonal antibodies against the F component of measles virus (Norrby, unpublished results), mumps virus (Örvell, 1984) or canine distemper virus (Örvell et al., 1984). In contrast to all these negative findings, one monoclonal antibody against the F protein of NDV has been reported to neutralize the infectivity of this virus (Russell et al., 1983). Monoclonal antibodies against the small glycoprotein of RS virus with ability to block infectivity of this virus have also been reported (Walsh and Hruska, 1983; Stott, personal communication).

3. Stability of antigenic characters: intratypic variations

Paramyxoviruses are monotypic. Individual serotypes are defined by use of polyclonal convalescent or hyperimmune sera. The antigenic stability of measles and mumps viruses is reflected in the epidemiology of the diseases they cause. After a generalized infection a lifelong protection against subsequent clinical disease is induced. Subclinical local reinfections may occur, but variants of measles and mumps virus able to cause renewed disease have never been encountered. Furthermore, propagation of these viruses in the presence of polyclonal antisera does not lead to the emergence of nonneutralizable mutants. The reason why paramyxoviruses are more antigenically stable than myxoviruses is not known. It should be noted that there seems to be no difference in the frequency of mutations in these two negative-strand RNA virus families. Portner et al. (1980) propagated Sendai, vesicular stomatitis and influenza A virus in the presence of neutralizing monoclonal antibodies. Resistant variants of all three viruses appeared with a frequency of $10^{-4.5}$ to $10^{-4.7}$.

Paramyxoviruses which are limited to more restricted respiratory mucosal replication, e.g. parainfluenza viruses and RS virus, may cause repeated infections. It has been suggested that this could be due to a waning immunity and not to changing antigenic properties of virus variants.

Availability of monoclonal antibodies has made it possible to identify variations in the epitopes of structural components of the stable monotypic paramyxoviruses. The occurrence of such variations will be discussed separately for peplomer antigens and antigens not available at the surface of virions.

3.1. Peplomer antigens

The surface antigens of the virus envelope represent the target for immune defence mechanisms. A selective immunological pressure involving these antigens therefore prevails under conditions of in vivo infections and can also be established in vitro for experimental purposes. Spontaneous epitope variations between homologous peplomers of virus variants have now been identified for several paramyxoviruses. The HN glycoprotein of mumps virus carries the antigenic structure of major importance in virus neutralization (Örvell, 1978a). Comparisons of mumps virus strains have revealed differences in their HA and NA activities (Klamm, 1980; Leprat and Aymard, 1979) and also in the molecular weight of the HN protein (McCarthy and Johnson, 1980), but no differences in antigenic characteristics have been demonstrable by use of polyclonal antisera. Two studies (Server et al., 1982; Örvell, 1984) employing monoclonal antibodies, however, have revealed that separate functional activities were blocked by different antibodies.

In recent studies of monoclonal antibodies against the HN protein of mumps virus (Örvell, unpublished) a subgroup of antibodies was found to react with egg-grown virus but not with virus produced in Vero cells. Since the same group of monoclonal antibodies also reacted with influenza A virus grown in eggs, they were probably directed against egg carbohydrate antigens. Although differences in glycosylation represent one source of host cell-induced modification of antigenic properties of virus components, other phenomena may also be influential as indicated by studies of influenza A virus (Schild et al., 1983). This study emphasized the importance of differing host cell tropisms of virus subpopulations. Passage of virus in different cell types gave a selection of virus with different properties.

Intratypic variations in paramyxoviruses have been examined by selection of parainfluenza type 1 strain 6/94 variants in the presence of 6 anti-HN monoclonal antibodies (Yewdell and Gerhard, 1982). Three of these antibodies had a sufficient neutralizing capacity to allow emergence of resistant variants.

A large number of paramyxoviruses have been isolated from birds and the most extensively studied representative of the group is Newcastle disease virus (NDV). NDV strains show similar reactivity in HI tests but differ in their ability to be neutralized (Schloer et al., 1975; Pennington, 1978; Avery and Niven, 1979). Monoclonal antibodies specific for the HN protein of NDV have been obtained in several laboratories (Russell and Alexander, 1983; Hoshi et al., 1983; Nishikawa et al., 1983). These antibodies were used to discriminate between different viral strains and to define a series of antigenic sites on the HN molecule.

Besides NDV, six groups of avian paramyxovirus types have been identified by HI tests (Alexander, 1980; Alexander et al., 1981; Yamane et al., 1982) and by immunodiffusion tests with the M protein (Kida and Yanagawa, 1981). In one of these studies (Yamane et al., 1982) it was noted that the HI cross-reactions between members of one group (type) in one case were unidirectional. This phenomenon deserves further study using monoclonal antibodies.

Measles virus is the prototype for a ubiquitous monotypic virus. Minor antigenic differences have been observed in neutralization tests comparing laboratory strains and fresh isolates from cases of acute measles (Albrecht et al., 1981) and subacute sclerosing panencephalitis (Payne and Baublis, 1973). Variation in epitopes of the H component has now been demonstrated by several groups by use of monoclonal antibodies (Birrer et al., 1981a; Giraudon and Wild, 1981a; ter Meulen et al., 1981; Trudgett et al., 1981; Sheshberadaran et al., 1983). Different assays including neutralization, hemagglutination inhibition, hemolysis inhibition, radioimmunoprecipitation and enzyme-linked immunosorbent tests were employed in these studies. In cases when a larger number of monoclonal anti-H antibodies were used it was noted that the differences concerned only a few epitopes and that the majority of monoclonal antibodies reacted equally well with different strains. Thus in the study by Sheshberadaran et al. (1983) a set of 11 different monoclonal antibodies operationally defined to react with 9 different epitopes was tested against 9 different lytically replicating strains of measles virus. Three epitope variations were identified. In the same study a comparison was also made of the F component of the 9 virus strains by determining their reactivity with a set of 11 monoclonal antibodies identifying at least 6 different epitopes. The F component appeared highly conserved since no variation in reactivity of different strains was found. Only one study (Birrer et al., 1981b) of induced variations in the measles virus H protein by propagation of virus in the presence of neutralizing antibodies has been presented. By use of this technique extensive changes of the H proteins have now been generated in this laboratory (Sheshberadaran, to be published).

Under conditions of viral persistence based on a nonproductive infection more extensive variations in epitope characteristics of both surface and internal components might be anticipated to occur (Holland et al., 1979). As long as these alterations do not interfere with the more limited requirements of the defective infection they would be allowed in the system. The relative mutation rate in lytic and persistent infections could be of importance in this context. Recently it was shown (Sato et al., 1984) that variants of Sendai virus resistant to either of three different monoclonal antibodies against the HN protein of the virus appeared with a frequency of $10^{-3.3}$ to $10^{-3.8}$. This frequency was higher than that seen in lytic infections in parallel experiments, and also by others (Portner et al., 1980). In one study of measles virus (ter Meulen et al., 1981) persistent subacute sclerosing panencephalitis (SSPE) virus infection was established by intracerebral inoculation of virus in hamsters followed by cocultivation of Vero cells with explanted brain tissue from these animals. Under these conditions the SSPE virus in the carrier cells had lost reactivity with 4 out of 6 monoclonal antibodies against the H protein. In studies of measles virus a defective neurotropic virus strain (HNT) persisting in Vero cells showed marked alterations of the H protein epitopes (Rammohan et al., 1983; Sheshberadaran et al., 1984).

Variants of paramyxoviruses with modified epitopes in the F protein have so far not been described. Only in the case of Newcastle disease virus has a direct neutralizing activity of monoclonal antibodies been described (Russell et al., 1983). Thus this system could offer interesting possibilities for selection of F epitope variants. In the case of other paramyxoviruses the capacity of monoclonal antibodies against the F protein to neutralize the virus in the presence of anti-immunoglobulin (cf. above) may have to be exploited in selection of variants. In view of the epitope stability of the F component of paramyxoviruses, so far as this has been studied, it would be interesting to evaluate the functional consequenses of changes in this component.

3.2. Internal components

The term 'internal component' is used to denote components, such as the matrix, nucleo, polymerase and large proteins, that are not exposed at the surface of virions. These components differ from the peplomer components in that they are not accessible to the immune defense system. The fact that these components serve specific structural or metabolic functions obviously imposes limitations on the variations which may be allowed in their structure. Only some studies have concerned variations in epitopes of internal structural components in paramyxovirus strains.

In the study of the reactions of monoclonal antibodies with structural proteins of mumps virus by Örvell (1984) seven different epitopes on the N and P proteins were found to be absent in one or more of 3 heterologous mumps virus strains. All of these epitopes were retained by the homologous strain, SBL-1, upon repeated passaging in eggs and Vero cells, but in the Enders, Kilham and RW strain some of these epitopes were missing. The Enders strain could only be distinguished from the Kilham strain when monoclonal antibodies against the P protein were used in ELISA tests. No variation has been observed in M protein epitopes of different strains of mumps. Similarly differences in Newcastle disease virus (NDV) strains were observed with antibodies to the N or P protein (Russell and Alexander, 1983). A separation of 40 NDV strains into 8 distinct antigenic groups was possible when the information on epitope characteristics of both surface and internal components was combined. It has been suggested that viruses in each group share biological and epizootological properties, but this correlation needs to be further evaluated.

In a short communication (Giraudon and Wild, 1981b) it was reported that 4 monoclonal antibodies against the measles virus Halle strain N protein antigen all reacted with 5 other strains, with the exception of one fresh isolate of virus which only reacted with 3 of the 4 antibodies. Results differing from the findings with NDV were obtained in studies of 9 lytically replicating measles virus strains of different origin (Sheshberadaran et al., 1983). Six different monoclonal antibodies against the N and P proteins distinguishing a minimum of 3 epitopes did not show any difference in their reactivity with the different strains. In contrast, marked variations were seen between the different strains when monoclonal antibodies against the M protein were employed. The monoclonal antibodies were separated into six groups and four different reaction patterns of strains were identified.

In further studies (Sheshberadaran and Norrby, unpublished results) another 9 lytically replicating strains were characterized. Altogether seven different reaction patterns were identified (Table 15.2). Attempts were made to correlate reaction patterns with virulence or pathogenicity features. Five out of 7 SSPE strains showed the same pattern of M epitopes. A similar pattern was also found in other isolates. Since all the 5 SSPE strains that shared M epitope patterns were derived from the same laboratory, it would appear necessary to study lytically replicating SSPE strains from other laboratories before drawing conclusions. In another separate study (Norrby, unpublished results) it was found that 30 consecutive limiting dilution passages of 4 strains with unique M epitope patterns did not lead to any detectable alteration in the properties of the M antigen. Although the N and P epitope patterns were stable in lytically replicating virus strains variations in epitopes were encountered in four different persistently infected cultures (Sheshberadaran et al., 1984). Thus it was found that the virus strains in two out of the four cultures, Mantooth and HNT, had lost 2 epitopes on the N protein and in the case of HNT two epitopes on the P protein. This appears to be another example of the more limited functional constringency on the structure of virus components under conditions of cell-associated persistent infections. For instance, the Mantooth strains showed the same characteristics as Mantooth strain material obtained from other laboratories in the F.R.G. and U.S.A.

In another study (Carter et al., 1983) the measles virus M protein of 3 lytically replicating virus strains, one SSPE virus isolate and virus in one persistenly infected cell line, CLu 106 (Norrby, 1967) were compared by competitive antibody binding. Only the virus strain in the persistently infected cells showed deviating characteristics. However, since the virus originally used to establish the persistent infection was not included in the study, the significance of this observation cannot be evaluated.

TABLE 15.2

Reaction of lytically replicating measles virus strains in radioimmunoprecipitation assay with 9 monoclonal antibodies identifying 6 epitopes on the M protein (Sheshberadaran et al., 1983)

Monoclonal antibody		LEC-KI, Woodfolk Schwarz, Moraten	Hu 2	MVO, MVP Hallé, McClellan, Dean, Zistev, Mantooth	LEC-WI	J.M.	Edmonston (DP, NIH), Fischer	Edmonston (V7), McI
Group	Clone							
1	10 EF10	+	+	+	+	+	+	+
2	16 BB2	+	+	+	+	+	+	+
3	19 AG10	+	+	+	−	+	+	+
4	19 CG6	+	+	±	+	−	−	−
	19 GF6							
5	19 DC5	+	−	−	+	+	−	−
6	19 DF10	+	−	+	−	−	+	−
	19 HC5							
	19 HF6							

The character and source of measles virus strains was the following. *Fresh isolates* from cases of acute measles – Woodfolk, MVO, MVP, Hu2 (Sheshberadaran et al., 1983), Edmonston V7 (6 passages in HEK cells and 7 passages in Vero cells; from Dr. P. Albrecht, Div. Virology, Natl. Centre Drugs and Biologics, Bethesda, MD, U.S.A.), McI (2 passages in RhMK cells and 6 passages in Vero cells; isolated by Dr. K. McIntosh and provided by Dr. P. Albrecht), JM (11 passages in Vero cells; from Dr. P. Albrecht). *Vaccine strains* – Schwarz, Moraten, *SSPE strains* – Hallé, McClelland, Dean, Zistev, Mantooth, Fischer (from Dr. J. Sever, Natl. Inst. Health, Bethesda, MD), LEC-VI (8 passages in Vero cells; from Dr. H. Koprowski, Wistar Institute, Philadelphia, PA). *Laboratory strains* – Edmonston NIK (From Dr. R. Lazzarini, Natl. Inst. Health, Bethesda, MD), Edmonston DP and LEC-KI (passaged repeatedly in Vero cells in this department).

4. Intertypic antigenic relationships between structural components

Cross-reactions between different paramyxoviruses were already observed in the early days of serological characterization of infections in man (Compans and Nakamura, 1978). Heterotypic antibody responses to infection with one type of parainfluenza virus is frequently observed in patients who have had prior infection with other types. Patients with mumps can also show a rise in the level of antibodies against parainfluenza viruses and NDV. Only a few studies have been published recently concerning the intertypic relationships between members of paramyxovirus genera. One of these concerned parainfluenza viruses and another morbilliviruses.

Four different serotypes of human parainfluenza viruses have been recognized. Parainfluenza viruses have also been isolated from other mammals. The virus SV5 has been extensively studied. Although originally identified as a virus of monkey origin, it has also been described in connection with infections in other animals and in man (Hsiung, 1972). Studies using monoclonal antibodies against the HN protein of human parainfluenza 3 and SV5 viruses have demonstrated the occurrence of specific epitopes in each serotype, as well as common epitopes, between human parainfluenza 1 and 3 and between human parainfluenza 2 and SV5 (Goswami and Russell, 1983). Cross-reactions between the HN proteins of these serotypes were not observed when polyclonal hyperimmune sera were employed. In the same study monoclonal anti-N antibodies against SV5 and human parainfluenza 3 components reacted equally well with human parainfluenza 2 and 1 (Sendai), respectively. The latter finding was a confirmation of data obtained in immunodiffusion tests by use of polyclonal hyperimmune sera (Goswami and Russell, 1982).

Measles and canine distemper viruses are closely related. Both homologous peplomer and internal components have been shown to share some antigen properties (Örvell and Norrby, 1974; Stephenson and ter Meulen, 1979; Hall et al, 1980; Gould et al., 1981). Comparative analysis of hyperimmune sera against H, F, M and N polypeptides from the two viruses isolated by preparatory SDS-polyacrylamide gel electrophoresis emphasized the sharing of immunological properties, in particular between homologous F and M proteins (Örvell and Norrby, 1980). The monoclonal antibodies against measles H and N proteins described by Birrer et al. (1981a) did not react with the Onderstepoort strain of distemper virus. In another report (Giraudon and Wild, 1981b) two out of 3 anti-N and 3 out of 4 anti-H measles monoclonal antibodies were described to react with canine distemper virus in immunofluorescence tests, but no neutralizing activity was found (Giraudon and Wild, 1981a) and only one antibody (N specific) showed detectable activity in radioimmunoprecipitation assay (RIPA) tests. Recently an extensive study of the capacity of monoclonal antibodies against measles virus components to react with distemper virus components and vice versa (distemper monoclonal antibodies prepared by Örvell et al., 1984) have been examined in ELISA, RIPA and immunofluorescence tests (Sheshberadaran et al., submitted). The observed difference between the properties of the two surface components is striking (Table 15.3). Monoclonal antibodies against the H component of either virus did not react with antigen of the other virus. In contrast, the structure of the F components was highly conserved in the two viruses. Concerning P and N components, a low to moderate degree of reciprocal cross-reaction was found and a certain strain dependence was also observed.

TABLE 15.3

The capacity of monoclonal antibodies against structural components of measles (LEC-KI strain) and distemper (Convac strain) virus to react with homologous components of the heterologous virus in immune fluorescence tests (Sheshberadaran et al., 1984).

Monoclonal antibody specificity		No. of monoclonal antibodies	No. of epitope specificities	No. of monoclonal antibodies reacting with heterologous virus
Virus	Protein			
Measles	H	11	≧9	0
	F	11	≧6	10
	M	9	6	5
	N	12	≧5	3
	P	6	≧3	2
Distemper	H equiv.[a]	8	≧6	0
	F_1+F_2 equiv.[b]	10	≧3	9
	N	39	≧18	21
	P	13	≧6	2

[a] 85-kDa glycoprotein (H equivalent).
[b] 41 and 18-kDa glycoproteins (F_1 and F_2 equivalents).

5. Immune responses to host cell proteins induced by paramyxoviruses

Like other enveloped viruses, paramyxovirions include host cell-derived material. Thus the carbohydrate moiety of peplomer glycoproteins derive from the host cell and cellular actin is also present inside the virus envelope. This presence of actin carries a certain relevance in relationship to the observed transient IgM antiactin antibody response found in measles infection (Haire, 1972). In an attempt to study the antiactin antibody response under conditions of experimental immunization with paramyxoviruses, hybridomas which did not produce antibodies against virus structural components were screened in immunofluorescence tests for their reactivity with cytoskeleton components of noninfected human fibroblast cultures (Fagraeus et al., 1983). Three hybridomas producing antibodies (IgM) with such a specificity were identified. All of them reacted with an epitope present on both actin and vimentin. In another study (Fujinami et al., 1983) a hybridoma derived from measles virus-immunized mice reacted with an intermediate filament, probably also vimentin. However, in this case the reaction appeared to be based on the occurrence of the same epitope on both cellular vimentin and on measles virus P protein.

Another example of such molecular mimicry has been encountered in this laboratory (Sheshberadaran and Norrby, 1984). Three out of nine monoclonal antibodies against the F protein of measles virus were found to give concomitant precipitation in RIPA tests of some nonviral polypeptides, in particular a 79-kDa protein. These antibodies also gave a weak immunofluorescence staining of noninfected cells. It was shown that the 79-kDa protein was not coprecipitated with viral H protein, but instead that it belonged to the group of cellular stress proteins. Thus, when noninfected cells were subjected to stress by a raise of incubation temperature or by treatment with 2-deoxyglucose, a large amount of 79-kDa cellular protein could be precipitated with the monoclonal antibodies against the F protein. Since the virus infection as such represents a situation of stress in cells it may lead to activation of an immune response against stress proteins. Phenomena of this kind may pertain to virus-induced autoimmunity. They may also have relevance to the appearance of the as yet unidentified C-reactive protein in connection with infections. Further studies within this area are warranted. Preliminary attempts to detect an antibody response to cellular stress proteins in measles virus infections have been negative (Sheshberadaran, unpublished results).

6. Implications for clinical virology of intra- and intertypic variations

The potential use of monoclonal antibodies in clinical virological analyses requires separate evaluation in different virus systems. A certain guidance in selection of reagents can be obtained from the accumulated experience gained in the epitope characterization of paramyxoviruses. In the absence of the pressure of immune selection the antigen properties of both surface and internal components of lytically replicating paramyxoviruses appear relatively stable, but in the presence of neutralizing monoclonal antibodies resistant variants readily emerge. Selection of monoclonal antibodies to be used for direct identification of viral antigen or virions in clinical specimens depends on the purpose of the test.

If the aim is to identify a strain of virus a strain-specific monoclonal antibody should be used. If instead type-specific identification is the goal a monoclonal antibody against a conserved epitope should be employed. In some cases it may be advisable to use a mixture of monoclonal antibodies. Finally there may be situations when intertype-specific monoclonal antibodies are most useful. The antigen target specificity of the antibody should be chosen to best serve the kind of test used. For example, if an immune fluorescence test with cells from the respiratory tract is to be examined it may be useful to use a monoclonal antibody against the N protein component, since this occurs in abundance. In an indirect immunofluorescence test for rapid detection of respiratory syncytial (RS) virus, a monoclonal antibody against the nucleocapsid was chosen (Bell et al., 1983).

In some situations monoclonal antibodies may be considered for anchoring of a selected antigen to the bottom of plastic wells for the purpose of quantification of antigen or for identification of specific antibodies against this antigen in sera. The efficacy of antigen anchoring by hybridoma antibodies varies depending upon their clonality and avidity. If there is a need for a polyclonal hyperimmune serum it should be kept in mind that monoclonal antibodies frequently can be effectively used for high degree of purification of individual virus structural components. Components purified by this method can then be used for preparation of highly specific hyperimmune sera. This approach has successfully been employed for characterization of measles virus H, F and M components (Varsanyi et al., 1984).

7. The role of different virus surface components in immune protection

Attempts to develop vaccines against paramyxoviruses have met with both successes and failures. Live vaccines against measles and mumps represent a safe way of inducing long-term immunity by way of innocuous infections. These vaccines are extensively used in many industrialized countries and are frequently combined into a triplex vaccine which includes attenuated live rubella virus. Simultaneously with the introduction of the live measles vaccine in the early 1960s, inactivated measles vaccines were also developed and tested. These inactivated vaccines did not provide long-term protection. In addition, exposure to wild strains of measles virus in many cases led to development of an atypical form of measles (Fulginiti et al., 1967). Similarly immunization with a formalin-inactivated RS virus product in some cases caused an accentuated form of disease in vaccinees exposed to wild virus (Kim et al., 1969). Furthermore, formalin-inactivated parainfluenza and mumps vaccines have not been found satisfactory in providing long-term protection, but in these cases no complications or aberrations have been noted of the form of disease upon virus exposure.

The relative role of the two surface antigens – the HN (or H in morbilliviruses) and F components – in inducing immune protection offers a means of explaining the varying experiences with different forms of vaccines. It was found that the F component of formalin-inactivated or Tween 80 ether-treated measles virus products did not retain complete immunizing capacity. Although antibodies against the F component could be detected in RIPA tests (Norrby, unpublished results), these antibodies did not have the capacity to give a direct inhibition of the hemolytic activity of the virus (Norrby et al., 1975; Norrby, 1975). However, the H component of the vaccines appeared fully immunogenic and induced antibodies which effectively could neutralize the virus in laboratory tests. Apparently the presence of hemagglutination-inhibiting (in vitro-neutralizing) antibodies did not suffice to provide protection in vivo. Similarly it was found that a formalin-inactivated mumps vaccine used for immunization of military recruits in Finland stimulated the production of antibodies blocking the H component but not the F component (hemolytic) activity (Norrby and Penttinen, 1978). It is possible that formalin-inactivated parainfluenza and RS virus products may also represent defective immunogens. The atypical form of disease seen in vaccinees after measles exposure and perhaps also the accentuated disease seen in the case of RS virus infections may be explained against the background of an immunity essentially directed against only one of the two types of surface components. Replication of virus will lead to a boosting of this immunity and can as a consequence set the stage for immune pathological complications.

The degree of in vivo protection provided solely by immunization with the F component of paramyxoviruses has not been defined. However, certain conclusions may be drawn from the experience of immunization of dogs with measles virus to provide protection against canine distemper. As discussed above, the F component in the two viruses is antigenically highly conserved whereas the H components have essentially distinct antigenic properties. It was found (Appel et al., 1984) that in this system the measles virus F immunity caused a markedly restricted replication of virulent distemper virus. The replication was associated with mild and transient symptoms and resulted in the production of distemper-neutralizing antibodies, but also of high titers of measles hemolyzing-inhibiting antibodies.

In other virus systems the use of monoclonal antibodies has allowed definition of the major antigenic site for induction of neutralizing antibodies. In the case of paramyxoviruses some information of this kind has accumulated, but the overall interpretation of the relevance of antigenic structures to immunity is complex since both the different surface components have to be considered. As discussed above, the capacity of antibodies against each of the two kinds of peplomers to neutralize infectivity in vitro varies with different paramyxoviruses. The reason for these variations is not known but the proportion of the two peplomers in the envelope could be one factor of importance. By use of specific hyperimmune sera against the surface components of SV5 it was shown that antibodies against HN and F proteins interfered with virus replication in different ways (Merz et al., 1980; Merz et al., 1981). Antibodies against the HN protein neutralized extracellular virus, but did not prevent the cell-to-cell spread of infection by fusion. In contrast, antibodies to the F component primarily interfered with the latter form of spread of infection. However, they could also neutralize virus infectivity (cf. above). Recently the corresponding observations were made in studies of rabbit hyperimmune sera against isolated envelope components of measles virus (Norrby et al., 1984).

The situation concerning RS virus may be unique among paramyxoviruses. Monoclonal antibodies against the large uncleaved glycoprotein did not neutralize the virus directly (Fernie et al., 1982), but gave some neutralization in the presence of complement (Walsh and Hruska, 1983), whereas monoclonal antibodies against the cleaved smaller glycoprotein gave a direct neutralization. Only antibodies with the latter specificity could block cell-to-cell fusion. The relative importance of antibodies against each of the two glycoproteins for protection against RS virus infection has not been defined. Ward et al. (1983) showed that children aged 6–12 months mobilized a relatively poor antibody response to the large glycoprotein relative to the response against the smaller cleaved glycoprotein in comparison with the situation in adults. The importance of this finding should be further evaluated. Studies employing polyclonal antisera specific for each of the two glycoproteins in RS virus are warranted.

Only limited studies of passive protection in vivo by specific monoclonal or polyclonal antisera against paramyxovirus surface components have been performed. We have shown that in the case of Sendai virus infections in mice both polyclonal (Örvell and Norrby, 1977) and monoclonal antibodies (Örvell and Grandien, 1982) against either H or F components can provide protection. The observation that the F antibodies were effective confirms observations in other systems that antibodies lacking neutralizing activity in vitro may give passive immune protection in animals. Further studies of this kind should be performed, preferably in situations of natural infections.

The currently available information does not allow formulation of a generalized strategy for development of inactivated vaccines against paramyxoviruses. Obviously the HN component alone cannot be used but admixture of F component is required. The proportions of components in a subunit peplomer vaccine requires consideration. In studies of measles and Sendai virus it has been experienced that immunization with whole noninfectious virus containing native F component primarily induces an antibody response to the hemagglutinin (Norrby and Gollmar, 1975; Örvell, 1976b; Örvell and Norrby, 1977). Thus it may be necessary to give the F component a certain advantage, for instance by increasing the proportion of this component in the aggregates. The possibility that immunization with the F component alone may provide satisfactory immunity deserves further study. Additional detailed molecular knowledge is required before recombinant DNA-produced proteins or synthetic peptides (Lerner, 1982; Norrby, 1983) can be used as immunogens. As yet no valid cloning of genes directing the synthesis of peplomer proteins of paramyxoviruses has been described. However, it seems likely that this will be accomplished in the near future.

8. Acknowledgement

The authors' work referred to in this review was supported by grants from the Swedish Medical Research Council (Project 16x-116) and from the National Swedish Board for Technical Development.

9. References

Albrecht, P., Herrmann, K. and Burns, G.R. (1981) *J. Virol. Methods* 3, 251–260.

Alexander, D.J. (1980) *Vet. Bull.* 50, 737–752.

Alexander, D.J., Hinshaw, V.S. and Collins, M.S. (1981) *Arch. Virol.* 68, 265–269.

Appel, M.J.G., Shek, W.R., Sheshberadaran, H. and Norrby, E. (1984) *Arch. Virol.* 82, 73–82.

Avery, R.J. and Niven, J. (1979) *Infect. Immun.* 26, 795–801.

Bell, D.M., Walsh, E.E., Hruska, J.F., Schnabel, K.C. and Hall, C.B. (1983) *J. Clin. Microbiol.* 17, 1099–1101.

Bellini, W.J., Silver, C.D. and McFarlin, D.E. (1983) *Arch. Virol.* 75, 87–101.

Bernard, J.P. and Northrop, R.L. (1974) *J. Virol.* 14, 183–186.

Bernstein, J.M. and Hruska, J.F. (1981) *J. Virol.* 38, 278–285.

Birrer, M.J., Bloom, B.R. and Udem, S. (1981a) *Virology* 108, 381–390.

Birrer, M.J., Udem, S., Nathenson, S. and Bloom, B.R. (1981b) *Nature* 293, 67–69.

Bowen, H.A. and Lyles, D.S. (1981) *J. Virol.* 37, 1079–1082.

Buetti, E. and Choppin, P.W. (1977) *Virology* 82, 493–508.

Campbell, J.J., Cosby, S.L., Scott, J.K., Rima, B.K., Martin, S.J. and Appel, M. (1980) *J. Gen. Virol.* 48, 149–159.

Carter, M.J., Willcocks, M.M., Löffler, S. and ter Meulen, V. (1982) *J. Gen. Virol.* 63, 113-120.

Carter, M.J., Willcocks, M.M., Löffler, S. and ter Meulen, V. (1983) *J. Gen. Virol.* 64, 1801–1805.

Cash, P., Wunner, W.H. and Pringle, C.R. (1977) *Virology* 82, 369–379.

Choppin, P.W. and Compans, R.W. (1975) *Comprehensive Virology* (H. Fraenkel-Conrat and R.R. Wagner Eds.) Vol. 4. Plenum Press, New York, pp. 95–178.

Choppin, P.W. and Scheid, A. (1980) *Rev. Infect. Dis.* 2, 40–61.

Compans, R.W. and Klenk, H.-D. (1979) *Comprehensive Virology* (H. Fraenkel-Conrat and R.R. Wagner, Eds.) Vol. 9. Plenum Press, New York. pp. 293–407.

Compans, R.W. and Nakamura, K. (1978) CRC Handbook Series in Clin. Lab. Sci., Section H: *Virology and Rickettsiology* (G.P. Hsiung and R.H. Green, Eds.) Vol. I, Part 2. CRC Press, West Palm Beach. pp. 361–385.

Dyke, S.F., Williams, W. and Seto, J.T. (1978) *J. Med. Virol.* 2, 143–152.

Fagraeus, A., Örvell, C., Norberg, R. and Norrby, E. (1983) *Exp. Cell. Res.* 145, 425–432.

Fernie, B.F., Cote, P.J. Jr. and Gerin, J.L. (1982) *Proc. Soc. Exp. Biol. Med.* 171, 266–271.

Fujinami, R.S., Oldstone, M.B.A., Wroblewska, Z., Frankel, M.E. and Koprowski, H. (1983) *Proc. Natl. Acad. Sci. USA* 80, 2346–2350.

Fulginiti, V.A., Eller, J.J., Downie, A.W. and Kempe, C.H. (1967) *J. Am. Med. Assoc.* 202, 1075–1080.

Gething, M.J., White, J.M. and Waterfield, M.D. (1978) *Proc. Natl. Acad. Sci. USA* 75, 2737–2740.

Giraudon, P. and Wild, T.F. (1981a) *J. Gen. Virol.* 54, 325–332.

Giraudon, P. and Wild, T.F. (1981b) *J. Gen. Virol.* 57, 179–183.

Giuffre, R.M., Tovell, D.R., Kay, C.M. and Tyrrell, D.L.J. (1982) *J. Virol.* 42, 963–968.

Goswami, K.K.A. and Russell, W.C. (1982) *J. Gen. Virol.* 60, 177–183.

Goswami, K.K.A. and Russell, W.C. (1983) *J. Gen. Virol.* 64, 1663–1672.

Gould, E.A., Armstrong, M., Shirodaria, P.V. and De'Ornellas, D. (1981) *J. Med. Virol.* 7, 51–60.

Haire, M. (1972) *Clin. Exp. Immunol.* 12, 335–341.

Hall, W.W., Lamb, R.A. and Choppin, P.W. (1979) *Proc. Natl. Acad. Sci. USA* 76, 2047–2051.

Hall, W.W., Lamb, R.A. and Choppin, P.W. (1980) *Virology* 100, 433–449.

Hardwick, J.M. and Bussell, R.H. (1978) *J. Virol.* 25, 687–692.

Hardwick, J.M. and Bussell, R.H. (1979) *Intervirology* 11, 300–306.

Heggeness, M.H., Smith, P.R. and Choppin, P.W. (1982) *Proc. Natl. Acad. Sci. USA* 79, 6232–6236.

Herrler, G. and Compans, R.W. (1982) *Virology* 119, 430–438.

Hewitt, J.A. and Nermut, M.V. (1977) *J. Gen. Virol.* 34, 127–136.
Holland, J.J., Grabau, E.A., Jones, C.L. and Semler, B.L. (1979) *Cell* 16, 495–504.
Hoshi, S., Mikami, T., Nagata, K., Onuma, M. and Izawa, H. (1983) *Arch. Virol.* 76, 145–151.
Hsiung, G.D. (1972) *Prog. Med. Virol.* 14, 241–274.
Hsu, C.H. and Kingsbury, D.W. (1982) *Virology* 120, 225–234.
Hsu, M.C., Scheid, A. and Choppin, P.W. (1979) *Virology* 95, 476–491.
Huang, A.S., Baltimore, D. and Bratt, M.A. (1971) *J. Virol.* 7, 389–394.
Iorio, R.M. and Bratt, M.A. (1983) *J. Virol.* 48, 440–450.
Jensik, S.C. and Silver, S. (1976) *J. Virol.* 17, 363–373.
Kida, H. and Yanagawa, R. (1981) *J. Gen. Virol.* 52, 103–111.
Kim, H.W., Chanchola, J.G., Brandt, C.D., Pyles, G., Chanock, R.M., Jensen, K. and Parrott, R.H. (1969) *Am. J. Epidemiol.* 89, 422–434.
Kingsbury, D.W., Bratt, M.A., Choppin, P.W., Hanson, R.P., Hosaka, Y., ter Meulen, V., Norrby, E., Plowright, W., Rott, R. and Wunner, W.H. (1978) *Intervirology* 10, 137–152.
Klamm, H. (1980) *Acta Virol.* 24, 127–131.
Kristensson, K. and Örvell, C. (1983) *J. Gen. Virol.* 64, 1673–1678.
Lamb, R.A. and Choppin, P.W. (1977) *Virology* 81, 382–397.
Lamb, R.A. and Choppin, P.W. (1978) *Virology* 84, 469–478.
Lamb, R.A., Mahy, B.W.J. and Choppin, P.W. (1976) *Virology* 69, 116–131.
Leprat, R. and Aymard, M. (1979) *Arch. Virol.* 61, 273–281.
Lerner, R.A. (1982) *Nature* 299, 592–596.
Lund, G.A. and Salmi, A.A. (1981) *J. Gen. Virol.* 56, 185–193.
Lyles, D.S. (1979) *Proc. Natl. Acad. Sci. USA* 76, 5621–5625.
Markwell, M.A.K. and Fox, C.F. (1980) *J. Virol.* 33, 152–166.
Marx, P.A., Portner, A. and Kingsbury, D.W. (1974) *J. Virol.* 13, 107–112.
Matsumoto, T. (1982) *Microbiol. Immunol.* 26, 285–320.
McCarthy, M. and Johnson, R.T. (1980) *J. Gen. Virol.* 46, 15–27.
Merz, D.C. and Wolinsky, J.S. (1981) *Virology* 114, 218–227.
Merz, D.C. and Wolinsky, J.S. (1983) *Virology* 131, 328–340.
Merz, D.C., Scheid, A. and Choppin, P.W. (1980) *J. Exp. Med.* 151, 275–289.
Merz, D.C., Scheid, A. and Choppin, P.W. (1981) *Virology* 109, 94–105.
Meulen, V. ter, Löffler, S., Carter, M.J. and Stephenson, J.R. (1981) *J. Gen. Virol.* 57, 357–364.
Miura, N., Ushida, T. and Okada, Y. (1982) *Exp. Cell. Res.* 141, 409–420.
Mountcastle, W.E. and Choppin, P.W. (1977) *Virology* 78, 463–474.
Nagai, Y. and Klenk, H.-D. (1977) *Virology* 77, 125–134.
Nagai, Y., Klenk, H.-D. and Rott, R. (1976) *Virology* 72, 494–508.
Naruse, H., Nagai, Y., Yoshida, T., Hamaguchi, M., Matsumoto, T., Isomura, S. and Suzuki, S. (1981) *Virology* 112, 119–130.
Nishikawa, K., Isomura, S., Suzuki, S., Watanabe, E., Hamaguchi, M., Yoshida, T. and Nagai, Y. (1983) *Virology* 130, 318–330.
Norrby, E. (1967) *Arch. Virol.* 20, 215–224.
Norrby, E. (1971) *Virology* 44, 599–608.
Norrby, E. (1975) *J. Biol. Stand.* 3, 375–380.
Norrby, E. (1983) *Arch. Virol.* 76, 163–177.
Norrby, E. and Gollmar, Y. (1975) *Infect. Immun.* 11, 231–239.
Norrby, E. and Penttinen, K. (1978) *J. Infect. Dis.* 138, 672–676.
Norrby, E. and Sievertsson, H. (1974) *Antimicrob. Agents Chemother.* 5, 426–430.
Norrby, E., Enders-Ruckle, G. and ter Meulen, V. (1975) *J. Inf. Dis.* 132, 262–269.
Norrby, E., Chen, S.-N., Togashi, T., Sheshberadaran, H. and Johnson, K.P. (1982) *Arch. Virol.* 71, 1–11.

Norrby, E., Varsanyi, T., Utter, G. and Appel, M. (1984) *In Modern Approaches to Vaccines* (R. Chanock and R. Lerner, Eds.) Cold Spring Harbor Laboratory Press, New York. pp. 21–25.
Örvell, C. (1976a) *Acta Pathol. Microbiol. Scand. B* 84, 441–450.
Örvell, C. (1976b) *Acta Pathol. Microbiol. Scand. B* 84, 451–457.
Örvell, C. (1978a) *J. Gen. Virol.* 41, 517–526.
Örvell, C. (1978b) *J. Gen. Virol.* 41, 527–539.
Örvell, C. (1980) *Arch. Virol.* 66, 193–206.
Örvell, C. (1984) *J. Immunol.* 132, 2622–2629.
Örvell, C. and Grandien, M. (1982) *J. Immunol.* 129, 2779–2787.
Örvell, C. and Norrby, E. (1974) *J. Immunol.* 113, 1850–1858.
Örvell, C. and Norrby, E. (1977) *J. Immunol.* 119, 1882–1887.
Örvell, C. and Norrby, E. (1980) *J. Gen. Virol.* 50, 231–245.
Örvell, C. Sheshberadaran, H. and Norrby, E. (1984) *J. Gen. Virol.* In press.
Ozawa, M., Asano, A. and Okada, Y. (1976) *FEBS Lett.* 70, 145–149.
Payne, F.E. and Baublis, J.V. (1973) *J. Infect. Dis.* 127, 505–511.
Peluso, R.W., Lamb, R.A. and Choppin, P.W. (1977) *J. Virol.* 23, 177–187.
Pennington, T.H. (1978) *Arch. Virol.* 56, 345–351.
Portner, A. (1981) *Virology* 115, 375–384.
Portner, A. and Kingsbury, D.W. (1976) *Virology* 73, 79–88.
Portner, A., Webster, R.G. and Bean, W.J. (1980) *Virology* 104, 235–238.
Rammohan, K.W., McFarland, H.F., Bellini, W.J., Ghuens, J. and McFarlind, D.E. (1983) *J. Infect. Dis.* 147, 546–550.
Richardson, C.D., Scheid, A. and Choppin, P.W. (1980) *Virology* 105, 205–222.
Rima, B.K. and Martin, S.J. (1979) *J. Gen. Virol.* 44, 135–144.
Rima, B.K., Martin, S.J. and Gould, E.A. (1979) *J. Gen. Virol.* 42, 603–608.
Rima, B.K., Roberts, M.W., McAdam, W.D. and Martin, S.J. (1980) *J. Gen. Virol.* 46, 501–505.
Robbins, S.J. and Bussell, R.H. (1979) *Intervirology* 12, 96–102.
Robinson, W.S. (1971) *J. Virol.* 8, 81–86.
Russell, P.H. and Alexander, D.J. (1983) *Arch. Virol.* 75, 243–253.
Russell, P.H., Griffiths, P.C., Goswami, K.K.A., Alexander, D.J., Cannon, M.J. and Russell, W.C. (1983) *J. Gen. Virol.* 64, 2069–2072.
Samson, A.C.R. and Fox, C.F. (1973) *J. Virol.* 12, 579–587.
Sato, H., Ogura, H., Tanaka, J. and Hatano, M. (1984) *J. Gen. Virol.* 65, 185–189.
Scheid, A. and Choppin, P.W. (1973) *J. Virol.* 11, 263–271.
Scheid, A. and Choppin, P.W. (1974) *Virology* 57, 475–490.
Scheid, A. and Choppin, P.W. (1976) *Virology* 69, 265–277.
Scheid, A. and Choppin, P.W. (1977) *Virology* 80, 54–66.
Scheid, A., Caliguiri, L.A., Compans, R.W. and Choppin, P.W. (1972) *Virology* 50, 640–652.
Scheid, A., Graves, M.C., Silver, S.M. and Choppin, P.W. (1978) In *Negative Strand Viruses and the Host Cell* (B.W.J. Mahy and R.D. Barry, Eds.). Academic Press, London, pp. 181–193.
Schild, G.C., Oxford, J.S., de Jong, J.C. and Webster, R.G. (1983) *Nature* 303, 706–709.
Schloer, G., Spalatin, J. and Hanson, R.P. (1975) *Am. J. Vet. Res.* 36, 505–508.
Seifried, A.S., Albrecht, P. and Milstein, J.B. (1981) *J. Virol.* 25, 781–787.
Server, A.C., Merz, D.C., Waxham, M.N. and Wolinsky, J.S. (1982) *Infect. Immun.* 35, 179–186.
Seto, J.T., Becht, H. and Rott, R. (1973) *Med. Microbiol. Immunol.* 159, 1–12.
Seto, J.T., Becht, H. and Rott, R. (1974) *Virology* 61, 354–360.
Shapshak, P., Graves, M.C. and Imagawa, D.T. (1982) *Virology* 122, 158–170.
Sheshberadaran, H. and Norrby, E. (1984) *J. Virol.* 52, 995–999.
Sheshberadaran, H., Chen, S.-N. and Norrby, E. (1983) *Virology* 128, 341–353.
Sheshberadaran, H., Norrby, E. and Rammohan, K.W. (1984) *Arch. Virol.* In press.

Shimizu, K. and Ishida, N. (1975) *Virology* 67, 427–437.
Shimizu, K., Shimizu, Y.K., Kohama, T. and Ishida, N. (1974) *Virology* 62, 90–101.
Stallcup, K.C., Wechsler, S.L. and Fields, B.N. (1979) *J. Virol.* 30, 166–176.
Stephenson, J.R. and ter Meulen, V. (1979) *Proc. Natl. Acad. Sci. USA* 76, 6601–6605.
Stone, H.O., Kingsbury, D.W. and Darlington, R.W. (1972) *J. Virol.* 10, 1037–1043.
Togashi, T., Örvell, C., Vartdal, F. and Norrby, E. (1981) *Arch. Virol.* 67, 149–157.
Tozawa, H., Watanabe, M. and Ishida, N. (1973) *Virology* 55, 242–253.
Trudgett, A., Gould, E.A., Armstrong, M., Mingioli, E.S. and McFarlin, D.E. (1981) *Virology* 109, 180–182.
Tyrrell, D.L.J. and Ehrnst, A. (1979) *J. Cell. Biol.* 81, 396–402.
Tyrrell, D.L.J. and Norrby, E. (1978) *J. Gen. Virol.* 39, 219–229.
Tyrrell, D.L.J., Rafter, D.J., Örvell, C. and Norrby, E. (1980) *J. Gen. Virol.* 51, 307–315.
Urata, D.M. and Seto, J.T. (1975/76) *Intervirology* 6, 108–114.
Varsanyi, T.M., Utter, G. and Norrby, E. (1984) *J. Gen. Virol.* 65, 355–366.
Walsh, E.F. and Hruska, J. (1983) *J. Virol.* 47, 171–177.
Wang, E., Wolf, B.A., Lamb, R.A., Choppin, P.W. and Goldberg, A.R. (1976) In *Cell Motility* (R. Goldman, T. Pollard and J. Rosenbaum, Eds.) Book A. Cold Spring Harbor Laboratory, New York, pp. 589–599.
Ward, K.A., Lambden, P.R., Ogilvie, M.M. and Watt, P.J. (1983) *J. Gen. Virol.* 64, 1867–1876.
Wechsler, S.L. and Fields, B.N. (1978) *Nature* 272, 458–460.
Wechsler, S.L., Weiner, H.L. and Fields, B.N. (1979) *J. Immunol.* 123, 884–889.
Yamane, N., Arikawa, J., Odigiri, T. and Ishida, N. (1982) *Microbiol. Immunol.* 26, 557–568.
Yewdell, J.W. and Gerhard, W. (1981) *Annu. Rev. Microbiol.* 35, 185–206.
Yewdell, J.W. and Gerhard, W. (1982) *J. Immunol.* 128, 2670–2675.
Yoshida, T., Nagai, Y., Yoshii, S., Maeno, K., Matsumoto, T. and Hoshino, M. (1976) *Virology* 71, 143–161.

Eds. M.H.V. van Regenmortel and A.R. Neurath
Immunochemistry of viruses. The basis for serodiagnosis and vaccines.

Antigenic structure of foot-and-mouth disease virus

16

F. BROWN

The Animal Virus Research Institute, Pirbright, Woking, Surrey GU24 ONF, U.K.

Foot-and-mouth disease is one of the most devastating diseases of farm animals. Cattle, sheep, pigs and goats are all affected and, although the disease rarely leads to death, productivity losses are usually estimated to be as high as 25%. In addition, indirect losses due to embargoes on trading can be crippling to a country whose economy depends on the export of farm products.

Control is by slaughter in those countries where the disease does not normally occur. In endemic areas such as Africa, Asia and South America, control is by vaccination. More than 1.5 billion doses of vaccine are injected each year.

Most vaccines are prepared by inactivating virus grown in tissue culture cells, using formaldehyde or an imine. Properly prepared, these vaccines are effective and their use in Western Europe during the last 20 years has led to the virtual eradication of the disease in that part of the world. The quality of the vaccines has gradually improved as the antigenic structure of the virus has become better understood. This knowledge has come from studies on the structure of the virus and its associated antigens.

However, a complicating factor in successful control of the disease is the considerable antigenic variation in the virus. Foot-and-mouth disease occurs as seven serotypes whose differences are such that recovery from infection with one serotype still leaves the animal fully susceptible to infection with the remaining serotypes. Moreover, variation within a serotype can be sufficient to leave animals immunized with vaccine of one strain of the virus susceptible to infection with other strains of the same serotype. An understanding of the chemistry underlying these antigenic differences is now being reached by applying the methods of recombinant DNA technology and nucleic acid sequencing. Using the information derived from this work, the amino acid sequences of major antigenic sites of viruses belonging to different serotypes have been determined and synthetic peptides corresponding to these sequences have been shown to possess immunogenic activity. The observations that have been made are of considerable practical importance and also provide interesting taxonomic information.

1. Virus-specific particles

Foot-and-mouth disease virus (FMDV) can be grown in a variety of tissue culture cells, including tongue epithelial cells, primary calf, pig and sheep kidney cells and bovine thyroid cells. Cell lines such as the pig kidney cell line IBRS-2 (De Castro, 1964) and the baby hamster kidney cell line BHK 21 (MacPherson and Stoker, 1962) have been used extensively and most of the work described here relates to experiments done with virus grown in these cells.

Four well-recognized virus-specified antigens are found in harvests produced in these cells: (*a*) the infectious virus particle; (*b*) the so-called empty particle; (*c*) the 12S protein subunit, and (*d*) the virus infection-associated (VIA) antigen. All these antigens have been purified and their composition is summarised in Table 16.1.

The virus particle consists of one molecule of single-stranded positive sense RNA with a molecular weight of approximately 2.6×10^6 and 60 copies of each of four proteins, VP1−4. The so-called empty particle is similar to the infectious particle in size but it lacks the RNA and the polypeptides VP2 and VP4 are covalently linked as the precursor VPO. The 12S particle, which is found in harvests of the

TABLE 16.1

Antigenic components of foot-and-mouth disease virus

Component	Sedimentation constant	Composition
Virus particle	146S	1 molecule ssRNA ($M_r = 2.6 \times 10^6$ Da) 60 copies of each of VP1–3 ($M_r \sim 24 \times 10^3$ Da) and VP4 ($M_r \sim 8 \times 10^3$ Da)
Empty particle	75S	60 copies each of VPO[a], VP1 and VP3
Protein subunit	12S	Pentamer of VP1–3
Virus infection-associated antigen	3.8S	RNA polymerase ($M_r \sim 56 \times 10^3$ Da)

[a] VPO comprises VP2 and VP4 covalently linked.

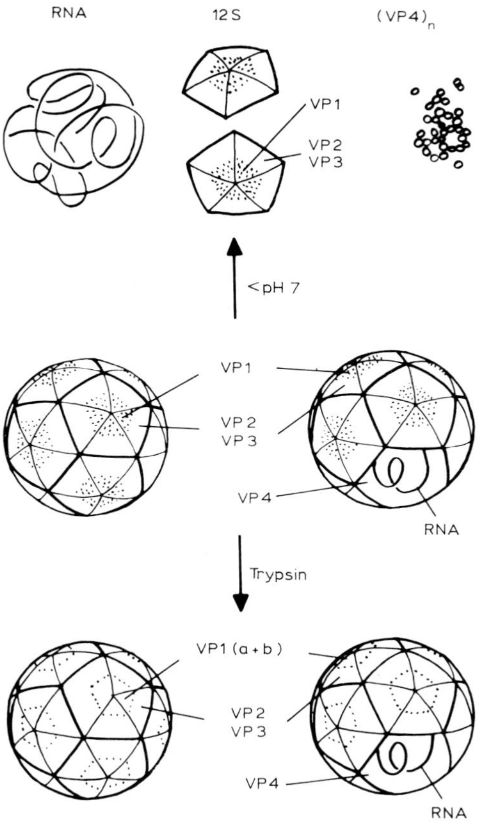

Fig. 16.1. Diagrammatic representation of the foot-and-mouth disease virus particle and the products obtained by cleavage below pH 7 and by hydrolysis with trypsin.

virus, can also be derived from the 146S infectious particle by heating to 56°C or by reducing the pH below 7. This treatment also releases the RNA in an infectious form (Fig. 16.1). The VIA antigen has been shown to be an inactive form of the RNA polymerase (Newman et al., 1979). Its presence in such large amounts in the virus harvest appears to be a unique feature of FMDV, because there are no similar reports with other picornaviruses. It is now recognised that one molecule of VIA antigen is present in the virus particle. The VIA antigen differs from the other antigens found in virus harvests in not being serotype-specific.

1.1. The virus particle

In addition to the RNA and four major structural polypeptides (VP1 — 4) there are small amounts, averaging one molecule per virus particle, of proteins with molecular weights of 56×10^3 and 38×10^3. The first of these minor components is the VIA antigen or RNA polymerase; the second is a copy of VPO, consisting of uncleaved VP2 and VP4. The molecular weights of VP1–3 of some isolates of the virus are now known with certainty because their compositions have been derived from the sequences of the corresponding regions of the RNA (Boothroyd et al., 1982). The precise molecular weight of VP4 is not known because the N-terminal amino acid is blocked, so that it has not been possible to determine the site at which it is cleaved from P16 (Fig. 16.2).

Electron microscopy of complexes of the virus particle with IgM and IgG antibody has provided useful information about the antigenic sites of the virus (Fig. 16.3). This work is described in detail in Section 3.

1.2. The empty particle

In common with most of the picornaviruses that have been examined in detail, FMDV harvests contain particles which are penetrated by phosphotungstic acid when they are prepared for examination in the electron microscope (Fig. 16.3). These particles are the same size as the full particles but can be separated from them because of their much lower sedimentation rate (75S compared with 146S). Failure to incorporate uridine during growth of the virus shows that the particles do not contain RNA. Moreover, polyacrylamide gel electrophoresis of the polypeptides shows that the 75S particles contain VP1 and VP3 but no VP2 and VP4. Instead they contain VPO, the uncleaved precursor of VP2 and VP4.

1.3. The 12S particle

A subviral particle was discovered by Bradish and his colleagues in 1952. Its structural relationship to the 146S virus particle was only established after the purification of the virus and analysis of its protein composition by polyacrylamide gel elec-

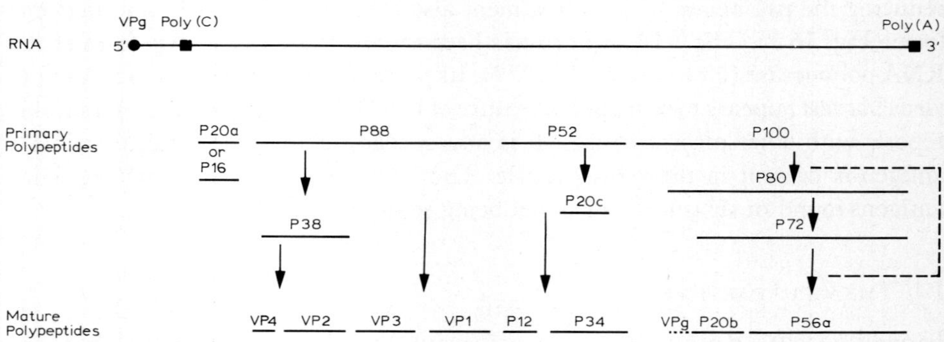

Fig. 16.2. Biochemical map of the foot-and-mouth disease virus genome.

trophoresis had been accomplished. It is now known that the 12S particle is a pentamer of polypeptides VP1 −3 (Vasques et al., 1979) which can be obtained from the 146S particle by lowering the pH of the suspending medium below 7 (Fig. 16.1). The low immunizing activity of the 12S particle compared with that of the 146S particle has provided important clues regarding the structural features which determine this biological activity. These are discussed in Section 3.

1.4. The virus infection-associated antigen

This protein, which was discovered by Cowan and Graves in 1966, is indistinguishable from the RNA polymerase in chemical tests but it does not possess any enzymatic activity. It differs from all the other virus-specified proteins found in infected cells in being cross-reactive in serological tests. It does not elicit neutralizing antibody when injected into experimental animals.

2. Antigenic variation

The virus occurs as seven serotypes: O, A, C, SAT 1, SAT 2, SAT 3 and Asia 1. The O, A and C serotypes occur in all parts of the world where the disease is found, while the other serotypes are confined to Africa (SAT 1–3) or Asia.

The difficulties encountered in vaccination because of the occurrence of the seven serotypes are multiplied by the considerable antigenic variation which occurs within a serotype. The failure to recognise the extent of this intratypic variation has led quite frequently to unsuccessful vaccinations over the last 20–30 years. However, the extent of this variation is now appreciated by control authorities and vaccine manufacturers. Consequently, in the face of a field outbreak, every effort is made to match the outbreak strain with the appropriate vaccine strain.

Early experiments to determine the biochemical basis for differences between serotypes were directed towards comparisons of (*a*) the RNAs of different isolates by hybridisation (Dietzschold et al., 1971; Robson et al., 1977), and (*b*) tryptic peptide maps of the virus-induced proteins (Robson et al., 1980). Both these approaches explored the differences between the entire genome of the viruses. There was good agreement between the results obtained by the RNA hybridization and protein analyses (Marquardt, 1982). The conclusion drawn was that there was considerably more variation at the 5′ end of the genome, i.e. in that part coding for the structural proteins. In particular, it was found that the structural proteins VP1 and VP3 showed more variation than VP2 and VP4. These observations have provided supporting evidence for the location of the immunogenic site of the virus obtained by studying the effect of proteolytic enzymes on the immunizing activity of the virus particle.

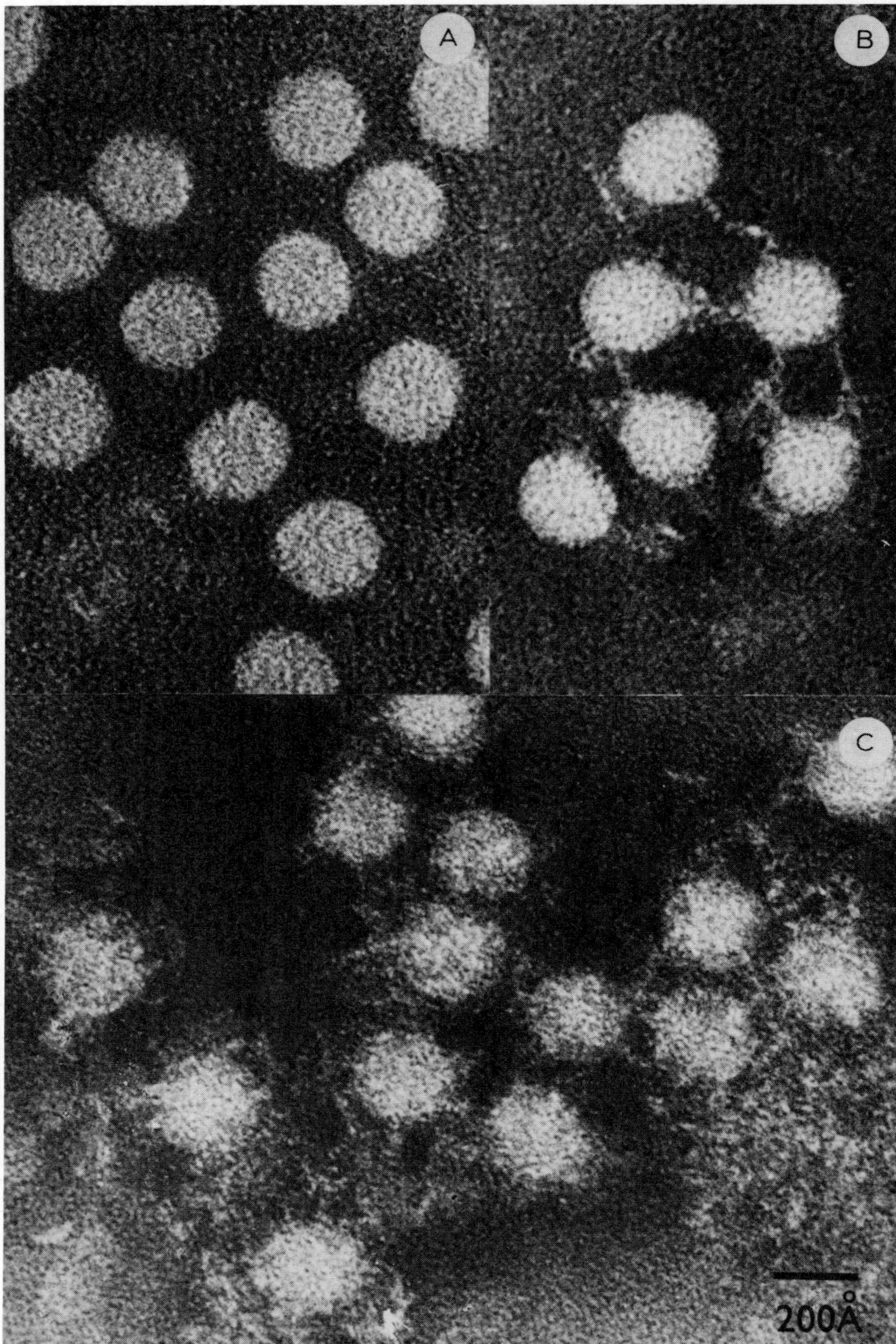
A
B
C
200Å

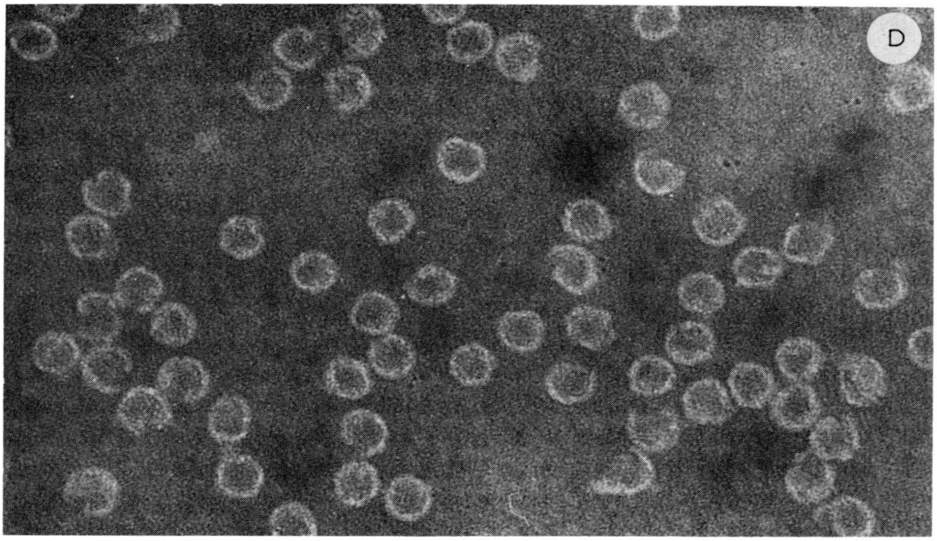

Fig. 16.3. Foot-and-mouth disease virus particles (A) complexed with IgM (B) and IgG (C) antibody molecules. Empty particles are shown in panel D.

3. Location of the immunogenic site

FMDV is unique among the picornaviruses, with the possible exception of Coxsackie A9 (Strohmaier et al., 1982; Matheka et al., 1962) in that its infectivity is reduced considerably by treatment with proteolytic enzymes (Brown et al., 1963). The loss of infectivity is caused by the failure of the treated virus to attach to susceptible cells. With some strains of the virus, the loss of infectivity is accompanied by a marked loss in immunizing activity (Wild et al., 1969).

The enzyme-treated particles are indistinguishable from the untreated particles in the electron microscope. However, complexing the two kinds of particle with IgM antibodies reveals interesting differences between them. Whereas complexes are formed between the intact virus particles and the IgM antibody, the treated particles do not form complexes. Moreover, the site of attachment of the IgM antibody molecules appears to be clearly defined as being at the apices of the virus particle (Fig. 16.3; Brown and Smale, 1970).

In contrast, IgG antibody molecules appear to attach at many positions on the surface of both the untreated and treated virus particles. However, absorption of the IgG antibody with trypsin-treated particles leaves a fraction which still neutralizes the virus and attaches to virus particles apparently at the same positions as those to which the IgM antibody molecules attach.

The only detectable effect of the enzyme on the virus particles is the cleavage of polypeptide VP1 (Wild et al., 1969; Burroughs et al., 1971; Fig. 16.1). None of the other three polypeptides are apparently affected by the enzyme, as judged by their mobility in polyacrylamide gel electrophoresis. The RNA of the treated virus has the same sedimentation characteristics and infectivity as that isolated from the untreated virus particles. The effect of the enzyme on VP1 in situ thus provided an important clue regarding the location of the major immunogenic site on the virus particle.

Confirmation that the major immunogenic site is located on VP1 came from experiments by Laporte and his colleagues on the activity of the individual proteins of the virus (Laporte et al., 1973). They found that only VP1 stimulated the production of neutralizing antibody. The level of activity produced was very low indeed compared with that evoked by the virus particle and several injections of VP1 were often required to obtain a measurable response. Nevertheless, the response has been confirmed in several laboratories (Bachrach et al., 1975; Kaaden et al., 1977; Meloen et al., 1979). Both pigs and cattle can be protected against challenge by two or three injections of about 500 μg of the isolated protein.

Biochemical mapping has established the position of VP1 on the virus genome (Sangar et al., 1977; Fig. 16.2). This allowed the portion of the genome coding for the protein to be cloned and expressed in *Escherichia coli* cells. Excellent yields were obtained by Kleid et al. (1981), as much as 17% of the bacterial protein being VP1. As with the protein isolated from virus particles, the activity of the biosynthesized VP1 was very low compared with that of virus particles. Nevertheless two injections at a 24-day interval of 250 μg of the chimeric protein protected cattle against challenge infection. This observation provided the first demonstration that a vaccine produced by genetic engineering can protect animals from a disease caused by a virus.

Using the approach introduced by Anderer with tobacco mosaic virus (Anderer, 1963), Strohmaier et al. (1982) cleaved VP1, either by hydrolysis of the isolated protein with cyanogen bromide or by proteolytic enzymes in situ, and then tested the separated fragments for biological activity (Fig. 16.4). The positions of the fragments on VP1 were established by comparing the sequences of their amino and carboxy termini with the complete sequence of VP1 derived from that of the RNA coding for it.

By comparing the overlapping peptides for their ability to elicit a neutralizing antibody response, Strohmaier's group predicted that residues 146–154 and 200–213 would be included in antigenic sites which possessed this activity (Fig. 16.4). This work was invaluable in providing the groundwork from which all subsequent developments have been made.

It would be expected in a highly variable virus that antigenic differences would correlate with amino acid sequence differences. Studies of sequences of the VP1 proteins of viruses from different serotypes and subtypes of the virus have revealed that, while much of the sequence is highly conserved, there are three positions where

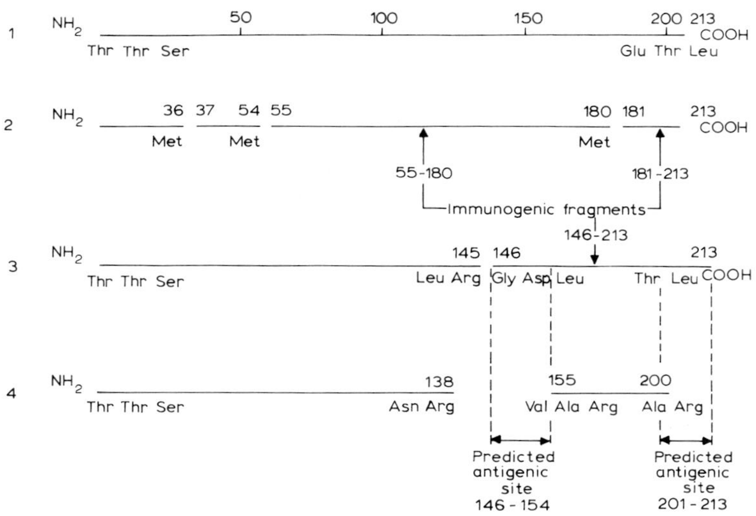

Fig. 16.4. Location of antigenic sites on VP1 of foot-and-mouth disease virus (adapted from Strohmaier et al., 1982). 1, Intact VP1. 2, VP1 cleaved at methionine residues with cyanogen bromide. 3, VP1 cleaved in situ at exposed arginine residues with mouse submaxillary gland protease. 4, VP1 cleaved in situ at exposed arginine and lysine residues with trypsin.

there is considerable variability. In comparisons of viruses from serotypes O, A and C it was found that the major variation occurred at residues 42–61, 131–160 and 193–204 (Fig. 16.5). Surface-orientated regions are likely to be hydrophilic in character and when plots of hydrophilic regions (Hopp and Woods, 1981) were superimposed on the amino acid sequences, it was found that the only regions of high variability which coincided with hydrophilic regions were at residues 131–169 and 193–204.

These ideas were tested by measuring the activity of several organically synthesized peptides in eliciting virus-neutralizing antibody and, in some instances, in protecting the animals against challenge infection. Early experiments with virus of serotype O by Bittle et al. (1982), in which peptides linked to keyhole limpet hemocyanin were injected into guinea pigs, showed that by far the greatest activity was found in a peptide containing residues 141–160. Much lower activities were found with peptides 151–160, 194–213 and 200–213, and peptide 146–154 did not elicit any neutralizing antibody. Although we have not investigated other FMDVs in the same detail, sufficient information has been obtained which indicates that region 141–160 is likely to be the immunodominant site, although more recent preliminary evidence indicates that region 131–160 may be more active. As was to be expected from its hydrophobic nature, peptide 41–60 was inactive and high-titer antibody produced against this peptide does not attach to virus particles.

Another technique for the prediction of potentially antigenic sites of VP1 was

```
O 1    TTSAGESADPVTTTVENYGGETQIQRRQHTDVSFIMDRFVKVTPQNQINI
A 10   TTTTGESADPVTTTVENYGGDTQVQRRHHTDVGFIMDRFVKIDSLSPTHV
A 12   TTATGESADPVTTTVENYGGETQVQRRHHTDVSFIMDRFVKIKSLNPTHV
A 24             VTTTVENTGGETQIQRRHHTDIGFIMDRFVKIQSLSPTHV
C 3    TTTTGESADPVTTTVENTGGETQIQRRHHTDVAFVLDRFVKVHVSGNQHT  50

O 1    LDLMQIPSHTLVGALLRASTYYFSDLEIAVKHEGDLTWVPNGAPEKALDN
A 10   IDLMQTHKHGIVGALLRAATYYFSDLEIVVRHDGNLTWVPNGAPEAALSN
A 12   IDLMQTHQHGLVGALLRAATYYFSDLEIVVRHDGNLTWVPNGAPEAALSN
A 24   IDLMQTHQHGLVGALLRATTYYFSDLEIVVRHEGNLTWVPNGAPESALLN
C 3    LDVMQVHKDSIVGALLRAATYYFSDLEIAVTHTGKLTWVPNGAPVSALDN  100

O 1    TTNPTAYHKAPLTRLALPHTAPHRVLATVYNGECRYNRNAVPNLRGDLQV
A 10   TSNPTAYNKAPFTRLALPYTAPHRVLATVYDGTNKTSASDS-RS-GDLGS
A 12   TGNPTAYNKAPFTRLALPYTAPHRVLATVYNGTNKYSASGSG-VRGDFGS
A 24   TSNPTAYNKAPFTRLALPYTAPHRVPATVYNGTSKYAVGGSGR-RGDMGS
C 3    TANPTAYHKGPLTRLALPYTAPHRVLATTYTGTTAYTASA--R-RGDLAH  150

O 1    LAQKVARTLPTSFNYGAIKATRVTELLYRMKRAETYCPRPLLAIHPTE-A
A 10   IAARVATQLPASFNYGAIQAQAIHELLVRMKRAELYCPKPLLAIKVTSQD
A 12   LAPRVARQLPASFNYGAIKAETIHELLVRMKRAELYCPRPLLAIEVSSQD
A 24   LAARVVKQLPASFNYGAIKADAIHELLVRMKRAELYC?RP?LVIEVSSQD
C 3    LAAAHARHLPTSFNFGAVKAETITELLVRMKRAELYCPRPVLPVQPTG-D  200

O 1    RHKQKIVAPVKQTL
A 10   RYKQKIIAPAKQLL
A 12   RHKQKIIAPGKQL
A 24   RHKQKVIAPAKQLL
C 3    RHKQPLIAPAKQLL
```

Fig. 16.5. Amino acid sequences of VP1 of five viruses belonging to three serotypes of foot-and-mouth disease virus.

used by Pfaff et al. (1982). They reasoned that a good candidate structure would be a strong α-helical region with hydrophilic and hydrophobic zones on opposite sides of the helix. Such a structure was found, using the Chou-Fasman prediction rules (Chou and Fasman, 1978) at residues 144–159 in the type O Kaufbeuren strain of the virus. However, our unpublished work indicates that a large helical content is not a conserved feature of this region from other FMDVs. Since this region is active for peptides from all the viruses we have tested, it seems that this method of prediction may be less valuable than the one we have used.

4. The chemical basis of antigenic variation

The considerable antigenic variation found with FMDV makes it a valuable system for studying the chemical basis for this variation. It has been mentioned in Section 2 that most of the variation between serotypes occurs at the 5′ end of the molecule. The

TABLE 16.2

Serological relationships between foot-and-mouth disease viruses belonging to serotypes A, O and C.

Serum	Log_{10} virus neutralized by 0.015 ml serum			
	A10	A12	01	C3
Antivirion A10	2.5	0.9	–	–
Antipeptide	2.7	0.5	–	–
Antivirion A12	0.7	3.5	–	–
Antipeptide	0.5	2.5	–	–
Antivirion 01	−0.2	–	3.1	−0.1
Antipeptide	1.1	–	4.3	−0.1

subsequent location of the immunodominant site at positions 141–160 on polypeptide VP1 has allowed more precise comparisons to be made of the chemical basis for antigenic variation. It is clear from Fig. 16.5 that there is more amino acid variation between serotypes than there is between subtypes of a single serotype. This would be expected from the serological relationships between the viruses. Examples illustrating the serological differences between viruses of serotypes A, O and C and two subtypes of serotype A are given in Table 16.2. These results illustrate clearly that serotype specificity is also found on residues 141–160.

A fortuitous observation has also enabled us to pinpoint even more precisely the chemical basis for antigenic variation. We found that the 141–160 peptide of the A12 virus derived from the nucleotide sequence obtained by the Plum Island-Genentech group by cloning (Kleid et al., 1981) did not neutralize the A12 virus currently being used in our laboratory. This was an unexpected result because suitably coupled peptides corresponding to three other viruses (serotype O subtype 1, serotype A subtype 10 and serotype C subtype 3) had elicited good levels of neutralizing antibody. However, the antibody to the A12 peptide neutralized the virus which had been used by the Plum Island-Genentech group for their cloning experiments. Our challenge virus had undergone a single passage in baby hamster kidney cells from virus isolated from cattle, whereas the Plum Island-Genentech virus, originally received from The Animal Virus Research Institute in 1954, had been passaged through different host cells, plaque picked and then cloned. We confirmed the sequence of the Plum Island-Genentech virus by use of the primer extension method. This takes advantage of the conservation of the sequence of the region of the genome coding for P52 which is located immediately to the 3′ side of VP1 (Fig. 16.2). Using a DNA primer corresponding to part of the P52 region, it was possible to obtain sequences of VP1 extending from the C-terminal amino acid at position 213 to the amino acid at position 120. However, the nucleotide sequence of our virus at positions corresponding to amino acids 148 and 153 was ambiguous, suggesting that our virus consisted of a mixture. Sequencing of the RNA of 10 plaque isolates confirmed this supposition. Of the 10 plaque isolates three had Ser at 148 and Leu at 153, five had Leu at 148 and Pro at 153, while two had Ser at both 148 and 153. The Plum Island-Genentech virus had Phe at 148 and Pro at 153. Otherwise all the viruses had a common sequence from 213 to 120 except that the Plum Island-Genentech virus had a Glu instead of a Val at position 171.

Serological analysis of the four viruses showed the importance of these amino acid positions in determining the specificity of the antigen-antibody reaction (Table 16.3). Moreover, the fact that the antibody elicited by the peptide mimics that produced by the virus particle is further evidence that the 141–160 peptide contains the immunodominant site.

The influence of the amino acids at positions 148 and 153 is also illustrated by the results obtained with two nonsense peptides. These had been synthesized when we were somewhat uncertain of the sequence of the RNA of our virus, before we had contemplated the possibilitiy that it was a mixture. Thus antiserum against a peptide with Leu at 148 instead of Ser did not neutralize the virus with Ser at that position. Similarly antisera against peptides with Leu at 153 instead of Pro did not neutralize the two viruses with Pro at that position.

Predictions of the shape of protein molecules have been the subject of much research ever since amino acid sequences became available. We have attempted similar predictions for the 141–160 peptide (Rowlands et al., 1983). It appears from comparison of the peptides for the four A12 viruses that these fall into two structural groups, those with Ser at 148 and either Leu or Ser at 153 on the one hand, and those with Leu or Phe at 148 and Pro at 153 on the other. The serological data in Table 16.3 correlate with these predictions. This is encouraging because it should allow us to design peptides with predictable antigenic properties.

The occurrence of deletions at residues 141–145 stimulated us to determine the effect of inserting an arginine residue at position 145 of the A10 sequence (where there is a deletion) so that it then matched the A12 sequence at that position. A second peptide was also tested in which, in addition to the inserted arginine at position 145, a proline was inserted at position 142. We found that the antibodies elicited by the 'artificial' peptides, besides neutralizing the homologous virus, also neutralized two of the A12 viruses (Table 16.4). This result leads us to the belief that appropriate manipulation of the sequence at positions 141–145 will enable us to tailor peptide vaccines which have a wider antigenic spectrum than the 'natural' peptides.

TABLE 16.3

Effect of changes at amino acid residues 148 and 153 on the serological relationship between four isolates of foot-and-mouth disease virus, serotype A12

Serum	Log$_{10}$ virus neutralized by 0.015 ml serum				
	148	Ser	Leu	Ser	Phe
	153	Leu	Pro	Ser	Pro
Antivirion 148 Ser		3.3	1.3	1.7	0.7
Antipeptide 153 Leu		2.5	1.1	1.5	0.9
Antivirion 148 Leu		0.9	3.7	0.9	2.1
Antipeptide 153 Pro		0.0	3.2	0.5	2.2
Antivirion 148 Ser		3.5	2.9	4.1	2.5
Antipeptide 153 Ser		1.5	0.1	2.6	1.5
Antivirion 148 Phe		1.3	2.5	1.1	3.7
Antipeptide 153 Pro		0.0	2.0	0.0	2.7
Antipeptide 148 Leu, 153 Leu		0.3	0.9	0.5	0.7
Antipeptide 148 Phe, 153 Leu		0.0	0.4	0.4	0.3

Antivirion sera were used at 1/200 dilution, antipeptide sera were used undiluted.

TABLE 16.4

Effect of inserting arginine and proline residues at positions 145 and 142 of the A10 VP1 sequence on its antigenic spectrum

Antipeptide serum	Log$_{10}$ virus neutralized by 0.015 ml serum					
		A10	A12	A12	A12	A12
	148		Ser	Leu	Ser	Phe
	153		Leu	Pro	Ser	Pro
A10		2.5	−0.1	0.9	0.5	≦0.5
A12 Phe 148, Pro 153		0.5	0.0	2.0	0.0	2.7
A10 Arg 145		2.7	2.1	≧2.5	0.7	0.1
A10 Pro 142, Arg 145		2.1	1.3	2.1	0.7	0.3

5. Newer vaccines

The encouraging results obtained in guinea pigs and rabbits in eliciting neutralizing antibodies prompted studies on the activity of the peptides in affording protection against challenge infection. In FMD, it is possible to obtain valuable preliminary data in guinea pigs because this species gives lesions similar to those found in cattle when virus is injected into the foot pad. Using the guinea pig model, Bittle et al. (1982) demonstrated that animals receiving the 141–160 peptide elicited levels of neutralizing antibody sufficient to protect them against challenge by injection of 10^5 ID_{50} into the tongue.

Even more encouraging are the results obtained by the Plum Island-Genentech group who have found (J.J. Callis, personal communication) that as little as 25 μg of a peptide, corresponding to a virus of serotype A, biosynthesized in *E. coli* cells as part of a fusion protein, will protect 100% of cattle against infection when the animals which had received the peptide are housed in contact with infected animals. It seems, therefore, that it will soon be possible to consider, in economic terms, whether it will be possible to replace the conventional vaccine with a totally synthetic product.

6. Conclusions

The recent observations with foot-and-mouth disease virus hold out great promise that an understanding of the chemical basis for immunogenicity and serological specificity is near at hand. Clearly the preliminary results with the limited number of virus isolates examined so far need to be extended to determine whether the ground rules which have been established with these viruses apply to other members of the genus. As more sequencing data become available it will become possible to correlate these with the mass of serological information on this virus obtained from both laboratory and field studies. Virus variation has been a major problem in the control of foot-and-mouth disease by vaccination. Now that the nucleotide sequences of the viruses can be obtained fairly readily, it should soon be possible to build up a more complete picture of the chemical basis for antigenic variation and apply this directly to field problems.

7. References

Anderer, F.A. (1963) *Z. Naturforsch.* 18b, 1010–1014.

Bachrach, H.L., Moore, D.M., McKercher, P.D. and Polatnick, J. (1975) *J. Immunol.* 115, 1636–1641.

Bittle, J.L., Houghten, R.A., Alexander, H., Shinnick, T.M., Sutcliffe, J.G., Lerner, R.A., Rowlands, D.J. and Brown, F. (1982) *Nature* 298, 30–33.

Boothroyd, J.C., Harris, T.J.R., Rowlands, D.J. and Lowe, P.A. (1982) *Gene* 17, 153–161.
Bradish, C.J., Brooksby, J.B., Dillon, J.F. and Norambuena, M. (1952) *Proc. Roy. Soc. London Ser. B* 140, 107–127.
Brown, F. and Smale, C.J. (1970) *J. Gen Virol.* 7, 115–127.
Brown, F., Cartwright, B. and Stewart, D.L. (1963) *J. Gen. Microbiol.* 31, 179–186.
Burroughs, J.N., Rowlands, D.J., Sangar, D.V., Talbot, P. and Brown, F. (1971) *J. Gen. Virol.* 13, 73–84.
Chou, P.Y. and Fasman, G.D. (1978) *Adv. Enzymol.* 47, 45–148.
Cowan, K.M. and Graves, J.H. (1966) *Virology* 30, 528–540.
De Castro, M.P. (1964) *Arch. Inst. Biol. S. Paulo*, 31, 63–78.
Dietzschold, B., Kaaden, O.R., Tokui, T. and Bohm, H.O. (1971) *J. Gen. Virol.* 15, 171–174.
Hopp, T.P. and Woods, K.R. (1981) *Proc. Natl. Acad. Sci. USA* 78, 3824–3828.
Kaaden, O.R., Adam, K.-H. and Strohmaier, K. (1977) *J. Gen. Virol.* 197, 397–400.
Kleid, D.G., Yansura, D., Small, B., Dowbenko, D., Moore, D.M., Grubman, M.J., McKercher, P.D., Morgan, D.O., Robertson, B.H. and Bachrach, H.L. (1981) *Science* 214, 1125–1129.
Laporte, J., Grosclaude, J., Wantyghem, J., Bernard, S. and Rouze, P. (1973) *C. R. Acad. Sci.* 276, 3399–3401.
MacPherson, I. and Stoker, M. (1962) *Virology* 16, 147–151.
Marquardt, O. (1982) *J. Gen. Virol.* 59, 283–294.
Matheka, H.D., Mayr, A. and Bögel, K. (1962) *Z. Bakt. Parasit. Infekt. Hyg.* 187, 137–143.
Meloen, R.H., Rowlands, D.J. and Brown, F. (1979) *J. Gen. Virol.* 45, 761–763.
Newman, J.F.E., Cartwright, B., Doel, T.R. and Brown, F. (1979) *J. Gen. Virol.* 45, 497–507.
Pfaff, E., Mussgay, M., Bohm, H.O., Schulze, G.E. and Schaller, H. (1982) *EMBO J.* 1, 869–874.
Robson, K.J.H., Harris, T.J.R. and Brown, F. (1977) *J. Gen. Virol.* 37, 271–276.
Robson, K.J.H., Doel, T.R., Gorman, B.M. and Brown, F. (1980) *J. Gen. Virol.* 46, 179–193.
Rowlands, D.J., Clarke, B.E., Carroll, A.R., Brown, F., Nicholson, B.H., Bittle, J.L., Houghten, R.A. and Lerner, R.A. (1983) *Nature* 306, 694–697.
Sangar, D.V., Black, D.N., Rowlands, D.J. and Brown, F. (1977) *J. Gen. Virol.* 35, 281–297.
Strohmaier, K., Franze, R. and Adam, K.-H. (1982) *J. Gen. Virol.* 59, 295–306.
Vasquez, C., Denoya, C.D., La Torre, J.L. and Palma, E.L. (1979) *Virology* 97, 195–200.
Wild, T.F., Burroughs, J.N. and Brown, F. (1969) *J. Gen. Virol.* 4, 313–320.

Eds. M.H.V. van Regenmortel and A.R. Neurath
Immunochemistry of viruses. The basis for serodiagnosis and vaccines.

Antigenic structure of poliovirus 17

EMILIO A. EMINI[1], BRADFORD A. JAMESON[2] and ECKARD WIMMER[2]

[1]Division of Virus and Cell Biology Research, Merck, Sharp and Dohme Research Laboratories, West Point, PA 19486 and [2]Department of Microbiology, State University of New York at Stony Brook, Stony Brook, NY 11794, U.S.A.

Poliovirus, a member of the Picornaviridae, exists in nature in three stable serotypes. The poliovirion is a naked icosahedron (Rueckert, 1976; Hogle, 1982) composed of twelve identical subunits. A subunit contains five copies of each of four structural proteins: VP1 (M_r = 34 kDa), VP2 (M_r = 30 kDa), VP3 (M_r = 27 kDa), and VP4 (M_r = 7 kDa) which are cleaved from a single precursor polyprotein (Holland and Kiehn, 1968; Jacobsen and Baltimore, 1968; Maizel and Summers, 1968; Kitamura et al., 1981). The intact virion usually contains 60 copies of VP1 and VP3, 58–59 copies of VP2 and VP4, as well as 1–2 copies of VP0, the uncleaved direct precursor of VP2 and VP4 (for a summary see Kitamura et al., 1980; Putnak and Phillips, 1981).

Cross-linking and surface labeling studies of intact virions (Lonberg-Holm and Butterworth, 1976; Beneke et al., 1977; Rothberg et al., 1978; Wetz and Habermehl, 1979) have established that VP1 is the predominantly exposed surface protein followed, respectively, by VP3 and VP2. VP4 is located in the interior of the virion in close association with the viral RNA (Wetz and Habermehl, 1982).

The genome of the virion is composed of a single-stranded 'messenger-sense' RNA molecule that is polyadenylated at the 3′ end (Yogo and Wimmer, 1972). A small protein, VPg (M_r = 2.4 kDa), is covalently linked to the 5′ end of the RNA (Lee et al., 1976; 1977; Nomoto et al., 1977). The viral mRNA is identical to virion RNA except that the 5′-terminal VPg is absent. It is translated into a single large polyprotein which is post-translationally cleaved to yield the above-mentioned structural proteins from the amino terminus and the nonstructural proteins from the carboxy terminus.

During viral morphogenesis, five copies of each of the structural proteins, VP0,

VP3 and VP1 self-assemble to produce a subunit (protomer) with a sedimentation value of 14S. The subunits further assemble to produce an 80S empty capsid, also called a procapsid, which, in turn, will associate with and encapsidate a single viral RNA molecule by an unknown mechanism. The incorporation of the viral RNA into the procapsid is associated with the cleavage of the VP0 structural precursors into the structural polypeptides VP2 and VP4, thereby producing the mature, highly stable virion (Rueckert, 1976; Putnak and Phillips, 1981).

1. Antigenic forms of the virion

In its native state, the poliovirion exists in an antigenic form that is commonly referred to as the D (or N; neutralizing) antigenic state. In this form, the virion sediments in velocity density gradients at 155S and is capable of eliciting a high-level neutralizing antibody response (Rueckert, 1976; Mandel, 1979). When subjected to mild heating (about 55°C in low salt) or to elevated pH (approximately 8.0), the virus undergoes an antigenic conversion (Le Bouvier, 1955; 1959) which is accompanied by the loss of VP4 from the virion (Breindl, 1971). The particle is now in the C (or H; heated) antigenic state. Usually these treatments also lead to the loss of the genome RNA. The resulting particle sediments at 80S and is capable of only occasionally eliciting a neutralizing antibody response, and then only to a low titer (Mandel, 1979). The viral procapsid, the morphogenetic empty capsid precursors consisting of VP0, VP1 and VP3, also exist in a predominantly C antigenic form when isolated from infected cells at pH 7.5 or above. However, the procapsids can be isolated as mostly D antigens if the extraction buffer has a pH of 6.0 (Rombaut et al., 1982). It is likely that the 70S poliovirus particles with D antigenic reactivity described by Ferguson et al. (1981) are such D-specific procapsids. Finally, while similar antigenicities are found between heat-treated virions and naturally occurring empty capsids, studies with monoclonal antibodies have shown that some antigenic determinants are not shared between the two (Emini et al., 1982b).

2. Neutralization antigenic sites

In the discussion that follows, we define an *epitope* as the conformation of a structural element at the surface of a protein to which a monospecific (e.g. monoclonal) antibody binds. The *structural element* may consist of a linear sequence of amino acids or of composite sequences formed by distant regions in the protein. Several epitopes may be clustered in a specific region of the protein. The epitopes in a cluster may or may not overlap functionally. Such a cluster of epitopes is defined as an *antigenic site*. If the binding of the antibody to a structural element results in the neutralization of the virus, we shall call these structural elements *neutralization epitopes* that belong to a neutralization antigenic site (Emini et al., 1984b; Wimmer et al., 1984; Jameson et al., 1984).

2.1. Studies with neutralizing antibodies raised to viral structural elements

As mentioned, the intact poliovirion is capable of consistently eliciting a high-level neutralizing antibody response. However, the infectious virion is the only poliovirus-specific structure with this capability. The morphogenetic precursors of the virion (the 80S naturally occurring empty capsids and the 14S subunits) give rise only to an occasional, weak neutralizing response (for a review see Mandel, 1979). Generally, each of the four viral structural proteins, when denatured and purified, is incapable of generating neutralizing antibodies, while nonneutralizing virion cross-reactive antibodies are produced. However, under certain circumstances, a consistent low-level neutralizing antibody response can be obtained using the purified structural proteins as immunogens (see below).

Purified, denatured VP1 will give rise to just such a response when inoculated into inbred Lewis rats (Chow and Baltimore, 1982). However, in outbred rabbits, the neutralizing antibody response is not as consistent. Only an occasional response is seen and no response is noted with any of the other structural proteins (Blondel et al., 1982a; Emini and Wimmer, unpublished results). Hence, it seems that the ability to respond to different epitopes on the purified capsid proteins is largely dependent upon the immune system of different animals.

The mode of presentation of the antigenic sites of capsid proteins is also important. VP1, VP2 and VP3 will give rise to a consistent neutralizing antibody response in rabbits when purified either from formalin-inactivated virus (Van der Marel et al., 1983) or from native virus by dissociation with formic acid and purification by reverse-phase high performance liquid chromatography (Dernick et al., 1983). Presumably some neutralization epitopes on all three proteins are preserved during these purification protocols, although the titer of the neutralizing sera is in all cases low compared to virus-induced sera. These experiments demonstrate, however, that more than one neutralization antigenic site is to be found on the surface of the intact virion and that these sites are located on all three of the surface-exposed structural proteins, VP1, VP2 and VP3.

Purified VP4 does not give rise to neutralizing antibodies in either Lewis rats or rabbits (Emini, unpublished results). However, a single rabbit inoculated with a VP4 preparation that had been apparently contaminated with degradation products of VP3 elicited antiviral neutralizing antibodies (Emini et al., 1983a). It was shown that the neutralizing activity in this anti-VP4/VP3 serum is due entirely to antibodies directed to a single site on the virion VP3. This antiserum proved especially interesting in subsequent studies on antibody-dependent virus neutralization (see Section 2.2.3).

2.2. Studies with neutralizing monoclonal antibodies

2.2.1. Location of neutralization antigenic sites

The studies described above all suggested that the neutralization epitopes of poliovirus are largely lost upon disruption of the intact virion. Hence, specific probes for these virion-specific epitopes were required in order to precisely define the locations and structures of the viral neutralization antigenic sites. To this end, a large number of neutralizing monoclonal antibodies were generated to several strains of poliovirus type 1 (Icenogle et al., 1981; Osterhaus et al., 1981; 1982; Brioen et al., 1982; Couillin et al., 1982; Emini et al., 1982b; 1983b; Humphrey et al., 1982; Blondel et al., 1983), type 2 (Osterhaus et al., 1981; 1982; Crainic et al., 1983), and type 3 (Ferguson et al., 1981; 1982; Osterhaus et al., 1981; 1982).

With a single exception, none of these antibodies was found capable of reacting with any of the purified viral structural proteins. The exception, antibody C3, was shown by Blondel et al. (1983) to bind to both native and heat-treated virions as well as to purified VP1.

The first direct experimental evidence that neutralization epitopes exist on VP1 was provided by Emini et al. (1982b), who chemically cross-linked F(ab) fragments of two independent neutralizing monoclonal antibodies to intact virus and identified the cross-links to have occurred on VP1. In fact, it appears that most of the neutralization antigenic sites are found on the predominantly exposed VP1. It should be emphasized again, however, that the other two capsid proteins, VP2 and VP3, also carry neutralization antigenic sites (Emini et al., 1983a; van der Marel et al., 1983; Dernick et al., 1983).

2.2.2. Variation and quantitation of the neutralizing antibody-binding sites

Neutralizing monoclonal antibodies and genetic variants resistant to neutralization by these antibodies provided a valuable tool in the study of the number and the locations of neutralizing antigenic sites. Emini et al. (1983b) reported that the neutralizing monoclonal antibodies isolated against poliovirus type 1 (Mahoney) by R. Crainic and his colleagues and by E. Emini fall into 6 groups, each group recognizing a different neutralization epitope. Recent studies indicate that other groups may exist (Bonin and Wimmer, unpublished results). As will be discussed below, these neutralization epitopes belong to two major neutralization antigenic sites on VP1 (Emini et al., 1983b). Interestingly, nearly all neutralizing monoclonal antibodies raised against poliovirus type 3 recognized only epitopes of a single neutralization antigenic site also located on VP1 (Minor et al., 1983). However, not all neutralization antigenic sites may be detectable by the hybridoma technique since some sites are more immunodominant than others (Emini et al., 1984b; Wimmer et al., 1984; Jameson et al., 1984).

In light of the stability of the three poliovirus serotypes, the high genetic variation of virus to variants resistant to neutralization by monoclonal antibodies is surprising. For all monoclonal antibodies, as well as for the neutralizing monospecific anti-VP4/VP3 serum, such variants arise on the order of 1 plaque-forming unit

(PFU) in 10^4 or 10^5 nonvariant PFU (Emini et al., 1982b; 1983a; 1983b). It seems that while the degree of variation of any single epitope is high, the poliovirion contains a large enough number of epitopes to render it antigenically stable with respect to neutralization by antiviral hyperimmune serum.

Neutralizing monoclonal antibodies have also been used to show that different strains of the same serotype bear different as well as identical neutralizing epitopes (Humphrey et al., 1982; Crainic et al., 1983). As a result, the antigenic variation which occurs upon passage of the virus in humans has been studied. In a very interesting study Crainic et al. (1983) demonstrated that neutralizing epitopes specific for avirulent strains of the virus undergo progressive conversion to epitopes characteristic of virulent strains upon replication in vaccinees.

2.2.3. Mechanisms of antibody-dependent neutralization

Monospecific neutralizing antibodies provided a means for studying in detail the mechanisms of antibody-dependent poliovirus neutralization. Using VP1-specific neutralizing monoclonal antibodies it was shown that at least 99% viral neutralization could be accomplished by binding a minimum of four neutralizing IgG molecules to the virion's surface (Icenogle et al., 1983; Emini et al., 1983b). Similar to hyperimmune serum (Mandel, 1979), neutralizing monoclonal antibodies generally do not neutralize by inhibiting adsorption of the virus onto or its penetration into the host cell (Emini et al., 1983b). However, virus-specific macromolecular synthesis does not occur (Emini et al., 1983b). Hence, neutralization may involve inhibition of viral RNA release from the penetrated virion. This model is supported by the finding that all neutralizing monoclonal antibodies tested alter the conformation of the virion, as evidenced by a shift in the virion's isoelectric point from $pI=7$ to $pI=4$ (Emini et al., 1983b). The conformational alteration and subsequent neutralization is dependent upon bivalent attachment of the IgG molecule across the virion's surface (Emini et al., 1983d). If, for example, the neutralized antibody-virus complex is treated with papain thereby converting the bivalent IgG molecules to monovalent F(ab) fragments that remain attached to the virion, the infectivity of the virus is restored. Concomitant with the recovery of the virus infectivity one observes that the pI of the F(ab)-virus complex is 7.0, i.e. the apparent conformational change of the virion has been reversed. Thus, nonneutralizing antivirion antibodies and the nonneutralizing F(ab) fragments of neutralizing antibodies fail to change the virion's conformation*. If quasi-bivalent binding of the F(ab) fragments in the infectious F(ab)-virus complex (8–12 molecules per virion) is restored by treatment of the complex with goat antimouse immunoglobulin antibodies, the infectivity is again lost and the pI is also reduced to 4.0 (Emini et al., 1983d). Nonneutralizing antibodies capable of binding to the virion can produce the same effect (Emini et al., 1983d).

* It should be kept in mind, however, that saturation of the virion's surface with F(ab) fragments leads to neutralization by sterical interference with binding (see Emini et al., 1982b).

The neutralizing monospecific anti-VP4/VP3 serum (see section 2.1) is the only exception so far to this scheme. These antibodies neutralize the virus without altering the virion's isoelectric point and can neutralize the virus as F(ab) fragments. Hence, not all neutralization antigenic determinants share the same neutralizing mechanism upon their interaction with antibody.

Yet another mechanism of neutralization, involving antibody-mediated polymerization, has been described (Brioen et al., 1983). Although this mode of neutralization may contribute to immune protection, it is only effective under conditions of antibody excess, a situation rarely found in vivo.

3. Nonneutralization antigenic sites

In addition to neutralization epitopes, the poliovirion, its morphogenetic precursors and purified structural proteins contain a large number of antigenic sites which will induce and bind nonneutralizing antibodies. Rabbit antisera prepared against each of the purified structural proteins, as mentioned previously, generally do not contain neutralizing antibodies. These antisera to purified VP1, VP2, or VP3 do, however, recognize sites found on C antigen-specific structures such as heat-inactivated (empty) particles. Unlike the neutralization epitopes, many of the epitopes of a purified capsid protein are found on the corresponding polypeptide of the heterotypic viruses (Blondel et al., 1982b; Wiegers and Dernick, 1983; van der Marel et al., 1983; Emini, Jameson and Wimmer, unpublished results). Thus, anti-VP1, anti-VP2, or anti-VP3 sera of poliovirus type 1 precipitate the corresponding capsid proteins of the other two serotypes. This broad cross-reactivity, however, is narrowed when the sera are used to immunoprecipitate C particles (heat-inactivated virus) of the three serotypes (Blondel et al., 1982). Similar results were obtained when polyclonal or monoclonal sera, C- or D-specific for poliovirus type 3, were tested for their ability to react with denatured capsid proteins in immunoblot assays (Thorpe et al., 1982). For example, the antibodies to poliovirion type 3 antigens readily recognized VP1 of all three serotypes. The same antibodies, however, recognized VP2 only of poliovirus type 3, and did not recognize VP2 or VP3 of the heterotypes.

Finally, at least one monoclonal antibody has been derived which has the interesting property of binding to and neutralizing the vaccine strain of poliovirus type 1, while it binds without neutralizing the virulent parent strain of the vaccine (Couillin et al., 1982; Emini et al., 1983b). Thus the epitope recognized by the monoclonal antibody on the Sabin strain is a neutralization epitope. The same antibody still presumably binds to the same site on the Mahoney strain but cannot prevent the initiation of the infectious cycle. Neutralization of infection of the complex between Sabin-specific antibodies and Mahoney virion can be forced if the complex is treated with goat antimouse immunoglobulin antibodies (Emini et al., 1983b).

4. Identification of a major neutralizing antigenic site

Minor et al. (1983), working with a strain of poliovirus type 3, derived nonneutralizable viral variants to a series of type 3-specific neutralizing monoclonal antibodies. The RNAs of these variants were analyzed by RNase T_1 oligonucleotide fingerprint mapping. Each RNase T_1 map was compared with that of the parent virus. Most of the variants exhibited a change in a single T_1 oligonucleotide spot. The nucleotide sequence of this oligonucleotide was found to map at 277–294 bases downstream from the start of the RNA region coding for VP1. Since the amino acid sequence of VP1 (see Section 5) had been deduced from the RNA nucleotide sequence, it was possible to identify the corresponding amino acids as being located 93–97 amino acid residues from the amino terminus of VP1. The site was further defined to an eight amino acid stretch extending from residue 93 to residue 100 in a study by Evans et al. (1983). The authors directly sequenced the RNAs of 16 variants through this site and determined which specific amino acid had been altered to generate the variant. In each case, a single residue change in this critical site was sufficient to yield a variant which could no longer be neutralized by a given monoclonal antibody. It was concluded that the region of amino acids 93–100 of VP1 harbors a major neutralization antigenic site of poliovirus type 3.

Similar T_1-oligonucleotide analyses of many type 1 variants yielded no change in the RNase T_1 fingerprints (Emini, Lynch and Wimmer, unpublished results). Although this result was disappointing it was not too surprising since the probability is low that a single point mutation among 7433 nucleotides becomes apparent by RNase T_1 fingerprinting. However, the region in the VP1 of type 1 homologous to the type 3 antigenic site was found to also be a neutralizing site of poliovirus type 1 by analysis of antisera prepared to specific peptides (see Section 5). This finding was supported by an investigation (van der Werf et al., 1983) in which a series of type 1 VP1 molecules, truncated from the carboxy to the amino terminus by genetic engineering, was studied for the ability to bind to the C3 neutralizing monoclonal antibody. As mentioned earlier, this neutralizing antibody exhibits the unique property of binding to purified VP1. Reactivity of the antibody with its attachment site was lost when amino acid residues 95–110 of VP1 were deleted.

5. Studies of the antigenicity of the poliovirion with antibodies to synthetic peptides

5.1. Peptides containing VP1 amino acid sequences

The complete nucleotide sequence of representative strains from all three poliovirus serotypes is known (Kitamura et al., 1981; Racaniello and Baltimore, 1981; Nomoto et al., 1982; Stanway et al., 1983; Toyoda et al., 1984). This information, along with terminal amino acid analyses of the viral proteins (Kitamura et al., 1981;

Semler et al., 1981; 1982; 1983; Emini et al., 1982a; Larsen et al., 1982), has been used to establish the complete amino acid sequence of each viral protein. Large regions of amino acid homologies are found among the three serotypes in both the structural and nonstructural proteins (Toyoda et al., 1984). Clustered areas of high interserotypic amino acid variation are present, however, in the structural proteins (Toyoda et al., 1984).

All attempts failed to localize a neutralization antigenic site in the poliovirus capsid protein VP1 by fragmenting the polypeptide and analyzing the peptide fragments for their ability to induce neutralizing antibodies. This strategy was very successful in the case of foot-and-mouth disease virus, an aphthovirus related to poliovirus (Strohmeyer et al., 1982). However, in the case of poliovirus both tryptic fragments and cyanogen bromide cleavage fragments of VP1 failed to elicit neutralizing antibody responses (Emini and Wimmer, unpublished results). We therefore reasoned that selectively chosen small synthetic peptides, freed from the constraints found in the larger peptide fragments or in the intact capsid protein, might assume conformations similar to those found in mature virus. Synthetic peptides have been shown in several systems (MS-2 bacteriophage, influenza virus, foot-and-mouth disease virus, hepatitis B surface antigen) to successfully elicit antibodies that bind to both the proteins and the virions from which their amino acid sequences are derived (for reviews see Arnon et al., 1983, Sutcliffe et al., 1983; Walter and Doolittle, 1983).

Emini et al. (1983c) submitted the amino acid sequence of the poliovirus type 1 (Mahoney strain) VP1 to computer analysis for regions of amino acids containing predominantly hydrophilic residues. Such regions have the highest probability of being located on the surface of a protein or macromolecular structure such as a virion. In addition, areas of amino acid variability among the three serotypes were also studied (Toyoda et al., 1984). Such regions are likely to have diverged as a result of selective pressure exerted by the human immune system.

Three regions in VP1 were identified and five synthetic peptides corresponding to the amino acid sequences of these regions were prepared (Fig. 17.1; Emini et al., 1983c). Each of the five peptides was tested in an enzyme-linked immunoadsorbent assay (ELISA) for its ability to be recognized by both neutralizing IgG monoclonal antibodies representing six different epitopes and rabbit hyperimmune antiviral serum. Four of the peptides (1–4) react positively in the assay, each exhibiting a different spectrum of reactivity which is dependent on amino acid composition and peptide length (Table 17.1; Emini et al., 1983c). These same four peptides also react positively with most of the rabbit hyperimmune sera tested. Peptide 5 is not positive with any of the sera, but nevertheless represents a neutralization antigenic site (see below).

Bovine serum albumin (BSA) conjugates of each of the five peptides were inoculated into rabbits. The conjugate of peptide 3 generally elicited a significant serotype-specific neutralizing antibody response (Emini et al., 1983c). As discussed

1 G L G Q M L E S M I D N T V R E T V G A A T S R D A L P N T
#5
31 E A S G P T H S K E I P A L T A V E T G A T N P L V P S D T
61 V Q T R H V V Q H R S R S E S S I E S F F A R G A C V T I M
#1 #4
91 T V D N P A S T T N K D K L F A V W K I T Y K D T V Q L R R
#2
#3
121 K L E F F T Y S R F D M E L T F V V T A N F T E T N N G H A
151 L N Q V Y Q I M Y V P P G A P V P E K W D D Y T W Q T S S N
181 P S I F Y T Y G T A P A R I S V P Y V G I S N A Y S H F Y D
211 G F S K V P L K D Q S A A L G D S L Y G A A S L N D F G I L
241 A V R V V N D H N P T K V T S K I R V Y L K P K H I R V W S
271 P R P P R A V A Y Y G P G V D Y K D G T L T P L S T K D L T
301 T Y

Fig. 17.1. Poliovirus type 1 (Mahoney strain) VP1 amino acid sequence. The amino acid residues comprising the peptides described in the text are underlined.

previously the VP1 region represented by this peptide is identical to the major neutralizing antigenic site identified by analysis of viral variants and deletion products of VP1. In addition, the conjugates of the remaining four peptides, while not directly eliciting neutralizing antibodies, are all capable of priming the immune system of the inoculated animal so that a single subsequent inoculation with a small dose of virus leads to a secondary high-level neutralizing IgG response. Poliovirus does not replicate in rabbits and the dose of 10^5 PFU of 'challenge' virus was too low to elicit by itself serum conversion. Hence, all three VP1 regions (Fig. 17.1) corre-

TABLE 17.1

Recognition of synthetic peptides by neutralizing monoclonal antibodies[a]

Epitope group[b]	Peptide				
	1	2	3	4	5
B	+	−	−	+	−
E	+	−	−	−	−
A	+	−	−	−	−
C	+	−	−	−	−
D	−	−	+	−	−
F	−	+	+	−	−

[a] +, Positive recognition by ELISA. −, ELISA reaction not significantly above background control (see Emini et al., 1983c).

[b] Epitope groupings established by Emini et al. (1983b).

sponding to the peptides used in these studies contain neutralizing antibody eliciting and binding sites. This includes the region represented in peptide 5. As mentioned, this peptide is not recognized in an ELISA by rabbit antibodies prepared against intact virus particles. We speculate that in rabbits this portion of the VP1 structural protein in the environment of the intact virion is not immunogenic.

A peptide identical to our peptide 3 (amino acids 93–103 of VP1 of poliovirus type 1 (Mahoney) was also found by Wychowski et al. (1983) to interact with a neutralizing monoclonal antibody (C3), although the ability of this peptide to elicit neutralizing antibodies was not reported in their study.

The region represented by peptide 1 is of interest because it is identical between serotypes 1 and 3 and bears only a single allelic amino acid difference with type 2. Accordingly, this peptide 'primes' for a neutralizing antibody response for all three serotypes. However, unlike the response with respect to types 1 and 2, the neutralizing response to type 3 was only transitory and the only observed secondary response produced antibodies capable of binding to but not of neutralizing poliovirus type 3 (Emini et al., 1983c). This region of the VP1, therefore, appears to represent a neutralization antigenic site in type 1 and 2 virions, but not in type 3 virions. In poliovirus type 3 this site may only rarely express a bona fide neutralization epitope.

5.2. Peptides containing VP2 and VP3 amino acid sequences

As mentioned earlier, both purified VP2 and VP3 are capable of occasionally giving rise to a neutralizing antibody response. The amino acid sequences of both proteins were analyzed for regions containing structural similarities to a major antigenic site in VP1. These similarities include (*a*) a hydrophilic amino acid sequence, (*b*) amino acid variability among the viral serotypes, and (*c*) a flanking 'bend'-like structure containing proline and several characteristic amino acid residues (Chou and Fasman, 1973; 1978). One such region was found in each of the two proteins (residues 162–173 in VP2 and residues 71–82 in VP3). Peptides corresponding to these regions were synthesized as described before, linked to bovine serum albumin via a tyrosine residue and analyzed for immunological properties. The VP3-specific peptide was found to be reactive by ELISA with the monospecific neutralizing anti-VP4/VP3 serum (Emini, Jameson and Wimmer, unpublished results). The BSA-linked conjugate of the VP2-specific peptide is capable of eliciting in rabbits a consistent low-level neutralizing antibody response that can be augmented by a subsequent administration of virus (Emini et al., 1984a).

The latter result is of special interest. Like peptide 5 of VP1, the VP2-specific peptide is not recognized in an ELISA by antibodies found in a number of rabbit hyperimmune antiviral sera. Hence, using the intact virion, rabbit antibodies to this site are not normally produced. However, use of the synthetic peptide 'forces' recognition of this site by the rabbit's immune system, thereby converting, in a sense, a less immunogenic site to one that is more dominant (Emini et al., 1984a).

6. Conclusions and perspectives

The antigenic manifestations of poliovirus have, to a large extent, evolved through selection by the ecological niche that the virus occupies in nature. The primary route of poliovirus infection is via the gastrointestinal system. As a result, the structure of the intact virion must withstand both low pH and attack by proteolytic enzymes. To meet these conditions, the capsid proteins fold into specific conformations and thereby express specific epitopes that may be favored to exist only in the intact virion. The epitopes expressed by the isolated purified capsid proteins or by morphological precursors are apparently quite different. Some epitopes are shared between purified capsid proteins and the intact virion, but it is interesting that these epitopes are predominantly of the nonneutralization variety and many are not serotype specific. Most of the serotype-specific neutralization epitopes are expressed only on the mature intact virion. Even the viral morphogenetic precursors (protomer and procapsid) have more epitopes in common with the purified proteins than with the virion. The structural proteins VP0, VP1 and VP3 that make up the procapsid undergo profound antigenic rearrangements when the genome RNA is encapsidated, and as a consequence the 60 VP0 precursor molecules are cleaved (by an unknown proteinase; see Hanecak et al., 1982) to yield VP2 and VP4 during formation of the poliovirion. This final step in the morphogenesis of poliovirus is directly responsible for giving rise to most of the virion's neutralization epitopes and, by analogy, possibly to most of the virion's functional sites, such as the attachment site to the host cell receptor.

As we have discussed, however, small peptides can mimic epitopes found on the intact virion. Studies with these peptides showed that several neutralization antigenic sites occupy a discrete number of specific sites. Additional sites will undoubtedly be discovered as more peptides are prepared, although it is impossible to even speculate how many of these sites exist as it is impossible to predict into how many neutralization epitopes the neutralization antigenic sites can fold. The three-dimensional structure of the poliovirion is unknown. The position of the neutralization antigenic sites and the conformational rearrangements occurring upon binding of neutralizing antibody to the virion remain fascinating problems to be resolved. Clearly, every model must accommodate the observation that neutralization by several groups of antibodies requires bivalent binding to VP1, and that this binding causes a shift in the p*I* of the virus particle. However, antibodies binding to other capsid proteins neutralize by a different mode of action.

It should be noted that the number of neutralization antigenic sites located in poliovirus type 1 (Emini et al., 1983c; 1984) is in contrast to the single site identified in type 3 virus by viral variant analysis (Minor et al., 1983; Evans et al., 1983; see earlier discussion). It is possible that a fundamental structural difference exists between the two serotypes. On the other hand, the monoclonal antibody spectrum used for generating the type 3 variants may have been limited in scope, i.e. not in number but in quality. At least some of the antibodies that have been used in the studies of type 3 virus are of the IgM class (Ferguson et al., 1982). This argues that these authors may have confined themselves to studying a primary immune response in which antibodies to other, less dominant neutralizing antigenic sites were not present.

Finally, the studies with the peptides also showed that the immune system of any individual animal only recognizes a limited set of antigenic sites and their epitopes on the surface of the virion. Antibodies to all the possible epitopes are not produced. It cannot be said at this time what determines the immunodominance of one neutralization antigenic site over another. The range of idiotypes that a given animal's immune system can express certainly plays a role in this determination, but other unknown factors must also be involved. It is extremely interesting that, with synthetic peptides, an immune response can be obtained (either directly or by priming) to an antigenic site that is not immunodominant. This raises a considerable potential for future study. Most neutralization epitopes vary to nonneutralizability at a high rate. It is theoretically possible to find a functionally critical antigenic site where such variation cannot occur (the virus-cell receptor binding site?) and which will lead to virus neutralization upon antibody binding. This site may also cross serotype boundaries. It appears that such an antigenic structure, if it exists, does not normally elicit an antibody response. The response may, therefore, be stimulated and/or generated by a synthetic peptide to yield a cross-serotype neutralizing antibody against which viral variants cannot arise. Future investigations aimed at this goal will be exciting indeed.

7. Acknowledgments

The authors thank Jutta Bonin, David Diamond and Richard Crowell for helpful discussions, and Akio Nomoto for generously communicating to us the sequences of all three poliovirus serotypes prior to publication. The work carried out by the authors was supported by NIH grants AI-15122, CA-28146 and AI-06485.

8. References

Arnon, R., Shapira, M. and Jacob, C.O. (1983) *J. Immunol. Methods* 61, 261–273.

Beneke, T.W., Habermehl, K.O., Diefenthal, W. and Buchholz, M. (1977) *J. Gen. Virol.* 34, 387–390.

Blondel, B., Crainic, R. and Horodniceanu, F. (1982a) *C. R. Acad. Sci.* 294, 91–94.

Blondel, B., Crainic, R., Akacem, O., Bruneau, P., Girard, M. and Horodniceanu, F. (1982b) *Virology* 123, 461–463.

Blondel, B., Akacem, O., Crainic, R., Couillin, P. and Horodniceanu, F. (1983) *Virology* 126, 707–710.

Breindl, M. (1971) *J. Gen. Virol.* 11, 147–156.

Brioen, P., DeKegel, D. and Boeye, A. (1983) *Virology* 127, 463–468.

Brioen, P., Sijens, R.J., Vrijsen, R., Rombaut, B., Thomas, A.A.M., Jackers, A. and Boeye, A. (1982) *Arch. Virol.* 74, 325–330.

Chou, P.Y. and Fasman, G. (1973) *J. Mol. Biol.* 74, 263–281.

Chou, P.Y. and Fasman, G.D. (1978) *Adv. Enzymol.* 47, 251–276.

Chow, M. and Baltimore, D. (1982) *Proc. Natl. Acad. Sci. USA* 79, 7518–7521.

Couillin, P., Crainic, R., Cabau, N., Horodniceanu, F. and Boue, A. (1982) *Ann. Virol.* 133E, 315–323.

Crainic, R., Couillin, P., Blondel, B., Cabau, N., Boue, A. and Horodniceanu, F. (1983) *Infect. Immun.* 41, 1217–1225.

Dernick, R., Heukeshoven, J. and Hilbrig, M. (1983) *Virology* 130, 243–246.

Emini, E.A., Elzinga, M. and Wimmer, E. (1982a) *J. Virol.* 42, 194–199.

Emini, E.A., Jameson, B.A., Lewis, A.J., Larsen, G.R. and Wimmer, E. (1982b) *J. Virol.* 43, 997–1005.

Emini, E.A., Dorner, A.J., Dorner, L.F., Jameson, B.A. and Wimmer, E. (1983a) *Virology* 124, 144–151.

Emini, E.A., Kao, S., Lewis, A.J., Crainic, R. and Wimmer, E. (1983b) *J. Virol.* 46, 466–474.

Emini, E.A., Jameson, B.A. and Wimmer, E. (1983c) *Nature* 304, 699–703.

Emini, E.A., Ostapchuk, P. and Wimmer, E. (1983d) *J. Virol.* 48, 547–550.

Emini, E.A., Jameson, B.A. and Wimmer, E. (1984a) *J. Virol.* 52, 719–721.

Emini, E.A., Jameson, B.A. and Wimmer, E. (1984b) In *Modern Approaches to Vaccines* (R. Chanock and R. Lerner, Eds.) Cold Spring Harbor Press. pp. 65–76.

Evans, D.M.A., Minor, P.D., Schild, G.C. and Almond, J.W. (1983) *Nature* 304, 459–462.

Ferguson, M., Schild, G.C., Minor, P.D., Yates, P.J. and Spitz, M. (1981) *J. Gen. Virol.* 54, 437–442.

Ferguson, M., Qi, Y.H., Magrath, D.I., Spitz, M. and Schild, G.C. (1982) *Lancet* ii, 122–124.

Hanecak, R., Semler, B.L., Anderson, C.W. and Wimmer, E. (1982) *Proc. Natl. Acad. Sci. USA* 79, 3973–3977.

Hogle, J.M. (1982) *J. Mol. Biol.* 160, 663–668.

Holland, J.J. and Kiehn, E.D. (1968) *Proc. Natl. Acad. Sci. USA* 60, 1015–1022.

Humphrey, D.D., Kew, O.M. and Feorino, P.M. (1982) *Infect. Immun.* 36, 841–843.

Icenogle, J., Gilbert, S.F., Grieves, J., Anderegg, J. and Rueckert, R. (1981) *Virology* 115, 211–215.

Icenogle, J., Shiwen, H., Duke, G., Gilbert, S., Rueckert, R. and Anderegg, J. (1983) *Virology* 127, 412–425.

Jacobsen, M.F. and Baltimore, D. (1968) *Proc. Natl. Acad. Sci. USA* 61, 77–84.

Jameson, B.A., Emini, E.A. and Wimmer, E. (1984) In *New Approaches to Vaccine Development* (World Health Organization) Schwabe and Co., Basel. In press.

Kitamura, N., Adler, C. and Wimmer, E. (1980) *Ann. N. Y. Acad. Sci.* 345, 183–201.

Kitamura, N., Semler, B.L., Rothberg, P.G., Larsen, G.R., Adler, C.J., Dorner, A.J., Emini, E.A., Hanecak, R., Lee, J.J., van der Werf, S., Anderson, C.W. and Wimmer, E. (1981) *Nature* 291, 547–553.

Larsen, G.R., Anderson, C.W., Dorner, A.J., Semler, B.L. and Wimmer, E. (1982) *J. Virol.* 41, 340–344.

Le Bouvier, G.L. (1955) *Lancet* i, 1013–1016.

Le Bouvier, G.L. (1959) *Br. J. Exp. Pathol.* 40, 605–620.

Lee, Y.F., Nomoto, A. and Wimmer, E. (1976) *Prog. Nucleic Acid Res. Mol. Biol.* 19, 89–95.

Lee, Y.F., Nomoto, A., Detjen, B.M. and Wimmer, E. (1977) *Proc. Natl. Acad. Sci. USA* 74, 59–63.

Lonberg-Holm, K. and Butterworth, B.E. (1976) *Virology* 71, 207–216.

Maizel, J.V. and Summers, D.F. (1968) *Virology* 36, 48–54.

Mandel, B. (1979) In *Comprehensive Virology* (H. Frankel-Conrat and R.R. Wagner, Eds.) Vol. 15. Plenum Publishing Corp., New York. pp. 37–121.

Minor, P.D., Schild, G.C., Bootman, J., Evans, D.M.A., Ferguson, M., Reeve, P. and Spitz, M. (1983) *Nature* 301, 674–679.

Nomoto, A., Detjen, B., Pozzatti, R. and Wimmer, E. (1977) *Nature* 268, 208–213.

Nomoto, A., Omata, T., Toyoda, H., Kuge, S., Horie, H., Kataoka, Y., Genba, Y., Nakano, Y. and Imura, N. (1982) *Proc. Natl. Acad. Sci. USA* 79, 5793–5797.

Osterhaus, A.D.M.E., van Wezel, A.L., van Steenis, B., Drost, G.A. and Hazendonk, T.G. (1981) *Intervirology* 16, 218–224.

Osterhaus, A.D.M.E., van Wezel, A.L., van Steenis, G., Hazendonk, T.G. and Drost, G. (1982) *Dev. Biol. Stand.* 50, 221–228.

Putnak, J.R. and Phillips, B.A. (1981) *Microbiol. Rev.* 45, 287–315.

Racaniello, V.R. and Baltimore, D. (1981) *Proc. Natl. Acad. Sci. USA* 78, 4887–4891.

Rombaut, B., Vrijsen, R., Brioen, P. and Boeye, A. (1982) *Virology* 122, 215–218.

Rothberg, P.G., Harris, T.J.R., Nomoto, A. and Wimmer, E. (1978) *Proc. Natl. Acad. Sci. USA* 75, 4868–4872.

Rueckert, R.R. (1976) In *Comprehensive Virology* (H. Fraenkel-Conrat and R.R. Wagner, Eds.) Vol. 6. Plenum Publishing Corp., New York. pp. 131–213.

Semler, B.L., Hanecak, R., Anderson, C.W. and Wimmer, E. (1981) *Virology* 114, 589–594.

Semler, B.L., Anderson, C.W., Hanecak, R., Dorner, L.F. and Wimmer, E. (1982) *Cell* 28, 405–412.

Semler, B.L., Hanecak, R., Dorner, L.F., Anderson, C.W. and Wimmer, E. (1983) *Virology* 126, 624–635.

Stanway, G., Cann, A.J., Hauptmann, R., Hughes, P., Clarke, L.D., Mountford, R.C., Minor, P.D., Schild, G.C. and Almond, J.W. (1983) *Nucleic Acids Res.* 11, 5629–5643.

Strohmeyer, K., Franze, R. and Adam, K.-H. (1982) *J. Gen. Virol.* 59, 295–306.

Sutcliffe, J.G., Shinnick, T.M., Green, N. and Lerner, R.A. (1983) *Science* 219, 660–666.

Thorpe, R., Minor, P.D., Mackay, A., Schild, G.C. and Spitz, M. (1982) *J. Gen. Virol.* 63, 487–492.

Toyoda, H., Kohara, M., Kataoka, Y., Suganuma, T., Omata, T., Imura, N. and Nomoto, A. (1984) *J. Mol. Biol.* 174, 561–585.

Van der Marel, P., Hazendonk, T.G., Henneke, M.A.C. and van Wezel, A.L. (1983) *Vaccine* 1, 17–22.

Van der Werf, S., Wychowski, C., Bruneau, P., Blondel, B., Crainic, R., Horodniceanu, F. and Girard, M. (1983) *Proc. Natl. Acad. Sci. USA* 80, 5080–5084.

Walter, G. and Doolittle, R.F. (1983) In *Genetic Engineering* (J.K. Setlow and A. Hollaender, Eds.) Vol. 5. Plenum Publishing Corp., New York. pp. 61–91.

Wetz, K. and Habermehl, K.O. (1979) *J. Gen. Virol.* 44, 525–534.

Wetz, K. and Habermehl, K.O. (1982) *J. Gen. Virol.* 59, 397–401.

Wiegers, K.J. and Dernick, R. (1983) *J. Gen. Virol.* 64, 777–785.

Wimmer, E., Jameson, B.A. and Emini, E.A. (1984) *Nature* 308, 19.

Wychowski, C., van der Werf, S., Siffert, O., Crainic, R., Bruneau, P. and Girard, M. (1983) *EMBO J.* 2, 2019–2029.

Yogo, Y. and Wimmer, E. (1972) *Proc. Natl. Acad. Sci. USA* 69, 1877–1882.

Eds. M.H.V. van Regenmortel and A.R. Neurath
Immunochemistry of viruses. The basis for serodiagnosis and vaccines.

Antigenic structure of the adenoviruses

18

ULF PETTERSSON[1] and GÖRAN WADELL[2]

[1] *Department of Medical Genetics, University of Uppsala, The Biomedical Centre, Box 589, S-751 23 Uppsala, and* [2] *Department of Virology University of Umeå, S-901 85 Umeå, Sweden*

1. Classification of human adenoviruses

The adenovirus family comprises a large number of serotypes (species) both of human and animal origin (Table 18.1). Two genera have been distinguished, Mastadenovirus and Aviadenovirus. The separation of adenoviruses into two genera was originally proposed because of the absence of immunologically cross-reactive proteins. The human adenoviruses have been studied most intensively. They have been subdivided into a number of subgenera (subgroups) based on biochemical, immunological and biological characteristics. Three different groups were identified based on hemagglutination properties (Rosen, 1958). Human adenoviruses have also been divided into highly oncogenic (A) weakly oncogenic (B) and nononcogenic on the basis of their oncogenicity for newborn hamsters (Huebner et al., 1967). The nononcogenic adenoviruses can, however, transform rodent cells in vitro. They have been divided into subgenera C and D based on the antigenic reactivity of their T antigen (McAllister et al., 1969). SDS-polyacrylamide gel electrophoresis of virion polypeptides has also been used as a more discriminative criterion for classification of human (Wadell, 1979; Wadell et al., 1980; 1981a) and simian adenoviruses (Tikchonenko, 1984).

The ultimate classification of adenoviruses will be based on nucleotide sequence differences. Information on the homology between human adenoviruses has been obtained by filter hybridization, liquid hybridization (Green, 1970; Green et al., 1979), and also by the use of restriction endonucleases. Today 41 human adenovirus serotypes (DeJong et al., 1983) divided into six subgenera have been identified (Wadell et al., 1980; Wadell, 1984). Table 18.2 outlines the classification of human adenoviruses.

TABLE 18.1

Adenovirus serotypes in man and animals (modified from Wigand et al., 1982)

Genus	Host	Number of members		Serotype
		Definite	Probable	
Mastadenovirus	Man	41	–	h 1–h 41
	Nonhuman primates	24	–	sim 1–sim 24
	Cattle	9	–	bos 1–bos 9
	Pig	4	1	sus 1–sus 4
	Sheep	5	–	ovi 1–ovi 5
	Horse	1	–	equ 1
	Dog	2	–	can 1
	Goat	–	1	cap
	Mouse	1	1	mus 1
	Peromyscus	–	1	
	Opossum	–	1	
	Three shrew (tupaia)	1	–	tup 1
Aviadenovirus	Fowl	9	3	gal 1–gal 9
	Turkey	2	–	mel 1–mel 2
	Goose	–	3	ans
	Pheasant	–	1	pha
	Duck	–	1	ana
	Frog	–	1	

Adenovirus serotypes are defined on the basis of their immunological distinctiveness, as determined by quantitative neutralization with animal antisera. A serotype exhibits either no cross-reaction or shows a homologous: heterologous titer ratio of 16 in both directions. If neutralization shows a certain degree of cross-reaction between two viruses in either of both directions (homologous: heterologous titer ratio of 8–16) distinctiveness of serotype is assumed if: (*a*) the hemagglutinins are unrelated, as shown by lack of cross-reaction in the hemagglutination-inhibition test; or (*b*) substantial biophysical/biochemical differences in the DNAs exist (Wigand et al., 1982).

Genetic variability within each adenovirus serotype can be studied by restriction enzyme analysis. Distinct viral entities and genome types have been identified by this procedure (Wadell et al., 1980; 1981b).

Analysis of the adenovirus proteins by SDS-PAGE (Maizel et al., 1968a; Maizel, 1971) has revealed that the particle contains at least 9 unique polypeptides (polypeptides II–IX and IIIa; see Table 18.3 and Fig. 18.1). Several minor polypeptides that are present in only a few copies per virus particle have also been observed.

Properties of human adenovirus subgenera A to F (modified from Wadell, 1984)

Sub-genus	Serotype	DNA				Apparent molecular mass of the major internal polypeptides (kDa)			Hemagglutination pattern[b]	Length of fibers (nm)	Oncogenicity in newborn hamsters
		Homology (%)		G+C (%)	Number of SmaI[a] fragments						
		Intra-generic	Inter-generic			V	VI	VII			
A	12,18,31	48–69	8–20	48	4–5	51.0–51.5 46.5–48.5[c]	25.5–26.0	18	IV	28–31	High (tumors in most animals in 4 months)
B[b]	3,7,11,14,16, 21,34,35	89–94	9–20	51	8–10	53.5–54.5	24	18	I	9–11	Weak (tumors in few animals in 4–18 months)
C	1,2,5,6	99–100	10–16	58	10–12	48.5	24	18.5	III	23–31	nil
D[d]	8,9,10,13,15, 17,19,20,22,23, 24,25,26,27,28, 29,30,32,33, 36,37,38,39	94–99	4–17	58	14–18	50.0–50.5[e]	23.2	18.2	II	12–13	nil
E	4		4–23	58	16–19	48	24.5	18	III	17	nil
F	40,41	ND[f]		ND	9–12	46.0–48.5	25.5	17	IV	28–33	nil

[a] The restricted DNA fragments were analyzed on 0.8–1.2% agarose slab gels. DNA fragments smaller than 400 bp were not resolved.
[b] I, Complete agglutination of monkey erythrocytes; II, complete agglutination of rat erythrocytes; III, partial agglutination of rat erythrocytes (fewer receptors); IV, agglutination of rat erythrocytes discernible only after addition of heterotypic antisera.
[c] Polypeptide V of ad31 was a single band of 48 kDa.
[d] Only DNA restriction and polypeptide analysis have been performed with ad32 – 39.
[e] Polypeptides V and VI of ad8 showed apparent molecular mass of 45 and 22 kDa respectively.
Polypeptide V of ad30 showed an apparent molecular mass of 48.5 kDa.
[f] Not determined.

TABLE 18.3

Molecular weights of polypeptides present in the adenovirus type 2 particle[a]

Polypeptide designation	Molecular mass (kDa)	Functional significance
II	108	Hexon
III	63	Penton base
IIIa	65	Present in vertex region
IV	62	Fiber
IVa_2	56	Maturation protein
V	42	Core protein
VI[b]	24	Hexon-associated
VII[b]	18.5	Major core protein
VIII[b]	17.5	Hexon-associated
IX	14.3	Hexon-associated
X–XII	5–6.5	Virion components

[a] Predicted from the DNA sequence.
[b] Polypeptide generated by cleavage from a larger precursor.

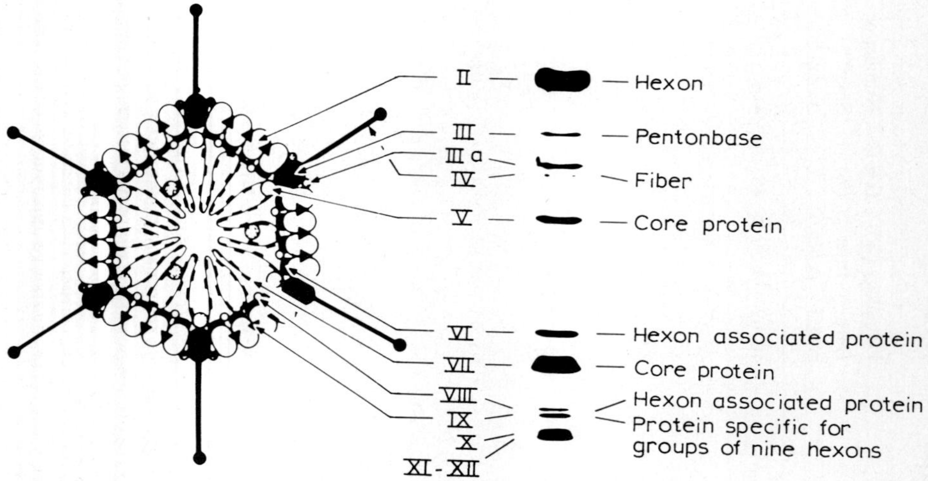

Fig. 18.1. Schematic diagram illustrating the architecture of the adenovirus particle. The SDS-PAGE pattern is also shown indicating all known virion polypeptides. The tentative locations of the different polypeptides in the virus capsid are also shown.

Other components, like polypeptides X–XII, are likely to represent cleavage products which remain when certain capsid polypeptides mature during virion assembly. Unless otherwise specified, all information given in this chapter concerns a member of subgenus C, mastadenovirus type 2 (ad2).

2. The major capsid proteins

Large quantities of soluble viral antigens are produced during adenovirus multiplication. These consist of three major components known as the hexons, pentons, and fibers, which are major components of the adenovirus capsid (Wilcox and Ginsberg, 1963a; Wilcox et al., 1963b; Valentine and Pereira, 1965; Norrby, 1966a). The nascent polypeptides assemble into multimeric structural proteins during or shortly after their synthesis (Horwitz et al., 1969; Velicer and Ginsberg, 1970) but only 1–5% of the fibers and penton bases and 20–30% of the hexons are used for assembly of virus particles (White et al. 1969; Everitt et al., 1971).

2.1. The hexon

Several methods for purification of adenovirus hexons from infected cells have been reported, most of which utilize DEAE-chromatography in combination with one or more additional purification steps (Valentine and Pereira, 1965; Hollinshead et al., 1967; Russell et al., 1967; Kjellén and Pereira, 1968; Maizel et al., 1968b; Levine and Ginsberg, 1967; Pettersson et al., 1967; Boulanger et al., 1969).

2.1.1. Structural and biochemical properties

Hexons have been described as spheres (Horne et al., 1959; Valentine and Pereira, 1965), or as cylindrical structures having a diameter of 8–10 nm and a central hole measuring 2.5 nm in diameter (Pettersson et al., 1967). Berger et al. (1978) described them as pyramids with a triangular top and a hexagonal base. A hole appears to extend from the apex to the base of the pyramid. The radius of gyration has been reported to be 4.3–4.9 nm in different studies (Devaux et al., 1982; Berger et al., 1978). Based on electron microscopic data several three-dimensional models of the hexon have been constructed (Nermut 1975; Nermut and Perkins, 1969; Nermut, 1980) and crystallographic studies have been reported (Burnett et al., 1979, Burnett, 1984).

Hexons from ad2 sediment at 12.9S (Table 18.2; Franklin et al., 1971a) and their molecular mass is 324 kDa as determined from the amino acid sequence of ad2 hexons (Jörnvall et al., 1981). They are composed of three identical polypeptide chains (Horwitz et al., 1970; Cornick et al., 1973; Boulanger and Puvion, 1974; Jörnvall et al., 1974a), each with a molecular mass of 108kDa (Jörnvall et al., 1981).

The ad2 hexon polypeptide consists of 967 animo acid residues (Jörnvall et al., 1981; Akusjärvi et al., 1984). Seven half-cystine/cysteine residues are present, none of which appears to be engaged in disulphide bond formation (Jörnvall et al., 1974a; Grütter and Franklin, 1974). One single cysteine residue is, however, accessible in the native molecule (residue 629) (Grütter and Franklin, 1974; Jörnvall and Philipson, 1980).

The hexon polypeptide carries a reactive serine residue as revealed by its affinity for diisopropylfluorophosphate (DFP) (Devaux and Boulanger, 1980). The N-terminus of ad2 hexons is blocked and has the sequence acetyl-Ala-Thr-Pro-Ser (Jörnvall et al., 1974b). No large clusters of hydrophobic residues are present in the hexon sequence, although charged residues are common. Residues with positive and negative charges are intermingled in the primary structure, except in a highly acidic region which includes fifteen dicarboxylic residues (positions 146–161 in ad2 hexon) and a more widespread region, with nine charged residues (positions 552–580). The former very acidic region appears to be exposed on the surface since it can be removed by limited tryptic digestion.

The complete amino acid sequence is so far only known for hexons from human types 2 and 5. The ad5 hexon polypeptide is 16 residues shorter than the ad2 counterpart (Kinloch et al., 1984). The best match between ad2 and ad5 hexons is obtained if a small number of deletions are introduced in the ad5 hexon relative to the ad2 hexon. The comparison reveals, moreover, that the hexon sequence can be divided into three separate regions; the amino acid sequence homology of the N-terminal region is 100% and that of the C-terminal region 98.6%. The intervening region is heterogeneous and contains well-conserved sequences intermingled with nonhomologous sequences (Fig. 18.2). The region between residues 131 and 320 probably contains type-specific antigenic determinants.

2.1.2. Immunological properties

Almost all adenoviruses contain a common antigenic specificity known as 'α' (Pereira et al., 1959; Wilcox and Ginsberg, 1963b). Hexons from avian and bovine subgroup II adenoviruses seem to be the only exceptions (Bürki et al., 1979; Wigand et al., 1982). More detailed immunological studies (Wilcox and Ginsberg, 1963b; Köhler, 1965; Norrby, 1969a; Pettersson, 1971) have revealed an additional type-specific component in the hexon designated 'ϵ' (Table 18.4). Using more sensitive techniques additional complexities of the antigenic determinants in the hexon molecule were revealed (Norrby 1969a; 1969b; Norrby and Wadell, 1969). Type, group, and intra- and intersubgenus specificities were discovered by cross-absorption and complement fixation tests. Ad12, which appears to have a distant relationship to other human adenoviruses, has hexons which share few antigenic determinants with hexons from other serotypes (Norrby and Wadell, 1969). Hexons from ad4, which is classified as the only member of subgenus E, are most closely related to hexons from subgenus B, and in particular to hexons from ad16 (Norrby and Wad-

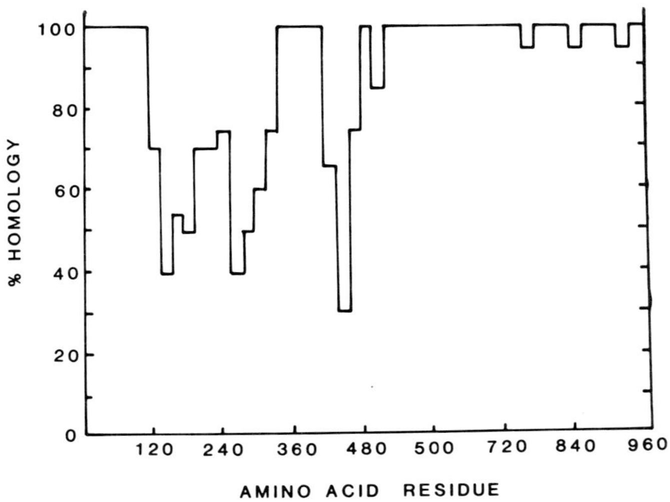

Fig. 18.2. Amino acid sequence homology between ad5 and ad2 hexon polypeptides. The percentage homology was calculated for 20-codon contiguous segments of the sequence. Adapted from Kinloch et al. (1984).

ell, 1969). Intermediate adenovirus strains which contain hexon specificities from one parental strain and fiber specificities from another parental strain have also been described, in particular among members of subgenera B and D (Norrby and Skaaret, 1968; Wigand and Fliedner, 1968). These strains are presumably naturally occurring recombinants between two adenovirus serotypes.

Type-specific hexon determinants have been demonstrated on the surface of the virion (Norrby et al., 1969a; Willcox and Mautner, 1976). The genus-specific determinant ('α'), in contrast, does not appear to be exposed on the surface since only homologous antisera give rise to a visible antibody interaction (Norrby et al., 1969a). Adenoviruses belonging to subgenus C seem to have few determinants exposed on the outside of the particle since homologous hexon antibodies fail to aggregate the virions as judged by electron microscopy (Wadell, 1972) and sucrose gradient centrifugation (Prage et al., 1970). An interesting observation is the finding that type-specific antibodies directed against the 'ϵ' determinant of the hexon interfere with the capacity of adenoviruses from subgenera A and B to agglutinate erythrocytes (Norrby and Wadell, 1972). This inhibition is presumably caused either by aggregation or by the attachment of antibodies to hexons in the peripentonal region, thereby preventing attachment to the red cell receptor because of steric hindrance (Norrby, 1969b; Norrby and Wadell, 1969).

The topography of the antigenic determinants on the hexon protein has been studied by the use of partial proteolysis. Treatment of purified ad2 hexons with trypsin introduces breaks in the polypeptide without changing the antigenic specificity (Pettersson, 1971). The trypsin-sensitive sites in the native hexon molecule have been identified in the ad2 hexon sequence and are located between the following residues: 142/143, 165/166, 286/287. The region between residues 140 and 290 must thus be exposed on the surface of the hexon (von Bahr-Lindström et al., 1982). Limited digestion with chymotrypsin, papain or subtilisin, in contrast, digests more than 50% of the hexon polypeptide, and parts of the group-specific determinant are destroyed under these conditions (Pettersson, 1971). The type-specific determinant ('ϵ') can only be removed by drastic proteolytic treatment, suggesting that this part of the hexon polypeptide has an unusually compact structure.

The use of monoclonal antibodies makes it possible to study the antigenic determinants of the hexon in more detail. Cepko et al. (1981) have prepared group-reacting monoclonal antibodies which precipitate native hexons as well as hexon nonamers*. They fail, however, to precipitate intact virions and unassembled hexon polypeptide chains. It appears that different sets of antibodies recognize the native hexon molecule and the unassembled hexon polypeptides. This has also been demonstrated by Ginsberg and coworkers (Ginsberg et al., 1974; Stinsky and Ginsberg, 1974; Ginsberg, 1979) who found that hexons which have been treated with strong denaturing agents induce different antibodies than do native hexons. The results of Cepko et al. (1981) show that certain determinants of the hexon are also unavailable in the intact virus particle, which confirms previous results obtained by electron microscopy (Norrby et al., 1969a). A collection of monoclonal antibodies against ad5 hexons has been isolated by Russell et al. (1981). These antibodies react with five apparently different epitopes on the hexon, confirming the immunological complexity of the hexon molecule previously demonstrated in cross-adsorption studies. One monoclonal antibody against ad5 hexons failed to react with ad2 hexons while reacting with ad1 and ad6 hexons, revealing a specificity which had been overlooked in previous studies. Monoclonal antibodies are likely to reveal a large number of antigenic determinants which exhibit different reactivities with hexons from different adenovirus species. In accord with this, Adam et al. (1984) have described a collection of 60 monoclonal antibodies against type 1 hexons which represent no less than 41 antigenic determinants. Such antibodies are likely to become important tools for detection and typing of adenoviruses. Cepko et al. (1983) have described an interesting monoclonal antibody which is extremely cross-reactive recognizing hexons from human, canine, porcine, bovine, murine and simian adenoviruses.

Two classes of ad2 hexons, a major and a minor some slowly migrating population have been identified after purification of native hexons by electrophoresis (Pet-

* Aggregates of nine hexons (nonamers) are released from the facets of the adenovirus capsid when it is disintegrated by certain agents (Pereira and Wrigley, 1974).

TABLE 18.4

Antigens associated with the major structural proteins (modified from Pettersson, 1983)

Protein	Corresponding polypeptide	Antigens		
		Designation	Specificity	Remarks
Hexon	II	α	Genus	Oriented towards the inside of the virion
		?	Inter- and intrasubgenus	
		ϵ	Serotype	Available at the surface of the virion
Penton base	III	β	Genus	Carries toxin activity
			Inter- and intrasubgenus	
Fiber	IV	γ	Serotype	Reacts with HI-antibody
		?	Intersubgenus	Shared between members of subgenera C and D
		δ	Intrasubgenus	Located at the proximal part of the fiber; only present in subgenera A C, D and E

tersson, 1971; Boulanger et al., 1978) Two populations of hexons have also been identified using ion exchange chromatography (Wadell, 1972). Analysis of the slowly migrating hexons by SDS-PAGE suggests that they are composed of a slightly shorter polypeptide, lacking approximately 50 residues at the N-terminal end, probably as the result of proteolytic trimming (Boulanger et al., 1978). The slowly migrating hexons are absent from virions and their biological function therefore remains to be determined.

2.2. The fiber

Fibers have been purified to homogeneity from the soluble antigen pool in cells infected with ad2 or ad5. The purification procedures include DEAE-chromatography in combination with other separation methods such as isoelectric focusing, hydroxylapatite chromatography and gel exclusion chromatography.

2.2.1. Structural and biochemical properties

The fiber has a highly asymmetric structure ($f/f_0 = 1.90$), consisting of a shaft with a terminal knob. The diameter of the shaft is estimated to be 2 nm and the diameter of the knob is 4 nm. The length of the fiber is a characteristic property for members of different adenovirus subgenera (Table 18.2). Subgenus B adenoviruses have the shortest fibers (10 nm in length) and subgenus C the longest (23–31 nm) (Norrby, 1969b). Ad4, the only member of subgenus (E), has fibers which are 17 nm in length (Wadell et al., 1967). Avian adenoviruses, except the agent of the so called egg-drop syndrome, have been found to have a peculiar morphology with two fibers of different length extending from different sites on the penton base (Laver et al., 1971; Gelderblom and Maichle-Lauppe, 1982). Extracts from adenovirus-infected cells contain dimers of fibers which are stable after several purification steps. It is, however, not yet known whether these dimers exist under physiological conditions or whether they contain an extra component which links the two units together (Wadell and Norrby 1969a; Norrby et al., 1969b).

Fibers from ad2 and ad5 sediment at about 6S (Levine and Ginsberg, 1967; Pettersson et al., 1968; Wadell et al., 1969; Devaux et al., 1982) and have a molecular mass of 156–207 kDa based on sedimentation equilibrium centrifugation or neutron scattering analysis (Sundquist et al., 1973; Dorsett and Ginsberg, 1975; Devaux et al., 1982). Ad2 fibers are composed of a single type of polypeptide chain with an estimated molecular mass of 62 kDa (Pettersson et al., 1968; Maizel et al., 1968b; Sundquist et al., 1973; Dorsett and Ginsberg, 1975), consisting of 581 residues (Hérissé et al., 1981). The size of the native fiber protein thus suggests that it either is a trimeric or a dimeric structure.

Green et al. (1983) have proposed a model for the fiber based on the predicted amino acid sequence of the fiber polypeptide (Fig. 18.3). According to this model the fiber protein contains a repeating motif of 15 residues. These repeats extend from residue 43 to residue 400 of the 581 residue-long fiber polypeptide. The model predicts that each 15-residue segment contains two short β-strands and two β-bends. The 44 strands together form a long (21 nm) narrow amphipathic β-sheet which could be stabilized by dimer formation to give the shaft of the fiber. The knob may arise from a dimer of the C-terminal 180-residue segment prediced to be an 8–10 stranded β-sandwich. According to this model the fiber protein is a dimeric unit consisting of two fiber polypeptides.

Ishibashi and Maizel (1974) originally discovered that the fiber is a glycoprotein containing two residues of *N*-acetylglycosamine in each polypeptide chain. The functional significance of the carbohydrate moiety remains unknown. A mutant of ad5 which fails to glycosylate fibers at the nonpermissive temperature (Ginsberg and Young, 1977; Cheng Chee-Sheung and Ginsberg, 1982) produces fibers with similar physical properties as fibers from wild type virus, although they do not seem to be immunologically reactive (Cheng Chee-Sheung and Ginsberg, 1982).

Nearly all studies on adenovirus fibers have so far been carried out with fibers

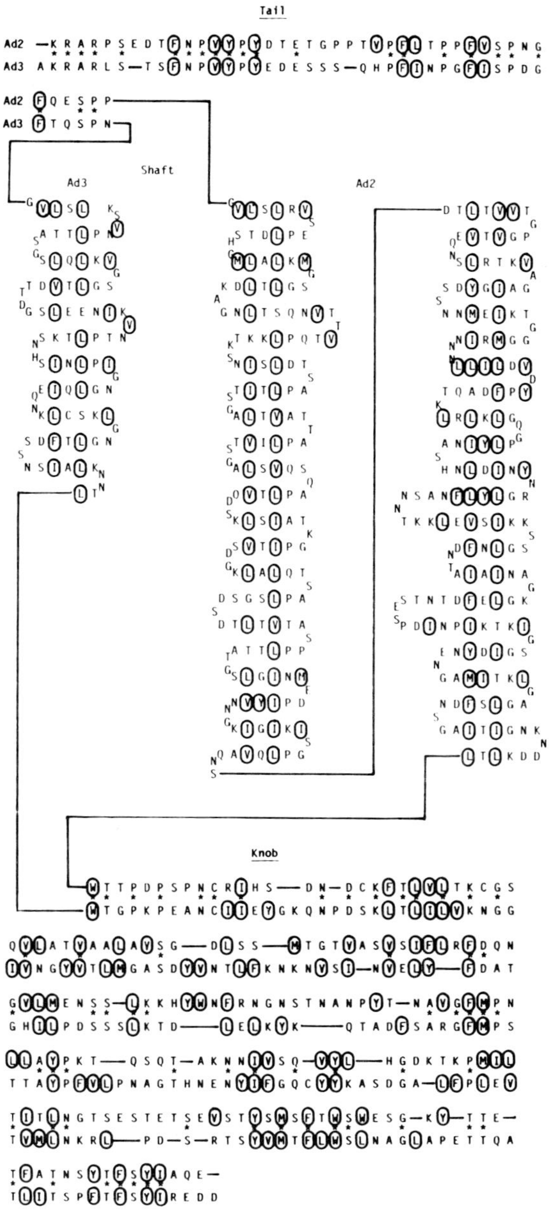

Fig. 18.3. A comparison between the predicted structures for the ad2 and ad3 fiber proteins (from Green et al., 1983 and Signäs et al., 1985). According to this model the fiber consists of a knob, a shaft and a tail. The main difference between ad2 and ad3 fiber is the length of the shaft.

from subgenus C members. Fibers from ad3, a subgenus B member, are composed of two polypeptides each with a molecular mass of 35 kDa (Signäs et al., 1985). The predicted structure of ad3 fibers supports the model proposed by Green et al. (1983) for ad2 fibers and the length of fibers from different subgenus members is apparently correlated to the size of the polypeptide subunit (Fig. 18.3).

2.2.2. Immunological properties

The fiber contains one type-specific determinant (γ) which resides in the knob (Norrby et al., 1969a). Serotypes belonging to subgenus B (Table 18.2), having the shortest fibers, contain exclusively the type-specific 'γ' determinant. For serotypes with longer fibers a positive correlation has been found between the length of the fiber and its antigenic complexity (Table 18.2). The long fibers present in virions from subgenera A, C and D have in their shafts additional antigenic determinants (δ) which are shared between fibers from members within the same subgenus. The long fibers of subgenus C members have in addition intersubgroup-specific determinants (Norrby, 1968; 1969b; Pettersson et al., 1968). The type specific 'γ' determinant present on the knob binds at least two IgG molecules (Norrby et al., 1969a) and these antibodies prevent the attachment of virions to red cell receptors (Pereira and Figueiredo, 1962) and also prevent the attachment of fibers to KB and HeLa cell receptors (Philipson et al., 1968). It is believed that the 'γ' determinants are located near the junction between the fiber and the penton base since they are masked in the intact penton structure (Pettersson and Höglund, 1969; Wadell and Norrby, 1969b).

2.3. THE PENTON

The pentons are composed of two structural units, the base and the fiber, and are more labile than the other major capsid components. They are consequently more difficult to purify in large quantities and have been subjected to less thorough studies than hexons and fibers.

2.3.1. Structural and biochemical properties

The penton bases have a morphology very similar to that of isolated hexons, having an average diameter of 8 nm (Valentine and Perera, 1965). Some investigators have observed a conical morphology (Laver et al., 1969; Wadell et al., 1969) whereas others have observed globular or polygonal structures sometimes showing a pentagonal outline (Pettersson and Höglund, 1969; Boudin et al., 1979). Pentons, like fibers, sometimes appear as aggregates when isolated from infected cells (Norrby 1966a; Wadell et al., 1969) and two types of multimers have been recognized. Dimers of pentons are present in cells infected with certain members of subgenera B and C. Other members of subgenera B and D as well as E produce aggregates consisting of 12 pentons in a dodecahedral arrangement, known as dodecons (Fig. 18.4).

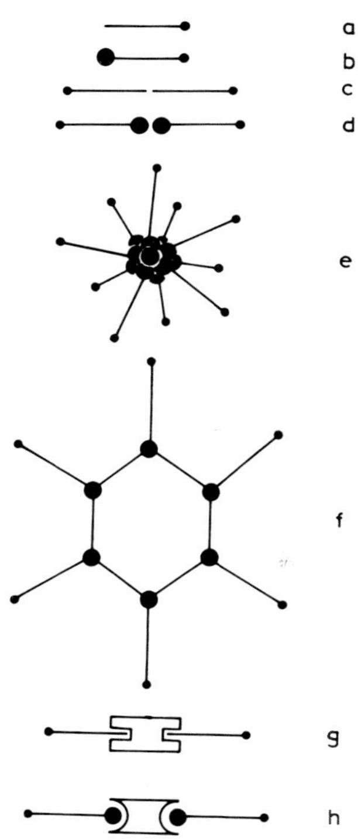

Fig. 18.4. Adenovirus haemagglutinins. A number of different viral components can cause haemagglutination. Fibers (a) and pentons (b) are unable to cause agglutination since they are monovalent units. Dimers of fibers (c) and dimers of pentons (d), on the other hand, are divalent and can hemagglutinate, as can dodecons (e) and complete virions (f), which are multivalent structures. It is possible to convert the monovalent units (pentons and fibers) into divalent haemagglutinins by the use of antibodies against the shaft of the fiber (g) or the base of the penton (h).

Pentons from ad2 and ad5 sediment at 10.5–11S (Pettersson and Höglund, 1969; Wadell et al., 1969). Devaux et al. (1982) have reported a molecular mass of 365 kDa for the complete penton structure, and the penton base is composed of identical polypeptide chains (Anderson et al., 1973; Everitt et al., 1973; Devaux et al., 1982). From the nucleotide sequence of the penton gene it can be predicted that the polypeptide has a molecular mass of 63.3 kDa (Roberts et al. 1984). Molecular mass determinations using neutron scattering techniques have given a molecular mass of 246 kDa for the base (Devaux et al., 1982) suggesting that it is a tetrameric structure. This is unexpected because of the pentagonal symmetry at the corners of the adenovirus icosahedron (Fig. 18.1). The fiber is joined to the penton base by non-covalent bonds which can be disrupted by treatment with guanidine hydrochloride (Norrby and Skaaret, 1967), formamide (Neurath et al., 1968), pyridine (Pettersson and Höglund, 1969) or deoxycholate (Boudin et al., 1979). Penton bases which have been dissociated from the fiber retain their antigenicity and their ability to induce cytopathic changes (see Section 2.3.3) (Pettersson and Höglund, 1969). Boudin et al. (1979) and Winters et al. (1970) have, moreover, shown that free vertex capsomers, sedimenting at 9.5S, exist in adenovirus infected cells. Penton bases and fibers are able to assemble in vitro into pentons (Boudin and Boulanger, 1982). Boudin and Boulanger (1981) have shown that the addition of antibodies against the fiber, particularly those directed against the 'δ' determinant, cause dissociation of the penton into its two constituents.

The penton bases are considerably more sensitive to the action of proteolytic enzymes than most other adenovirus capsid proteins. The base is completely destroyed at high enzyme concentrations, liberating free fibers, whereas the antigenic as well as the morphological properties remain unchanged after treatment with low concentrations of trypsin (Pettersson and Höglund, 1969; Wadell and Norrby, 1969b).

2.3.2. Immunological properties

The penton base carries an antigenic determinant, known as 'β', which is common to pentons from all adenovirus serotypes. A sensitive hemagglutination enhancement assay for serological adenovirus detection has been designed on the basis of this determinant (Norrby et al., 1970). In addition, inter- and intrasubgenus specific determinants have been found (Wadell and Norrby 1969b) and these can be demonstrated in special types of haemagglutination assays. Immunodiffusion, on the other hand, reveals primarily the intrasubgenus specific antigen. The cytopathic effect, caused by penton (see Section 2.3.3), can be neutralized by homologous antisera and, depending on the respective subgenus, by antisera against members of other subgenera (Wadell and Norrby, 1969b; Döring et al., 1972). Pentons from serotype 12 appear to carry few determinants which cross-react with pentons from other serotypes (Wadell and Norrby, 1969b).

2.3.3. Biological functions

Adenovirus-infected cells show cytopathic changes during the course of the lytic infection. Two different kinds of cytopathic effects can be distinguished; the first type is apparent 8–24 hours after infection and the second type appears 2–5 days after infection in conjunction with production of progeny virus (Pereira and Kelly, 1957). It was later shown (Everett and Ginsberg, 1958; Pereira, 1958; Rowe et al., 1958) that the early effect is linked to the pentons (Valentine and Pereira, 1965) and that a penton concentration of 0.1 μg/10^6 cells is required to induce a clearly demonstrable effect. The morphological changes do not seem to be accompanied by perturbation of the macromolecular synthesis and the effect is fully reversible by washing with fresh medium (Pereira, 1958; Rowe et al., 1958). The cytopathic effect of the penton is nullified by mild proteolytic degradation although the antigenic properties of the penton base are not affected by this treatment (Pettersson and

Höglund, 1969; Wadell and Norrby, 1969b). The cytopathic effect of pentons is neutralized by antipenton sera but not by sera against fibers (Pettersson and Höglund, 1969; Wadell and Norrby, 1969b) which suggests that the penton base alone is responsible for the phenomenon. In accord with this, it has been found that isolated penton bases are able to induce early cytopathic changes (Pettersson and Höglund, 1969; Winters et al., 1970; Boudin et al., 1979).

An endonuclease activity has been found to be associated with pentons (Burlingham et al., 1971). It appears, however, that the endonuclease is a cellular component which remains tightly associated with pentons even after several purification steps (Reif et al., 1977a; 1977b).

2.4. The major core proteins

Two major polypeptides, designated polypeptides V and VII, are associated with adenovirus cores (Maizel et al., 1968a; Laver et al., 1968; Russell et al., 1971; Prage and Pettersson, 1971; Everitt et al., 1973; Anderson et al., 1973). Polypeptide VII is the most abundant core protein and accounts for about 10% of the protein mass of the virion. There are approximately 1070 copies in each particle and the weight ratio of VII to DNA is approximately 1:1 (Corden et al., 1976; Sung et al., 1977). Anderson et al. (1973) discovered that polypeptide VII is synthesized as a precursor, pVII, which is cleaved into the mature core protein during virion maturation. Polypeptide VII has an estimated molecular mass of 18.5 kDa and the proteolytic cleavage of pVII results in the loss of one methionine-containing tryptic peptide (Anderson et al., 1973; Sung et al., 1977; Aleström et al., 1984). The predicted amino acid sequence shows that protein VII contains both tyrosine and tryptophan residues and has alanine as its N-terminal amino acid (Aleström et al. 1984). The protein is very basic, containing approximately 24% arginine. With regard to its amino acid composition it resembles the arginine-rich histones that have a high arginine and alanine content, but differs from these in containing tryptophan and lower amounts of lysine. The purified protein interacts with adenovirus DNA in a specific manner, forming a compact structure in which the DNA appears to be organized in the same way as in cores isolated from virus particles. The complex between DNA and polypeptide VII is extremely stable (Black and Center, 1979). The major core protein, like other basic proteins, is weakly antigenic and special care must be taken to reveal its immunological reactivity (Prage and Pettersson, 1971). Core proteins from ad2 and ad3 are immunologically related although there are antigenic differences between them.

Little is known about polypeptide V. It is moderately basic and has a predicted molecular mass of 41.6 kDa. There are about 180 molecules of polypeptide V per virus particle. Polypeptide V seems to mature without proteolytic trimming and it appears to be less tightly bound to the DNA than protein VII (Brown et al., 1975).

3. Minor polypeptides in the adenovirus capsid

In addition to the major capsid components (hexons, pentons, fibers, and core proteins) several other adenovirus polypeptides have been revealed by SDS-PAGE (Maizel et al., 1968a; 1968b; Everitt et al., 1973) as illustrated in Fig. 18.1.

3.1. Protein IIIa

There are approximately 60 copies of polypeptide IIIa in the virus particle (Boudin et al., 1980; Devaux et al., 1982), probably associated with the vertex region. The stoichiometry implies the existence of five polypeptide IIIa molecules for each penton structure. Antibodies against polypeptide IIIa aggregate virions, indicating that the antigenic determinants are exposed on the surface of the virion (Everitt et al., 1975).

Other observations (Everitt et al., 1975) indicate that polypeptide IIIa, although accessible on the surface, might extend through the capsid and have contacts with components in the interior of the virion. Boudin et al. (1980) have reported that polypeptide IIIa suffers proteolytic cleavage during virion maturation. The precursor (pIIIa) is only slightly larger and has an estimated molecular mass of 67 kDa.

Polypeptide IIIa appears to be produced in excess during the infectious cycle (Lemay et al., 1980). It sediments at 6.0S and the native molecule has an estimated molecular mass of 65.4 kDa. A value of 66 kDa has been determined by SDS-PAGE, suggesting that the native protein exists as a monomer. Polypeptide IIIa is remarkably uniform in size between different subgenus members (Wadell et al., 1980). The mature protein IIIa appears to have glycine as the N-terminal amino acid whereas the N-terminus of the precursor is blocked by acetylation (Lemay et al., 1980).

Polypeptide IIIa is a phosphoprotein, and a protein kinase activity which is associated with adenovirus particles appears to preferentially phosphorylate this polypeptide (Akusjärvi et al., 1978; Blair and Russell, 1978).

Antisera prepared against purified polypeptide IIIa precipitate similar polypeptides from cells infected with adenovirus types 2, 3 and 5, and no type-specific antigenic determinants have yet been detected on protein IIIa by immunoprecipitation. Although it was originally claimed that antibodies against polypeptide IIIa have neutralizing capacity (Everitt and Philipson, 1974) more recent data (Boudin et al., 1980) suggest that antibodies against protein IIIa do not play a major role in adenovirus neutralization.

3.2. Protein VI and its precursor polypeptide

Polypeptide VI from ad2 has an estimated molecular mass of 24 kDa (Maizel et al., 1968a; Everitt et al., 1973; Anderson et al., 1973) and there are about 450 molecules per virion. It is synthesized as a precursor, polypeptide pVI. Polypeptide pVI of ad2 consists of 249 amino acids and has a molecular mass of 27.1 kDa (Akusjärvi and Persson, 1981). During virion maturation 33 amino acids are removed from the N-terminal end of pVI (Akusjärvi and Persson, 1981), the cleavage taking place between a glycine and an alanine residue. The location in the capsid of the polypeptide VI has not yet been determined unambigously, although it appears to be associated with the facets of the adenovirus icosahedron. The basis for this assumption is that polypeptide VI is released together with the hexons when the capsid is broken by repeated cycles of freezing and thawing (Everitt et al., 1973). Enzymatic labeling of the exterior of the capsid demonstrates, however, that polypeptide VI cannot be labeled from the outside (Everitt et al., 1975). Cross-linking experiments have also revealed an association between polypeptides V and VI in the virion, supporting the notion that it is located on the interior side of the capsid (Everitt et al., 1975). Studies by Russell and Precious (1982) indicate that polypeptide VI is a DNA-binding protein. This observation, together with the results discussed above, points towards a more intimate relationship with the core than with the outer capsid. It is, in fact, possible that polypeptide VI is a core protein which is less tightly bound to the DNA than polypeptides VII and V. The immunological properties of protein VI are largely unknown, although it is clear that the molecular mass of polypeptide VI differs among members of different subgenera (Wadell, 1980).

3.3. Protein VIII

Polypeptide VIII has a molecular mass of 13 kDa as estimated by SDS-PAGE (Maizel et al., 1968a; Anderson et al., 1973; Everitt et al., 1973). It is synthesized as a considerably larger precursor, pVIII. Nothing is known about the fate of the parts of the precursor which are eliminated during cleavage. Polypeptide VIII, when present in the virion, is associated with hexons, although less firmly than polypeptide IX.

Polypeptide pVIII consists of 227 amino acids, with a calculated molecular mass of 24.7 kDa (Galibert et al., 1979; Hérissé et al., 1980).

3.4. Protein IX

Protein IX is a minor constituent of the adenovirus particle (Maizel et al., 1968a; Everitt et al., 1973; Anderson et al., 1973). In the virus particle, polypeptide IX has been found to be associated with those hexons which constitute the facets of the adenovirus icosahedron (Maizel et al., 1968b; Everitt et al., 1973). The bond between polypeptide IX and the hexon appears to be firm and proteolytic treatment of the hexon nonamers leaves polypeptide IX antigenically intact. Unlike many other minor adenovirus polypeptides, it does not mature through proteolytic trimming. The amino acid sequence is known for polypeptide IX from ad2 (Aleström et al., 1980; Anderson and Lewis, 1980), ad5 (Maat et al., 1980), ad3 (Engler, 1981), ad7 (Dijkema et al., 1981) and ad12 (Bos et al., 1981). Polypeptide IX from ad2 comprises 139 amino acids and has a predicted molecular mass of 14.3 kDa. The N-terminal amino acid is acetylated and the amino acid composition reveals high contents of serine, alanine and leucine. Substantial homologies are found when polypeptide IX from different serotypes are compared, particularly in the N-terminal parts (Engler, 1981). Antibodies have been raised against the purified polypeptide IX and these have been used to study the immunological properties of the polypeptide. The results (Boulanger et al., 1979) show that protein IX preparations from different serotypes cross-react, although type-specific components appear to be present as well (Boulanger et al., 1979).

3.5. Polypeptides X–XII

Maizel et al. (1968a) reported that the adenovirus particle contains a fast migrating polypeptide, designated X. Anderson et al. (1973) were later able to resolve this component into three polypeptide components which range in molecular mass from 5 to 6.5 kDa. It is questionable whether they are unique translation products and at least some of them may represent degradation products from other capsid proteins.

Hosokawa and Sung (1976) have described a virion polypeptide, designated 'μ', which is associated with adenovirus cores. Polypeptide 'μ' was estimated to have a molecular mass of 4–5 kDa and may be identical to one of polypeptides X–XII. It has been estimated that approximately 125 copies of the 'μ' polypeptide are present in the adenovirus particle and a biochemical analysis has revealed a very unusual amino acid composition with 69% basic residues.

3.6. The viral protease

Several of the internal adenovirus polypeptides, including the terminal protein which is covalently linked to the viral DNA, are processed during viral assembly. Cleavage is caused by a viral endopeptidase. Its gene is located on the immediate 3′ side of the hexon gene and encodes a 23-kDa polypeptide (Liang et al., 1983). The protease appears to be a chymotrypsin-like neutral protease (Bhatti and Weber 1979) and cleaves at the sequence Met-X-Gly-Gly-Ala/Val (Aleström et al., 1984).

4. Adenoviral proteins synthesized at intermediate times after infection

Certain adenovirus polypeptides are produced in very low amounts before the onset of viral DNA replication, whereas their synthesis increased considerably late after infection. Polypeptides IX and IVa2 appear to belong to this category. The properties of polypeptide IX have been described in Section 3.4. Polypeptide IVa2 was originally recognized as a minor component of the virion (Maizel et al., 1968a; Everitt et al., 1973; Anderson et al., 1973), having a molecular mass of 50 kDa (Everitt et al., 1973; Anderson et al., 1973). Persson et al. (1979) have proposed that it serves as a maturation protein during adenovirus morphogenesis. It is phosphorylated (d'Halluin et al., 1978) and appears to be a DNA-binding protein (Russell and Precious, 1982). A possible function for polypeptide IVa2 may therefore be to interact with the viral DNA when it enters the preformed capsid. A comparison between the sequences of polypeptide IVa2 from ad5 and ad7 reveals that the N-terminal portion is poorly conserved whereas the C-terminal portions exhibit a very high degree of homology (Dijkema et al., 1981; Engler and van Bree, 1982; Gingeras et al., 1982). Polypeptide IVa2 from adenovirus types 2 and 5 appear to be identical (Gingeras et al., 1982).

5. Late nonstructural adenovirus proteins

In addition to the capsid components a prominent polypeptide has been observed late after infection, designated '100 kDa' because of its estimated molecular mass (Russell and Skehel, 1972; Anderson et al., 1973; Everitt et al., 1973). A 33-kDa adenovirus-specific polypeptide has also been identified which may be nonstructural.

5.1. The 100-kDa polypeptide

This polypeptide is a phosphoprotein. A close physical association has been observed between the 100-kDa and hexon polypeptides and it was postulated that the 100-kDa polypeptide is required for proper assembly of hexon trimers (Cepko and Sharp, 1982). It has, moreover, been demonstrated that antibodies against the 100-kDa protein prevent the assembly of trimeric hexon in cell extracts (Oosterom-Dragon and Ginsberg, 1981).

Monoclonal antibodies have been prepared against the 100-kDa polypeptide from ad5 (Russell et al., 1981). These antibodies react with the corresponding polypeptide in cells infected with ad2 but not in cells infected with ad3. The results thus suggest that the 100-kDa polypeptide reacts in a subgenus-specific manner.

5.2. The 33-kDa phosphoprotein

A 33-kDa phosphoprotein is present late after adenovirus infection (Russell and Blair, 1977; Levinson et al., 1977; Axelrod, 1978). It is encoded by a mRNA which completely overlaps with the mRNA for the 100-kDa protein. The proteins have completely unrelated sequences since different translational reading frames are utilized (Oosterom-Dragon and Anderson, 1982). No information exists as to the function of the 33-kDa polypeptide.

6. Early adenovirus proteins

A large number of virus-specific polypeptides are synthesized early after an adenovirus infection. These polypeptides are presumed to play catalytic roles in connection with the replication of the viral DNA and the synthesis of viral proteins. The early proteins contribute little to the total pool of viral antigens in the infected cell. Some of them have, however, also played an important role as antigens since they are detectable in sera from animals, bearing adenovirus-induced tumors. A class of the early adenovirus proteins are therefore known as T-antigens. Members of immunologically different adenovirus subgenera induce T-antigens which have been used for classification of human adenoviruses (McAllister et al., 1969; Gerna et al., 1982). Hyperimmune sera against early antigens from members of subgenera A–E cross-react with early antigens of ad4, whereas early antigens of all other tested adenoviruses showed a restricted cross-reactivity (Gerna et al., 1982).

7. Adenovirus antigens and their detection

The adenovirus genome encodes some 50 different polypeptides. An adenovirus infection thus gives rise to a large number of antigenically different viral proteins, each of which carries several epitopes. From a diagnostic point of view the major structural proteins, the hexon, the penton and the fiber are of greatest significance. Since the discovery of the adenovirus family, the hexon antigen has been an important immunological tool for adenovirus detection. Traditionally complement fixation assays were utilized to detect the group-specific reactivity of hexons. Other methods which are used for detection of hexons in infected cells or released from purified viral particles include single radial immunodiffusion (Pereira et al., 1972), immunoelectrophoresis (Bohn and Wigand, 1983), crossed immunoelectrophoresis (Boulanger and Puvion, 1974), radioimmunoassays (Halonen et al., 1980), ELISA tests (Kornadt et al., 1982; Roggendorf et al., 1982) and solid phase immunoelectron microscopy (Svensson and von Bonsdorff, 1982; Svensson et al., 1983). When the hexons were characterized in greater detail it became apparent that they are very complex from an immunological point of view. Early studies revealed the presence of type-specific components; the so-called 'ϵ' specificity (Wilcox and Ginsberg, 1963a; Köhler, 1965; Norrby and Wadell, 1969; Pettersson, 1971). Monoclonal antibodies will become a valuable immunological tool to obtain broadly cross-reacting hexon antibodies as well as type-specific ones, and encouraging results have already been reported (Cepko et al., 1981; Russell et al., 1981; Cepko et al., 1983; Adam et al., 1984).

The fiber component has been shown to interact with the receptors present on the surface of erythrocytes and infected cells. The 'γ' determinant of the fiber is type-specific and the adenovirus capsid thus contains two major type-specific components, the 'γ' specificity of the fiber and the 'ϵ' specificity of the hexon. The former is identified in hemagglutination inhibition assays and the latter in neutralization tests (see Section 7.2). The long fibers from members of subgroups A, C, D and E contain additional antigenic specificities known as 'δ'. Moreover, subgenus C fibers show inter- as well as intrasubgenus reactivity. The antigenic specificities which have been identified on the fiber so far consist of several different epitopes, and monoclonal antibodies are likely to reveal additional complexity.

A group-specific 'β' antigen is present on the vertex capsomers. A hemagglutination enhancement test for serological detection of adenoviruses, based on this reactivity, was described by Norrby et al. (1970).

Antigenic extinction tests have revealed that the lowest dose of antigen capable of inducing a primary immune response in guinea pigs is 1 ng and 100 ng for ad5 hexons and fibers respectively. This means that 10^7 virions are needed to elicit and immune response (Haase et al., 1972).

Different adenovirus species sometimes display very peculiar patterns of cross-reactivity; this is particularly true for ad4. This virus shows immunological cross-reactivity at the hexon level with ad16 of subgenus B (Norrby and Wadell, 1969b) and at the vertex capsomer level with several members of subgenus B (Wadell and Norrby, 1969b). The ad4 fiber shows immunological cross-reactivity with fibers of members of subgenera C and D (Wadell and Norrby, 1969b). Hyperimmune sera against ad4 virions cross-react with exposed epitopes on the surface of virions of ad40, a member of subgenus F (Svensson et al., 1983).

It is interesting that the ad4 early antigens also display a very broad cross-reactivity (Gerna et al., 1982).

The relationship between adenoviruses of different host species is sometimes closer than between adenoviruses of the same host species. Simian adenovirus 20 (S5), for instance, cross-reacts to a larger extent with human subgenus C members than with simian adenovirus 15 (S3). Pentons of simian adenovirus 15 (S3), on the other hand, cross-react with pentons of human adenoviruses belonging to subgenus B (Norrby et al., 1971).

7.1. Quantitative measurements of adenovirus infectivity

Three major assays are utilized to titrate adenovirus infectivity. The plaque assay is an important tool which, in combination with specific antisera, can be used for identification of adenovirus species. The disadvantage of the plaque assay is that it is cumbersome and that the specific infectivity of most adenovirus serotypes is very low, sometimes requiring 10^4 particles to yield one plaque (Green et al., 1967; Wigand et al., 1983). The plaques usually appear after an incubation period of one week. A less time-consuming assay is the fluorescent focus assay. In this test specific adenovirus sera are utilized to reveal adenovirus antigens inside the infected cell. The most important antigen in this assay is the hexon. Using polyclonal antisera prepared against adenovirus particles, the test identifies the 'α' specificity present on any human adenovirus serotype. The test can be carried out within 48 hours and is more sensitive than the plaque assay. The third assay which is used to titrate adenovirus infectivity is the endpoint dilution test, which is based on the appearance of cytopathic changes. Although very sensitive it has the disadvantage of being time consuming and requires two weeks or more to complete.

Nucleic acid hybridization offers an interesting alternative for adenovirus detection (Stålhandske et al., 1983). Solid-phase immunoelectron microscopy is also a useful procedure for detecting adenoviruses (Svensson and von Bonsdorff, 1982; Svensson et al., 1983).

7.2. Neutralization of adenovirus infectivity

Inhibited expression of the viral genome as indicated by the lack of formation of viral progeny is taken as the criterion for neutralization of viral infectivity (Wadell, 1972).

Neutralization of adenoviruses is a ternary reaction involving the virus particle and antibodies, as well as the infected cell. Blocking of the interaction between the fiber and the cell-associated receptor is not sufficient to obtain neutralization. It is known that antibodies can bind to the surface of the virion without causing neutralization (Kjellén and Pereira, 1968; Wadell, 1972).

The significance of different viral antigens in the neutralization reaction seems to vary among the different subgenus members. Hexons of all studied adenoviruses induce neutralizing antibodies. Soluble hexons of subgenera B, D and E induce antisera with powerful neutralizing capacity (Norrby, 1969a; 1969b; Wadell, 1972). Hexons from subgenus C members, on the other hand, are much less potent in inducing neutralizing antibodies (Pettersson et al., 1967; Pettersson, 1971; Wadell, 1972).

Subgroup C hexons also induce antibodies with reduced ability to aggregate virions as measured by ultracentrifugation (Prage et al., 1968) and immunoelectron microscopy (Wadell, 1972). This suggests that subgenus C members have fewer hexon determinants exposed on the surface, which may account for the poor neutralizing effect of these hexon antibodies. Pettersson (1971) and Wadell (1972) have described two populations of ad2 hexons differing in charge and ability to induce neutralizing antibodies although they induce type-specific complement-fixing antibodies to the same degree (Wadell, 1972).

Antisera against ad2 fibers (30 nm) do not neutralize whereas antisera against ad3 fibers (10 nm) have a limited neutralizing capacity (Norrby, 1969). Antisera against dodecons of subgenera D and E show a low degree of neutralization, whereas sera against ad3 dodecons display a neutralizing capacity close to that of hexons (Norrby, 1968a; Norrby and Wadell, 1972).

Taken together, these results suggest that hexons are of critical importance for induction of neutralizing antibodies against adenoviruses although other components in the vertex region also seem to play a role.

7.3. Hemagglutination

Hemagglutination by adenoviruses was first demonstrated by Rosen (1958). Subsequently it was shown that almost all human adenovirus serotypes can agglutinate erythrocytes (Rosen, 1960; Rosen et al., 1962; Bauer and Wigand, 1963; Schmidt et al., 1965; Norrby, 1969b; DeJong et al., 1983). When the structural proteins of the virion were isolated and morphologically defined it became possible to show that the type-specific determinant 'γ' of the fiber is the component which interacts with the red cell surface (Pereira and Figueiredo, 1962; Valentine and Pereira, 1965; Norrby, 1969b; Pettersson et al., 1968; Norrby et al., 1969a). Specific receptors are present on the erythrocyte membrane of certain animal species. A broad variability is seen in the reactivity of receptors on different erythrocytes. This variability can sometimes be demonstrated within one human adenovirus serotype. For instance the adenovirus 19A genome type, which is responsible for epidemic outbreaks of keratoconjunctivitis (Wadell and deJong, 1980), agglutinates guinea pig and dog erythrocytes (Kemp et al., 1983), whereas the adenovirus 19 prototype is devoid of this hemagglutination capacity. The number of receptors for different adenovirus serotypes varies for different subgenera. Rat erythrocytes carry a higher number of receptors for members of subgenus D than for members of subgenus C (Wadell et al., 1969), whereas monkey erythrocytes carry 5 times more receptors for ad3 (subgenus B) than rat erythrocytes for ad2 (Inouye and Norrby, 1973).

Since intact virions carry several vertex projections they can establish a bridge between the erythrocytes and give a complete hemagglutination pattern. Dodecons, like dimers of pentons and fibers, can in the same way give rise to complete hemagglutination (Fig. 18.4). Pentons or fibers alone can, in contrast, only establish a monovalent link with the cell and agglutination can only be detected when antibodies are used to bridge the receptor-associated structural units (Fig. 18.4; Pereira and Figueiredo, 1962; Norrby, 1966a; Norrby 1969b). Heterotypic antibodies must be used to demonstrate hemagglutination by penton and fiber monomers since the type-specific determinant of the fiber is responsible for the attachment to the erythrocyte receptor (Pereira and Figueiredo, 1962; Norrby and Skaaret, 1967). This method of revealing adenovirus hemagglutination by heterotypic antisera has been called the *hemagglutination enhancement* reaction (Fig. 18.4). The short fibers of subgenus B viruses fail to hemagglutinate in this test since they only contain type-specific determinants (Norrby and Skaaret, 1967) and are therefore unable to form divalent aggregates after interaction with heterotypic antibodies. Antibodies directed against the 'γ' specificity of the fiber can be assayed by *hemagglutination inhibition* tests. Specific antibody consumption tests have been designed in order to assay for the group, inter- and intrasubgenus-specific determinants on penton bases and fibers. The penton base may thus be assayed by *hemagglutination enhancement consumption* (HEC) tests (Norrby and Skaaret, 1967) and free fibers from subgenus B viruses which lack subgenus specificity can be assayed by the *hemagglutination inhibition consumption* (HIC) test (Norrby and Skaaret, 1967). A *virion hemagglutination inhibition* (VHI) test which detects exposed epitopes on the surface of virions has also been described (Norrby et al., 1970). With the aid of hemagglutination techniques it has been possible to elucidate the detailed immunological relationship between the capsid components from adenovirus species representing subgenera A, B, C, D and E (Norrby, 1968; 1969b; Wadell, 1970; Wadell and Norrby, 1969b). Based on their ability to agglutinate different red cells, Rosen (1960) proposed a subgroup classification for human adenoviruses (see Table 18.2). Subgenus I (B) viruses agglutinate monkey erythrocytes whereas subgenus II (D) viruses agglutinate rat erythrocytes. Subgenus III (C + E) viruses also agglutinate rat erythrocytes but with a partial pattern (a partial hemagglutination pattern is a pattern in which a fraction of the erythrocytes has sedimented without being agglutinated and therefore form a ring in a less dense lattice). The partial agglutination is probably due to a competition between the mono- and multivalent virus components for the receptors on the erythrocytes (Wadell, 1969). Members of subgenus IV (A), which include the highly oncogenic adenoviruses, were originally thought to lack hemagglutination capacity but were later shown to agglutinate rat erythrocytes with a partial pattern (Schmidt et al., 1965; Norrby et al., 1969b). Further subdivision of the human adenoviruses has been suggested based on the types of erythrocytes which the different serotypes within one subgenus agglutinate (Norrby, 1968; 1969b; Hierholzer, 1973).

8. Acknowledgements

The authors are greatly indebted to Marianne Gustafson and Ingegärd Schiller for outstanding secretarial help and to Hannu Ukkonen for the artwork. Work in the authors' laboratories has been supported by grants from The Swedish Medical Research Council, The Swedish Cancer Society and The National Swedish Board for Technical Development.

9. References

Adam, E., Erdei, J., Lengyel, A., Berencsi, G., Fachet, J. and Nasz, I. (1984) *Intervirology* In press.
Akusjärvi, G. and Persson, H. (1981) *J. Virol.* 38, 469.
Akusjärvi, G., Philipson, L. and Pettersson, U. (1978) *Virology* 87, 276.
Akusjärvi, G., Aleström, P., Pettersson, M., Lager, M., Jörnvall, H. and Pettersson, U. (1984) *J. Biol. Chem.* 259, In press.
Aleström, P., Akusjärvi, G., Perricaudet, M., Mathews, M.B., Klessig, D.F. and Pettersson, U. (1980) *Cell* 19, 671.
Aleström, P., Akusjärvi, G., Lager, M., Liang, Y.-K. and Pettersson, U. (1984) *J. Biol. Chem.* 259, In press.
Anderson, C.W. and Lewis, J.B. (1980) *Virology* 104, 27.
Anderson, C.W., Baum, P.R. and Gesteland, R.F. (1973) *J. Virol.* 12, 241.
Axelrod, N. (1978) *Virology* 87, 366.
Bauer, H. and Wigand, R. (1963) *Z. Hyg. Infektionskr.* 149, 96.
Berger, J., Burnett, R.M. and Franklin, R.M. (1978) *Biochim. Biophys. Acta* 535, 233.
Bhatti, A.R. and Weber, J. (1979) *Virology* 96, 478.
Black, B.C. and Center, M.S. (1979) *Nucl. Acids Res.* 6, 2329.
Blair, G.E. and Russell, W.C. (1978) *Virology* 86, 157.
Bohn, K.A. and Wigand, R. (1983) *Zentralbl. Bakteriol. Parasitenkd. Infektionskr. Hyg. Abt. 1: Orig. Reihe A* 255, 196–201.
Bos, J.L., Polder, L.J., Bernards, R., Schrier, P.I., van den Elsen, P.J., van der Eb, A.J. and van Ormondt, H. (1981) *Cell* 27, 121.
Boudin, M.-L. and Boulanger, P. (1981) *Virology* 113, 781.
Boudin, M.-L. and Boulanger, P. (1982) *Virology* 116, 589.
Boudin, M.L., Moncany, M., D'Halluin, J.C. and Boulanger, P. (1979) *Virology* 92, 125.
Boudin, M.L., D'Halluin, J.C., Cousin, C. and Boulanger, P. (1980) *Virology* 101, 144.
Boulanger, P.A. and Puvion, G. (1974) *Eur. J. Biochem.* 43, 465.
Boulanger, P.A., Flamencourt, P. and Biserte, G. (1969) *Eur. J. Biochem.* 10, 116.
Boulanger, P., Devaux, C. and Lemay, P. (1978) *Virology* 84, 456.
Boulanger, P., Lemay, P., Blair, E.G. and Russell, W.C. (1979) *J. Gen. Virol.* 44, 783.
Brown, D.T., Westphal, M., Burlingham, B.T., Winterhoff, U. and Doerfler, W. (1975) *J. Virol.* 16, 366.
Burlingham, B.T., Doerfler, W., Pettersson, U. and Philipson, L. (1971) *J. Mol. Biol.* 60, 45.
Burnett, R.M. (1984) *Biological Macromolecules and Assembilas* (A. McPherson, Ed.) Vol. 1. John Wiley and Sons, New York. pp. 337–385.
Burnett, R.M., Grütter, M.G., Markovic, Z. and White, J.L. (1979) *J. Supramol. Struct.* Suppl. 3, 92.
Bürki, F., Hinaidy, B. and Böckmann, J. (1979) *Microbiologica* 2, 65.
Cepko, C.L. and Sharp, P.A. (1982) *Cell* 31, 407.
Cepko, C.L., Changelian, P.A. and Sharp, P.A. (1981) *Virology* 110, 385.
Cepko, C.L., Whetsone, C.A. and Sharp, P.A. (1983) *J. Clin. Microbiol.* 17, 360.
Cheng Chee-Sheung, C. and Ginsberg, H.S. (1982) *J. Virol.* 42, 932.
Corden, J., Engelking, H.M. and Pearson, G.D. (1976) *Proc. Natl. Acad, Sci. USA* 73, 401.
Cornick, G., Sigler, P.B. and Ginsberg, H.S. (1973) *J. Mol. Biol.* 73, 533–537.
DeJong, J.C., Wigand, R., Kidd, A.H., Wadell, G., Kapsenberg, J.G., Muzerie, C.J., Wermenbol, A.G. and Firtzlaff, R.G. (1983) *J. Med. Virol.* 11, 215.
Devaux, C. and Boulanger, P. (1980) *Virology* 102, 94.
Devaux, C., Zulauf, M., Boulanger, P. and Jacrot, B. (1982) *J. Mol. Biol.* 156, 927.
D'Halluin, J.-C., Martin, G.R., Torpier, G. and Boulanger, P.A. (1978) *J. Virol.* 26, 357.

Dijkema, R., Maat, J., Dekker, B.M.M., van Ormondt, H. and Boyer H.W. (1981) *Gene* 13, 375.
Döring, N., Nguyen, C.X. and Wigand, R. (1972) *Med. Microbiol. Immunol.* 157, 325.
Dorsett, P.H. and Ginsberg, H.S. (1975) *J. Virol.* 15, 208.
Engler, J.A. (1981) *Gene* 13, 387.
Engler, J.A. and van Bree, M. (1982) *Gene* 19, 71.
Everett, S.F. and Ginsberg, H.S. (1958) *Virology* 6, 770.
Everitt, E. and Philipson, L. (1974) *Virology* 62, 253.
Everitt, E., Sundquist, B. and Philipson, L. (1971) *J. Virol.* 8, 742.
Everitt, E., Sundquist, B., Pettersson, U. and Philipson, L. (1973) *Virology* 52, 130.
Everitt, E., Lutter, L. and Philipson, L. (1975) *Virology* 67, 197.
Franklin, R.M., Pettersson, U., Åkervall, K., Strandberg, B. and Philipson, L. (1971a) *J. Mol. Biol.* 57, 383.
Galibert, F., Herissé, J. and Courtois, G. (1979) *Gene* 6, 1.
Gelderblom, H. and Maichle-Lauppe, I. (1982) *Arch. Virol.* 72, 289.
Gerna, G., Cattaneo, E., Grazia Revello, M. and Battaglia, M. (1982) *J. Infect. Dis.* 145, 678–682.
Gingeras, T.R., Sciaky, D., Gelinas, R.E., Bing-Dong, J., Yen, C.E., Kelly, M.M., Bullock, P.A., Parsons, B.L., O'Neill, K.E. and Roberts, R.J. (1982) *J. Biol. Chem.* 257, 13475.
Ginsberg, H.S. (1979) In *Comprehensive Virology* (H. Fraenkel-Conrat and R.R. Wagner, Eds.) Vol. 13. Plenum Press, New York. pp. 409–457.
Ginsberg, H.S. and Young, C.S.H. (1977) In *Comprehensive Virology* (H. Fraenkel-Conrat and R.R. Wagner, Eds.) Vol. 9. Plenum Press, New York. pp. 27–86.
Ginsberg, H.S., Ensinger, M.J., Kaufman, R.S., Mayer, A.J. and Lundholm, U. (1974) *Cold Spring Harbor Symp. Quant. Biol.* 39, 419.
Green, M. (1970) *Annu. Rev. Biochem.* 39, 701–756.
Green, M., Pina, M. and Kimes, R.C. (1967) *Virology* 31, 562–565.
Green, M., Mackey, J.K., Wold, W.S.M. and Rigden, P. (1979) *Virology* 93, 481–492.
Green, N.M., Wrigley, N.G., Russell, W.C., Martin, S.R. and McLachlan, A.D. (1983) *EMBO J.* 2, 1357.
Grütter, M. and Franklin, R.M. (1974) *J. Mol. Biol.* 89, 163.
Halonen, P., Sarkkinen, H., Arstila, P., Hjertsson, E. and Torfason, E. (1980) *J. Clin. Microbiol.* 11, 614–617.
Haase, A.T., Mautner, V. and Pereira, H.G. (1972) *J. Immunol.* 108, 483.
Hérissé, J., Courtois, G. and Galibert, F. (1980) *Nucl. Acids Res.* 8, 2173.
Hérissé, J., Rigolet, M., Dupont de Dinechin, S. and Galibert, F. (1981) *Nucl. Acids Res.* 9, 4023.
Hierholzer, J.C. (1973) *Inf. Dis.* 128, 541.
Hollinshead, A.C., Bunnag, B. and Alford, T.C. (1967) *Nature* 215, 397.
Horne, R.W., Brenner, S., Waterson, A.P. and Wildy, P. (1959) *J. Mol. Biol.* 1, 84.
Horwitz, M.S., Scharff, M.D. and Maizel, J.V. Jr. (1969) *Virology* 39, 682.
Horwitz, M.S., Maizel, J.V. and Scharff, M.D. (1970) *J. Virol.* 6, 569.
Hosokawa, K. and Sung, M.T. (1976) *J. Virol.* 17, 924.
Huebner, R.J. (1967) In *Perspectives in Virology* (M. Pollard, Ed.) Vol. 5. Academic Press, New York, pp. 147–166.
Inouye, S. and Norrby, E. (1973) *Arch. Gesamte Virusforsch.* 42, 388–398.
Ishibashi, M. and Maizel, J.V. (1974) *Virology* 58, 345.
Jörnvall, H. and Philipson, L. (1980) *Eur. J. Biochem.* 104, 237.
Jörnvall, H., Pettersson, U. and Philipson, L. (1974a) *Eur. J. Biochem.* 48, 179.
Jörnvall, H., Ohlsson, H. and Philipson, L. (1974b) *Biochem. Biophys. Res. Commun.* 56, 304.
Jörnvall, H., Akusjärvi, G., Aleström, P., van Bahr-Lindström, H., Pettersson, U., Appella, E., Fowler, A.V. and Philipson, L. (1981) *J. Biol. Chem.* 256, 6181.
Kemp, M.C., Hierholzer, J.C., Cabbradilla, C.P. and Obijeski, J.F. (1983) *J. Infect. Dis.* 148, 24.

Kinloch, R., Mackay, N. and Mautner, V. (1984) *J. Biol. Chem.* 259, 6431–6436.
Kjellén, L. and Pereira, H.G. (1968) *J. Gen. Virol.* 2, 177.
Köhler, K. (1965) *Z. Naturforsch.* 20b, 747.
Kornadt, C.-U., Wigand, R., Adrian, T.H. and Roggendorf, M. (1982) *Zentralbl. Bakteriol. Parasitenkd. Infektionskr. Hyg. Abt. 1: Orig Reihe A* 253, 285–296.
Laver, W.G., Pereira, H.G., Russell, W.C. and Valentine, R. (1968) *J. Mol. Biol.* 37, 379.
Laver, W.G., Wrigley, N.G. and Pereira, H.G. (1969) *Virology* 39, 599.
Laver, W.G., Younghusband, H.B. and Wrigley, N.G. (1971) *Virology* 45, 598.
Lemay, P., Boudin, M.-L., Milleville, M. and Boulanger, P. (1980) *Virology* 101, 131.
Levine, A.J. and Ginsberg, H.S. (1967) *J. Virol.* 1, 747.
Levinson, A.D., Postel, E.H. and Levine, A.J. (1977) *Virology* 79, 144.
Liang, Y.-K., Akusjärvi, G., Aleström, P., Pettersson, U., Tremblay, M. and Weber, J. (1983) *J. Mol. Biol.* 167, 217.
Maat, J., van Beveren, C.P. and van Ormondt, H. (1980) *Gene* 10, 27.
Maizel, J.V. Jr. (1971) In *Methods in Virology* (K. Maramosch and H. Koprowski, Eds.) Vol. V. Academic Press, New York. pp. 179–246.
Maizel, J.V. Jr., White, D.O. and Scharff, M.D. (1968a) *Virology* 36, 115.
Maizel, J.V. Jr., White, D.O. and Scharff, M.D. (1968b) *Virology* 36, 126.
McAllister, R.M., Nicolson, M.O., Reed, G., Kern, J.I., Gilden, R.V. and Huebner, R.J. (1969) *J. Natl. Cancer Inst.* 43, 917–923.
Nermut, M.V. (1975) *Virology* 65, 480.
Nermut, M.V. (1980) *Arch. Virol.* 64, 175.
Nermut, M.V. and Perkins, W.J. (1979) *Micron* 10, 247.
Neurath, A.R., Rubin, B.A. and Stasny, J.T. (1968) *J. Virol.* 2, 1086.
Norrby, E. (1966a) *Virology* 28, 236.
Norrby, E. (1966b) *Virology* 30, 608.
Norrby, E. (1968) *Curr. Top. Microbiol. Immunol.* 43, 1.
Norrby, E. (1969a) *Virology* 37, 565.
Norrby, E. (1969b) *J. Gen. Virol.* 5, 221.
Norrby, E. and Skaaret, P. (1967) *Virology* 32, 489.
Norrby, E. and Skaaret, P. (1968) *Virology* 36, 201.
Norrby, E. and Wadell, G. (1969) *J. Virol.* 4, 663.
Norrby, E. and Wadell, G. (1972) *Virology* 48, 757.
Norrby, E., Marusyk, H.L. and Hammarskjöld, M.-L. (1969a) *Virology* 38, 477.
Norrby, E., Wadell, G. and Marusyk, H. (1969b) *Arch. Gesamte Virusforsch.* 28, 239.
Norrby, E., Van der Veen, J. and Espmark, Å. (1970) *Proc. Soc. Exp. Biol. Med.* 134, 889–895.
Norrby, E., Marusyk, R. and Wadell, G. (1971) *Can. J. Microbiol.* 17, 1227.
Oosterom-Dragon, E. and Anderson, C.W. (1983) *J. Virol.* 45, 251.
Oosterom-Dragon, E.A. and Ginsberg, H.S. (1981) *J. Virol.* 40, 491.
Pereira, H.G. (1958) *Virology* 6, 601.
Pereira, H.G. and Figueiredo, M.V.T. de (1962) *Virology* 18, 1.
Pereira, H.G. and Kelly, B. (1957) *J. Gen. Microbiol.* 17, 517.
Pereira, H.G. and Wrigley, N.G. (1974) *J. Mol. Biol.* 85, 617.
Pereira, H.G. Allison, A.C. and Farthing, C. (1959) *Nature* 183, 895.
Pereira, H.G., Machado, R.D. and Schild, G.C. (1972) *J. Immun. Methods* 2, 121–128.
Persson, H., Mathisen, B., Philipson, L. and Pettersson, U. (1979) *Virology* 93, 198.
Pettersson, U. (1971) *Virology* 43, 123.
Pettersson, U. and Höglund, S. (1969) *Virology* 39, 90.
Pettersson, U., Philipson, L. and Höglund, S. (1967) *Virology* 33, 575.
Pettersson, U., Philipson, L. and Höglund, S. (1968) *Virology* 35, 204.

Philipson, L., Lonberg-Holm, K., and Pettersson, U. (1968) *J. Virol.* 2, 1064.
Prage, L. and Pettersson, U. (1971) *Virology* 45, 364.
Prage, L., Pettersson, U. and Philipson, L. (1968) *Virology* 36, 508.
Prage, L., Pettersson, U., Höglund, S., Lonberg-Holm, K. and Philipson, L. (1970) *Virology* 42, 341.
Reif, U.K., Winterhoff, U. and Doerfler, W. (1977a) *Eur. J. Biochem.* 73, 327.
Reif, U., Winterhoff, U., Lundholm, U., Philipson, L. and Doerfler, W. (1977b) *Eur. J. Biochem.* 73, 313.
Roberts, R.J., O'Neill, K.E. and Yen, C.E. (1984) *J. Biol. Chem.* 259, In press.
Roggendorf, M., Wigand, R., Deinhardt, F. and Frösner, G.G. (1982) *J. Virol. Meth.* 4, 27.
Rosén, L. (1958) *Virology* 5, 574.
Rosén, L. (1960) *Am. J. Hyg.* 71, 120.
Rosén, L., Hovis, J.F. and Bell, J.A. (1962) *Proc. Soc. Exp. Biol. Med.* 110, 710.
Rowe, W.P., Hartley, J.W., Roizmann, B. and Levy, H.B. (1958) *J. Exp. Med.* 108, 713.
Russell, W.C. and Blair, G.E. (1977) *J. Gen. Virol.* 34, 19.
Russell, W.C. and Precious, B. (1982) *J. Gen. Virol.* 63, 69.
Russell, W.C. and Skehel, J.J. (1972) *J. Gen. Virol.* 15, 45.
Russell, W.C., Valentine, R.C. and Pereira, H.G. (1967) *J. Gen. Virol.* I, 509.
Russell, W.C., McIntosh, K. and Skehel, J.J. (1971) *J. Gen. Virol.* 11, 35.
Russell, W.C., Patel, G., Precious, B., Sharp, I. and Gardner, P.S. (1981) *J. Gen. Virol.* 56, 393.
Schmidt, N.J., King, C.J. and Lennette, E.H. (1965) *Proc. Soc. Exp. Biol. Med.* 118, 208.
Signäs, C., Akusjärvi, G. and Pettersson, U. (1985) *J. Virol.* In press.
Stinski, M.F. and Ginsberg, H.S. (1974) *Intervirology* 4, 226.
Stålhandske, P., Hyypiä, T., Gadler, H., Halonen, P. and Pettersson, U. (1983) *In Current Topics in Microbiology and Immunology* (P.A. Bachmann, Ed.) Vol. 104. Springer-Verlag, Berlin. pp. 299–306.
Sundquist, B., Pettersson, U. and Philipson, L. (1973) *Virology* 51, 252.
Sung, M.T., Lischwe, M.A., Richards, J.C. and Hosokawa, K. (1977) *J. Biol. Chem.* 252, 4981.
Svensson, L. and von Bonsdorff, C.H. (1982) *J. Med. Virol.* 10, 243.
Svensson, L., Wadell, G., Uhnoo, I., Johansson, M. and von Bonsdorff, C.H. (1983) *J. Gen. Virol.* 64, 2517–2520.
Tikchonenko, T.I. (1984) *Curr. Top. Microbiol. Immunol.* 110, 169–189.
Valentine, R.C. and Pereira, H.G. (1965) *J. Mol. Biol.* 13, 13.
Wadell, G. (1969) *Proc. Soc. Exp. Biol. Med.* 132, 413.
Wadell, G. (1970) Thesis, Karolinska Institute, Stockholm, Sweden.
Wadell, G. (1972) *J. Immunol.* 108, 622.
Wadell, G. (1979) *Intervirology* 11, 47–57.
Wadell, G. (1984) *Curr. Top. Microbiol. Immunol.* 110, 191–220.
Wadell, G. and DeJong, J.C. (1980) *Infect. Immun.* 27, 292–296.
Wadell, G. and Norrby, E. (1969) *J. Virol.* 4, 671.
Wadell, G., Norrby, E. and Schönning, U. (1967) *Arch. Gesamte Virusforsch.* 21, 234–242.
Wadell, G., Norrby, E. and Skaaret, P. (1969) *Arch. Gesamte Virusforsch.* 26, 33.
Wadell, G., Hammarskjöld, M.-L., Winberg, G., Varsanyi, T.M. and Sundell, G. (1980) *Ann. N.Y. Acad. Sci.* 354, 16–42.
Wadell, G., Sundell, G. and deJong, J.C., (1981a) *J. Med. Virol* 7, 119–125.
Wadell, G., DeJong, J.C. and Wolontis, S. (1981b) *Infect. Immun.* 34, 368–372.
Velicer, L.F. and Ginsberg, H.S. (1970) *J. Virol.* 5, 338.
White, D.O., Scharff, M.D. and Maizel, J.V. Jr. (1969) *Virology* 38, 395.
Wigand, R. and Fliedner, D. (1968) *Arch. Gesamte Virusforsch.* 24, 235–256.
Wigand, R., Bartha, A., Dreizin, R.S., Esche, H., Ginsberg, H.S., Green, M., Hierholzer, J.C., Kalter, S.S., McFerran, J.B., Pettersson, U., Russell, W.C. and Wadell, G. (1982) *Intervirology* 18, 169.

Wigand, R., Gelderblom, H., Özel, M., Distler, H. and Adrian, T.H. (1983) *Arch. Virol.* 76, 307–319.
Wigand, R., Baumeister, H.G., Maass, G., Kühn, J. and Hammer, H.J. (1983) *J. Med. Virology* 11, 233–240.
Wilcox, W.C. and Ginsberg, H.S. (1963a) *J. Exp. Med.* 118, 295.
Wilcox, N. and Mautner, V. (1976) *J. Immunol.* 116, 19.
Wilcox, W.C. and Ginsberg, H.S. (1963b) *Proc. Soc. Exp. Biol. Med.* 114, 37.
Wilcox, W.C., Ginsberg, H.S. and Andersson, T.F. (1963) *J. Exp. Med.* 118, 307.
Winters, W.D., Brownstone, A. and Pereira, H.G. (1970) *J. Gen. Virol.* 9, 105.
Von Bahr-Lindström, H., Jörnvall, H., Althin, S. and Philipson, L. (1982) *Virology* 118, 353.

Eds. M.H.V. van Regenmortel and A.R. Neurath
Immunochemistry of viruses. The basis for serodiagnosis and vaccines.

Antigenic structure of human hepatitis viruses

19

A.R. NEURATH[1] and S.B.H. KENT[2]

[1]The Lindsley F. Kimball Research Institute of the New York Blood Center, New York, NY 10021 and [2]Division of Biology, California Institute of Technology, Pasadena, CA 91125, U.S.A.

Many distinct viruses may cause human liver disease (Howard and Simpson, 1983). Of these, only hepatitis A and B viruses (HAV and HBV), the δ-agent and at least two additional viruses, classified as 'non-A, non-B,' recognize the liver as the primary target organ. Until now, only antigens associated with HAV, HBV and the δ-agent have been characterized. For this reason, the subject of this chapter will be limited to these three infectious agents.

HAV is a member of the family of *Picornaviridae* (Gust et al., 1983), HBV is a member of a new virus group with the proposed name of *Hepadnaviridae* (Robinson, 1980; Melnick, 1982) and the δ-agent has unique properties distinct from viruses so far classified (Rizzetto, 1983).

The properties of HAV, HBV and the δ-agent have been described in detail in a series of recent reviews (Deinhardt and Gust, 1982; Popper and Schaffner, 1982; Zuckerman, 1982, 1983; Deinhardt and Deinhardt, 1983; Vyas and Blum, 1983; Chisari, 1984; Rizzetto et al., 1984; Shih, 1985). This chapter will selectively focus on the immunochemistry of viral structural components from the vantage point of the recent information concerning the genetic organization and sequence (if available) of the respective viral nucleic acids.

1. Hepatitis B virus (HBV)

1.1. Genetic organization of the HBV genome

The common properties of *Hepadnaviridae* are:

(*a*) Virion ultrastructure: the viruses are 40–50 nm in diameter, have a lipoprotein envelope and an ~ 27 nm nucleocapsid core; incomplete forms (~ 22 nm spherical particles and tubules), representing envelope lipoprotein, usually predominate over complete virions in infected plasma.

(*b*) Circular double-stranded DNA about 3.2 kilobases long, having a single-stranded region and a nick in the longer DNA strand.

(*c*) DNA polymerase within the nucleocapsid that can repair the single-stranded region in the DNA.

(*d*) A protein covalently linked to the 5′ end of the longer (L) DNA strand.

(*e*) A protein kinase within the nucleocapsid core.

(*f*) DNA replication through an RNA intermediate utilizing reverse transcription;

(*g*) Similar assembly of viral proteins (antigens).

(*h*) Partial DNA and protein sequence homology and serological cross-reactivity. (Summers et al., 1978; Werner et al., 1979; Gerlich and Robinson, 1980; Gerlich et al., 1980; Marion et al., 1980; 1983; Mason et al., 1980; 1982; Feitelson et al., 1981; 1982a; Galibert et al., 1981; 1982; Ganem et al., 1982; Melnick, 1982; Ogston et al., 1982; Summers and Mason, 1982; Robinson, 1983; Stannard et al., 1983; Mandart et al., 1984).

(*i*) Conservation of secondary structure of viral proteins even in the absence of primary amino acid sequence homology (Schaeffer and Sninsky, 1984; Sninsky et al., 1984).

The proteins of HBV have been studied in greater detail than those of the other known members of the Hepadnaviridae family, i.e. woodchuck, ground squirrel and duck hepatitis viruses (WHV, GSHV and DHV) and the newly discovered viruses of tree squirrels and kangaroos (Cossart and Keirnan, 1984; Feitelson et al., 1984). For this reason, and because of the worldwide medical significance of hepatitis B, HBV was selected as the subject for this review.

The precise number of base pairs in HBV DNA varies for different antigenic subtypes of HBV and is between 3182 and 3221. The DNA is circular and only partially double stranded. One strand (denoted L) has a nick at a fixed position; the other strand (S) has a variable length gap with a favored minimum length of 650 to 700 nucleotides (Delius et al., 1983). HBV DNA corresponding to several antigenic subtypes has been cloned and sequenced (Galibert et al., 1979; Pasek et al., 1979; Valenzuela et al., 1980; Fujiyama et al., 1983; Ono et al., 1983). A computer search for termination and AUG initiation codons on putative RNA transcripts from both L and S strands of HBV DNAs, read in the 3 possible reading frames, lead to the identification of 4 potential protein products (PP) translated from RNA transcripts of the DNA L strand with the number of amino acid residues (distinct for different serologic subtypes) given in parentheses: PP1 (183 or 185), PP2 (389, 399 or 400), PP3 (154) and PP4 (832, 843 or 845; with one of the subtypes, this coding region contained a stop codon; Ono et al., 1983). Proteins corresponding to PP1, PP2 and PP3 have been experimentally identified in HBV and its subviral components (see

Section 1.2). On the other hand, RNA transcripts of the DNA S strand could code only for one protein consisting of ≧ 94 amino acids (PP5; Valenzuela et al., 1980; Fujiyama et al., 1983). This protein has not yet been identified experimentally.

The structure and proposed genetic organization of HBV DNA is schematically presented in Fig. 19.1.

HBV DNA may become integrated into host DNA. Deletions, rearrangements, insertions and duplications of HBV DNA sequences have been reported (Dejean et al., 1983; Koike et al., 1983; 1984; Figus et al., 1984; Freytag von Loringhoven et al., 1984; Fung et al., 1984; Koch et al., 1984; Koshy et al., 1984; Rogler et al., 1984; Rutter, 1984). These alterations could possibly result in new kinds of gene products not obvious from the genetic organization of HBV DNA (Fig. 19.1).

1.2. Structural proteins of HBV

The following HBV-specific antigens have been identified in plasma of HBV-infected humans or chimpanzees, in infected liver cells or in human hepatoma cell lines containing HBV DNA:

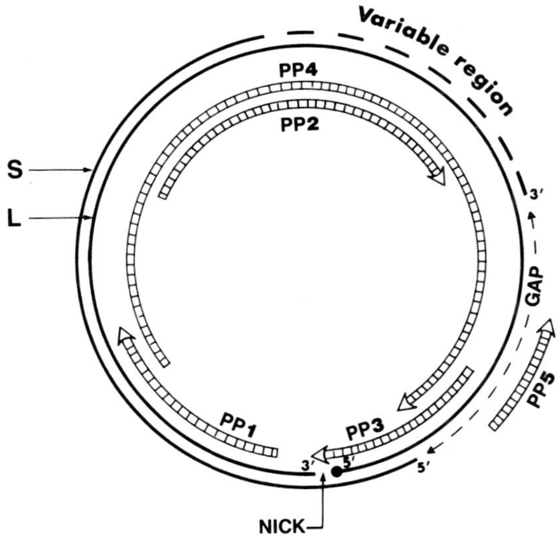

Fig. 19.1 Schematic diagram of the structure and genetic organization of HBV DNA. Hatched arrows indicate the position and length of the potential protein products (PP) coded for by HBV DNA. The full circle at the 5′ end of the L strand represents the protein covalently linked to DNA. PP1, PP2, PP3 and PP4 are coded for by the RNA transcript of the HBV DNA L strand; PP5 is coded for by the RNA transcript of the S strand in which the gap has to be filled in for potential transcription to occur. Transcription of mRNAs from different strands of DNA from overlapping regions of the genome has been observed for other small DNA viruses (from Griffin, 1980).

(*a*) Hepatitis B surface antigen (HBsAg), which represents the lipoprotein envelope component(s) of HBV and also occurs in the form of subviral ~ 22-nm diameter particles and tubular forms ~ 20-nm wide and up to ~ 200-nm long.

(*b*) Hepatitis B core antigen (HBcAg), which represents the major antigen of the HBV nucleocapsid.

(*c*) Hepatitis B e antigen (HBeAg), which is present in plasma in various physical forms and also arises from in vitro dissociation of the HBV nucleocapsid (Neurath and Strick, 1979; Takahashi et al., 1979).

(*d*) A nuclear antigen detected in hepatoma cells which contains integrated HBV DNA (Wen et al., 1983).

Membrane antigens specific for human hepatocellular carcinoma have been identified (Shouval, 1983). However, it has not yet been established whether these antigens are coded for by HBV DNA.

Antigen-antibody systems corresponding to HBV protein kinase, DNA polymerase and the protein covalently attached to HBV DNA, respectively, have not yet been identified.

The different experimentally detected HBV proteins may be assigned to distinct PPs predicted from DNA sequence data on the basis of: (*a*) expression of defined HBV DNA fragments (inserted into appropriate expression vectors) in prokaryotic or eukaryotic host cells transfected with the recombinant DNAs, (*b*) (partial) amino acid sequencing of HBV proteins and alignment of the experimentally determined sequences with those predicted for PPs, and (*c*) synthesis of polypeptides corresponding to segments of distinct PPs, preparation of antipeptide antisera and their utilization for detection and purification of proteins in virus-infected tissues or in virus particles (Sutcliffe et al., 1980; Walter et al., 1980; Lerner, 1982).

PP1 + PP3. Using the first approach, the major protein of the HBV nucleocapsid (HBcAg) was assigned to PP1 (Burrell et al., 1979; Pasek et al., 1979; Edman et al., 1981; Hardy et al., 1981; Stahl et al., 1982; Will et al., 1984). Its molecular mass is ~ 21 kDa and its amino acid sequence for several antigenic subtypes of HBV is shown in Fig. 19.2. The assignment of HBcAg (and HBeAg) to PP1 was confirmed by data on amino acid composition and partial sequencing (Yamaki et al., 1982; Takahashi et al., 1983). However, polyacrylamide gel electrophoresis of HBV nucleocapsids revealed several distinct protein components (data summarized by Feitelson et al., 1982b). The observed 15–19-kDa components are proteolytic cleavage products of the major 21–22-kDa gene product. The larger minor components (with molecular masses between ~ 25 and ~ 43 kDa) contain additional amino acid sequences not present in the major protein, as suggested by tryptic peptide mapping. In DHV DNA the genes for proteins analogous to PP1 and PP3 are fused and the major protein of the DHV nucleoprotein has a molecular mass of ~ 35 kDa (Mandart et al., 1984). For this reason it is likely that a protein corresponding to PP3 (designated as protein X) is a component of the HBV nucleocapsid and that the larger protein components observed are translation products of spliced RNA tran-

```
                            1                   2                   3                   4                   5                   6
         1 2 3 4 5 6 7 8 9 0 1 2 3 4 5 6 7 8 9 0 1 2 3 4 5 6 7 8 9 0 1 2 3 4 5 6 7 8 9 0 1 2 3 4 5 6 7 8 9 0 1 2 3 4 5 6 7 8 9 0 1 2

1 ayw    M D I D P Y K E F G A T V E L L S F L P S D F F P S V R D L L D T A S A L Y R E A L E S P E H C S P H H T A L R Q A I L C W
2 adr    M D I D P Y K E F G A S V E L L S F L P S D F F P S I R D L L D T A S A L Y R E A L E S P E H C S P H H T A L R Q A I L C W
3 adyw   M D I D P Y K E F G A T V E L L S F L P S D F F P S V R D L L D T A A A L Y R D A L E S P E H C S P H H T A L R Q A I L C W
4 adw2   M D I D P Y K E F G A T V E L L S F L P S D F F P S V R D L L D T A S A L Y R E A L E S P E H C S P H H T A L R Q A I L C W
5 adw    M D I D P Y K E F G A T V E L L S F L P S D F F P S V R D L L D T A S A L Y R E A L E S P E H C S P H H T A L R Q A I L C W

         M D I D P Y K E F G A   V E L L S F L P S D F F P S   R D L L D T A   A L Y R   A L E S P E H C S P H H T A L R Q A I L C W


         6             7                   8                   9                   10                  11                  12
         3 4 5 6 7 8 9 0 1 2 3 4 5 6 7 8 9 0 1 2 3 4 5 6 7 8 9 0 1 2 3 4 5 6 7 8 9 0 1 2 3 4 5 6 7 8 9 0 1 2 3 4 5 6 7 8 9 0 1 2 3 4

1 ayw    G E L M T L A T W V G V N L E D P A S R D L V V S Y V N T N M G L K F R Q L L W F H I S C L T F G R E T V I E Y L V S F G V
2 adr    G E L M N L A T W V G S N L E D P A S R E L V V S Y V N V N M G L K I R Q L L W F H I S C L T F G R E T V L E Y L V S F G V
3 adyw   G D L M T L A T W V G T N L E D P A S R D L V V S Y V N T N V G L K F R Q L L W F H I S C L T F G R E T V L E Y L V S F G V
4 adw2   G E L M T L A T W V G N N L E D P A S R D L V V N Y V N T N V G L K I R Q L L W F H I S C L T F G R E T V L E Y L V S F G V
5 adw    G E L M T L A T W V G N N L Q D P A S R D L V V N Y V N T N M G L K I R Q L L W F H I S C L T F G R E T V L E Y L V S F G V

         G   L M   L A T W V G   N L   D P A S R   L V V   Y V N   N   G L K   R Q L L W F H I S C L T F G R E T V   E Y L V S F G V


         12        13                  14                  15                  16                  17                  18
         5 6 7 8 9 0 1 2 3 4 5 6 7 8 9 0 1 2 3 4 5 6 7 8 9 0 1 2 3 4 5 6 7 8 9 0 1 2 3 4 5 6 7 8 9 0 1 2 3 4 5 6 7 8 9 0 1 2 3 4 5

1 ayw    W I R T P P A Y R P P N A P I L S T L P E T T V V R R R - - G R S P R R R T P S P R R R R S Q S P R R R R S Q S R E S Q C
2 adr    W I R T P P A Y R P P N A P I L S T L P E T T V V R R R - - G R S P R R R T P S P R R R R S Q S P R R R R S Q S R E S Q C
3 adyw   W I R T P P A Y R P P N A P I L S T L P E T T V V R R R - - G R S P R R R T P S P R R R R S Q S P R R R R S Q S R E S Q C
4 adw2   W I R T P P A Y R P P N A P I L S T L P E T T V V R R R D R G R S P R R R T P S P R R R R S P S P R R R R S Q S R E S Q C
5 adw    W I R T P P A Y R P P N A P I L S T L P E T T V V R R R D R G R S P R R R T P S P R R R R S Q S P R R R R S Q S R E S Q C

         W I R T P P A Y R P P N A P I L S T L P E T T V V R R R     G R S P R R R T P S P R R R R S   S P R R R R S Q S R E S Q C
```

Fig.19.2. Amino acid sequences of the major protein of the HBV nucleocapsid (HBcAg) as deduced from sequences of HBV DNA. Sequences for 5 HBV subtypes are presented. The bottom line shows amino acid residues common to all five subtypes.

scripts corresponding to the gene for PP1 and to portions of or to the total of the gene for PP3. The molecular mass of PP3 is ~ 16.5 kDa; its amino acid sequence is shown in Fig. 19.3. Using antisera to a synthetic peptide analogue of PP3, it was possible to detect a protein in cells containing integrated HBV DNA (Moriarty et al., 1984). Thus, a protein corresponding to PP3 is undoubtedly a bona fide translation product of HBV DNA.

HBeAg, discovered by Magnius (1975), is present in the serum of some HBV carriers in free form or associated with IgG (Neurath and Strick, 1977; Takahashi et al., 1978a; Neurath et al., 1979) or with lipoproteins (Tedder and Bull, 1979). The assignment of HBeAg to PP1 has been demonstrated by MacKay et al. (1981) by conversion of HBcAg (produced in transfected *Escherichia coli*) into HBeAg by proteolytic cleavage in the presence of detergents and reducing agents. HBcAg preparations produced in *E. coli* contain particles similar in appearance to empty HBV nucleocapsids isolated from plasmas or livers of HBV-infected humans and chimpanzees (Cohen and Richmond, 1982). The loss of HBcAg antigenicity and the appearance of HBeAg antigenicity coincide with the morphological disintegration of HBV nucleocapsids. The protein moieties of HBcAg and HBeAg appear identical (Budkowska et al., 1979; Ohori et al., 1980; 1984). Thus, proteolytic cleavage, which leads to conversion of HBcAg into HBeAg (Takahashi et al., 1979), is not obligatory for this change in immunologic specificities. In summary, HBcAg and HBeAg apparently represent neotopes and cryptotopes, respectively, associated with HBV nucleocapsids.

PP4. Proteins corresponding to PP4 (molecular mass ~ 94 kDa; amino acid sequence shown in Fig. 19.4) have not been identified so far in cells transfected with recombinant DNA containing the whole genome of HBV, with recircularized, cloned HBV DNA or in HBV-infected cells. PP4 shares regions of homology with retrovirus reverse transcriptase (including the enzyme of human adult T cell leukemia virus; Toh et al., 1983). For this reason, it may represent the DNA polymerase and reverse transcriptase of HBV. Therefore, it has been designated as 'P' protein. Interestingly, PP4 has several potential N-glycosylation sites (Asn-X-Ser (Thr); Fig. 19.4), unlike the PP1 and PP3 proteins of the HBV nucleocapsids which lack any such sites.

PP2. Transfection of cells with recombinant DNAs containing an HBV DNA fragment corresponding to the gene for PP2 resulted in synthesis of HBsAg (Pourcel et al., 1982; Stratowa et al., 1982; Lamb et al., 1983; Siddiqui, 1983; Stenlund et al., 1983; Wang et al., 1983) in the form of ~ 22-nm particles morphologically similar to subviral particles observed in serum of HBV-infected humans and chimpanzees. The expected molecular mass of PP2 is 43.5 – 44 kDa (not considering saccharide chains expected on at least some of the several potential N-glycosylation sites; Figs. 19.5 and 19.6). However, polyacrylamide gel electrophoresis of the synthesized HBsAg consistently revealed two major components of about 25 and 29 kDa, and the detection of an ~ 44 kDa (minor) component was reported only by Pourcel et al.

```
                          1                   2                   3                   4                   5                   6
        1 2 3 4 5 6 7 8 9 0 1 2 3 4 5 6 7 8 9 0 1 2 3 4 5 6 7 8 9 0 1 2 3 4 5 6 7 8 9 0 1 2 3 4 5 6 7 8 9 0 1 2 3 4 5 6 7 8 9 0 1 2

1 ayw   M A A R L C C Q L D P A R D V L C L R P V G A E S C G R P F S G S L G T L S S P S P S A V P T D H G A H L S L R G L P V C A
2 adr   M A A R V C C Q L D P A R D V L C L R P V G A E S R G R P V S G P F G A L P S P S S S A V P A D H G A H L S L R G L P V C A
3 adw2  M A A R L Y C Q L D P S R D V L C L R P V G A E S R G R P L S G P L G T L S S P S P S A V P A D H G A H L S L R G L P V C A
4 adw   M A A R L Y C Q L D P S R D V L C L R P V G A E S R G R P L S G P L G T L S S P S P S A V P A D H G A H L S L R G L P V C A

        M A A R     C Q L D P   R D V L C L R P V G A E S   G R P   S G     G   L   S P S   S A V P   D H G A H L S L R G L P V C A
        ' ' ' '     ' ' ' ' '   ' ' ' ' ' ' ' ' ' ' ' ' '   ' ' '   ' '     '   '   ' ' '   ' ' ' '   ' ' ' ' ' ' ' ' ' ' ' ' ' ' '

        6             7                   8                   9                   10                  11                  12
        3 4 5 6 7 8 9 0 1 2 3 4 5 6 7 8 9 0 1 2 3 4 5 6 7 8 9 0 1 2 3 4 5 6 7 8 9 0 1 2 3 4 5 6 7 8 9 0 1 2 3 4 5 6 7 8 9 0 1 2 3 4

1 ayw   F S S A G P C A L R F T S A R R M E T T V N A H Q I L P K V L H K R T L G L S A M S T T D L E A Y F K D C L F K D W E E L G
2 adr   F S S A G P C A L R F T S A R R M E T T V N A H Q V L P K V L H K R T L G L S A M S T T D L E A Y F K D C V F K D W E E L G
3 adw2  F S S A G P C A L R F T S A R C M E T T V N A H Q I L P K V L H K R T L G L P A M S T T D L E A Y F K D C V F K D W E E L G
4 adw   F S S A G P C A L R F T S A R C M A T T V N A H Q I L P K V L H K R T L G L P A M S T T D L E A Y F K D C V F K D W E E L G

        F S S A G P C A L R F T S A R   M   T T V N A H Q   L P K V L H K R T L G L   A M S T T D L E A Y F K D C   F K D W E E L G
        ' ' ' ' ' ' ' ' ' ' ' ' ' ' '   '   ' ' ' ' ' ' '   ' ' ' ' ' ' ' ' ' ' ' '   ' ' ' ' ' ' ' ' ' ' ' ' ' '   ' ' ' ' ' ' ' '

        12        13                  14                  15
        5 6 7 8 9 0 1 2 3 4 5 6 7 8 9 0 1 2 3 4 5 6 7 8 9 0 1 2 3 4

1 ayw   E E I R L K V F V L G G C R H K L V C A P A P C N F F T S A
2 adr   E E I R L K V F V L G G C R H - - - - - - - - - N F F T S A
3 adw2  E E I R L K V F V L G G C R H K L V C A P A P C N F F T S A
4 adw   E E I R L M I F V L G G C R H K L V C A P A P C N F F T S A

        E E I R L     F V L G G C R H                   N F F T S A
        ' ' ' ' '     ' ' ' ' ' ' ' '                   ' ' ' ' ' '
```

Fig. 19.3. Amino acid sequences of X protein. For further explanation see legend to Fig. 19.2.

```
                          1                   2                   3                   4                   5                   6
        1 2 3 4 5 6 7 8 9 0 1 2 3 4 5 6 7 8 9 0 1 2 3 4 5 6 7 8 9 0 1 2 3 4 5 6 7 8 9 0 1 2 3 4 5 6 7 8 9 0 1 2 3 4 5 6 7 8 9 0 1 2

1 ayw   M P L S Y Q H F R R L L L L D D - - E A G P L E E E L P R L A D E G L N R R V A E D L N L G N L N V S I P W T H K V G N F T
2 adw2  M P L S Y Q H F R K L L L L D D G T E A G P L E E E L P R L A D A D L H R R V A E D L N L G N L N V S I P W T H K V G N F T
3 adw   M P L S Y Q H F R K L L L L D D G T E A G P L E E E L P R L A D A D L N R R V A E D L N L G N L N V S I P W T H K V G N F T
4 adr   M P L S Y Q H F R K L L L L D D - - E A G P L E E E L P R L A D E G L N R R V A E D L N L G N L N V S I P W T H K V G N F T

        M P L S Y Q H F R   L L L L D D     E A G P L E E E L P R L A D     L   R R V A E D L N L G N L N V S I P W T H K V G N F T
        ' ' ' ' ' ' ' ' '   ' ' ' ' ' '     ' ' ' ' ' ' ' ' ' ' ' ' ' '     '   ' ' ' ' ' ' ' ' ' ' ' ' ' ' ' ' ' ' ' ' ' ' ' ' ' '


        6             7                   8                   9                  10                  11                  12
        3 4 5 6 7 8 9 0 1 2 3 4 5 6 7 8 9 0 1 2 3 4 5 6 7 8 9 0 1 2 3 4 5 6 7 8 9 0 1 2 3 4 5 6 7 8 9 0 1 2 3 4 5 6 7 8 9 0 1 2 3 4

1 ayw   G L Y S S T V P V F N P H W K T P S F P N I H L H Q D I I K K C E Q F V G P L T V N E K R R L Q L I M P A R F Y P K V T K Y
2 adw2  G L Y S S T V P I F N P E W Q T P S F P K I H L Q E D I I N R C Q Q F V G P L T V N E K R R L K L I M P A R F Y P T H T K Y
3 adw   G L Y S S T V P I F N P E W Q T P S F P K I H L Q E D I I N R C Q Q F V G P L T V N E K R R L K L I M P A R F Y P T H T K Y
4 adr   G L Y S S T V P V F N P E W Q T P S F P H I H L Q E D I I N R C Q Q Y V G P L T V N E K R R L K L I M P A R F Y P N L T K Y

        G L Y S S T V P   F N P   W   T P S F P   I H L     D I I     C   Q   V G P L T V N E K R R L   L I M P A R F Y P     T K Y
        ' ' ' ' ' ' ' '   ' ' '   '   ' ' ' ' '   ' ' '     ' ' '     '   '   ' ' ' ' ' ' ' ' ' ' ' '   ' ' ' ' ' ' ' ' '     ' ' '


       12        13                  14                  15                  16                  17                  18
        5 6 7 8 9 0 1 2 3 4 5 6 7 8 9 0 1 2 3 4 5 6 7 8 9 0 1 2 3 4 5 6 7 8 9 0 1 2 3 4 5 6 7 8 9 0 1 2 3 4 5 6 7 8 9 0 1 2 3 4 5 6

1 ayw   L P L D K G I K P Y Y P E H L V N H Y F Q T R H Y L H T L W K A G I L Y K R E T T H S A S F C G S P Y S W E Q D L Q H G - -
2 adw2  L P L D K G I K P Y Y P D Q V V N H Y F Q T R H Y L H T L W K A G I L Y K R E T T R S A S F C G S P Y S W E Q E L Q H G R L
3 adw   L P L D K G I K P Y Y P D Q V V N H Y F Q T R H Y L H T L W K A G I L Y K R E T T R S A S F C G S P Y S W E Q E L Q H - - -
4 adr   L P L D K G I K P Y Y P E H A V N H Y F K T R H Y L H T L W K A G I L Y K R E T T R S A S F C G S P Y S W E Q E L Q H G R L

        L P L D K G I K P Y Y P       V N H Y F   T R H Y L H T L W K A G I L Y K R E T T   S A S F C G S P Y S W E Q   L Q H
        ' ' ' ' ' ' ' ' ' ' ' '       ' ' ' ' '   ' ' ' ' ' ' ' ' ' ' ' ' ' ' ' ' ' ' ' '   ' ' ' ' ' ' ' ' ' ' ' ' '   ' ' '


       18    19                  20                  21                  22                  23                  24
        7 8 9 0 1 2 3 4 5 6 7 8 9 0 1 2 3 4 5 6 7 8 9 0 1 2 3 4 5 6 7 8 9 0 1 2 3 4 5 6 7 8 9 0 1 2 3 4 5 6 7 8 9 0 1 2 3 4 5 6 7 8

1 ayw   - - - - - - - - - A E S F H Q Q S S G I L S R P P V G S S L Q S K H R K S R L G L Q S Q Q G H L A R R Q Q G R S W S I R A G
2 adw2  V I K T S Q R H G D E S F C S Q S S G I L S R S S V G P C I R S Q L K Q S R L G L Q P R Q G R L A S S Q P S R S G S I R A K
3 adw   - - - - S Q R H G D E S F C S Q P S G I P S R S S V G P C I R S Q L N K S R L G L Q P H Q G P L A S S Q P G R S G S I R A R
4 adr   V F Q T S T R H G D E S F C S Q S S G I L S R S P V G P C V R S Q L K Q S R L G L Q P Q Q G S M A R G K S G R S G S I R A R

                            E S F     Q   S G I   S R     V G         S         S R L G L Q     Q G     A           R S   S I R A
                            ' ' '     '   ' ' '   ' '     ' '         '         ' ' ' ' ' '     ' '     '           ' '   ' ' ' '


       25                    26                  27                  28                  29                  30                  31
        9 0 1 2 3 4 5 6 7 8 9 0 1 2 3 4 5 6 7 8 9 0 1 2 3 4 5 6 7 8 9 0 1 2 3 4 5 6 7 8 9 0 1 2 3 4 5 6 7 8 9 0 1 2 3 4 5 6 7 8 9 0

1 ayw   F H P T A R R P F G V E P S G S G H T T N F A S K S A S C L H Q S P V R K A A Y P A V S T F E K H S S S G H A V E F H N L P
2 adw2  A H P S T R R Y F G V E P S G S G H I D H S V N N S S S C L H Q S A V R K A A Y S H L S T S K R Q S S S G H A V E F H C L P
3 adw   A H P S T R R Y F G V E P S G S G H I D H S V N N S S S C L H Q S A V R K A A Y S H L S T S K R Q S S S G H A V E F H C L A
4 adr   V H P T T R R S F G V E P S G S G H I D N S A S S T S S C L H Q S A V R K T A Y S H L S T S K R Q S S S G H A V E F H N I P

          H P     R R   F G V E P S G S G H                   S C L H Q S   V R K   A Y       S T         S S S G H A V E F H
          ' '     ' '   ' ' ' ' ' ' ' ' ' '                   ' ' ' ' ' '   ' ' '   ' '       ' '         ' ' ' ' ' ' ' ' ' '


       31                32                  33                  34                  35                  36                  37
        1 2 3 4 5 6 7 8 9 0 1 2 3 4 5 6 7 8 9 0 1 2 3 4 5 6 7 8 9 0 1 2 3 4 5 6 7 8 9 0 1 2 3 4 5 6 7 8 9 0 1 2 3 4 5 6 7 8 9 0 1 2

1 ayw   P N S A R S Q S E R P V F P C W W L Q F R N S K P C S D Y C L S L I V N L L E D W G P C A E H G E H H I R I P R T P S R V T
2 adw2  P N S A G S Q S Q G S V S S C W W L Q F R N S K P C S E Y C L S H L V N L R E D W G P C D E H G E H H I R I P R T P A R V T
3 adw   P S S A G S Q S Q G S V S S C W W L Q F R N S K P C S E Y C L S H L V N L R E D W G P C D D H G E H H I R I P R T P A R V T
4 adr   P S S A R S Q S E G P I F S C W W L Q F R N S K P C S D Y C L T H I V N L L E D W G P C T E H G E H N I R I P R T P A R V T

        P   S A   S Q S             C W W L Q F R N S K P C S   Y C L       V N L   E D W G P C     H G E H   I R I P R T P   R V T
        '   ' '   ' ' '             ' ' ' ' ' ' ' ' ' ' ' ' '   ' ' '       ' ' '   ' ' ' ' ' '     ' ' ' '   ' ' ' ' ' ' '   ' ' '


       37            38                  39                  40                  41                  42                  43
        3 4 5 6 7 8 9 0 1 2 3 4 5 6 7 8 9 0 1 2 3 4 5 6 7 8 9 0 1 2 3 4 5 6 7 8 9 0 1 2 3 4 5 6 7 8 9 0 1 2 3 4 5 6 7 8 9 0 1 2 3 4

1 ayw   G G V F L V D K N P H N T A E S R L V V D F S Q F S R G N Y R V S W P K F A V P N L Q S L T N L L S S N L S W L S L D V S A
2 adw2  G G V F L V D K N P H N T A E S R L V V D F S Q F S R G I S R V S W P K F A V P N L Q S L T N L L S S N L S W L S L D V S A
3 adw   G G V F L V D K N P H N T A E S R L V V D F S Q F S R G I T R V S W P K F A V P N L Q S L T N L L S S N L S W L S L D V S A
4 adr   G G V F L V D K N P H N T T E S R L V V D F S Q F S R G S T H V S W P K F A V P N L Q S L T N L L S S N L S W L S L D V S A

        G G V F L V D K N P H N T   E S R L V V D F S Q F S R G       V S W P K F A V P N L Q S L T N L L S S N L S W L S L D V S A
        ' ' ' ' ' ' ' ' ' ' ' ' '   ' ' ' ' ' ' ' ' ' ' ' ' ' '       ' ' ' ' ' ' ' ' ' ' ' ' ' ' ' ' ' ' ' ' ' ' ' ' ' ' ' ' ' ' '


       43        44                  45                  46                  47                  48                  49
        5 6 7 8 9 0 1 2 3 4 5 6 7 8 9 0 1 2 3 4 5 6 7 8 9 0 1 2 3 4 5 6 7 8 9 0 1 2 3 4 5 6 7 8 9 0 1 2 3 4 5 6 7 8 9 0 1 2 3 4 5 6

1 ayw   A F Y H L P L H P A A M P H L L V G S S G L S R Y V A R L S S N S R I L N N Q H G T M P D L H D Y C S R N L Y V S L L L L Y
2 adw2  A F Y H I P L H P A A M P H L L I G S S G L S R Y V A R L S S N S R I N N N Q Y G T M Q N L H D S C S R Q L Y V S L M L L Y
3 adw   A F Y H I P L H P A A M P H L L I G S S G L S R Y V A R L S S N S R I N N N Q Y G T M Q N L H D S C S R Q L X V S L M L L Y
4 adr   A F Y H I P L H P A A M P H L L V G S S G L P R Y V A R L S S T S R N I N H Q H G A M Q D L H D S C S R N L Y V S L L L L Y

        A F Y H   P L H P A A M P H L L   G S S G L   R Y V A R L S S   S R     N   Q   G   M     L H D   C S R   L Y V S L   L L Y
        ' ' ' '   ' ' ' ' ' ' ' ' ' ' '   ' ' ' ' '   ' ' ' ' ' ' ' '   ' '     '   '   '   '     ' ' '   ' ' '   ' ' ' ' '   ' ' '
```

```
        49    50                  51                  52                  53                  54                  55
        7 8 9 0 1 2 3 4 5 6 7 8 9 0 1 2 3 4 5 6 7 8 9 0 1 2 3 4 5 6 7 8 9 0 1 2 3 4 5 6 7 8 9 0 1 2 3 4 5 6 7 8 9 0 1 2 3 4 5 6 7 8

1 ayw   Q T F G R K L H L Y S H P I I L G F R K I P M G V G L S P F L L A Q F T S A I C S V V R R A F P H C L A F S Y M D D V V L G
2 adw2  K T Y G W K L H L Y S H P I V L G F R K I P M G V G L S P F L L A Q F T S A I C S V V R R A F P H C L A F S Y M D D V V L G
3 adw   K T Y G W K L H L Y S H P I V L G F R K I P M G V G L S P F L L A Q F T S A I C S V V R R A F P H C L A F S Y M D D V V L G
4 adr   K T F G R K L H L Y S H P I I L G F R K I P M G V G L S P F L L A Q F T S A I C S V V R R A F P H C L A F S Y M D D V V L G

          T   G   K L H L Y S H P I   L G F R K I P M G V G L S P F L L A Q F T S A I C S V V R R A F P H C L A F S Y M D D V V L G


          56                  57                  58                  59                  60                  61                  62
        9 0 1 2 3 4 5 6 7 8 9 0 1 2 3 4 5 6 7 8 9 0 1 2 3 4 5 6 7 8 9 0 1 2 3 4 5 6 7 8 9 0 1 2 3 4 5 6 7 8 9 0 1 2 3 4 5 6 7 8 9 0

1 ayw   A K S V Q H L E S L F T A V T N F L L S L G I H L N P N K T K R W G Y S L N F M G Y V I G C Y G S L P Q E H I I Q K I K E C
2 adw2  A K S V Q H R E S L Y T A V T N F L L S L G I H L N P N K T K R W G Y S L N F M G Y I I G S W G T L P Q D H I V Q K I K H C
3 adw   A K S V Q H R E S L Y T A V T N F L L S L G I H L N P N K T K R W G Y S L N F M G Y I I G S W G T L P Q D H I V Q K I K H C
4 adr   A K S V Q H L E S L F T S I T N F L L S L G I H L N P N K T K R W G Y S L N F M G Y V I G S W G T L P Q E H I V L K L K Q C

        A K S V Q H   E S L   T     T N F L L S L G I H L N P N K T K R W G Y S L N F M G Y   I G     G   L P Q   H I     K   K   C


        62                63                  64                  65                  66                  67                  68
        1 2 3 4 5 6 7 8 9 0 1 2 3 4 5 6 7 8 9 0 1 2 3 4 5 6 7 8 9 0 1 2 3 4 5 6 7 8 9 0 1 2 3 4 5 6 7 8 9 0 1 2 3 4 5 6 7 8 9 0 1 2

1 ayw   F R K L P I N R P I D W K V C Q R I V G L L G F A A P F T Q C G Y P A L M P L Y A C I Q S K Q A F T F S P T Y K A F L C K Q
2 adw2  F R K L P V N R P I D W K V C Q R I V G L L G F A A P F T Q C G Y P A L M P L Y A C I Q A K Q A F T F S P T Y K A F L S K Q
3 adw   F R K L P V N R P I D W K V C Q R I V G L L G F A A P F T Q C G Y P A L M P L Y A C I Q A K Q A F T F S P T Y K A F L S K Q
4 adr   F R K L P V N S P I D W K V C Q R I V G L L G F A A P F T Q C G Y P A L M P L Y A C I Q S K Q A F T F S P T Y K A F L C K Q

        F R K L P   N   P I D W K V C Q R I V G L L G F A A P F T Q C G Y P A L M P L Y A C I Q   K Q A F T F S P T Y K A F L   K Q


        68            69                  70                  71                  72                  73                  74
        3 4 5 6 7 8 9 0 1 2 3 4 5 6 7 8 9 0 1 2 3 4 5 6 7 8 9 0 1 2 3 4 5 6 7 8 9 0 1 2 3 4 5 6 7 8 9 0 1 2 3 4 5 6 7 8 9 0 1 2 3 4

1 ayw   Y L N L Y P V A R Q R P G L C Q V F A D A T P T G W G L V M G H Q R M R G T F S A P L P I H T A E L L A A C F A R S R S G A
2 adw2  Y M N L Y P V A R Q R P G L C Q V F A D A T P T G W G L A I G H Q R M R G T F V A P L P I H T A E L L A A C F A R S R S G A
3 adw   Y M N L Y P V A R Q R P G L C Q V F A D A T P T G W G L A I G H Q R M R G T F V A P L P I H T A E L L A A C F A R S R S G A
4 adr   Y L N L Y P V A R Q R S G L C Q V F A D A T P T G W G L A I G H R R M R G T F V A P L P I H T A E L L A A C F A R S R S G A

        Y   N L Y P V A R Q R   G L C Q V F A D A T P T G W G L     G H   R M R G T F   A P L P I H T A E L L A A C F A R S R S G A


        74        75                  76                  77                  78                  79                  80
        5 6 7 8 9 0 1 2 3 4 5 6 7 8 9 0 1 2 3 4 5 6 7 8 9 0 1 2 3 4 5 6 7 8 9 0 1 2 3 4 5 6 7 8 9 0 1 2 3 4 5 6 7 8 9 0 1 2 3 4 5 6

1 ayw   N I I G T D N S V V L S R K Y T S F P W L L G C A A N W I L R G T S F V Y V P S A L N P A D D P S R G R L G L S R P L L R L
2 adw2  K L I G T D N S V V L S R K Y T S F P W L L G C T A N W I L R G T S F V Y V P S A L N P A D D P S R G R L G L S R P L L R L
3 adw   K L I G T D N S V V L S R K Y T S F P W L L G C T A N W I L R G T S F V Y V P S A L N P A D D P S R G R L G L S R P L L R L
4 adr   K L I G T D N S V V L S R K Y T S F P W L L G C A A N W I L R G T S F V Y V P S A L N P A D D P S R G R L G L Y R P L L L L

            I G T D N S V V L S R K Y T S F P W L L G C   A N W I L R G T S F V Y V P S A L N P A D D P S R G R L G L   R P L L   L


        80    81                  82                  83                  84
        7 8 9 0 1 2 3 4 5 6 7 8 9 0 1 2 3 4 5 6 7 8 9 0 1 2 3 4 5 6 7 8 9 0 1 2 3 4 5

1 ayw   P F R P T T G R T S L Y A D S P S V P S H L P D R V H F A S P L H V A W R P P
2 adw2  P F Q P T T G R T S L Y A V S P S V P S H L P V R V H F A S P L H V A W R P P
3 adw   P F Q P T T G R T S L Y A V S P S V P S H L P V R V H F A S P L H V A W R P P
4 adr   P F R P T T G R T S L Y A V S P S V P S H L P D R V H F A S P L H V A W R P P

        P F   P T T G R T S L Y A   S P S V P S H L P   R V H F A S P L H V A W R P P
```

Fig. 19.4. Amino acid sequences of P protein. For further explanation see legend to Fig. 19.2.

(1982) and Stenlund et al. (1983). Additional 34 and 37-kDa minor components were described by Stratowa et al. (1982) and Wang et al. (1983). Similarly, 22-nm HBsAg particles isolated from serum and from a hepatoma cell line with integrated HBV DNA contained 25 and 29-kDa major components as well as the minor, larger molecular mass components mentioned before (Marion et al., 1979; Stibbe and Gerlich, 1982; 1983; Feitelson et al., 1983; Machida et al., 1983). The 44-kDa component is more prominently represented in tubular forms of HBsAg and in the

```
                            1                   2                   3                   4                   5                     6
          1 2 3 4 5 6 7 8 9 0 1 2 3 4 5 6 7 8 9 0 1 2 3 4 5 6 7 8 9 0 1 2 3 4 5 6 7 8 9 0 1 2 3 4 5 6 7 8 9 0 1 2 3 4 5 6 7 8 9 0 1

1 ayw     M E N I T S G F L G P L L V L Q A G F F L L T R I L T I P Q S L D S W W T S L N F L G G T T V C L G Q N S Q S P T S N H S
2 adw2    M E N I T S G F L G P L L V L Q A G F F L L T R I L T I P Q S L D S W W T S L N F L G G S P V C L G Q N S Q S P T S N H S
3 adw     M E N I T S G F L G P L L V L Q A G F F L L T R I L T I P Q S L D S W W T S L N F L G G S P V C L G Q N S Q S P T S N H S
4 adr     M E N T T S G F L G P L L V L Q A G F F L L T R I L T I P Q S L D S W W T S L N F L G G A P T C P G Q N S Q S P T S N H S
5 adyw    M E N I T S G F L G P L L V L Q A G F F L L T R I L T I P Q S L D S W W T S L N F L G G T T V C L G Q N S Q S P I S N H S

          M E N   T S G F L G P L L V L Q A G F F L L T R I L T I P Q S L D S W W T S L N F L G G       C   G Q N S Q S P   S N H S
          ' ' '   ' ' ' ' ' ' ' ' ' ' ' ' ' ' ' ' ' ' ' ' ' ' ' ' ' ' ' ' ' ' ' ' ' ' ' ' ' '       '   ' ' ' ' ' ' '   ' ' ' '

          6             7                   8                   9                  10                  11                  12
          3 4 5 6 7 8 9 0 1 2 3 4 5 6 7 8 9 0 1 2 3 4 5 6 7 8 9 0 1 2 3 4 5 6 7 8 9 0 1 2 3 4 5 6 7 8 9 0 1 2 3 4 5 6 7 8 9 0 1 2 3

1 ayw     T S C P P T C P G Y R W M C L R R F I I F L F I L L L C L I F L L V L L D Y Q G M L P V C P L I P G S S T T S T G P C R T
2 adw2    T S C P P I C P G Y R W M C L R R F I I F L F I L L L C L I F L L V L L D Y Q G M L P V C P L I P G S T T T S T G P C K T
3 adw     T S C P P I C P G Y R W M C L R R F I I F L F I L L L C L I F L L V L L D Y Q G M L P V C P L I P G S T T T S T G P C K T
4 adr     T S C P P I C P G Y R W M C L R R F I I F L F I L L L C L I F L L V L L D Y Q G M L P V C P L L P G T S T T S T G P C K T
5 adyw    T S C P P T C P G Y R W M C L R R F I I F L F I L L L C L I F L L V L L D Y Q G M L P V C P L I P G S S T T S T G S C R T

          T S C P P   C P G Y R W M C L R R F I I F L F I L L L C L I F L L V L L D Y Q G M L P V C P L   P G     T T S T G   C   T
          ' ' ' ' '   ' ' ' ' ' ' ' ' ' ' ' ' ' ' ' ' ' ' ' ' ' ' ' ' ' ' ' ' ' ' ' ' ' ' ' ' '   ' '     ' ' ' ' '   '   '

         12        13                  14                  15                  16                  17                  18
          5 6 7 8 9 0 1 2 3 4 5 6 7 8 9 0 1 2 3 4 5 6 7 8 9 0 1 2 3 4 5 6 7 8 9 0 1 2 3 4 5 6 7 8 9 0 1 2 3 4 5 6 7 8 9 0 1 2 3 4 5

1 ayw     M T T A Q G T S M Y P S C C C T K P S D G N C T C I P I P S S W A F G K F L W E W A S A R F S W L S L L V P F V Q W F V G
2 adw2    T T P A Q G N S M F P S C C C T K P T D G N C T C I P I P S S W A F A K Y L W E W A S V R F S W L S L L V P F V Q W F V G
3 adw     T T P A Q G N S K F P S C C C T K P T D G N C T C I P I P S S W A F A K Y L W E W A S V R F S W L S L L V P F V Q W F V G
4 adr     T I P A Q G T S M F P S C C C T K P S D G N C T C I P I P S S W A F A R F L W E W A S V R F S W L S L L V P F V Q W F V G
5 adyw    T T P A Q G I S M Y P S C C C T K P S D G N C T C I P I P S S W A F G K F L W E W A S A R F S W L S L L V P F V Q W F V G

                A Q G   S     P S C C C T K P   D G N C T C I P I P S S W A F       L W E W A S   R F S W L S L L V P F V Q W F V G
                ' ' '   '     ' ' ' ' ' ' ' '   ' ' ' ' ' ' ' ' ' ' ' ' ' ' '       ' ' ' ' ' '   ' ' ' ' ' ' ' ' ' ' ' ' ' ' ' ' '

         18    19                  20                  21                  22
          7 8 9 0 1 2 3 4 5 6 7 8 9 0 1 2 3 4 5 6 7 8 9 0 1 2 3 4 5 6 7 8 9 0 1 2 3 4 5 6

1 ayw     S P T V W L S V I W M M W Y W G P S L Y S I L S P F L P L L P I F F C L W V Y I
2 adw2    S P T V W L S A I W M M W Y W G P S L Y S I V S P F I P L L P I F F C L W V Y I
3 adw     S P T V W L S A I W M M W Y W G P S L Y S I V S P F I P L L P I F F C L W V Y I
4 adr     S P T V W L S V I W M M W Y W G P S L Y N I L S P F L P L L P I F F C L W V Y I
5 adyw    S P I V W L S V I W M M W Y W G P S L Y S I L S P F L P L L P I F F C L W A Y I

          S P   V W L S   I W M M W Y W G P S L Y   I   S P F   P L L P I F F C L W   Y I
          ' '   ' ' ' '   ' ' ' ' ' ' ' ' ' ' ' '   '   ' ' '   ' ' ' ' ' ' ' ' ' '   ' '
```

Fig. 19.5. Amino acid sequences of the translational product of the S gene as deduced from sequences of HBV DNA. For further explanation see legend to Fig. 19.2. Additional data obtained from amino acid sequencing (Peterson et al., 1984a) are summarized in the following table:

Subtype	Position						
	3	4	125	127	131	134	224
adw			T	P	N	F	V
adr	S	T	T	P	N	F	V
ayw			T	P	T	Y	V

The apparent discrepancies between some data obtained from direct sequencing of S-protein and those deduced from DNA sequencing may be ascribed to either the failure to more precisely characterize the subtype of S-protein which was sequenced, or to variability of amino acid sequences of S-proteins belonging to the same subtype. The subtype *adyw* (Pasek et al., 1979) probably represents a phenotypic mixture; there is no evidence for a single S-gene coding for an *ady* subtype (Peterson et al., 1984c). Out of 3 potential N-glycosylation sites at positions 3, 59 and 146, only the latter is actually glycosylated.

envelope of HBV (Shih et al., 1980b; Gerlich et al., 1984; Neurath et al., 1984d).

Partial sequencing of the major 25 and 29 kDa components and alignment of the sequences with those of PP2 revealed that these two components represent the non-glycosylated and glycosylated forms of the same protein corresponding to the C-terminal 226 amino acids of PP2 (Charnay et al., 1979; Peterson, 1981). This protein has been designated S protein (Fig. 19.5), and the corresponding DNA as S gene. The N-terminal half of PP2 (lacking the S protein) has been designated as pre-S protein (Fig. 19.6) and the corresponding DNA as pre-S gene.

The 34 and 37-kDa components appear to differ from each other only in the extent of glycosylation, and correspond to S protein with 55 additional amino acids (coded for by a portion of the pre-S gene) on its N-terminal region (Machida et al., 1983; Stibbe and Gerlich, 1983; Michel et al., 1984; Neurath et al., 1984a; Persing et al., 1984). Transfection of cells with recombinant DNAs containing the S gene leads to synthesis of S protein in the form of ~ 22-nm particles, of free polypeptide, or of a fusion protein with prokaryotic enzymes (Charnay et al., 1980; Edman, 1981; Moriarty et al., 1981; Liu et al., 1982; Valenzuela et al., 1982; Crowley et al., 1983; Fujisawa et al., 1983; Hitzeman et al., 1983; Lamb et al., 1983; Miyanohara et al., 1983; Smith et al., 1983).

Transfection of *E. coli* with recombinant DNA containing the pre-S gene (lacking the last 41 trinucleotides corresponding to the C-terminal end of the pre-S protein) resulted in the synthesis of an antibody-eliciting protein which specifically reacted with an ~ 45-kDa protein present in cells containing integrated HBV DNA and in HBV particles (Sninsky, personal communication). This protein corresponds to the entire PP2 (Neurath et al., 1984d).

A protein corresponding to pp5 (Fig. 19.12.) has not yet been identified.

The clinical significance of the presence of HBV-specific antigens or their corresponding antibodies in human sera has been described in detail (Overby et al., 1981; Deinhardt and Gust, 1982; Vyas and Blum, 1983). A simplified overview concerning the interpretation of diagnostic test results is presented in Table 19.1. The future availability of additional serological markers for HBV (see below) is expected to further improve the diagnosis of hepatitis B.

1.3. HBV SUBTYPES

Serological reactions involving the use of human or animal antibodies to HBsAg (anti-HBs) and of HBsAg, either in the form of sera from HBV-infected individuals or as purified ~ 22-nm spherical particles, revealed 4 major antigenic subtypes which could be further subdivided (as indicated in parentheses) to yield the following subtypes: *adw* (*adw*2, *adw*4); *adr* (*adr* q^-, *adr* q^+); *ayw* (*ayw1*, *ayw*2, *ayw*3, *ayw*4) and *ayr*. The frequency of these HBV subtypes is distinct in different geogra-

```
                          1                   2                   3                   4                   5                   6
        1 2 3 4 5 6 7 8 9 0 1 2 3 4 5 6 7 8 9 0 1 2 3 4 5 6 7 8 9 0 1 2 3 4 5 6 7 8 9 0 1 2 3 4 5 6 7 8 9 0 1 2 3 4 5 6 7 8 9 0 1 2

1 ayw   G H H I L G N K I Y S M G Q N L S T S N P L G F F P D H Q L D P A F R A N T A N P D W D F N P N K D T W P D A N K V G A G A
2 adyw  G H H I L G N K S Y S M G Q N L S T S N P L G F F P D H Q L D P A F R A N T N N P D W D F N P N K D T W P D A N K V G A G A
3 adw2  M G G W S S K P R K G M G T N L S V P N P L G F F P D H Q L D P A F G A N S N N P D W D F N P V K D D W P A A N Q V G V G A
4 adw   L G N K S Y S I R K G M G T N L S V P N P L G F L P D H Q L D P A F G A N S T N P D W D F N P I K D H W P A A N Q V G V G A
5 adr   M G G W S S K P R Q G M G T N L S V P N P L G F F P D H Q L D P A F G A N S N N P D W D F N P N K D Q W P E A N Q V G A G A

                              M G   N L S     N P L G F   P D H Q L D P A F   A N     N P D W D F N P   K D   W P   A N   V G   G A
                              ' '   ' ' '     ' ' ' ' '   ' ' ' ' ' ' ' ' '   ' '     ' ' ' ' ' ' ' '   ' '   ' '   ' '   ' '   ' '

        6             7                   8                   9                   10                  11                  12
        3 4 5 6 7 8 9 0 1 2 3 4 5 6 7 8 9 0 1 2 3 4 5 6 7 8 9 0 1 2 3 4 5 6 7 8 9 0 1 2 3 4 5 6 7 8 9 0 1 2 3 4 5 6 7 8 9 0 1 2 3 4

1 ayw   F G L G F T P P H G G L L G W S P Q A Q G I L Q T L P A N P P P A S T N R Q S G R Q P T P L S P P L R N T H P Q A M Q W N S
2 adyw  F G L G F T P P H G G L L G W S P Q A Q G I M Q T L P A N P P P A S T N R Q S G R Q P T P L S P P L R T T H P Q A M H W N S
3 adw2  F G P R L T P P H G G I L G W S P Q A Q G I L T T V S T I P P P A S T N R Q S G R Q P T P I S P P L R D S H P Q A M Q W N S
4 adw   F G P G L T P P H G G I L G W S P Q A Q G I L T T V S T I P P P A S T N R Q S G R Q P T P I S P P L R D S H P Q A M Q W N S
5 adr   F G P G F T P P H G G L L G W S P Q A Q G I L T T V P A A P P P A S T N R Q S G R Q P T P I S P P L R D S H P Q A M Q W N S

        F G       T P P H G G   L G W S P Q A Q G I     T         P P P A S T N R Q S G R Q P T P   S P P L R     H P Q A M   W N S
        ' '       ' ' ' ' ' '   ' ' ' ' ' ' ' ' ' '     '         ' ' ' ' ' ' ' ' ' ' ' ' ' ' ' '   ' ' ' ' '     ' ' ' ' '   ' ' '

        12        13                  14                  15                  16                  17
        5 6 7 8 9 0 1 2 3 4 5 6 7 8 9 0 1 2 3 4 5 6 7 8 9 0 1 2 3 4 5 6 7 8 9 0 1 2 3 4 5 6 7 8 9 0 1 2 3 4

1 ayw   T T F H Q T L Q D P R V R G L Y F P A G G S S S G T V N P V L T T A S P L S S I F S R I G D P A L N
2 adyw  T T F H Q T L Q D P R V R G L Y F P A G G S S S G T V N P V P T T T S P I S S I F S R I G D P A L N
3 adw2  T A F H Q T L Q D P R V R G L Y L P A G G S S S G T V N P A P N I A S H I S S I S A R T G D P V T N
4 adw   T A L H Q A L Q D P R V R G L Y L P A G G S S S G T V N P A P N I A S H I S S I S A R T G D P V T I
5 adr   T T F H Q A L L D P R V R G L Y F P A G G S S S G T V N P V P T T A S P I S S I F S R T G D P A P N

        T     H Q   L   D P R V R G L Y   P A G G S S S G T V N P           S     S S I     R   G D P
        '     ' '   '   ' ' ' ' ' ' ' '   ' ' ' ' ' ' ' ' ' ' ' '           '     ' ' '     '   ' ' '
```

Fig. 19.6. Amino acid sequence of the translation product of the pre-S gene. For further explanation see legend to Fig. 19.2. Potential glycosylation sites common to all subtypes are at positions 15 and 123. Experimental evidence suggests such a site at the latter position (Stibbe and Gerlich, 1983).

TABLE 19.1

Simplified interpretation of the presence of various serological markers for diagnosis of hepatitis B.

HBsAg	HBeAg	Anti-HBe	Anti-HBc[a]	Anti-HBs	Interpretation
+	+	−	−	−	Incubation period or early acute period during hepatitis B
+	+	−	+	−	Acute hepatitis B or chronic carrier
+	−	+	+	−	Late during hepatitis B infection or chronic state
−	−	+	+	+	Convalescent from acute hepatitis B infection
−	−	−	+	+	Recovered from past hepatitis B infection
−	−	−	−	+	Immunized without infection, repeatedly exposed to HBsAg without infection, or recovered from past hepatitis B infection
−	−	−	+	−	Recovered from past hepatitis B infection with undetectable anti-HBs, early convalescent or chronic infection

[a]The immunological class of anti-HBc (IgM or IgG antibody) and its titer may differentiate between onset of disease or acute disease (anti-HBc is predominantly IgM) and the asymptomatic carrier state (anti-HBc is predominantly IgG). Persistence of IgM anti-HBc and of high titers of IgG anti-HBc seem to be generally associated with continuing viral replication. IgM anti-HBc is a marker for fulminant hepatitis B, especially in the absence of HBsAg.
HBcAg serves as an additional marker indicating high levels of virus replication. However, HBcAg is usually sequestered within HBV particles present in plasma. It can be detected after disrupting the envelope of HBV (Neurath et al., 1982b).

phical areas (Courroucé-Pauty et al., 1983). In this scheme, the letter *a* denotes antigenic specificities common to all HBV subtypes; the letters *d/y* and *w/r*, respectively, denote pairs of mutually exclusive specificities characteristic for the antigenic subtypes; number 1–4 further subdivide the *w* specificity; and the presence or absence of *q* determinants further subdivides the *adr* subtypes. The antigenic specificites *a*, *d* and *y* were also detected on the surface of HBV particles (Hess et al., 1977). Immunization of experimental animals with the major and minor polypeptide components of HBsAg suggested that the *a, d* and *y* specificities exist on each of the constituent components of HBsAg (Dreesman et al., 1975; Gold et al., 1976;

Sanchez et al., 1981). Therefore, the corresponding antigenic determinants appear to be characteristic for the S protein. The *a* specificity appears to comprise a group of distinct antigenic determinants (Courouçé et al., 1976; Cote et al., 1982; Kennedy et al., 1983c; Pillot et al., 1984; Wands et al., 1984).

Antibodies to at least some of these *a* determinants appear to provide protection against infection by any of the HBV subtypes (Purcell and Gerin, 1978; Szmuness et al., 1982). A failure to respond to these *a* determinants after infection by one HBV subtype may later result in reinfection by another HBV subtype. This is true despite the presence of antibodies specific for the subtype determinants on HBV which are causing the primary infection, but which are absent on HBV which cause the secondary infection (Koziol et al., 1976; Le Bouvier et al., 1976; Sasaki et al., 1976; Tabor et al., 1976; Moraczewska et al., 1978; Swenson et al., 1983). Therefore, in order to provide protection against HBV, a vaccine has to elicit an antibody response to common group-specific *a* determinants of HBV (Jilg et al., 1984). A small proportion of people appear nonresponsive to these determinants despite a response to subtype-specific determinants (McAuliffe et al., 1982; Gold et al., 1974). This problem may, perhaps, be overcome by including antigens corresponding to several HBV subtypes in a vaccine. Further issues concerning vaccination against HBV will be discussed in the section dealing with synthetic peptides.

The use of monoclonal antibodies provides an improved tool in the study of subtype specificities of HBsAg, and its application provides evidence for previously unidentified subtype specificities (Shih et al., 1980a; Kalil et al., 1981; Wands et al., 1981; Wands and Zurawski, 1981; Boniolo et al., 1982; Goodall et al., 1982; Kennedy et al., 1983c; Tedder et al., 1983; Courouçé et al., 1984). Monoclonal antibodies may also allow the detection of HBsAg determinants within immune complexes containing HBsAg (Shafritz et al., 1982).

The great genetic heterogeneity of HBV, suggested by the antigenic variability of protein S, is further supported by the multitude of restriction endonuclease cleavage patterns of DNAs isolated from distinct HBV subtypes, as well as by differences in HBV isolates belonging to the same subtype (Siddiqui et al., 1979). The variability in amino acid sequences of S proteins belonging to distinct HBV subtypes, the genomes of which have been cloned and sequenced, is evident from Fig. 19.5. However, such variability is also evident for all other recognized gene products of HBV (Figs. 19.2 and 19.6) as well as for the putative proteins PP3 and PP4 (Figs. 19.3 and 19.4). Despite the predicted differences in amino acid sequence and the availability of monoclonal antibodies (Imai et al., 1982; Takahashi et al., 1983), antigenic variations in HBcAg or HBeAg have not yet been demonstrated experimentally. Data on the intertypic variations in primary sequence of the distinct HBV DNA gene products are summarized in Table 19.2. The smallest percentage of amino acid replacements occur within the internal nucleocapsid of HBV (HBcAg), while the highest variability is predicted for the pre-S protein. Antigenic determinants on HBsAg, and on the HBV envelope which corresponds to the pre-S protein,

TABLE 19.2

Variations in amino acid sequence between HBV DNA gene products belonging to distinct antigenic subtypes

Subtype	Gene product	Subtype				
		ayw	*adr*	*adyw*	*adw2*	*adw*
ayw	S	—	17	7	16	17
	Core	—	8	6	8	7
	Pre-S	—	33	8	47	50
	X	—	20		9	12
	P	—	111		126	123
adr	S	8	—	20	16	17
	Core	4	—	11	11	10
	Pre-S	19	—	33	23	31
	X	13	—		20	23
	P	13	—		89	95
adyw	S	3	9	—	18	19
	Core	3	6	—	9	10
	Pre-S	5	19	—	47	51
adw2	S	7	7	8	—	1
	Core	4	6	5	—	3
	Pre-S	27	13	27	—	14
	X	6	13		—	3
	P	15	11		—	20
adw	S	8	8	8	0	—
	Core	4	5	5	2	—
	Pre-S	29	18	29	8	—
	X	8	15		2	—
	P	15	11		2	—

The upper right-hand half of this table indicates the *number* of amino acid differences between subtypes; the lower left-hand half indicates the *percentage* of amino acid differences, expressed as round numbers.

have only been recognized very recently (Gerlich et al., 1984; Neurath et al., 1984a). Variations in their immunologic specificities are very likely, considering the amino acid sequence data, and will have to be considered in future efforts to classify HBV subtypes.

Despite the multitude of amino acid replacements within the sequence of each of the HBV DNA gene products, their secondary structures remain essentially conserved as indicated, for example, by profiles of relative hydrophilicity (Figs. 19.7–11). Of particular interest is the higher hydrophilicity of the pre-S protein as com-

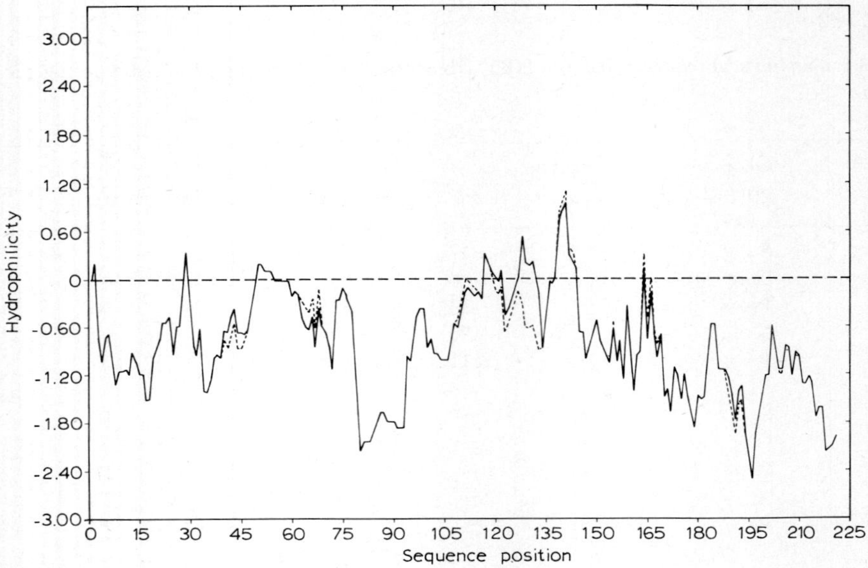

Fig. 19.7. Profiles of relative hydrophilicity (Hopp and Woods, 1981) corresponding to the amino acid sequence of the S gene product (subtype *adw*). Profiles for other subtypes are similar: the dotted lines represent subtype *ayw*.

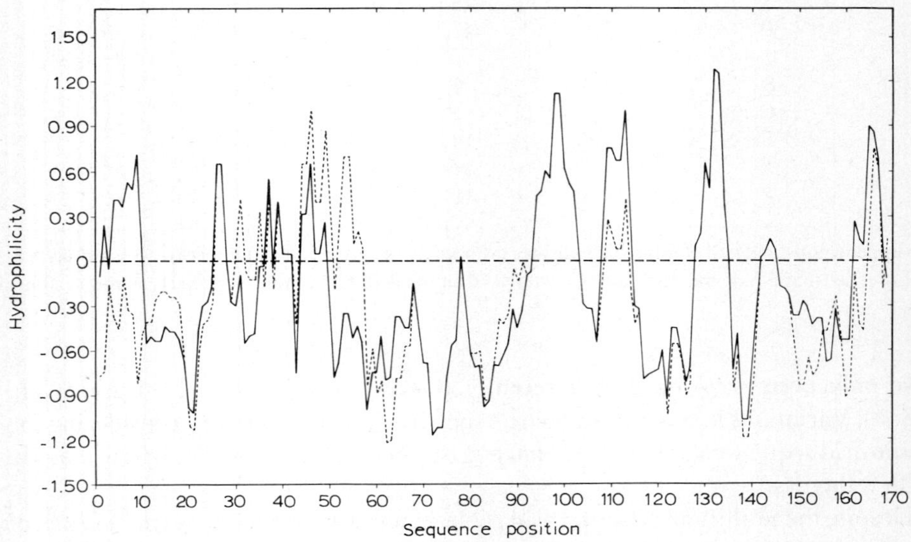

Fig. 19.8. Profiles of relative hydrophilicity corresponding to the amino acid sequence of the pre-S gene product (subtype *adw*). Dotted lines indicate subtype *ayw*.

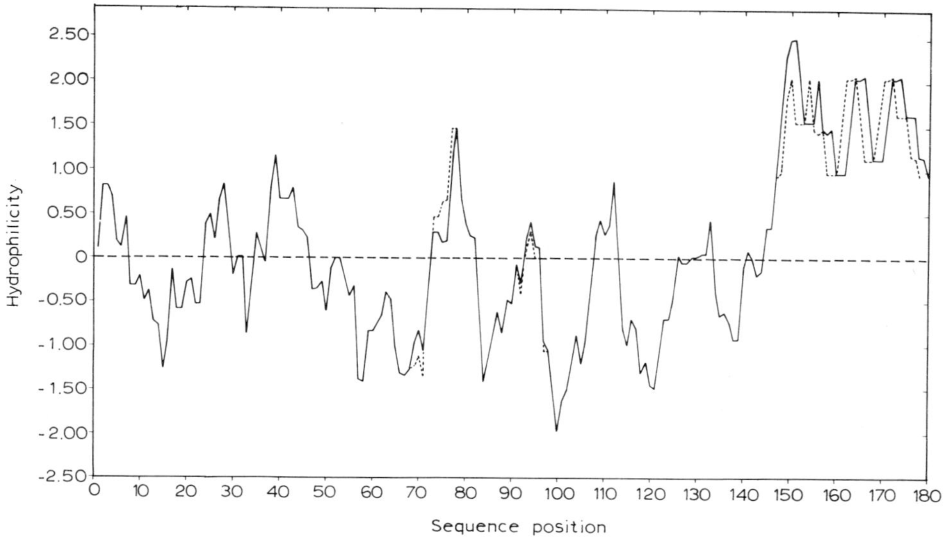

Fig. 19.9. Profiles of relative hydrophilicity corresponding to the amino acid sequence of the C gene product (subtype *adw*). Dotted lines represent subtype *ayw*.

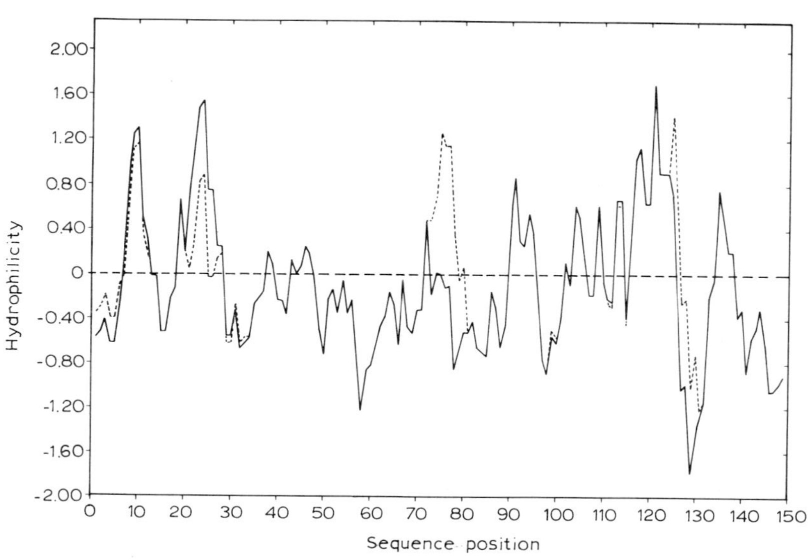

Fig. 19.10. Profiles of relative hydrophilicity corresponding to the amino acid sequence of the X-gene product (subtype *adw*). Dotted lines represent subtype *ayw*.

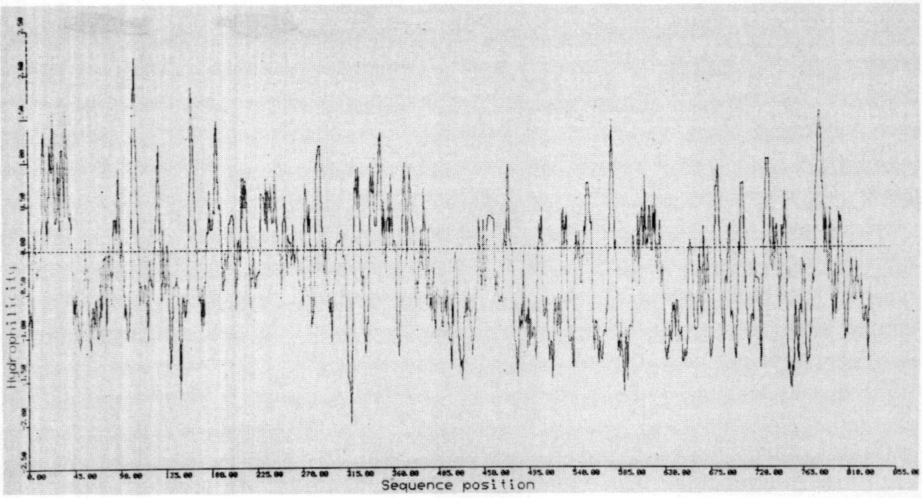

Fig. 19.11. Profiles of relative hydrophilicity corresponding to the amino acid sequence of the P-gene product (subtype *adw*).

pared with the S protein (Figs. 19.7 and 19.8). The hydrophilic nature of the pre-S protein and the higher mutation rate within the pre-S gene compared to the S gene makes it likely, in analogy with other viruses (for example, see Chapter 14 of this book), that the pre-S protein contains antigenic determinants important in the process of virus neutralization (see below).

1.4. Genetic regulation of antibody responses to HBV

The distinct clinical manifestations associated with HBV infections have been ascribed to variations in immune responsiveness, and the genetic background of these differences in host responses were recognized in early studies on hepatitis B (Blumberg et al., 1970; Dudley et al., 1972). A relationship between HLA type, immune responsiveness to HBsAg, and the development of chronic HBV infection has been recognized (Rubinstein et al., 1982). Individual variations in responses to distinct antigenic determinants of HBsAg among vaccinated humans were observed (see above).

The rating of immunogenicities of distinct HBsAg determinants varies with different animal species (Imai et al., 1974; Gerin et al., 1975). Inbred strains and H-2 congenic mice were used as a model system to evaluate genetic factors that regulate the antibody response to HBsAg determinants (Milich and Chisari, 1982; Milich et al., 1983a; 1983b; 1984; Neurath et al., 1983). These studies revealed that antibody responses to the *a, d* and *y* determinants are regulated by genes mapping to the murine major histocompatibility complex (MHC). Haplotypes conferring high, intermediate and low responsiveness as well as nonresponsiveness to HBsAg were distinguished. Studies with Fl hybrids between good responder and nonresponder mouse strains, and with these Fl hybrids backcrossed to nonresponder parental strains, revealed that responsiveness to HBsAg is inherited as a dominant trait. Interestingly, the subtype-specific anti-*d* and anti-*y* responses preceded the anti-*a* response. In some low responder strains, in which no primary anti-HBs response could be demonstrated, a secondary response to subtype-specific determinants but not to *a* determinants was detected – observations reminiscent of findings in humans, as mentioned before. Further studies (Milich et al., 1984) revealed that humoral anti-HBs responses are regulated by at least two immune response (*Ir*) genes, one in the I-A subregion of MHC and another in the I-C subregion, defined as *Ir-HBs-1* and *Ir-HBs-2*, respectively. *Ir-HBs-1* regulates the primary antibody responses to all determinants of the S protein, while *Ir-HBs-2* affects the responses to the *d* or *y* determinants. Strains having only the *Ir-HBs-2* gene fail to produce anti-*a* antibodies, and subtype-specific antibodies are detected only after secondary or tertiary immunization. In the presence of the *Ir-HBs-1* gene, the influence of *Ir-HBs-2* appears already detectable after primary immunization, and is additive rather than exclusive.

The lack of an antibody response to the S protein may be overcome by linking HBsAg to a carrier macromolecule, for example sheep red blood cells (Milich et al., 1983b). The IgG antibody response in that case appears to be subtype specific and is inhibited by preceding immunization with unconjugated HBsAg. No data are available at present on the genetic regulation of antibody responses to determinants on the pre-S protein, which is an integral part of the HBV envelope (Neurath et al., 1984d) or to determinants on HBcAg and HBeAg. Such antibodies may play an essential role in the outcome of HBV infections and in protection against hepatitis B. Therefore, immunogenetic studies concerning host responses to these HBV components, as well as an expansion of investigations of HLA-dependent responses to HBsAg, are needed for the design of strategies in the prevention of hepatitis B and its sequelae.

1.5. Generation of HBV-specific lymphocytes and their effector functions

Both antibodies and cytotoxic T cells play a role in host defense against HBV and in the pathogenesis of hepatitis B infections (Chisari et al., 1978; Chisari and Nowoslawski, 1982; Dienstag, 1984). The production of antibodies to HBsAg by B lymphocytes is regulated by T lymphocytes (Roberts et al., 1975). For these reasons, any scrutiny of the antigenic structure of HBV and of its relevance to immunity against hepatitis B must involve (*a*) an analysis of cellular correlates of the immune response to HBV, and (*b*) investigations concerning the recognition by HBV-specific T cells of HBV structural components and of HBV-infected hepatocytes.

Cell-mediated immunity to HBsAg has been demonstrated in both humans and experimental animals immunized with HBsAg or with HBsAg-derived polypeptides (Vyas et al., 1972; Cabral et al., 1975; 1978; Kremastinou et al., 1981). HBsAg-specific T cells were demonstrated in the circulation of some recipients of a hepatitis B vaccine and in asymptomatic HBsAg carriers. A great variation in detectability of these cells among individuals may be observed, and no correlation with the level of anti-HBs antibodies could be found (Celis et al., 1984; Hellstrom et al., 1984). These studies suggest that T cell responses are not exclusively specific for HBsAg subtypes, and that these cells may recognize sequential rather than 'conformational' epitopes on HBsAg.

Studies with inbred and congenic strains of mice demonstrated unequivocally that the anti-HBs antibody non-responsiveness is correlated with defects in generation of HBsAg-specific T (helper) cells in the presence of a fully competent HBsAg-specific B cell repertoire. The MHC-restricted response to HBsAg is thus mediated through T cell responses. Common and subtype-specific determinants appear to be recognized as separate entities by T cells. Both subtype-specific B and T cells appear to be able to recognize similar or at least adjacent regions on HBsAg. On the other hand, the subtype-independent determinants on HBsAg seem to be distinct for T and for B cells. Thus, T cells do not recognize those HBsAg determinants which are serologically defined by anti-*a* antibodies, but react with common 'carrier determinants' on HBsAg (Milich et al., 1983b; Milich and Chisari, 1984).

HBsAg on hepatocyte membranes might serve as a target for cell-mediated cytotoxic effects during the early or acute stages of hepatitis B (Alberti et al., 1984). More recent findings (Mondelli et al., 1982; 1984) implicate HBcAg as a target on hepatocytes and HBcAg-specific cytotoxic T lymphocytes in lysis of liver cells in which HBV replicates. Depending on other factors (see Section 1.6), this may lead to elimination of infected cells and subsequent recovery, or to chronic hepatitis. These findings are in accord with earlier observations indicating the membrane expression of HBcAg (Huang and Neurath, 1979) and the specific attachment of anti-HBc to HBV-infected hepatocytes (Trevisan et al., 1982). The latter antibodies might interfere with cytotoxic T cells which eliminate infected hepatocytes, thus modulating cell-mediated liver damage. Such interference may contribute to establishment of chronic infection.

The exposure on membranes of infected cells of components which are internal in the mature virion and the occurrence of cytotoxic T cells specific for these components is not unique for HBV (Braciale, 1977; Bennink et al., 1982; Kees and Krammer, 1984). Antibodies to surface components of virions and cytotoxic T cells directed against internal components may thus, together, provide a defense mechanism with dual (or multiple) antigenic specificities. In this context, it is of utmost importance that chimpanzees immunized with HBcAg are completely or partially protected against subsequent challenge with HBV (Murray et al., 1984; Iwarson et al., 1984). Such chimpanzees develop anti-HBc prior to challenge, but unfortunately results concerning elicitation of cytotoxic T cells with HBcAg specificity were not reported.

The delineation of antigenic determinants on the S and pre-S proteins (see below) and on other gene products of HBV DNA is expected to promote progress in defining the specificity of cells of the immune system generated by exposure to HBV and to establish their role in defense and pathogenesis.

1.6. Attempts to define antigenic determinants on HBV structural proteins

The contributory role of an immune response to HBcAg (HBeAg) in protection against hepatitis B has emerged only recently. Therefore, little attention has been given so far to defining in greater detail the antigenic determinants on HBcAg (HBeAg). Studies using monoclonal antibodies indicated multiple antigenic determinants on HBcAg and HBeAg (Imai et al., 1982; Takahashi et al., 1983), and sequence data (Table 19.2, Fig.19.2) suggest subtype-dependent variations in the amino acid sequence of HBcAg, possibly causing antigenic variability.

Most immunochemical studies of HBV have been focused on defining antigenic determinants on envelope components of HBV known to elicit immunity to HBV. In some of these studies (summarized by Peterson et al., 1984a), the complexity of the HBV envelope proteins was ignored and components larger than S protein (in its glycosylated and nonglycosylated forms) were declared to represent aggregates or contaminants, based on the erroneous conclusions of Koistinen (1980), and despite evidence to the contrary (see Section 1.2).

Results of experiments aimed at delineation of antigenic determinants on each of the S and pre-S proteins, based on the experimental approaches described in Chapter 1, will be summarized here. The use of synthetic peptide analogues for defining epitopes on HBV DNA gene products will be discussed separately in Section 1.7.

The smallest structural unit having an antigenicity and immunogenicity comparable to that of the S protein assembled within HBsAg particles is a dimer consisting of one molecule each of glycosylated and nonglycosylated S protein (Mishiro et al., 1980; Skelly et al., 1981). The dimeric protein carries both the common and the subtype-specific antigenic determinants. Reduction of disulfide bonds results in dissociation of the dimer and in a concomitant drastic decrease of antigenicity with respect to all determinants mentioned. This is in agreement with earlier reports concerning the decrease of antigenicity caused by cleavage of disulfide bonds within intact HBsAg, although disulfide bond-independent determinants were also convincingly demonstrated with both polyclonal (Imai et al., 1974) and monoclonal antibodies (Pillot et al., 1984). The location(s) of the latter determinants remains to be determined. Neither carbohydrate chains nor lipids are required for expression of full antigenicity or immunogenicity (Neurath et al., 1978; McAleer et al., 1984). Thus, a dimer of the S protein containing the appropriate intra- and intersubunit cystine links may represent the minimum structure that carries *all* antigenic determinants (corresponding to the S protein) detected on the HBV envelope and on HBsAg. Experiments involving tryptic digestion and the use of a photoactivable hydrophobic probe indicated that regions within the N-terminal and C-terminal parts of *both* glycosylated and nonglycosylated forms of the S protein (Fig. 19.5) are buried inside the lipid of HBsAg particles, while domains roughly spanning residues 122–150 are exposed outside the particles (Peterson, 1981; Gavilanes et al., 1982). These experimental findings are in disagreement with a model of the HBsAg structure which proposes that the same hydrophilic domains are alternatively inside and outside the lipid bilayer (Tiollais and Wain-Hobson, 1984) – a model difficult to reconcile with vectorial discharge of nascent polypeptides (Sabatini et al., 1983).

Proteolytic cleavage of the isolated glycosylated S protein by protease from *Bacillus amyloliquefaciens* and pronase P resulted in the release of a glycopeptide containing amino acid residues 141–155 and the single carbohydrate chain (corresponding to glycosylated S protein) attached to residue 146 (Fig. 19.5) (Machida et al., 1982). Antisera prepared against this glycopeptide reacted with HBsAg subtypes *adw, adr* and *ayw*, indicating that sequence 141–155 encompasses an epitope of a group-specific *a* determinant. This finding agrees with earlier suggestions that one or both lysine residues 141 and 160 (common for subtypes *ad* and *ay*; Fig. 19.5) are part of a dominant HBsAg epitope (Neurath et al., 1981; Vyas, 1981). Stretches of amino acid sequences of S protein common to all subtypes (except hydrophobic regions hidden in the interior of HBsAg; see Fig. 19.7), which may potentially carry *a* epitopes, are evident from Fig. 19.5. The amino acid sequence differences possibly causing antigenic differences between subtypes *d* and *y* are at residues 45, 46, 68, 122, 134, 159 and 168; and those possibly involved in *w* and *r* determinant changes are at locations 4, 45, 47, 110, 113, 126, 160 and 207. In attempts to assign antigenic variations to differences in amino acid sequence, one should consider, however, that some amino acid changes may not result in alterations of antigenic specificity. On the other hand, amino acid replacements outside a sequence comprising an epitope may possibly lead to changes in antigenic specificity.

Some insight concerning the contribution of distinct amino acid residues to specific antigenic determinants may be gained by following changes, caused by chemical modifications, in the reactivity of S protein with determinant-specific (preferably monoclonal) antibodies. Conversion of all lysine residues into *N*-dimethylysine by reductive methylation (i.e. residues 122, 141, 160 for subtype *ad*, and 141 and 160 for subtype *ay*) abolished the reactivity of HBsAg *ad* with monoclonal anti-*d* antibodies, while the reactivity of HBsAg *ay* with monoclonal anti-*y* antibodies remained unaltered, not only after reductive methylation but also following chemical modification of the arginine at position 122 (Paul and Peterson, 1984; Peterson et al., 1984b). These data suggest that Lys 122 is essential for the *d* determinant and that *y* and *d* determinants do not correspond to equivalent amino acid sequences, differing only at residue 122. However, conclusions based on the lack of observed changes in binding of monoclonal antibodies to chemically modified HBsAg can be valid only if it is known that these antibodies bind to determinants which include the modified amino acid residues. For example, modification of all lysine residues in either subtypes *ad* or *ay* did not affect the binding of 3 distinct, but undefined, monoclonal anti-*a* antibodies to the modified HBsAg (Peterson et al., 1984b). However, in another study the binding of polyclonal anti-HBs and of monoclonal anti-*a* antibodies, reacting with a synthetic peptide comprising residues 135–155 (see Section 1.7), was drastically diminished after reductive methylation of the peptide, suggesting the involvement of Lys 141 in one of the *a* determinants of the S protein (Neurath et al., 1981; Neurath et al., 1984e).

The first suggestion that the envelope of HBV carries antigenic determinants which are not present (or present at low levels) on the ~ 22-nm spherical subviral particles came from the electron microscopic studies of Moodie et al. (1974), which revealed immune complexes containing HBV particles but no ~ 22-nm subviral particles. These results were confirmed by additional studies in which either sera from HBV-infected humans or sera from animals hyperimmunized with HBV particles and tubular forms of HBsAg were used to precipitate HBV particles (Neurath et al., 1976; Alberti et al., 1978). These initial observations were made at a time when (*a*) the dogma 'one gene – one protein' had not yet been modified to allow overlapping genes in different reading frames (Sanger et al., 1977) and when the HBV genome was considered too small to have a coding capacity for proteins in addition to HBsAg, HBcAg, HBeAg and DNA polymerase; and (*b*) the variability in the antigenic and protein composition of the ~ 22-nm spherical subviral particles, depending on the stage of HBV infection (Neurath et al., 1980; Stibbe and Gerlich, 1982), had not yet been recognized. Therefore, these 'additional' antigenic determinants were erroneously ascribed to HBeAg, although it was recognized that 'e antigen' did not represent a single entity (Neurath et al., 1976). Using a single anti-HBe-positive serum each and the techniques of immunodiffusion and immunoprecipitation, Takahashi et al. (1978b) and Gerin et al. (1978) failed to detect any reactivity with HBV. The work of Vnek et al. (1979) and Neurath et al. (1980), who employed radioimmunoassays, confirmed the presence in some anti-HBe-positive sera of antibodies reacting with the HBV envelope.

However, these 'additional' HBV surface determinants have been located within the pre-S gene-coded HBV proteins (see Sections 1.2 and 1.7). Antibodies reacting with antigenic determinants located within residues 120 and 145 of the pre-S protein were detected in sera of humans who had recovered from hepatitis B. These antibodies specifically recognize the ~ 34 and ~ 37-kDa components in HBsAg and the HBV envelope (Neurath et al., 1984a) and react with intact HBsAg and HBV. They also react with the ~ 45-kDa component of HBV, as do antibodies recognizing determinants nearer to the N-terminal of the pre-S protein (Gerlich et al., 1984; Neurath et al., 1984d; Sninsky et al., personal communication).

Amino acid sequence differences in the pre-S protein, possibly contributing to antigenic differences between subtypes *d* and *y*, are at residues 14, 18, 19, 35, 38, 51, 54, 57, 65, 86, 88, 91, 108, 114, 115, 168 and 173; and those possibly involved in *w* and *r* determinant changes are at locations 51, 54, 91, 132 and 173 (Fig. 19.6). In accordance with this, a synthetic peptide encompassing residues 120–145 elicited antibodies recognizing both *d* and *y* subtypes of HBsAg (Neurath et al., 1984a).

The importance of cell-mediated and humoral immune responses to HBV envelope determinants, distinct from those corresponding to the S protein, in the early stage of hepatitis B, as well as in the outcome of the infection, has been suggested by Alberti et al. (1984) and Mondelli et al. (1984). Such antibodies appear to be absent in chronic active hepatitis, and this defect is assumed to permit the infectious cycle to proceed.

In this respect, it is of importance to note that a hepatitis B vaccine, the production of which involves pepsin treatment of HBsAg, although highly efficient in apparently healthy adults, has had low immunogenicity and no protective effect in hemodialysis patients and in other immunosuppressed individuals. Other vaccines produced without pepsin treatment do not seem to have this defect (Desmyter, 1984; Stevens et al., 1984; Taylor et al., 1984). Although treatment with pepsin does not affect the apparent morphological integrity of HBsAg and its reactivity with antibodies to the S protein, it does lead to loss of reactivity with antibodies to pre-S protein determinants (Neurath et al., 1984d).

1.7. Polypeptides produced by chemical synthesis or by expression vectors containing cloned HBV DNA fragments coding for viral epitopes

Peptide analogues of the translational products of HBV DNA have been synthesized with the following goals in mind:

(*a*) Delineation of antibody binding sites on HBV proteins.

(*b*) Definition of antigenic determinants involved in common and subtype specificities of HBV.

(*c*) Characterization of sites recognized by receptors on HBV-specific T cells.

(*d*) Detection of HBV DNA gene products not yet recognized by standard serological techniques utilizing convalescent sera or sera from HBV carriers.

(*e*) Determination of epitopes on HBV-proteins reacting with defined antibody idiotypes.

(*f*) Preparation of reagents applicable to clinical diagnosis of HBV infections.

(*g*) Prophylaxis of hepatitis B.

In principle, these analogues can be prepared by two distinct approaches:

(*a*) Chemical synthesis of peptides.

(*b*) Biosynthesis in eukaryotic or prokaryotic cells transfected with recombinant DNA expression vectors containing selected defined subgenomic HBV DNA fragments (Sninsky et al., 1984). This route appears to represent the method of choice for preparation of protein domains encompassing more than ~ 50 amino acids and for studies aimed at the characterization of antibody or T cell binding sites by localized mutagenesis (Lathe et al., 1983) of the HBV DNA fragments.

Before describing results obtained so far with peptide analogues of HBV proteins, some aspects of their preparation and the raising of antisera to them must be discussed. These aspects involve: selection of peptides for synthesis, choice of carrier-adjuvant combinations, detection and characterization of antibodies elicited by peptides, selection of peptides as candidates for hepatitis B vaccines and definition of diagnostic tests that will potentially predict recovery from HBV infection, and/or indicate immunity to HBV.

1.7.1. Selection of peptides for synthesis

Two issues have to be considered in selection of peptides for synthesis.

Finding the appropriate amino acid sequence within the viral protein sequence This can be accomplished by a 'brute force' approach consisting of synthesis of a number of peptides (possily having partially overlapping sequences) consecutively covering the total sequence of the protein (Kazim and Atassi, 1980; Bittle et al., 1982). Attempts to replace this cumbersome approach lead to the application of methods which predict the localization of antigenic determinants based on displaying the hydrophilic character of proteins (Hopp and Woods, 1981; Kyte and Doolittle, 1982; Ch. 1). Profiles of relative hydrophilicity corresponding to amino acid sequences of distinct HBV DNA gene products served as useful guides in selecting the corresponding synthetic analogues. These profiles are shown in Figs. 19.7–12. However, prediction of the localization of antigenic determinants based simply on determination of the sequences with highest hydrophilicity is not generally applicable to *all* proteins (Baron and Baltimore, 1982; Juillerat et al., 1984). Joint histograms which combine the prediction of α-helices, β-pleated sheets and reverse turns with hydrophilicity profiles and the helix wheel representation method (Schiffer and Edmundson, 1967; Pfaff et al., 1982) appear to have superior predictive potential, especially if combined with information concerning segments with the highest virus subtype-specific amino acid sequence variability.

Estimation of the minimum optimal number of amino acid residues in the synthetic analogue Studies concerned with elicitation of antiprotein antibodies with synthetic peptides indicate that the length of the peptide should not be limited to the number of expected amino acid residues comprising an epitope, but should be increased to enhance the probability that the synthetic peptide analogue will have a similar conformation to that which it has as a segment of the intact protein (Pfaff et al., 1982; Van Eldik et al., 1983). The chances of accomplishing this are higher with peptides that correspond to either the C-terminal or N-terminal portion of the protein. Fortunately, with appropriately designed synthetic peptides, the frequency of evoking protein-reactive antibodies is several orders of magnitude greater than expected from the predicted number of conformations the peptide could acquire in

A

```
                           1                   2                   3                   4                   5
            1 2 3 4 5 6 7 8 9 0 1 2 3 4 5 6 7 8 9 0 1 2 3 4 5 6 7 8 9 0 1 2 3 4 5 6 7 8 9 0 1 2 3 4 5 6 7 8 9 0 1 2 3 4 5 6 7 8

1 ayw       M R R A E V K R S A H G P A D E K A Q T G S P R K E R C A P W S V G T A D G E G D E R V P S D P E K G R P Q D S A P
2 adw       M Q R A E V K R S A H G P A D E K A Q T G R P R K E R C A P W S A G T A D G E G D E R V P S G P E R G R P R D S A P
3 adw2      M Q R A E V K R S A H G P A D E K A Q T G R P R K E R C A P W S A G T A D G E G D E R V P S G P E R G R P R D S A P
4 adr       M R R A E B K R S A H G P A D E K A Q T G R P R K E R C A P W S A G T A E E E G D G R A P N G P E T G R P R D S A P
5 adr4      M R R A E V K R S A H G P A D E K A Q T K R P R K E R C A P W S A G T A D E E G D G R V P N G P E T G R P R D S A P

            M   R A E V K R S A H G P A D E K A Q T     P R K E R C A P W S   G T A     E G D   R   P     P E   G R P   D S A P

              6                   7                   8                   9
            9 0 1 2 3 4 5 6 7 8 9 0 1 2 3 4 5 6 7 8 9 0 1 2 3 4 5 6 7 8 9 0 1 2 3 4 5

1 ayw       T G R K Q R T S R A G S S W Q H S L A A M E T M Y I C G I G Q Q S Y Q S R
2 adw       T G R K Q R T S R E G S S W Q Y S L A A M E M M Y I S A R G R Q N C Q F R
3 adw2      T G R K Q R T S R E G S S W Q Y S L A A M E T M Y I S A R G R Q N C Q F R
4 adr       T G R R Q R T S R A G S S W Q H T L A A M E R R C I S E R G Q Q S C R F R
5 adr4      T G R R Q R T S R A G S S W Q H T L A A M E R R C I S E R G Q Q S C R C R

            T G R   Q R T S R   G S S W Q H S L A A M E       I       G   Q         R
```

B

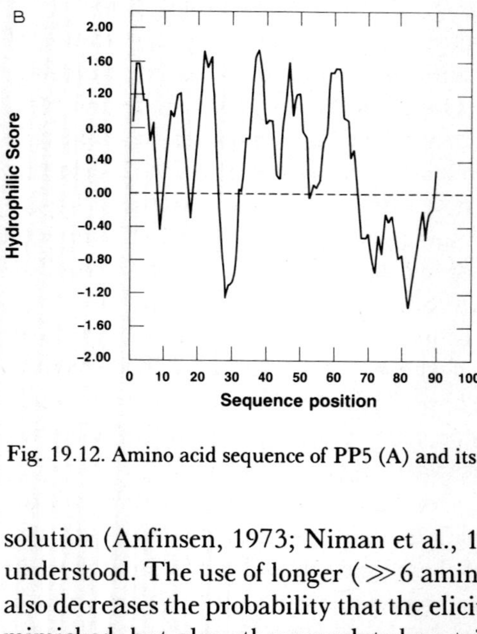

Fig. 19.12. Amino acid sequence of PP5 (A) and its profile of relative hydrophilicity (B).

solution (Anfinsen, 1973; Niman et al., 1983), for reasons which are not yet well understood. The use of longer ($\gg$6 amino acid residues) peptides as immunogens also decreases the probability that the elicited antibodies will recognize, not only the mimicked, but also other unrelated proteins (Nigg et al., 1982; Walter and Werchau, 1982; Kabsch and Sander, 1984).

1.7.2. Choice of carrier-adjuvant combinations

Although free peptides administered with adjuvants or incorporated into liposomes may elicit antibodies (Dreesman et al., 1982; Sanchez et al., 1982; Young et al., 1983; Shih et al., 1984), their level can usually be increased when peptides linked to appropriate carriers are used for immunization. However, potential problems associated with the use of protein carriers, such as keyhole lympet hemocyanin (KLH), tetanus toxoid, or other highly immunogenic carriers have rarely been appreciated. Immunization with KLH may suppress the antibody responses to haptens attached to it. This suppression is especially evident after secondary or tertiary immunizations with KLH conjugates (Herzenberg et al., 1983), and occurs while antibody titers against the carrier are still rising and exceeding the antihapten antibody titers by several orders of magnitude (Neurath et al., 1984d). This phenomenon probably explains the frequently observed lack of significant antipeptide booster responses beyond secondary immunizations (Gerin et al., 1983; Hopp, 1984; Neurath et al., 1984c; 1984d). These problems may be overcome by immunization with synthetic peptides covalently linked to the surface of liposomes, stabilized by cross-linking and having receptors for T cells (Neurath et al., 1984a; 1984b). Such fully synthetic conjugates elicit high levels of antibodies without any additional adjuvant (Neurath et al., 1984d). Results superior to those obtained with KLH conjugates have also been reported for a peptide linked to dipalmityl lysine (Hopp, 1984).

1.7.3. Detection of and characterization of peptide-elicited antibodies

The methodology of measuring antibodies elicited by synthetic peptides and directed against the peptide itself, and against the protein containing the same peptide sequence, is based on principles described in Chapters 5–10 and 25. In addition, fluorescence enzyme-linked immunoassays, in which peptides linked to β-galactosidase were used, have been described (Neurath et al., 1982a; Neurath et al., 1984c).

For the development of diagnostic tests or vaccines based on synthetic peptides, it appears essential to know (*a*) the proportion of antibodies recognizing the intact protein within the repertoire of antipeptide antibodies, and (*b*) the association constants for the antibody-peptide and antibody-protein reactions which predetermine their applicability to distinct immunoassays (Peterfy et al., 1983). Few studies addressed to this problem have been published so far. For example, ~ 30% and ~ 0.5% of antibodies elicited by S(110–137)* and S(135–155), respectively, recognized intact HBsAg (Neurath et al., 1982a; Alexander and Lerner, 1984). Similar association constants (*K*) for the reaction of a protein with antiprotein and with antipeptide antibodies, respectively, have been reported for calmodulin (Van Eldik et al., 1983). The *K* value for the reaction between HBsAg and anti-S(135–155) was $4 \times 10^5 M^{-1}$ (Neurath et al., 1982a), i.e. several orders of magnitude less than expected for 'good' anti-HBs (Wands et al., 1984). Results of assays measuring the inhibition of the HBsAg-anti-HBs reaction by HBsAg and synthetic peptides

---

* In the nomenclature for peptide analogues of HBV proteins, S, pre-S, C, X, P denotes the S, pre-S, X, and P proteins, respectively. Numbers in parentheses indicate the first and last residues in the synthetic peptide and correspond to the sequence positions shown in Figs. 19.2–6.

revealed that S(139–147) and cyclic S(122–137) inhibited the reaction, respectively, $\sim 10^7$ and $\sim 10^4$ less efficiently than intact HBsAg (Bhatnagar et al., 1982; Ionescu-Matiu et al., 1983). Using homologous antipeptide antibodies instead of anti-HBs in such reactions, the quantity of S(139–147), S(49–68) and S(69–79) in comparison with the amount of HBsAg required for the same degree of inhibition was $\sim 10^3$, $\sim 10^0$ and $\sim 10^0$, respectively (Bhatnagar et al., 1982; Neurath et al., 1984c). All these comparisons are based on molar concentrations and on the assumption that each HBsAg particle contains $\sim 10^2$ copies of S protein. The possibility that the reaction between HBsAg and antibodies to peptides encompassing the glycosylation site (residue S146) may be prevented by carbohydrate chains attached to S protein has to be considered in light of recent observations with influenza virus hemagglutinins (Skehel et al., 1984).

1.7.4. Selection of peptides for hepatitis B vaccines

Antibodies elicited by small peptides obviously have a narrow specificity. Therefore, the selection of an appropriate region of a viral protein is crucial for preparation of synthetic analogues capable of inducing protection against virus infection. The protective antibodies are likely to correspond to antibodies involved in in vitro neutralization of virus infectivity. However, as discussed in Sections 1.5 and 1.6, an immune response to antigenic determinants on internal HBV proteins, which can play no role in in vitro virus neutralization, also contributes to defense against HBV infection. As discussed in Chapter 5, the mechanisms whereby virus neutralization is accomplished are multiple. Therefore, there is good reason for exploring a variety of synthetic peptide analogues of distinct HBV DNA gene products as candidates for synthetic vaccines. Such peptides must be (*a*) highly immunogenic; (*b*) induce antibodies having a high association constant with the pertinent HBV protein; (*c*) elicit an increase in antibody levels upon booster immunizations; (*d*) induce relevant cell-mediated immune responses to HBV and (*e*) the antipeptide antibodies must interfere with an early event essential for HBV replication.

Requirement *e* is difficult to demonstrate experimentally because tissue culture systems in which HBV replicates are not available. Therefore, it may be necessary instead to: (*a*) carry out direct efficacy trials in chimpanzees, (*b*) set up model experiments with nonhuman hepadnaviruses and synthetic peptide analogues of their protein constituents; and (*c*) explore those facets of HBV-liver cell interactions which are amenable to experimentation – for example the adsorption of HBV to these cells – and their inhibition by antibodies. Peptides selected for use as vaccines must elicit subtype-independent protection, or alternately, multiple peptides corresponding to each subtype must be identified and used.

1.7.5. Definition of new diagnostic tests

Reagents derived from synthetic peptide analogues of the S protein and of HBcAg (if they fulfill requirements *a–c* above), especially if they prove suitable for HBV

TABLE 19.3

Synthetic peptide analogues of HBV proteins and of potential translational products of HBV DNA

Designation of peptide[a]	Subtype of HBV selected for synthesis[b]	Determined serologic specificity	Reference[d]
S(1–17)	ayw	common[c]	1,2
S(1–20[21])	ayw		3
S(11–17)	ayw[e]		1,2
S(22–36[35])	ayw[e]	common[c]	1,2,5
S(31–36)	ayw[e]		1,2
S(38–52)	ayw		1,2
S(21–47)	ayw		3
S(47–52)	ayw		1,2
S(45–68)	adw		4
S(48–81)	ayw	common[c]	1,2,5
S(48–81)	ayw		3
S(65–81)	ayw		1,2
S(69–79)	adw[e]		4
S(95–110)	ayw	common[c]	1,2
S(104–110)	ayw		1,2
S(110–137)	ayw	predominantly *y*	1,2,5
S(110–137)	adw	predominantly *d*	1,5
S(125–137)	ayw	predominantly *y*	1,2,5
S(125–137)	adw	predominantly *d*	1,5
S(107–137)	adw	predominantly *d*	1
Cyclic S(122[117]–137)	ayw	*a* and *y*	6–9
Linear S(122[117]–137)	ayw	*y*	6–9
S(110–151)	ayw		3
S(135–155)	ayw	*a*	10,11
S(138–149) (141–146)	adw (not adr)	*a*, *d* ?	12,13
S(139–147) (139–158)	adw (not adr)	*a*	14,15
Cyclic S(139–147)	adw (not adr)	*a*	18
S(156–185)	ayw		3
S(212[221]–226)	ayw		1,2
pre-S(120–145)	adw	contains common determinants	16
pre-S(12–32)	adw		19
X(138[139]–154)	?	?	17

[a] See footnote on page 351.
[b] See amino acid sequence data in Figs. 19.3, 19.5 and 19.6.
[c] Shared by HBV subtypes.
[d] 1, Alexander and Lerner (1984). 2, Lerner et al. (1981). 3, Shih et al. (1984). 4, Neurath et al. (1984c). 5, Gerin et al. (1983). 6, Dreesman et al. (1982). 7, Dreesman (1984). 8, Sanchez et al. (1982). 9, Ionescu-Matiu et al. (1983). 10, Neurath et al. (1982a). 11, Neurath et al. (1984e). 12, Hopp (1981). 13, Prince et al. (1982). 14, Bhatnagar et al. (1982). 15, Vyas et al. (1983). 16, Neurath et al. (1984a). 17, Moriarty et al. (1984). 18, Howard et al. (1984). 19, Neurath et al. (1984d).
[e] Sequences for all subtypes of S protein are common (Fig. 19.5).

subtyping, may contribute to the battery of tests already established for diagnosis of hepatitis B (see Table 19.1). However, the most important contribution of peptides produced by chemical synthesis or by cloning and expression of 'epitope genes' for diagnostic purposes may emerge from studies on the pre-S, X and P proteins. The detection of antibodies reacting with HBV, which appear before any antibodies to the S protein can be demonstrated, and which seem to be essential for the outcome of the disease (see Section 1.6) is a matter of first priority. Evidence indicating that these antibodies recognize determinants on the pre-S protein is accumulating (Neurath et al., 1984a; 1984d; Kent et al., 1984).

1.7.6. Delineation and specificity of antibody and T cell binding sites on HBV proteins

The peptide analogues of the S protein synthesized so far are summarized in Table 19.3 together with their properties. Only those peptides which elicited an anti-HBs response or reacted with antibodies to the S protein are listed. These peptides collectively span the whole amino acid sequence of the S protein except the hydrophobic segments 82–94 and 186–211 (Fig. 19.7). The data indicate the localization of the *d/y* subtype-specific determinants within residues S(125–137), suggesting that a replacement of phenylalanine by tyrosine is sufficient for a *d*→*y* subtype alteration (Fig. 19.5). However, modifications of amino acids outside residues 125–137 may affect the interaction of the S protein with subtype-specific antibodies, since methylation of Lys 122 leads to a loss of reactivity with anti-*d* (see Section 1.6). It seems intriguing that the hexadecamer S(122–137) reacts not only with subtype-specific antibodies, but also with anti-*a* antibodies after cyclization by means of a covalent link between residues 124 and 137. This cyclic peptide inhibited the reaction between idiotypic determinants on anti-HBs (anti-*a*) and anti-idiotypic antibodies, thus indicating an association of the particular idiotypic determinants with the anti-*a* antibody combining site and further supporting the *a* specificity of this peptide (Kennedy et al., 1983b). An additional *a*-specific epitope appears to be located on peptides encompassing the sequence S(139–147). This agrees with the reported essential role of Lys 141 in HBsAg and S(135–155) antigenicity and their reactivity with monoclonal anti-*a* antibodies (Rao and Vyas, 1974; Neurath et al., 1981; Neurath et al., 1984e). Interestingly, cyclization of S(139–147) by a disulfide bond between residues 139–147 enhances the binding of anti-*a* antibodies (Howard et al., 1984). Additional sequences, which may contribute to common antigenic determinants of the S protein, are evident from Table 19.3 and Fig. 19.5. Possibly because of differences in the conformation of the peptides and the corresponding sequences within the S protein, the serological specificities of some peptides may not be definable according to conventional identification schemes adopted for HBsAg (Shih, 1984).

In humans, HBV elicits not only antibodies to the S protein, but also antibodies which specifically react with the 34 and 37-kDa components of HBsAg (see Section

1.2) (Neurath et al., 1984a; 1984d). The latter antibodies recognize determinants within the C-terminal portion of the pre-S protein (pre-S(120–174)). A synthetic peptide, pre-S(120–145) (Figs. 19.6 and 19.8) proved to be highly immunogenic and elicited in rabbits high levels of apparently group-specific antibodies to HBsAg and to the HBV envelope. These antibodies were capable of neutralizing HBV. Furthermore, rabbit antibodies recognizing HBV particles, but not the S protein, reacted with pre-S(120–145) even after a 10^6-fold dilution. This suggests that pre-S(120–145) contains an immunodominant neutralizing HBV determinant.

Fusion proteins containing the N-terminal portion of the pre-S protein, produced using bacterial expression vectors, were recognized by antibodies present in sera of humans infected with HBV. Antibodies elicited by the fusion proteins specifically reacted with an ~ 45-kDa protein presumably corresponding to the translational product of the tandem pre-S and S genes of HBV DNA (Sninsky et al., 1984; Neurath et al., 1984d).

Antibodies against the synthetic peptide X(138[139]–154), corresponding to the hydrophobic C-terminal portion of X protein (Figs. 19.3 and 19.10), reacted with an antigen produced by a human hepatoma cell line containing integrated HBV DNA (Moriarty et al., 1984).

It has been suggested that HBsAg epitopes recognized by specific T and B cells are distinct. Synthetic peptide analogues offer the opportunity to locate these distinct epitopes within the sequence of the S protein. A series of such peptides were screened in HBsAg-specific T cell proliferative assays using lymph node cells of high responder mice immunized with HBsAg. Peptides S(38–52), S(47–52) and S(95–109) stimulated the proliferation of T cells which had been primed in vivo with HBsAg of either subtype *ad* or *ay*. On the other hand, S(107–137) and S(110–137) only stimulated T cells primed with the homologous HBsAg subtype (See Table 19.3). These peptides, known from earlier studies to elicit a good anti-HBs response (Gerin et al., 1983), did not require a carrier to induce humoral immunity. Thus, multiple T cell epitopes appear to be exposed on the surface of HBsAg. The dominant T cell epitope recognized by a particular mouse strain depends on the I region genotype of the strain. The common dominant T cell-specific epitope of all HBsAg subtypes appears to encompass residues 38–52 (Milich and Chisari, 1984), displaying considerable variability in amino acid sequence within distinct subtypes (Fig. 19.5).

Several of the synthetic peptides listed in Table 19.3 (S(110–137); cyclic S(122–137) and pre-S(120–145)), all of which induce antibodies to HBsAg without a carrier, have been tested with some success in chimpanzees for immunogenicity and/or protection against HBV infection. It is hoped that with the right carrier-adjuvant combination, synthetic hepatitis B vaccines will become feasible.

Recent results indicate that human as well as animal antibodies to an *a* epitope of HBsAg share idiotypic determinants (Kennedy et al., 1982; 1983b; 1983d; Kennedy and Dreesman, 1983). The immune response of mice to HBsAg can be enhanced by prior injection of anti-idiotypic antibodies (Kennedy et al., 1983a; Kennedy and Dreesman, 1984). When administered with an adjuvant, anti-idiotypic antibody alone induced anti-HBs (Kennedy et al., 1984). These results suggest that human anti-idiotypic antibodies may have potential for use in active immunization or in modulating immune responses to HBV. In addition, anti-idiotypic antibodies and synthetic peptide analogues inhibiting the idiotype-anti-idiotype reaction may contribute to the identification and characterization of cell receptors for HBV and of the complementary domains on the HBV envelope, following similar accomplishments with other virus-cell systems (Kauffman et al., 1983; Noseworthy et al., 1983).

2. The delta agent

Humans or chimpanzees already infected with HBV are susceptible to infection with another agent, named delta (δ) (Rizzetto, 1983; Verme et al., 1983). This superinfection exacerbates the liver disease. Simultaneous infections with both HBV and δ may also occur. Simultaneous infections with both HBV and δ may also occur. WHV-infected woodchucks are also susceptible to δ infection. Individuals or experimental animals either not infected with or immune to HBV (WHV) are resistant to infection by the δ-agent. This agent is unique since it has neither the properties of a virus nor of a viroid. It is defective and absolutely dependent on HBV (WHV) replication. It seems to correspond to particles ~ 36 nm in diameter having an outer lipoprotein coat serologically identical to the envelope of HBV, surrounding a nucleoprotein consisting of 'δ-antigen' and a single-stranded RNA 1.75 kilobases long (Rizzetto et al., 1980; Bonino et al., 1984). These 36-nm particles are present in the serum of infected individuals or animals. The structural organization of the nucleoprotein has not been described, but defined structures resembling HBcAg were not detected within the ~ 36-nm particles. Originally discerned in nuclei of infected cells by immunofluorescence, δ-antigen in its monomeric form has a molecular mass of 68 kDa. The δ-RNA has been reverse transcribed, cloned and sequenced (Denninston et al., 1984). Translational reading frames have been identified in δ-RNA which could potentially code for δ-antigen. δ-RNA appears not to have any sequence homology with HBV DNA. The molecular basis of HBV (WHV) helper function in replication of the δ-agent has not yet been reported. Although chronic double infections with HBV and the δ-agent may occur, δ infections in HBV carriers appear transient and result in subsequent immunity to δ infection, despite the apparent absence of antibodies to HBV envelope components and thus to the surface of the δ-agent. However, IgM and later IgG antibodies to δ-antigen appear in the course of δ infection. In the case of chronic δ infections, IgM antibodies persist (Smedile et al., 1982). The role of anti-δ in immunity to the agent has not yet been explained. Detailed information concerning the structure and immunochemistry of δ-antigen is not available at this time.

3. Hepatitis A virus (HAV)

HAV particles are icosahedral and consist of 32 capsomers. The HAV genome is a single-stranded polyadenylated RNA, about 7.5 kilobases long, functioning as a mRNA. The protein moiety of mature virions contains 4 polypeptides with molecular masses of 30–33 kDa (VP1); 24–27 kDa (VP2); 21–23 kDa (VP3) and 7–14 kDa (VP4). Small amounts of a larger polypeptide (~ 40 kDa; VP0) are also present. Unlike with poliovirus, VP4 appears to be present at less than equimolar quantities in relation to VP1, VP2 and VP3 (see Chapter 17) (Siegl et al., 1981; Tratschin et al., 1981; Coulepis et al., 1982; Ticehurst et al., 1983). As with other picornaviruses, VP1 appears to represent the most exposed protein component within assembled virions, whereas VP3 and especially VP2 are more hidden. None of the separated polypeptides react with antibodies to intact virions (anti-HAV; Gerlich and Frösner, 1983). Studies in which polyclonal anti-HAV was used suggested a close antigenic relatedness between distinct HAV isolates, indicating that a vaccine prepared with any HAV strain should provide universal protection against hepatitis A (Lemon and Binn, 1983a). Attenuated live vaccines are being prepared for that purpose (Provost et al., 1982). On the other hand, genetic variation between HAV isolates has been directly demonstrated by oligonucleotide mapping (Weitz and Siegl, 1984) and has also been suggested from restriction endonuclease cleavage maps of cloned HAV cDNA fragments (Ticehurst et al., 1983). Studies on the time course of anti-HAV development following infection suggested a close relationship between virus-neutralizing antibodies and anti-HAV detectable by radioimmunoassay or ELISA tests (Lemon and Binn, 1983b; Flehmig and Zahn, 1984). Tests for IgM anti-HAV proved to be the method of choice for diagnosis of HAV infections (Deinhardt, 1982).

The application of monoclonal anti-HAV (MacGregor et al., 1983; Dawson et al., 1984) for immunochemical studies in combination with cloning and sequence data on the HAV genome (Von der Helm et al., 1981; Ticehurst et al., 1983) is expected to lead to rapid definition of epitopes on HAV particles and their constituent polypeptides. Results obtained with poliovirus (see Chapter 17) may serve as a model for such investigations.

4. Acknowledgements

We are grateful to all colleagues who graciously responded to our appeal to provide us with their published and unpublished results: Drs. A. Alberti, C.J. Burrell, R.P. Beasley, A.G. Coulepis, A.M. Courouc鬬, F. Deinhardt, J.L. Dienstag, B. Flehmig, W. Gerlich, N.M. Gough, F.B. Hollinger, C.R. Howard, B.B. Knowles, W.T. London, M. Rizzetto, J.W.-K. Shih, D. Shouval, J.J. Sninsky, P. Tiollais, G.N. Vyas, A.J. Zuckerman and their respective collaborators. We thank Dr. D.G. George for computer searches and M. Halpern for editorial assistance.

5. References

Alberti, A., Diana, S., Sculard, G.H., Eddleston, A.L.W.F. and Williams, R. (1978) *Brit. Med. J.* 1056–1058.

Alberti, A., Trevisan, A., Fattovich, G. and Realdi, G. (1984) In *Advances in Hepatitis Research* (F.V. Chisari, Ed.) Masson, New York. pp. 134–143.

Alexander, H. and Lerner, R.A. (1984) In *Advances in Hepatitis Research* (F.V. Chisari, Ed.) Masson, New York. pp. 230–237.

Anfinsen, C.B. (1973) *Science* 181, 223–230.

Baron, M.H. and Baltimore, D. (1982) *J. Virol.* 43, 969–978.

Bennink, J.R., Yewdell, J.W. and Gerhard, W. (1982) *Nature* 296, 75–76.

Bhatnagar, P.K., Papas, E., Blum, H.E., Milich, D.R., Nitecky, D., Karels, M.J. and Vyas, G.N. (1982) *Proc. Natl. Acad. Sci. USA* 79, 4400–4404.

Bittle, J.L., Hougten, R.A., Alexander, H., Shinnick, T.M., Sutcliffe, J.G. and Lerner, R.A. (1982) *Nature* 298, 30–33.

Blumberg, B.S., Sutnick, A.I. and London, W.T. (1970) *Am. J. Med.* 48, 1–8.

Bonino, F., Hoyer, B., Shih, J.W.-K., Rizzetto, M., Purcell, R.H. and Gerin, J.L. (1984) *Infect. Immun.* 43, 1000–1005.

Boniolo, A., Dovis, M. and Matteja, R. (1982) *J. Immunol. Meth.* 49, 1–15.

Braciale, T.J. (1977) *J. Exp. Med.* 146, 673–689.

Budkowska, A., Kalinowska, B. and Nowoslawski, A. (1979) *J. Immunol.* 123, 1415–1416.

Burrell, C.J., MacKay, P., Greenaway, P.J., Hofschneider, P.H. and Murray, K. (1979) *Nature* 279, 43–47.

Cabral, G.A., Chairez, R., Marciano-Cabral, F., Suarez, M., Dreesman, G.R., Melnick, J.L. and Hollinger, F.B. (1975) *Infect. Immun.* 12, 564–570.

Cabral, G.A., Marciano-Cabral, F., Funk, G.A., Sanchez, Y., Hollinger, F.B., Melnick, J.L. and Dreesman, G.R. (1978) *J. Gen. Virol.* 38, 339–350.

Celis, E., Kung, P.C. and Chang, T.W. (1984) *J. Immunol.* 132, 1511–1516.

Charnay, P., Gervais, M., Louise, A., Galibert, F. and Tiollais, P. (1980) *Nature* 286, 893–895.

Charnay, P., Mandart, E., Hampe, A., Fitoussi, F., Tiollais, P. and Galibert, F. (1979) *Nucl. Acids Res.* 7, 335–346.

Chisari, F.V. (Ed.) (1984) *Advances in Hepatitis Research*, Masson, New York.

Chisari, F.V. and Nowoslawski, A. (1982) In *Viral Hepatitis* (W. Szmuness, H.J. Alter and J.E. Maynard, Eds.) Franklin Institute Press, Philadelphia. pp. 779–794.

Chisari, F.V., Routenberg, J.A., Anderson, D.S. and Edgington, T.S. (1978). In *Viral Hepatitis* (G.N. Vyas, S.N. Cohen and R. Schmid, Eds.) Franklin Institute Press, Philadelphia. pp. 245–266.

Cohen, B.J. and Richmond, J.E. (1982) *Nature* 296, 677–678.

Cossart, Y.E. and Keirnan, E. (1984) In *Viral Hepatitis and Liver Disease* (G.N. Vyas, J.L. Dienstag and J.H. Hoofnagle, Eds.) Grune and Stratton, Orlando FL. p. 647.

Cote, P.J. Jr., Dapolito, G.M., Shih, J.W.-K. and Gerin, J.L. (1982) *J. Virol.* 42, 135–142.

Coulepis, A.G., Locarnini, S.A., Westaway, E.G., Tannock, G.A. and Gust, I.D. (1982) *Intervirology* 18, 107–127.

Couroucé, A.M., Holland, P.V., Müller, J.Y. and Soulier, J.P. (1976) *HBs Antigen Subtypes* S. Karger, Basel.

Couroucé, A.M., Lee, H., Drouet, J., Canavaggio, M. and Soulier, J.P. (1984) In press.

Couroucé-Pauty, A.-M., Plançon, A. and Soulier, J.P. (1983) *Vox Sang.* 44, 197–211.

Crowley, C.W., Liu, C.-C. and Levinson, A.D. (1983) *Mol. Cell. Biol.* 3, 44–55.

Dawson, G.J., Mushahawar, I.K. and Decker, R.H. (1984) In *Viral Hepatitis and Liver Disease* (G.N. Vyas, J.L. Dienstag and J.H. Hoofnagle, Eds.) Grune and Stratton, Orlando FL. p. 662.

Deinhardt, F. (1982) In *Progress in Liver Diseases* (H. Popper and F. Schaffner Eds.) Vol. VII. Grune and Stratton, New York. pp. 451–467.

Deinhardt, F. and Deinhardt J. (Eds.) (1983) *Viral Hepatitis: Laboratory and Clinical Science.* Marcel Dekker, New York.

Deinhardt, F. and Gust, I.D. (1982) *Bull. WHO* 60, 661–691.

Dienstag, J.L. (1984) In *Advances in Hepatitis Research* (F.V. Chisari, Ed.) Masson, New York. pp. 163–167.

Dejean, A., Brechot, C., Tiollais, P., Wain-Hobson, S. (1983) *Proc. Natl. Acad. Sci. USA* 80, 2505–2509.

Delius, H., Gough, N.M., Cameron, C.H. and Murray, K. (1983) *J. Virol.* 47, 337–343.

Denninston, K.J., Wells, F., Engle, R., Hoyer, B. and Gerin, J.L. (1984) In *Viral Hepatitis and Liver Disease* (G.N. Vyas, J.L. Dienstag and J.H. Hoofnagle, Eds.) Grune and Stratton, Orlando FL. p. 696.

Desmyter, J. (1984) In *Viral Hepatitis and Liver Disease* (G.N. Vyas, J.L. Dienstag and J.H. Hoofnagle, Eds.) Grune and Stratton, Orlando FL. p. 709.

Dreesman, G.R. (1984) In *Advances in Hepatitis Research* (F.V. Chisari, Ed.) Masson, New York. pp. 216–222.

Dreesman, G.R., Chairez, R., Suarez, M., Hollinger, F.B., Courtney, R.J. and Melnick, J.L. (1975) *J. Virol.* 16, 508–515.

Dreesman, G.R., Sanchez, Y., Ionescu-Matiu, I., Sparrow, J.T., Six, H.R., Peterson, D.L., Hollinger, F.B. and Melnick, J.L. (1982) *Nature* 295, 158–160.

Dudley, F.J., Fox, R.A. and Sherlock, S. (1972) *Lancet* i, 191–194.

Edman, J.C., Hallewell, R.A., Valenzuela, P., Goodman, H.M. and Rutter, W.J. (1981) *Nature* 291, 503–506.

Feitelson, M.A., Marion, P.L. and Robinson, W.S. (1981) *J. Virol.* 39, 447–454.

Feitelson, M.A., Marion, P.L. and Robinson, W.S. (1982a) *J. Virol.* 43, 741–748.

Feitelson, M.A., Marion, P.L. and Robinson, W.S. (1982b) *J. Virol.* 43, 687–696.

Feitelson, M.A., Marion, P.L. and Robinson, W.S. (1983) *Virology* 130, 76–90.

Feitelson, M., Millman, I., Simmons, H. and Halbherr, T. (1984) In *Viral Hepatitis and Liver Disease* (G.N. Vyas, J.L. Dienstag and J.H. Hoofnagle, Eds.) Grune and Stratton, Orlando FL. p. 648.

Figus, A., Fung, Y.-K.T., Blum, H.E., Vyas, G.N. and Varmus, H.E. (1984) In *Viral Hepatitis and Liver Disease* (G.N. Vyas, J.L. Dienstag and J.H. Hoofnagle, Eds.) Grune and Stratton, Orlando FL. p. 632.

Flehmig, B. and Zahn, J. (1984) *J. Inf. Dis.* In press.

Fowler, M.J.F., Monjardino, J., Tsiquaye, K.N., Zuckerman, A.J. and Thomas, H.C. (1984) *J. Med. Virol.* 13, 83–92.

Freytag von Loringhoven, A., Koshy, R., Koch, S. and Hofschneider, P.H. (1984) In *Viral Hepatitis and Liver Disease* (G.N. Vyas, J.L. Dienstag and J.H. Hoofnagle, Eds.) Grune and Stratton, Orlando FL. p. 628.

Fujisawa, Y., Ito, Y., Sasada, R., Ono, Y., Igarashi, K., Marumoto, R., Kikuchi, M. and Sugino, Y. (1983) *Nucl. Acids. Res.* 11, 3581–3591.

Fujiyama, A., Miyanohara, A., Nozaki, C., Yoneyama, T., Ohtomo, N. and Matsubara, K. (1983) *Nucl. Acids. Res.* 11, 4601–4610.

Fung, Y.K.T., Lai, C.L., Todd, D., Ganem, D. and Varmus, H.E. (1984) In *Viral Hepatitis and Liver Disease* (G.N. Vyas, J.L. Dienstag and J.H. Hoofnagle, Eds.) Grune and Stratton, Orlando FL. p. 633.

Galibert, F., Mandart, E., Fitoussi, F., Tiollais, P. and Charnay, P. (1979) *Nature* 281, 646–650.

Galibert, F., Chen, T.N. and Mandart, E. (1981) *Proc. Natl. Acad. Sci. USA.* 78, 5315–5319.

Galibert, F., Chen, T.N. and Mandart, E. (1982) *J. Virol.* 41, 51–65.

Ganem, D., Greenbaum, L. and Varmus, H.E. (1982) *J. Virol.* 44, 374–383.

Gavilanes, F., Gonzales-Ros, J.M. and Peterson, D.L. (1982) *J. Biol. Chem.* 257, 7770-7777.

Gerin, J.L., Faust, R.M. and Holland, P.V. (1975) *J. Immunol.* 115, 100–105.

Gerin, J.L., Shih, J.W.-K., McAuliffe, V.J. and Purcell, R.H. (1978) *J. Gen. Virol.* 38, 561–566.

Gerin, J.L., Alexander, H., Shih, J.W.-K., Purcell, R.H., Dapolito, G., Engle, R., Green, N., Sutcliffe, J.G., Shinnick, T.M. and Lerner, R.A. (1983) *Proc. Natl. Acad. Sci. USA*. 80, 2365–2369.

Gerlich, W.H. and Robinson, W.S. (1980) Cell 21, 801–809.

Gerlich, W.H., Feitelson, M.A., Marion, P.L. and Robinson, W.S. (1980) *J. Virol.* 36, 787–795.

Gerlich, W.H. and Frösner, G.G. (1983) *Med. Microbiol. Immunol.* 172, 101–106.

Gerlich, W.H., Goldmann, U., Heermann, K.H., Schwartz, W. and Baumgarten, H. (1984) In *Viral Hepatitis and Liver Disease* (G.N. Vyas, J.L. Dienstag and J.H. Hoofnagle, Eds.) Grune and Stratton, Orlando FL. p. 711.

Gold, J.W.M., Alter, H.J., Holland, P.V., Gerin, J.L. and Purcell, R.H. (1974) *J. Immunol.* 112, 1100–1106.

Gold, J.W.M., Shih, J.W.-K., Purcell, R.H. and Gerin, J.L. (1976) *J. Immunol.* 117, 1404–1406.

Goodall, A.H., Meek, F.L., Waters, J.A., Miescher, G.C., Janossy, G. and Thomas, H.C. (1982) *J. Immunol. Meth.* 52, 167–174.

Griffin, B.E. (1980) In *DNA Tumor Viruses* (J. Tooze, Ed.) Cold Spring Harbor Laboratory, New York. pp. 61–123.

Gust, I.D., Coulepis, A.G., Feinstone, S.M., Locarnini, S.A., Moritsugu, Y., Najera, R. and Siegl, G. (1983) *Intervirology* 20, 1–7.

Hardy, K., Stahl, S. and Küpper, H. (1981) *Nature* 293, 481–483.

Hellstrom, U., Sylvan, S. and Lundbergh, P. (1984) In *Viral Hepatitis and Liver Disease* (G.N. Vyas, J.L. Dienstag and J.H. Hoofnagle, Eds.) Grune and Stratton, Orlando FL. p. 687.

Herzenberg, L.A., Tokuhisa, T. and Hayakawa, K. (1983) *Ann. Rev. Immunol.* 1, 609–632.

Hess, G., Shih, W.-K., Kaplan, P.M. and Gerin, J.L. (1977) *J. Immunol.* 119, 1542–1544.

Hitzeman, R.A., Chen, C.Y., Hagie, F.E., Patzer, E.J., Liu, C.-C., Estell, D.A., Miller, J.V., Yaffe, A., Kleid, D.G., Levinson, A.D. and Oppermann, D. (1983) *Nucl. Acids. Res.* 11, 2745–2762.

Hopp, T.P. (1981) *Mol. Immunol.* 18, 869–872.

Hopp, T.P. (1984) *Mol. Immunol.* 21, 13–16.

Hopp, T.P. and Woods, K.R., (1981) *Proc. Natl. Acad. Sci. USA* 78, 3824–3828.

Howard, C.R. and Simpson, D.I.H. (1983) In *Viral Hepatitis: Laboratory and Clinical Science* (F. Deinhardt and J. Deinhardt, Eds.) Marcel Dekker, New York. pp. 139–155.

Howard, C.R., Brown, S., Hogben, D.N., Zuckerman, A.J., Murray-Lyon, I.M. and Steward, M.W. (1984) In *Viral Hepatitis and Liver Disease* (G.N. Vyas, J.L. Dienstag and J.H. Hoofnagle, Eds.) Grune and Stratton, Orlando FL. pp. 561–572.

Huang, S.-N. and Neurath, A.R. (1979) *Lab. Invest.* 40, 1–17.

Imai, M., Gotoh, A., Nishioka, K., Kurashina, S., Miyakawa, Y. and Mayumi, M. (1974) *J. Immunol.* 112, 416–419.

Imai, M., Nomura, M., Gotanda, T., Sano, T., Tachibana, K., Miyamoto, H., Takahashi, K., Toyama, S., Miyakawa, Y. and Mayumi, M. (1982) *J. Immunol.* 128, 69–72.

Ionescu-Matiu, I., Kennedy, R.C., Sparrow, J.T., Culwell, A.R., Sanchez, Y., Melnick, J.L. and Dreesman, G.R. (1983) *J. Immunol.* 130, 1947–1951.

Iwarson, S., Tabor, E., Thomas, H., Snoy, P. and Gerety, R.J. (1984) In *Viral Hepatitis and Liver Disease* (G.N. Vyas, J.L. Dienstag and J.H. Hoofnagle, Eds.) Grune and Stratton, Orlando FL. p. 679.

Jilg, W., Delhoune, C., Deinhardt, F., Roumeliotou-Karayannis, A.J., Papaevangelou, G.J., Mushahwar, I.K. and Overby, L.R. (1984) *J. Med. Virol* 13, 171–178.

Juillerat, M.A., Barkas, T. and Tzarto, S.J. (1984) *FEBS Lett.* 168, 143–148.

Kabsch, W. and Sander, C. (1984) *Proc. Natl. Acad. Sci. USA* 81, 1075–1078.

Kalil, J., Crevat, D., Fellous, M., Drouet, J., Couroucé, A.M. and Ropars, C. (1981) *Ann. Immunol. Inst. Pasteur* 132C, 319–326.

Kees, U. and Krammer, P.H. (1984) *J. Exp. Med.* 159, 365–377.

Kauffman, R.S., Noseworthy, J.H., Nepom, J.T., Finberg, R., Fields, B.N. and Greene, M.I. (1983) *J. Immunol.* 131, 2539–2541.

Kazim, A.L. and Atassi, M.Z. (1980) *Biochem. J.* 191, 261–264.

Kennedy, R.C. and Dreesman, G.R. (1983) *J. Immunol.* 130, 385–389.

Kennedy, R.C. and Dreesman, G.R. (1984) *J. Exp. Med.* 159, 655–665.

Kennedy, R.C., Sanchez, Y., Ionescu-Matiu, I., Melnick, J.L. and Dreesman, G.R. (1982) *Virology* 122, 219–221.

Kennedy, R.C., Adler-Storthz, K., Henkel, R.D., Sanchez, Y., Melnick, J.L. and Dreesman, G.R. (1983a) *Science* 221, 853–855.

Kennedy, R.C., Dreesman, G.R., Sparrow, J.T., Culwell, A.R., Sanchez, Y., Ionescu-Matiu, I., Hollinger, F.B. and Melnick, J.L. (1983b) *J. Virol.* 46, 653–655.

Kennedy, R.C., Ionescu-Matiu, I., Adler-Storthz, K., Henkel, R.D., Sanchez, Y. and Dreesman, G.R. (1983c) *Intervirology* 19, 176–180.

Kennedy, R.C., Ionescu-Matiu, I., Sanchez, Y. and Dreesman, G.R. (1983d) *Eur. J. Immunol* 13, 232–235.

Kennedy, R.C., Melnick, J.L. and Dreesman, G.R. (1984) *Science* 223, 930–931.

Kent, S.B.H., Neurath, A.R. and Strick, N. (1984) In *Cold Spring Harbor Symp. on Modern Approaches to Vaccines* (R. Chanock and R. Lerner, Eds.) Cold Spring Harbor Laboratory, Cold Spring Harbor. In press.

Koch, S., Koshy, R., Freytag von Loringhoven, A. and Hofschneider P.H. (1984) In *Viral Hepatitis and Liver Disease* (G.N. Vyas, J.L. Dienstag and J.H. Hoofnagle, Eds.) Grune and Stratton, Orlando FL. p. 628.

Koike, K., Kobayashi, M., Mizusawa, H., Yoshida, E., Yaginuma, K. and Taira, M. (1983) *Nucl. Acids Res.* 11, 5391–5402.

Koike, K., Mizusawa, H., Kobayashi, M., Taira, M., Yoshida, E. and Yaginuma, K. (1984) In *Viral Hepatitis and Liver Disease* (G.N. Vyas, J.L. Dienstag and J.H. Hoofnagle, Eds.) Grune and Stratton, Orlando FL. p. 627.

Koistinen, V.U. (1980) *J. Virol.* 35, 20–23.

Koshy, R., Burgelt, E. and Hofschneider, P.H. (1984) In *Viral Hepatitis and Liver Disease*(G.N. Vyas, J.L. Dienstag and J.H. Hoofnagle, Eds.) Grune and Stratton, Orlando FL. p. 628.

Koziol, D.E., Alter, H.J., Kirchner, J.P. and Holland, P.V. (1976) *J. Immunol.* 117, 2260–2262.

Kremastinou, T., Howard, C.R., Skelly, J. and Zuckerman, A.J. (1981) *Intervirology* 16, 176–183.

Kyte, J. and Doolittle, R.F. (1982) *J. Mol. Biol.* 157, 105–132.

Lamb, O., Rall, L., Truett, M., Shaul, Y., Standring, D.N., Valenzuela, P. and Rutter, W.J. (1983) *J. Virol.* 48, 271–280.

Lathe, R.F., Lecocq, J.P. and Everett, R. (1983) In *Genetic Engineering* (R. Williamson, Ed.) Vol. 4. Academic Press, New York. pp. 31–49.

Le Bouvier, G.L., Capper, R.A., Williams, A.E., Pelletier, M. and Katz, A.J. (1976) *J. Immunol.* 117, 2262–2264.

Lemon, S.M. and Binn, L.N. (1983a) *Inf. Immun.* 42, 418–420.

Lemon, S.M. and Binn, L.N. (1983b) *J. Infect. Dis.* 148, 1033–1039.

Lerner, R.A. (1982) *Nature* 299, 592–596.

Lerner, R.A., Green, N., Alexander, H., Liu, F.-T., Sutcliffe, J.G. and Shinnick, T.M. (1981) *Proc. Natl. Acad. Sci. USA* 78, 3403–3407.

Liu, C.-C., Yansura, D. and Levinson, A.D. (1982) *DNA* 1, 213–221.

Machida, A., Kishimoto, S., Ohnuma, H., Miyamoto, H., Baba, K., Oda, K., Nakamura, T., Funatsu, G., Miyakawa, Y. and Mayumi, M. (1982) *Mol. Immunol.* 19, 1087–1093.

Machida, A., Kishimoto, S., Ohnuma, H., Miyamoto, H., Baba, K., Oda, T., Nakamura, T., Miyakawa, Y., Mayumi, M. (1983) *Gastroenterology* 85, 268–274.

MacGregor, A.M., Kornitschuk, M., Hurrell, J.G.R., Lehmann, N.I., Coulepis, A.G., Locarnini, S.A. and Gust, I.D. (1983) *J. Clin. Microbiol.* 18, 1237–1243.

MacKay, P., Lees, J. and Murray, K. (1981) *J. Med. Virol.* 8, 237–243.

Magnius, L.O. (1975) *Clin. Exp. Immunol.* 20, 209–216.

Mandart, E., Kay, A. and Galibert, F. (1984) *J. Virol.* 49, 782–792.

Marion, P.L., Knight, S.S., Feitelson, M.A., Oshiro, L.S. and Robinson, W.S. (1983) *J. Virol.* 48, 534–541.

Marion, P.L., Oshiro, L.S., Regnery, D.C., Scullard, G.H. and Robinson, W.S. (1980) *Proc. Natl. Acad. Sci. USA* 77, 2941–2945.

Marion, P.L., Salazar, F.H., Alexander, J.J. and Robinson, W.S. (1979) *J. Virol.* 32, 796–802.

Mason, W.S., Aldrich, C., Summers, J. and Taylor, J.M. (1982) *Proc. Natl. Acad. Sci. USA* 79, 3997–4001.

Mason, W.S., Seal, G. and Summers, J. (1980) *J. Virol.* 36, 829–836.

McAleer, W.J., Buynak, E.B., Maigetter, R.Z., Wampler, D.E., Miller, W.J. and Hilleman, M.R. (1984) *Nature* 307, 178–180.

McAuliffe, V.J., Purcell, R.H., Gerin, J.L. and Tyeryar, F.J. (1982) In *Viral Hepatitis* (W. Szmuness, H.J. Alter and J.E. Maynard, Eds.) Franklin Institute Press, Philadelphia. pp. 425–435.

Melnick, J.L. (1982) Intervirology 18, 105–106.

Michel, M.L., Sobczak, E., Malpiece, Y., Tiollais, P. and Streek, R. (1984) In *Viral Hepatitis and Liver Disease* (G.N. Vyas, J.L. Dienstag and J.H. Hoofnagle, Eds.) Grune and Stratton, Orlando FL. p. 659.

Milich, D.R. and Chisari, F.V. (1982) *J. Immunol.* 129, 320–325.

Milich, D.R. and Chisari, F.V. (1984) In *Advances in Hepatitis Research* (F.V. Chisari, Ed.) Masson, New York. pp. 91–109.

Milich, D.R., Alexander, H. and Chisari, F.V. (1983a) *J. Immunol.* 130, 1401–1407.

Milich, D.R., Leroux-Roels, G.G. and Chisari, F.V. (1983b) *J. Immunol.* 130, 1395–1400.

Milich, D.R., Leroux-Roels, G.G., Louie, R.E. and Chisari, F.V. (1984) *J. Exp. Med.* 159, 41–56.

Mishiro, S., Imai, M., Takahashi, K., Machida, A., Gotanda, T., Miyakawa, Y. and Mayumi, M. (1980) *J. Immunol.* 124, 1589–1593.

Miyanohara, A., Toh-e, A., Nozaki, C., Hamada, F., Ohtomo, N. and Matsubara, K. (1983) *Proc. Natl. Acad. Sci USA* 80, 1–5.

Mondelli, M., Naumov, N. and Eddleston, A.L.W.F. (1984) In *Advances in Hepatitis Research* (F.V. Chisari, Ed.) Masson, New York. pp. 144–151.

Mondelli, M., Vergani, G.M., Alberti, A., Vergani, D., Portmann, B., Eddleston, L.W.F. and Williams, R. (1982) *J. Immunol.* 129, 2773–2778.

Moodie, J., Stannard, L.M. and Kipps, A. (1974) *J. Gen. Virol.* 24, 375–379.

Moraczewska, Z., Madalinski, K. and Holland, P.V. (1978) *Intervirology* 9, 189–192.

Moriarty, A.M., Hoyer, B.H., Shih, J.W.-K., Gerin, J.L. and Hamer, D.H. (1981) *Proc. Natl. Acad. Sci. USA* 78, 2606–2610.

Moriarty, A.M., Alexander, H., Thornton, G.B. and Lerner, R.A. (1984) In *Viral Hepatitis and Liver Disease* (G.N. Vyas, J.L. Dienstag and J.H. Hoofnagle, Eds.) Grune and Stratton, Orlando FL. p. 629.

Murray, K., Bruce, S.A., Hinnen, A., Wingfield, P., van Erd, P.M.C.A., de Reus, A. and Schellekens, H. (1984) *EMBO J.* 3, 645–650.

Neurath, A.R. and Strick, N. (1977) *Proc. Natl. Acad. Sci. USA* 74, 1702–1706.

Neurath, A.R. and Strick, N. (1979) *J. Gen. Virol.* 42, 645–649.

Neurath, A.R., Trepo, C., Chen, M. and Prince, A.M. (1976) *J. Gen. Virol.* 30, 277–285.

Neurath, A.R., Strick, N. and Huang, C.Y. (1978) *Intervirology* 10, 265–275.

Neurath, A.R., Strick, N., Szmuness, W., Stevens, C.E. and Harley, E.J. (1979) *J. Gen. Virol.* 42, 493–504.

Neurath, A.R., Strick, N. and Huang, C.-Y. (1980) *J. Gen. Virol.* 48, 53–61.

Neurath, A.R., Strick, N. and Oleszko, W.R. (1981) *J. Virol. Meth.* 3, 115–125.

Neurath, A.R., Kent, S.B.H. and Strick, N. (1982a) *Proc. Natl. Acad. Sci. USA* 79, 7871–7875.

Neurath, A.R., Strick, N., Baker, L. and Krugman, S. (1982b) *Proc. Natl. Acad. Sci. USA* 79, 4415–4419.

Neurath, A.R., Stark, D., Strick, N. and Sproul, P. (1983) J. Med. Virol. 12, 227–236.

Neurath, A.R., Kent, S.B.H. and Strick, N. (1984a) *Science* 224, 392–395.

Neurath, A.R., Kent, S.B.H. and Strick, N. (1984b) *J. Gen. Virol.* 65, 1009–1014.

Neurath, A.R., Kent, S.B.H. and Strick, N. (1984c) *Virus Res.* 1, 321–331.

Neurath, A.R., Strick, N. and Kent, S.B.H. (1984d) in *Cold Spring Harbor Symp. on Modern Approaches to Vaccines* (R. Chanock and R. Lerner, Eds.) Cold Spring Harbor Laboratory, Cold Spring Harbor. In press.

Neurath, A.R., Strick, N. and Kent, S.B.H. (1984e) *J. Virol Methods.* In press.

Nigg, E.A., Walter, G. and Singer, S.J. (1982) *Proc. Natl. Acad. Sci. USA* 79, 5939–5943.

Niman, H.L., Houghten, R.A., Walker, L.E., Reisfeld, R.A., Wilson, I.A., Hogle, J.M. and Lerner, R.A. (1983) *Proc. Natl. Acad. Sci. USA* 80, 4949–4953.

Noseworthy, J.N., Fields, B.N., Dichter, M.A., Sobotka, C., Pizer, E., Perry, L.L., Nepom, J.T. and Greene, M.I. (1983) *J. Immunol.* 131, 2533–2538.

Ogston, C.W., Jonak, G.J., Rogler, C.E., Astrin, S.M. and Summers, J. (1982) *Cell* 29, 385–394.

Ohori, H., Yamaki, M., Onodera, S., Yamada, E. and Ishida, N. (1980) *Intervirology* 13, 74–82.

Ohori, H., Shimizu, N., Yamada, E., Onodera, S. and Ishida, N. (1984) *J. Gen. Virol.* 65, 405–414.

Ono, Y., Onda, H., Sasada, R., Igarashi, K., Sugino, Y. and Nishioka, K. (1983) *Nucl. Acids, Res.* 11, 1747–1757.

Overby, L.R., Ling, C.-M., Decker, R.H., Mushahwar, I.K. and Chau, K. (1981) In *Viral Hepatitis* (W. Szmuness, H.J. Alter and J.E. Maynard, Eds.) The Franklin Institute Press, Philadelphia. pp. 169–182.

Pasek, M., Goto, T., Gilbert, W. Zink, B., Schaller, H., MacKay, P., Leadbetter, G. and Murray, K. (1979) *Nature* 282, 575–579.

Paul, D.A. and Peterson, D.L. (1984) In *Viral Hepatitis and Liver Disease* (G.N. Vyas, J.L. Dienstag and J.H. Hoofnagle, Eds.) Grune and Stratton, Orlando FL. p. 690.

Persing, D.H., Varmus, H.E. and Ganem, D. (1984) In *Viral Hepatitis and Liver Disease* (G.N. Vyas, J.L. Dienstag and J.H. Hoofnagle, Eds.) Grune and Stratton, Orlando FL. p. 687.

Peterfy, F., Kuusela, P. and Mäkelä, O. (1983) *J. Immunol.* 130, 1809–1813.

Peterson, D.L. (1981) *J. Biol. Chem.* 256, 6975–6983.

Peterson, D.L., Gavilanes, F., Paul, D.A. and Achord, D.T. (1984a) In *Advances in Hepatitis Research* (F.V. Chisari, Ed.) Masson, New York. pp. 30–39.

Peterson, D.L., Paul, D.A., Lam, J., Tribby, I.I.E. and Achord, D.T. (1984b) *J. Immunol.* 132, 920–927.

Peterson, D.L., Purcell, R.H. and Paul, D.A. (1984c) In *Viral Hepatitis and Liver Disease* (G.N. Vyas, J.L. Dienstag and J.H. Hoofnagle, Eds.) Grune and Stratton, Orlando FL. p. 689.

Pfaff, E., Mussgay, M., Böhm, H.O., Schulz, G.E. and Schaller, H. (1982) *EMBO J.* 1, 869–874.

Pillot, J., Geneste, C., Riottot, M.M. and Petit, M.A. (1984) In *Viral Hepatitis and Liver Disease* (G.N. Vyas, J.L. Dienstag and J.H. Hoofnagle, Eds.) Grune and Stratton, Orlando FL. p. 691.

Popper, H. and Schaffner, F. (Eds.) (1982) *Progress in Liver Diseases.* Vol. 7. Grune and Stratton, New York.

Pourcel, C., Sobzack, E., Dubois, M.-F., Gervais, M., Drouet, J. and Tiollais, P. (1982) *Virology* 121, 175–183.

Prince, A.M., Ikram, H. and Hopp, T.P. (1982) *Proc. Natl. Acad. Sci. USA* 79, 579–582.

Provost, P.J., Banker, F.S., Giesa, P.A., McAleer, W.J., Buynak, E.B. and Hilleman, M.R. (1982) *Proc. Soc. Exp. Biol. Med.* 170, 8–14.

Purcell, R.H. and Gerin, J.L. (1978) In *Viral Hepatitis* (G.N. Vyas, S.N. Cohen and R. Schmid, Eds.) Franklin Institute Press, Philadelphia. pp. 491–505.

Rao, K.R. and Vyas, G.N. (1984) *J. Gen. Virol.* 54, 571–573.

Rizzetto, M. (1983) *Hepatology* 3, 729–737.

Rizzetto, M., Hoyer, B., Canese, M.G., Shih, J.W.-K., Purcell, R.H. and Gerin, J.L. (1980) *Proc. Natl. Acad. Sci. USA* 77, 6124–6128.

Rizzetto, M., Verme, G. and Bonino, F. (Eds.) (1984) *Viral Hepatitis and Delta Infection*. Alan R. Liss, New York.

Roberts, I.M., Bernard, C.C., Vyas, G.N. and Mackay, I.R. (1975) *Nature* 254, 606–607.

Robinson, W.S. (1980) Anu. N.Y. Acad. Sci. 354–371.

Robinson, W.S. (1983) In *Viral Hepatitis: Laboratory and Clinical Science* (F. Deinhardt and J. Deinhardt Eds.) Marcel Dekker, New York. pp. 57–116.

Rogler, C.E., Shafritz, D.A. and Summers, J. (1984) In *Viral Hepatitis and Liver Disease* (G.N. Vyas, J.L. Dienstag and J.H. Hoofnagle, Eds.) Grune and Stratton, Orlando FL. p. 630.

Rubinstein, P., Walker, M., Krassner, J., Carrier, C., Carpenter, C., Stevens, C., Szmuness, W. and Baker, L. (1982) *Abstracts of the Eighth Annual Meeting of the American Association for Clinical Histocompatibility Testing*, San Francisco, California. p. A13.

Rutter, W.J., Ziemer, M., Ou, J., Shaul, Y., Laub, D., Garcia, P. and Standring, D.N. (1984) In *Viral Hepatitis and Liver Disease* (G.N. Vyas, J.L. Dienstag and J.H. Hoofnagle, Eds.) Grune and Stratton, Orlando FL. pp. 67–86.

Sabatini, D.D., Kreibich, G., Morimoto, T. and Adesnik, N. (1983) In *Plasma Protein Secretion by the Liver* (H. Glaumann, T. Peters, Jr. and C. Redman, Eds.) Academic Press, New York. pp. 237–284.

Sanchez, Y., Ionescu-Matiu, I., Dreesman, G.R., Hollinger, F.B. and Melnick, J.L. (1981) *Virology* 114, 71–80.

Sanchez, Y., Ionescu-Matiu, I., Sparrow, J.T., Melnick, J.L. and Dreesman, G.R. (1982) *Intervirology* 18, 209–213.

Sanger, F., Air, G.M., Barrell, B.G., Brown, N.L., Coulson, A.R., Fiddes, J.C., Hutchison, C.A. III, Slocombe, P.M. and Smith, M. (1977) *Nature* 265, 687–695.

Sasaki, T., Ohkubo, Y., Yamashita, Y., Imai, M., Miyakawa, Y. and Mayumi, M. (1976) *J. Immunol.* 117, 2258–2259.

Schaeffer, E. and Sninsky, J.J. (1984) *Proc. Natl. Acad. Sci. USA* 81, 2902–2906.

Schiffer, M. and Edmundson, A.B. (1967) *Biophys. J.* 7, 121–135.

Shafritz, D.A., Lieberman, H.M., Isselbacher, K.J. and Wands, J.R. (1982) *Proc. Natl. Acad. Sci. USA* 79, 5675–5679.

Shih, J.W.-K. (1984) In *Hepatitis B* (R.J. Gerety, Ed.) In press.

Shih, J.W.-K., Cote, P.J. Jr., Dapolito, G.M. and Gerin, J.L. (1980a) *J. Virol. Meth.* 1, 257–273.

Shih, J.W.-K., Hess, G., Kaplan, P.M. and Gerin, J.L. (1980b) *J. Virol. Meth.* 1, 47–59.

Shih, J.W.-K., Yajima, H., Gerety, R.J. and Liu, T.Y. (1984) In *Viral Hepatitis and Liver Disease* (G.N. Vyas, J.L. Dienstag and J.H. Hoofnagle, Eds.) Grune and Stratton, Orlando FL. p. 659.

Shouval, D., Carlson, R., Eilat, D. and Wands, J.R. (1983) *European Association for the Study of the Liver*. Proceedings of the 18th Meeting, Southampton, England.

Siddiqui, A. (1983) *Mol. Cell. Biol.* 3, 143–146.

Siddiqui, A., Sattler, F. and Robinson, W.S. (1979) *Proc. Natl. Acad. Sci. USA* 76, 4664–4668.

Siegl, G., Frösner, G.G., Gauss-Müller, V., Tratschin, J.-D. and Deinhardt, F. (1981) *J. Gen. Virol.* 57, 331–341.

Skehel, J.J., Stevens, D.J., Daniels, R.S., Douglas, A.R., Knossow, M., Wilson, I.A. and Wiley, D.C. (1984) Proc. Natl. Acad. Sci. USA 81, 1779–1783.

Skelly, J., Howard, C.R. and Zuckerman, A.J. (1981) *Nature* 290, 51–54.

Smedile, A., Lavarini, C., Crivelli, O., Raimondo, G., Fassone, M. and Rizzetto, M. (1982) *J. Med. Virol.* 9, 131–138.

Smith, G.L., Mackett, M. and Moss, B. (1983) *Nature* 302, 490–495.

Sninsky, J.J., Zraly, C.B., Meyers, M.L., Loewy, Z.G., Wong, D.T. and Schaeffer, E. (1984) In *Advances in Hepatitis Research* (F.V. Chisari, Ed.) Masson, New York. pp. 21–29.

Stahl, S., MacKay, P., Magazin, M., Bruce, S.A. and Murray, K. (1982) *Proc. Natl. Acad. Sci. USA* 79, 1606–1610.

Stannard, L.M., Hantz, O. and Trepo, C. (1983) *J. Gen. Virol.* 64, 975–980.

Stenlund, A., Lamy, D., Moreno-Lopez, J., Ahola, H., Pettersson, U. and Tiollais, P. (1983) *EMBO J.* 2, 669–673.

Stevens, C.E., Alter, H.J., Taylor, P.E., Zang, E.A., Harley, E.J. and Szmuness, W. (1984) *N. Engl. J. Med* 311, 496–501.

Stibbe, W. and Gerlich, W.H. (1982) *Virology* 123, 436–442.

Stibbe, W. and Gerlich, W.H. (1983) *J. Virol.* 46, 626–628.

Stratowa, C., Doehmer, J., Wang, Y. and Hofschneider, P.H. (1982) *EMBO J.* 1, 1573–1578.

Summers, J. and Mason, W.S. (1982) *Cell* 29, 403–415.

Summers, J., Smolec, J.M. and Snyder, R. (1978) *Proc. Natl. Acad. Sci. USA* 75, 4533–4537.

Sutcliffe, J.G., Shinnick, T.M., Green, N., Liu, F.-T., Niman, H.L. and Lerner, R.A. (1980) *Nature* 287, 801–805.

Swenson, P.D., Escobar, M.R., Carithers, R.L. Jr. and Sobieski, T.J. III. (1983) *J. Clin. Microbiol.* 18, 305–309.

Szmuness, W., Stevens, C.E., Harley, E.J., Zang, E.A., Alter, H.J., Taylor, P.E., DeVera, A., Chen, G.T.S. and Kellner, A. (1982) *New Engl. J. Med.* 307, 1481–1486.

Tabor, E., Gerety, R.J., Smallwood, L.A. and Barker, L.F. (1976) *J. Immunol.* 117, 2038–2040.

Takahashi, K., Akahane, Y., Gotanda, T., Mishiro, T., Imai, M., Miyakawa, Y. and Mayumi, M. (1979) *J. Immunol.* 122, 275–279.

Takahashi, K., Imai, M., Miyakawa, Y., Iwakiri, S. and Mayumi, M. (1978a) *Proc. Natl. Acad. Sci. USA* 75, 1952–1956.

Takahashi, K., Machida, A., Funatsu, G., Nomura, M., Usuda, S., Aoyagi, S., Tachibana, K., Miyamoto, H., Imai, M., Nakamura, T., Miyakawa, Y. and Mayumi, M. (1983) *J. Immunol.* 130, 2903–2907.

Takahashi, K., Yamashita, S., Imai, M., Miyakawa, Y. and Mayumi, M. (1978b) *J. Gen. Virol.* 38, 431–436.

Taylor, P.E., Stevens, C.E. and Alter, H.J. (1984) In *Viral Hepatitis and Liver Disease* (G.N. Vyas, J.L. Dienstag and J.H. Hoofnagle, Eds.) Grune and Stratton, Orlando FL. p. 682.

Tedder, R.S. and Bull, F.G. (1979) *Clin. Exp. Immunol.* 35, 380–389.

Tedder, R.S., Guarascio, P., Yao, J.L., Lord, R.B. and Eddleston, A.L.W.F. (1983) *J. Hyg.* 90, 135–142.

Ticehurst, J.R., Racaniello, V.R., Baroudy, B.M., Baltimore, D., Purcell, R.H. and Feinstone, S.M. (1983) *Proc. Natl. Acad. Sci. USA* 80, 5885–5889.

Tiollais, P. and Wain-Hobson, S. (1984) In *Advances in Hepatitis Research* (F.V. Chisari, Ed.) Masson, New York. pp. 9–20.

Toh, H., Hayashida, H. and Miyata, T. (1983) *Nature* 305, 827–829.

Tratschin, J.D., Siegl. G., Frösner, G.G. and Deinhardt, F. (1981) *J. Virol.* 38, 151–156.

Trevisan, A., Realdi, G., Alberti, A., Ongaro, G., Pornaro, E. and Meliconi, R. (1982) *Gastroenterology* 82, 218–222.

Valenzuela, P., Medina, A., Rutter, W.J., Ammerer, G. and Hall, B.D. (1982) *Nature* 298, 347–350.

Valenzuela, P., Quiroga, M., Zaldivar, J., Gray, P. and Rutter, W.J. (1980) In *Animal Virus Genetics* (B.N. Fields and R. Jaenisch, Eds.) Academic Press, New York. pp. 57–70.

Van Eldik, L.J., Fok, K.-F., Erickson, B.W. and Watterson, D.M. (1983) *Proc. Natl. Acad. Sci. USA* 80, 6775–6779.
Verme, G., Bonino, F. and Rizzetto, M. (Eds.) (1983) *Viral Hepatitis and Delta Infection*. Alan R. Liss, New York.
Vnek, J., Prince, A.M., Trepo, C., Williams, A.E., Mushahwar, I.K., Ling, C.M. and Overby, L.R. (1979) *J. Med. Virol.* 4, 187–199.
Von der Helm, K., Winnacker, E.L., Deinhardt, F., Frösner, G.G., Gauss-Müller, V., Bayerl, B., Scheid, R. and Siegl, G. (1981) *J. Virol. Meth.* 3, 37–43.
Vyas, G.N. (1981) In *Hepatitis B Vaccine* (P. Maupas and P. Guesry, Eds.) Elsevier, Amsterdam. pp. 227–237.
Vyas, G.N. and Blum, H.E. (1983) In *Medical Virology II* (L.M. de la Maza and Peterson, E.M. Eds.) Elsevier Biomedical, Amsterdam. pp. 121–146.
Vyas, G.N., Rao, K.R. and Ibrahim, A.B. (1972) *Science* 178, 1300–1301.
Vyas, G.N., Bhatnagar, P.K., Blum, H.E., Expose, J. and Heldebrandt, C.M. (1983) In *Developments in Biological Standardization* (G. Papaevangelou and W. Hennessen, Eds.) Vol. 54. S. Karger, Basel. pp. 93–102.
Walter, G. and Werchau, H. (1982) *J. Cell Biochem.* 19, 119–125.
Walter, G., Scheidtmann, K.-H., Carbone, A., Laudano, A.P. and Doolittle, R.F. (1980) *Proc. Natl. Acad. Sci. USA* 77, 5197–5200.
Wands, J.R., Ben-Porath, E. and Wong, M.A. (1984) In *Viral Hepatitis and Liver Disease* (G.N. Vyas, J.L. Dienstag and J.H. Hoofnagle, Eds.) Grune and Stratton, Orlando FL. pp. 543–559.
Wands, J.R. and Zurawski, V.R. Jr. (1981) *Gastroenterology* 80, 225–232.
Wands, J.R., Carlson, R.I., Schoemaker, N., Isselbacher, K.J. and Zurawski, V.R. Jr. (1981) *Proc. Natl. Acad. Sci. USA* 78, 1214–1218.
Wands, J.R., Ben-Porath, E.P. and Wong, M.A. (1984) In *Advances in Hepatitis Research* (F.V. Chisari, Ed.) Masson, New York. pp. 110–122.
Wang, Y., Stratowa, C., Schaefer-Ridder, M., Doehmer, J. and Hofschneider, P.H. (1983) *Mol. Cell. Biol.* 3, 1032–1039.
Weitz, M. and Siegl, G. (1984) In *Viral Hepatitis and Liver Disease* (G.N. Vyas, J.L. Dienstag and J.H. Hoofnagle, Eds.) Grune and Stratton, Orlando FL. p. 704.
Wen, Y.-M., Mitamura, K., Merchant, B., Tang, Z.Y. and Purcell, R.H. (1983) *Infect. Immun.* 39, 1361–1367.
Werner, B.G., Smolec, J.M., Snyder, R. and Summers, J. (1979) *J. Virol.* 32, 314–322.
Will, H., Cattaneo, R., Pfaff, E., Kuhn, C., Roggendorf, M. and Schaller, H. (1984) *J. Virol.* 50, 335–342.
Yamaki, M., Ohori, H., Onodera, S., Ishida, N. and Maeda, H. (1982) *Biochim. Biophys. Acta* 706, 165–173.
Young, C.R., Schmitz, H.E. and Atassi, M.Z. (1983) *Mol. Immunol.* 20, 567–570.
Zuckerman, A.J. (1982) In *The Liver Annual* (I.M. Arias, M. Frenkel and J.H.P. Wilson, Eds.) Elsevier/North-Holland, Amsterdam. pp. 122–153.
Zuckerman, A.J. (1983) In *The Liver Annual* (I.M. Arias, M. Frenkel and J.H.P. Wilson, Eds.) Elsevier/North-Holland, Amsterdam. pp. 149–169.

Eds. M.H.V. van Regenmortel and A.R. Neurath
Immunochemistry of viruses. The basis for serodiagnosis and vaccines.

Antigenic structure of rhabdoviruses

20

W.H. WUNNER, B. DIETZSCHOLD and T.J. WIKTOR

The Wistar Institute, 36 Street at Spruce, Philadelphia, PA 19104, U.S.A.

Rhabdoviruses infect a wide variety of vertebrate, invertebrate and plant species. Presently, more than 80 viruses are included in the family Rhabdoviridae, based on morphologic, structural, and morphogenic characteristics (Brown et al., 1979; Murphy and Harrison, 1979). The antigenic relationship among these viruses, however, particularly those infecting vertebrates, has been more tenuous and has not allowed major subgrouping of viruses except those belonging to the Vesiculovirus and Lyssavirus genera; all others remain unclassified or are grouped vaguely into small virus serogroups. Eight serologically related vertebrate rhabdoviruses comprise the Vesiculovirus genus: vesicular stomatitis (VS) Indiana (prototype species), Cocal, Argentina, Brazil (Alagoas), VS New Jersey, Chandipura, Piry, and Isfahan. Six antigenically related viruses are included in the Lyssavirus genus: rabies (prototype species of this genus), Lagos bat, Mokola, Duvenhage, Kotonkan, and Obodhiang. Up to now, investigations of the antigenic structure of rhabdoviruses have centered primarily on the two prototype viruses, VS virus and rabies virus, both of which share a major portion of all structural, chemical and biological studies that have been carried out on rhabdoviruses. The key to the antigenic relationships among these viruses clearly lies in understanding the similarities and differences in basic chemical structure between their corresponding antigens. To this end, molecular cloning of viral gene sequences has been the focus of recent studies to determine the amino acid sequence of virus antigens.

In this chapter, developments in the analysis of antigenic structure of rhabdoviruses are described based on studies of VS and rabies viruses. In particular, the description of antigenic determinants on viral glycoproteins, as revealed by monoclonal antibody analysis, is emphasized. This technique allows characterization of these antigens which is necessary to understand their interactions with the immune system of the infected host, their interactions with cell surface receptors, and their effect on pathogenicity in the host. Monoclonal antibodies also permit differentiation of virus strains based on antigenic composition. In addition, these structural characteristics should provide a better understanding of the evolutionary diversity of rhabdoviruses.

1. Classification of rhabdovirus proteins

Most rhabdoviruses contain five structural proteins: virion transcriptase (L), glycoprotein (G), nucleoprotein (N), nucleocapsid-associated phosphoprotein (NS), and matrix protein (M). The nomenclature was devised primarily for VS virus, but all other rhabdoviruses that have been studied appear to conform to this classification (Wagner et al., 1972). Originally, it was determined that three viral proteins (G, M_1 and M_2) were associated with the viral envelope of rabies virus-like rhabdoviruses. However, upon reexamining rabies virus proteins, the M_1 protein was found to be associated with nucleocapsid proteins and was also identified as a phosphoprotein (Zaidis et al., 1979; Dietzschold et al., 1979; Cox et al., 1981). The M_1 protein of rabies virus is therefore analagous to NS protein of VS virus. The location of these structural proteins in the characteristic bullet-shaped rhabdovirion of higher animals is shown schematically in Fig. 20.1.

The largest protein in the virion, L ($\sim$190 kDa), is most likely the catalytic subunit of the endogenous transcriptase activity which is associated with the nucleocapsid core (Bishop and Roy, 1972; Emerson and Wagner, 1972). It is the product of the virus-specific 30S L-mRNA (Morrison et al., 1974; Freeman et al., 1977;

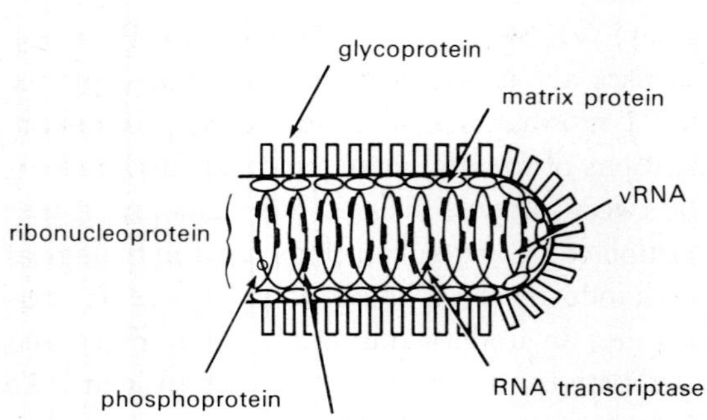

Fig. 20.1. Schematic diagram of a bullet-shaped rhabdovirus illustrating the location of structural elements in this particle. Shown are the glycoprotein peplomers projecting from the membrane envelope and matrix protein lying beneath the membrane. The ribonucleoprotein helix contains a single nonsegmented strand of genomic RNA linked with nucleoprotein and together with the RNA transcriptase and phosphoprotein comprises the nucleocapsid core.

Pennica et al., 1980), and there are approximately 30–60 copies per virion (Bishop and Roy, 1972; Madore and England, 1977).

The next largest protein, G, is the major surface antigen and the only viral protein which is glycosylated. The G protein (~ 69 kDa) can vary in size between species depending on the percent carbohydrate composition as well as on the number of amino acids per molecule. The carbohydrate composition can be affected by the host cell in which the virus is grown (Etchison and Holland, 1974; McSharry et al., 1978) and can account for as much as 9 kDa of the G protein molecular mass (Dietzschold, 1977). The glycoprotein is the basic unit of the surface projections of the virus and accounts for approximately one-third of the total virus protein, with ~ 1800 protein molecules per virion (Bishop and Roy, 1972; Madore and England, 1977). It is the product of the 18S G-mRNA (Morrison et al., 1974; Freeman et al., 1977; Pennica et al., 1980).

The major protein of the nucleocapsid structure, N, is also a major virus component of all rhabdoviruses studied (Cartwright et al., 1970a; Knudsen and MacLeod, 1972; Sokol et al., 1971). Approximately 1800–2300 molecules of N protein (~ 50 kDa) translated from the 16S N-mRNA are noncovalently bound to the virion RNA forming a helical ribonucleoprotein (RNP) complex (Bishop and Roy, 1972; Madore and England, 1977). The RNP helix is tightly bound within the nucleocapsid and surrounded by a lipoprotein envelope through which the surface projections of G protein extend to the exterior of the virion. The N protein of rabies virus is phosphorylated, as are the N proteins of several rabies-related viruses and some fish rhabdoviruses, unlike the N protein of VS virus (Sokol and Koprowski, 1975). The nucleocapsid structure also contains the L and NS proteins which, together with the RNP, constitute an active complex responsible for the transcription and replication of the virion RNA (Bishop and Roy, 1972; Cartwright et al., 1970b; Schneider and Diringer, 1976; Mellon and Emerson, 1978).

The NS protein (~ 40 kDa) is the major phosphoprotein in most rhabdoviruses (Cox et al., 1981; Sokol and Koprowski, 1975). It is a minor component (2–11%) of the virion (Bishop and Roy, 1972; Madore and England, 1977), but represents a significantly greater proportion of the virus-specific proteins synthesized in virus-infected cells; hence the original designation 'NS' for 'nonstructural' (Wagner et al., 1972). The NS protein, which is translated from 12S mRNA, seems to be a multifunctional phosphoprotein that participates in viral RNA transcription and genome replication (Szilagyi and Pringle, 1979; Evans et al., 1979). It is possible that the different functions of NS protein are modulated by phosphorylation and dephosphorylation of different specific sites on the protein (Clinton et al., 1978a; Kingsford and Emerson, 1980).

The second protein associated with the viral envelope is the nonglycosylated membrane protein, M. The M protein is the smallest protein (~ 29 kDa) and is thought to be located on the inner leaflet of the lipid bilayer, where it may interact with the cytoplasmic domain (C-terminal portion) of the membrane-inserted G protein. The association of M and G proteins is necessary for successful viral maturation and budding from the cellular membranes (Printz and Wagner, 1971; Lefay, 1971). The M protein also interacts with the RNP structure (Newcomb and Brown, 1981; Newcomb et al., 1982), but only when M protein is associated with the cellular membrane and not while either nucleocapsids or M protein are free in the cytoplasm of infected cells (Knipe et al., 1977). The M protein is synthesized from 12S mRNA on free polysomes (Morrison and Lodish, 1975). As a result of its interaction with the RNP structure, the M protein may play a role in the modulation of transcription (Clinton et al., 1978b; Martinet et al., 1979; Wilson and Lenard, 1981).

2. Primary structures of vesicular stomatitis virus proteins

Knowledge of the complete primary sequence of the viral proteins is crucial to the basic understanding of molecular structure in general and for the determination of antigenic structure of immunologically important proteins in particular. The amino acid sequences of all five of the VS virus proteins have been determined from cDNA clones that contain the complete coding sequences either from each of the messenger RNAs or from direct cDNA copies of the genome. Nucleotide sequences of the mRNAs encoding the G, N, NS and M proteins were determined from sequences of full-length or nearly full-length cDNA copies that were generated by reverse transcribing in vitro synthesized mRNAs from the San Juan strain of the Indiana serotype of VS virus (Rose and Gallione, 1981; Gallione et al., 1981). A cDNA clone of the complete L protein gene sequence was derived essentially from three large genomic cDNA copies of the Mudd Summers' strain of the Indiana serotype (Schubert et al., 1984). Single extended open reading frames for translation of the corresponding monocistronic mRNAs established the coding sequence for each of the specific proteins. These were located as beginning with the translation initiation ATG codon in the cDNA near the 5′-terminus of the analogous mRNA sequence and ending with a termination codon (one of three specific codons: TAA, TAG or TGA) in the cDNA near the 3′ end of the corresponding mRNA molecules. The amino acid sequences of the proteins are then deduced from these nucleotide sequences and these define the polypeptide structure which can be correlated with the biologic functions of the proteins.

2.1. Virion transcriptase (L)

The L protein is derived from the 5′ half of the rhabdovirus genome and represents more than 60% of the total coding capacity of the virus genome. Schubert and his colleagues (1984) determined the nucleotide sequence of genomic cDNA copies of this massive L gene of VS virus and found that it codes for a basic protein of 2109 amino acids (241.012 kDa). The rest of the genome (the 3′ half) is made up with the remaining four genes arranged in the following order: 3′-N, NS, M, G, L-5′ (Ball and White, 1976; Ball, 1977; Flamand and Delagneau, 1978).

2.2. Glycoprotein (G)

Rose and Gallione (1981) determined the nucleotide sequence of VS virus-specific G-mRNA which predicted a polypeptide sequence of 511 amino acids including the amino-terminal signal sequence of 16 amino acids that is cleaved from the mature protein during post-translational processing. The calculated molecular mass of the mature G protein minus its carbohydrate moiety is 55.057 kDa. The VS virus G protein sequence contains two putative carbohydrate attachment sites, identified by the sequences Asn-Ser-Thr and Asn-Gly-Thr (Neuberger et al., 1972), at asparagine residues 163 and 320 (numbered from the amino-terminal lysine) of the mature protein. A particularly hydrophobic domain of 20 amino acids extending from residues 447 to 466 is presumed to be the transmembrane segment of the mature protein that spans the lipid bilayer of the viral envelope. The 28 amino acids extending from residue 467 to the carboxy-terminal lysine are a mixture of charged and uncharged residues constituting an hydrophilic domain that presumably interacts with the M protein inside the viral envelope. The remaining 446 amino acids of G protein, located on the amino-terminal side of the transmembrane segment, represent 90% of the molecule and constitute the characteristic surface projections of the virion.

2.3. Nucleoprotein (N)

Gallione et al. (1981) deduced the sequence of the N protein of VS virus from the nucleotide sequence determined for the virus-specific N-mRNA. The calculated molecular mass of 47.355 kDa for the predicted 422 amino acids of N protein agrees with the molecular mass of N protein determined from its migration in sodium dodecyl sulfate-polyacrylamide gels (Wagner et al., 1972; Wunner and Pringle, 1972).

2.4. Nucleocapsid-associated phosphoprotein (NS)

The amino acid sequence of one of the two smallest proteins was predicted from the nucleotide sequence of two cDNA clones derived from the NS-mRNA (Gallione et al., 1981). The nucleotide sequence determined for the VS virus NS-mRNA encoded a protein of 222 amino acids; a calculated molecular mass of 25.110 kDa. This molecular mass, however, does not conform with that previously estimated from the migration of NS in sodium dodecyl sulfate-polyacrylamide gel electrophoresis (SDS-PAGE), which puts it in the range of 40–50 kDa (Knipe et al., 1975).

Reasons for this discrepancy are not clear, particularly as phosphorylation of NS protein alone cannot explain its anomalous migration on gels (Gallione et al., 1981). Gallione et al. (1981) suggested that a deficiency of sodium dodecyl sulfate binding to an unusually large domain containing 18 negatively charged amino acids and no positively charged residues may account for the anomalous electrophoretic mobility of NS protein, but there may be alternative explanations which have not been considered. The specific location of phosphorylation sites in NS protein may have to be determined before molecular conformation and relative binding of sodium dodecyl sulfate is understood (Clinton et al., 1978a).

2.5. Matrix protein (M)

The amino acid sequence of the other small size protein was deduced from the nucleotide sequence of a cDNA clone derived from the M-mRNA (Rose and Gallione, 1981). It is comprised of 229 amino acids (26.064 kDa). This agrees with the molecular mass of VS virus M protein estimated from its migration in sodium dodecylsulfate-polyacrylamide gels (Wagner et al., 1972; Wunner and Pringle, 1972). The M protein sequence does not reveal any nonpolar regions that might promote an association of M protein with the lipid bilayer of the plasma membrane (Rose and Gallione, 1981). The protein is, however, rich in basic amino acids, especially the predicted amino-terminal sequence of M protein, which contains 8 lysine residues within the first 19 amino acids. The suggestion that this basic domain plays a role in the regulation of viral RNA transcription, perhaps by interacting with genomic RNA (Rose and Gallione, 1981), clearly deserves further investigation.

3. Primary structure of rabies virus glycoprotein (G)

The first of the rabies virus genes to be cloned and sequenced was the G gene. This protein, which forms the 10 nm long peplomers on the external surface of the virus membrane, is responsible for the induction and binding of virus-neutralizing (VN) antibodies and for conferring immunity against lethal infection with rabies virus (Crick and Brown, 1969; 1970; Wiktor et al., 1973; Cox et al., 1977). To define the antigenic structure of rabies virus G protein, the nucleotide sequence of the virus-specific G-mRNA of rabies virus (ERA strain) was determined from a cloned cDNA copy. From the nucleotide sequence (Fig. 20.2), a polypeptide 524 amino acids long was deduced, which included a signal sequence of 19 amino acids. Residue 20 (Lys) and the next 5 amino acids in the predicted sequence, as well as the carboxy-terminal sequence Thr-Arg-Leu-COOH, which were identified by direct amino acid sequence analysis of the purified rabies virus (ERA strain) glycoprotein (Lai and Dietzschold, 1981), confirmed the correct reading frame of the cDNA sequence. It is of interest that the sequence of the first 19 amino acids predicted from the rabies virus G protein cDNA coding sequence and the signal peptide sequence of 16 amino acids preceding the amino terminus of the mature VS virus G protein are both highly hydrophobic sequences, but not highly homologous (Rose et al., 1982). Thus, the two amino-terminal leader sequences appear to function similarly in the vectorial translocation of the nascent proteins across the membrane of the endoplasmic reticulum (Lingappa et al., 1978) predominantly because of the hydrophobic character maintained in these sequences. In contrast, proteolytic cleavage of the leader peptide within the lumen of the membrane vesicles may involve recognition of the highly conserved sequence within the first four residues of the amino terminus of the mature G proteins.

The deduced sequence of the mature rabies virus G protein has an uninterrupted hydrophobic domain of 22 amino acids (shown in Fig. 20.2), bounded by lysine 439 (numbered from the amino-terminal lysine) and two arginine residues at positions 462 and 463 near the carboxy-terminus, that match the presumptive transmembrane domain in VS virus G protein (Rose et al., 1982). Extending from the putative transmembrane domain to the carboxy-terminal leucine 505 of the rabies virus G protein sequence is a relatively long stretch of 44 charged and uncharged residues compared with 29 residues in the cytoplasmic domain of VS virus G protein. This apparent hydrophilic cytoplasmic domain of the rabies glycoprotein may provide a site for stronger interaction with the viral M protein and nucleocapsid proteins. The deduced amino acid sequence also contains carbohydrate acceptor sites as defined by the sequences Asn-X-Ser and Asn-X-Thr; three of these are located on the amino-terminal side of the proposed transmembrane segment at asparagines 37, 247 and

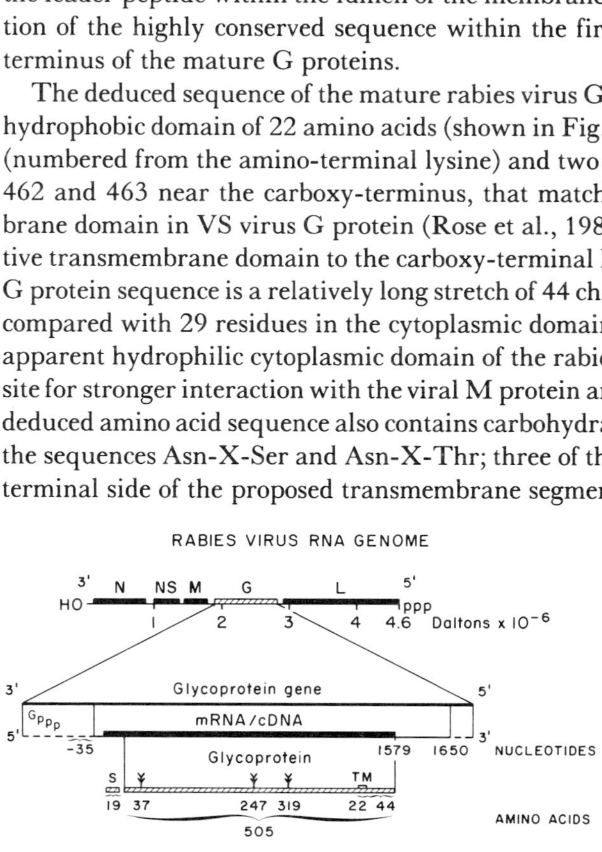

Fig. 20.2. The glycoprotein gene of rabies virus ERA strain is fourth in the order of the five structural genes, designated N, NS, M, G and L, of the RNA genome and codes for the surface glycoprotein of the virus. The glycoprotein gene sequence was copied from mRNA into cDNA. The sequence of a polypeptide 524 amino acids long was deduced from the cDNA nucleotide sequence. The cloned cDNA sequence of 1650 nucleotides is missing 35 nucleotides from the 5′ terminus of the glycoprotein mRNA. The signal peptide (S) contains 19 residues and is cleaved from the mature glycoprotein which contains 505 residues. The putative transmembrane (TM) domain comprises 22 residues and 44 residues complete the polypeptide to the carboxy terminus. Three potential carbohydrate acceptor sites (») are located at residues 37, 247 and 319.

319 (Fig. 20.2). Two residues in the amino acid sequence predicted from the nucleotide sequence of the ERA strain G-cDNA clone described by Anilionis et al. (1981) happen to be copied from a mutated message which resulted in base changes in the cloned cDNA sequence compared with the genomic nucleotide sequence coding for the mature glycoprotein (Wunner and Smith, unpublished results). The cDNA clone registers a leucine at position 8 instead of proline, and leucine is substituted for valine at position 399. Proline 8 was also detected by direct amino acid sequence analysis of cyanogen bromide peptides of the virion G protein (Dietzschold et al., 1982).

The extent of overall sequence homology between the predicted amino acid sequence of the rabies and VS virus G proteins was not as great as that seen for other family-related structural G proteins (Rose et al., 1982). Using a computer program which essentially considers all possible alignments of the two sequences, the number of identities between the rabies virus ERA strain and VS virus Indiana serotype G protein sequences in the optimium alignment revealed a 20% identity. Certain regions of the proteins show greater homology than others, such as a sequence of 24 amino acids preceding and including the putative second and third glycosylation sites of the VS and rabies virus G proteins, respectively. In this region the identity is greater than 50% and includes exact aligment of the putative glycosylation sites (Rose et al., 1982). It is expected that the strong conservation of the sequence in this portion of the molecule reflects on the correlation of structure to function in these proteins, although the nature of that function remains undetermined.

Not surprisingly, the homology of the predicted amino acid sequences for G proteins between two strains of rabies virus, challenge virus standard (CVS) and ERA, is about 90% at the amino acid level (Yelverton et al., 1983). The extent of nucleotide sequence homology corresponding to these two proteins would be expected to be less because of degeneracy in the genetic code. The homology between these two predicted amino acid sequences conforms with the closely related antigenic characteristics of the two virus strains. With the exception of one of the three putative carbohydrate acceptance sites, the structural features of the two rabies virus G proteins including the 14-amino acid signal peptide and 22-amino acid hydrophobic transmembrane segment are in perfect register (Yelverton et al., 1983) between amino-terminal lysine 1 and carboxy-terminal leucine 505. It should be noted that 9 of the predicted 22 amino acids in the transmembrane domain differ between the G protein of CVS and ERA rabies virus strains, but none of the changes alter the characteristic hydrophobicity of this region. The location of the second carbohydrate acceptor sequence Asn-X-Thr in the predicted rabies virus G protein sequence of CVS strain is changed from position 247, as in G protein of ERA strain virus (Anilionis et al., 1981) to asparagine 204 (Yelverton et al., 1983). This occurred due to a double mutation, AGT to AAC, in the nucleotide sequence of the G protein gene of CVS strain virus in the codon for residue 204 (Ser to Asn) and a single mutation, AAT to GAT, in the codon for residue 247 (Asn to Asp) (Wunner and Smith, unpublished results). The function of this putative carbohydrate acceptance site for carbohydrate attachment in the CVS strain G protein is discussed further below with regard to possible antigenic variation within the G protein molecule.

4. Identification of immunogenic and antigenic domains on the rabies virus G protein

Rabies virus, because of its enzootic nature and life-threatening potential, is the only rhabdovirus which has been intensively targeted for vaccine development. The virus, which continues to be endemic in most areas of the world, causes an acute central nervous system disease which is normally fatal in man and domestic and wild animals. The G protein of rabies virus is the hemagglutinin and also the major immunogenic subunit responsible for the induction of virus neutralizing (VN) and cellular immune responses in immunized animals and man (for review see Wunner et al., 1983). The possibility that only a portion of the mature viral antigen may be required to induce the protective immunity against lethal challenge infection has intensified the need for a greater understanding of the rabies virus glycoprotein structure.

4.1. Structural requirements for immunogenic activity of rabies virus G protein

The immunogenic activity of purified native G protein and of smaller fragments of G protein, both naturally occurring and derived by chemical cleavage of the rabies virus G protein, has been compared in a variety of ways to determine the structural basis of VN antibody production following immunization. A naturally occurring fragment of G protein is shed from rabies virus-infected cells as a soluble antigen (G_s antigen). The G_s antigen has been compared with the virion-associated G protein following purification by immunoadsorption chromatography (Dietzschold et al., 1983a). By SDS-PAGE analysis, G_s antigen exhibited an apparent molecular mass of 61 kDa, compared to 67 kDa for G protein. The two proteins, however, were indistinguishable in their ability to bind monoclonal VN antibodies that recognized five distinct antigenic sites on the ERA strain virus G protein (Dietzschold et al., 1983a). The antigenic similarity of G_s antigen to G protein was not enough to confer comparable protection against lethal challenge infection with rabies virus in immunized mice. In these experiments mice were immunized with different amounts of G or G_s proteins. It was shown that 33% of mice immunized with as little as 0.24 μg of G protein survived the challenge with a lethal dose of rabies virus, whereas none of the mice immunized with the highest concentration (30 μg) of G_s antigen survived the challenge. The VN antibody titer following immunization with G_s antigen was also 15 times lower than the VN antibody titer developed by immunization with G protein, suggesting that the immunogenic potential of G_s antigen is markedly reduced.

Since the difference in molecular mass of G_s and G proteins might account for the disparate immunogenicities, the structural relationship between G_s antigen and G protein was determined (Dietzschold et al., 1983a). It was found that a carboxy-terminal portion of G protein equivalent to 58 amino acid residues was missing in G_s antigen, leaving G_s antigen with the carboxy-terminal sequence Ser-Ala-Gly-Ala-COOH, terminating at position 447. Thus, the determined carboxy-terminal region of the rabies G_s antigen contains the first 8 amino acids of the putative transmembrane segment (Anilionis et al., 1981), an insufficient number of residues to anchor the G_s antigen into the plasma membrane. This finding could explain why the antigenic sites which bound VN antibodies were unaffected, since all sites were located on the amino-terminal side of the transmembrane segment. Moreover, since the 58 amino acids missing in G_s antigen accounted for the estimated 6 kDa difference between G_s antigen and G protein determined by gel electrophoresis, no difference in glycosylation was expected between the two antigens. A similar relationship exists between the G_s and G proteins of VS virus (Little and Huang, 1978; Irving and Ghosh, 1982). This implies that the low immunogenicity of G_s antigen is probably the result of the loss of the hydrophobic transmembrane segment or carboxy-terminal fragment present in the G protein. Thus, because of a difference in presentation to the immune system, G_s is incapable of inducing an adequate humoral or cell-mediated response in immunized mice to confer protection comparable to the effect of G protein.

The rabies virus G protein has been chemically cleaved into smaller peptide fragments of known amino acid sequence in order to determine whether the peptides induce VN antibodies, and to locate the amino acids responsible for immunogenic activity in the intact molecule. Eight peptides which resulted from cyanogen bromide (CNBr) cleavage were designated Cr1, Cr2, Cr2A, Cr3, Cr4, Cr5, Cr6 and Cr7 and were mapped by matching peptide amino acid sequences to positions 1–44, 244–291, 386–452, 292–323, 103–178, 324–376, 188–236 and 57–102, respectively, of the predicted sequence (Dietzschold et al., 1982). Peptides Cr2 and Cr2A were not resolved by SDS-PAGE and migrated as a single band which could only be differentiated after sequencing (Dietzschold et al., 1982). The individual CNBr peptides, purified by SDS-PAGE, were injected with Freund's adjuvant into mice and VN antibody titers induced by the peptides were evaluated. Three CNBr fragments Cr1, Cr3 and Cr4 induced antibodies which exhibited significant but low VN activity, producing titers of 180, 270 and 510, respectively, compared to 9000 for native G protein (Dietzschold et al., 1982). Two suggestions have been presented to explain why higher titers of VN activity were not achieved: firstly, absence of high molecular weight carriers and secondly, loss of conformational structure and reduction of intramolecular disulfide bridging during peptide purification might account for the low VN titers obtained (Wunner et al., 1983). Alternatively, an important site on the molecule might have been cleaved by CNBr. Precisely how the immunodominant site(s) on rabies virus G protein is affected by conformational changes or by breaking of disulfide bonds has not been determined from these experiments. Clearly, knowledge of the three-dimensional structure of rabies virus G protein would be particularly advantageous in this analysis. Currently, efforts to determine the regions that constitute the antigenic determinant(s) on G protein are beginning to reveal probable locations of complementary surfaces in antigen-antibody complexes. These studies and their significance to the analysis of antigenic drift in rabies virus antigens are described below.

Another area in which peptide fragments derived from viral G protein may be gainfully employed is in the analysis of the activation of T lymphocyte-mediated immunity by rabies virus. In attempts to investigate the minimal molecular requirements for eliciting antirabies virus T lymphocyte specificities, the CNBr peptides and chemically synthesized peptides containing partial rabies virus ERA strain G protein sequences were tested in a murine splenic T-cell proliferation assay (Macfarlan et al., 1984). The proliferative response of rabies virus-primed T lymphocytes from A/J mice in vitro was strongest with the Cr1, Cr2 + CR2A and Cr3 peptides added individually to the culture. Peptides Cr4, Cr5, Cr6 and Cr7 did not stimulate proliferation of these cells. The interesting observation in this particular case is that while peptide Cr4 induced a VN antibody response in mice, it failed to stimulate T-cell proliferation in vitro. T cells primed in vivo with virus no longer respond to peptide Cr4 in the secondary proliferation assay, but do respond when this peptide is isolated under nonreducing conditions (Macfarlan et al., 1984). The determinant in peptide Cr4 which induces VN antibody is distinct from the reduction-sensitive T-cell determinant present on the intact glycoprotein and peptide Cr4 isolated under nonreducing conditions. The fact that reduced peptide Cr4 invokes a T cell-dependent VN antibody response following primary immunization may mean that unfolding of the peptide has created a new T-cell determinant which is not present on the native G protein molecule (Macfarlan, personal communication).

When synthetic peptides were used to stimulate rabies virus-primed T lymphocytes in vitro (Macfarlan et al., 1984b), only certain synthetic peptides, derived from the amino terminal end of G protein, stimulated proliferation. Peptide R21, containing residues 18–44 of the predicted amino acid sequence, was as effective as G protein itself, although larger molar concentrations of R21 were required to produce the same result (Macfarlan et al., 1984). Peptide R20, which contained residues 32–44 also stimulated proliferation, but to a lesser extent. Both peptides share amino acid sequences with peptide Cr1, but other peptides from this region, including R1 (residues 1–24), R2 (residues 24–34) and R25 (residues 10–31), did not stimulate proliferation. These data suggest that a major T-cell determinant may be located towards the carboxy-terminal end of Cr1 in the amino terminal region of the

G protein molecule. Using monoclonal antibodies to determine the phenotype of T lymphocytes responding to stimulation, it was shown that peptides R21 and R20, in contrast to rabies virus G protein, effected stimulation of a population of Lyt 2^+ cells (Macfarlan et al., 1984b). Lyt 2 antigen is expressed by both cytotoxic and suppressor T lymphocytes (Cantor and Boyse, 1975). Furthermore, preliminary experiments have demonstrated that supernatants from R21-stimulated cultures contain interleukin 2, demonstrating that at least some of the T lymphocytes have helper function (Macfarlan et al., 1984a). Further experiments will be required to identify the immunogenic determinant for each of the immunoreactive T-cell subpopulations responding in vitro and in vivo to rabies virus antigen.

4.2. Mapping antigenic domains in rabies virus G protein

The antigenic relationships among rhabdoviruses in general and rabies viruses in particular have been determined chiefly on the basis of serological studies using specific hyperimmune sera in complement fixation, immunofluorescent antibody staining and plaque reduction (neutralization) tests. As a result, antigenic classification of rhabdoviruses is defined by major and minor serogroups (Shope, 1975; Bauer and Murphy, 1975; Tesh et al., 1977; Tignor et al., 1977; Tesh et al., 1983). For example, at one time rabies virus was considered to be a single antigenic 'species' until this concept of antigenic unity was challenged and rabies virus was identified as a distinguishable member of a group of serologically related viruses (Shope et al., 1970). The application of monoclonal antibody technology to the study of antigenic relationships among rhabdoviruses confirmed the differentiation between rabies and rabies-related viruses (Wiktor et al., 1980). Indeed, monoclonal antibodies have revealed extensive antigenic variation occurring in both the G and N proteins of a number of laboratory and street virus strains (Flamand et al., 1980a; 1980b). The distinct possibility that vaccine treatment failures might be correlated with antigenic differences between the strain used for vaccine production and the strain of virus present in the animal population (Wiktor and Koprowski, 1980) has impelled an extensive and detailed comparative analysis of the antigenic determinants found in the G protein of rabies virus strains.

The antigenic structure of the rabies virus G protein has been analyzed using VN monoclonal antibodies and laboratory-selected mutant viruses. This technique has allowed the description of at least 12 unique epitopes (sites reacting with monoclonal antibody) on the G protein molecule of the laboratory-adapted CVS strain virus (Lafon et al., 1983) and up to 17 epitopes on the ERA strain virus G protein molecule (Wiktor et al., 1983; Lafon et al., 1983). Mutant viruses with single or possibly double amino acid substitutions in the G protein were obtained by titrating cloned parental CVS-11 or ERA strain virus in the presence of an excessive amount of a VN monoclonal antibody (Wiktor and Koprowski, 1980). Mutations in the G protein result in antigenic change if the altered amino acid sequence affects the epitope either directly or indirectly. The sequence change need not be in the actual site of the antibody combining site. In a three-dimensional structure, conformational change may result from a mutation occuring upstream or downstream from the epitope. The high frequency of variant virus occurence (10^{-4} to 10^{-5}) reported (Wiktor and Koprowski, 1980; Lafon et al., 1983) is compatible with the idea that the selected neutralization-resistant variants of rabies virus represent single point mutations affecting sole epitopes in the antigenically altered G proteins. The fact that double variants affecting two independent epitopes in G protein were not selected (frequency of occurrence should theoretically be equal to the product of the individual mutation rates, i.e. 10^{-8} to 10^{-10}) (Wiktor and Koprowski, 1980) supports the concept of independent mutability of epitopes. It is possible, however, that nonessential mutations which do not affect an antibody combining site may occur elsewhere in the molecule. Evidence for this has been found in one of several virus variants selected with monoclonal antibody 194-2 (Dietzschold et al., 1984).

Since the three-dimensional structure of rabies virus G protein (or any other rhabdovirus G protein) is not yet known, the precise topographic locations of the different epitopes are difficult to identify. We can therefore only define the relationships among epitopes in a functional or structural sense by comparing reactivities of closely related antigens with a panel of monoclonal antibodies. Comparative antigenic analysis of various selected G protein variants of CVS-11 and ERA strains of rabies virus using 22 monoclonal antibodies capable of neutralizing CVS-11 virus and 25 monoclonal antibodies capable of neutralizing ERA strain virus has permitted the delineation of three functionally independent antigenic sites which are common to both strains and two additional independent sites on the ERA strain virus G protein (Fig. 20.3) (Lafon et al., 1983; 1984; Wiktor et al., 1983). One of the three common antigenic sites (site I) is represented by a single epitope which appears to be topographically isolated from the other major sites. None of the antibodies competed with antibody 509-6 for binding to site I, and reciprocally 509-6 did not compete with any other antibody (Lafon et al., 1983). Epitopes delineating antigenic sites II and III, on the other hand, are at least operationally linked within each site (epitope cluster) since an individual mutation in one epitope concomitantly modifies certain other epitopes within the same antigenic site.

To correlate the operational antigenic map with the physical antigenic topography of rabies virus G protein, physical mapping studies have been initiated to determine the structural properties of the epitopes themselves. In certain instances it has been possible to predict which amino acids are capable of acting either as contact residues in epitopes of rabies virus G protein or to influence the structure of a distal epitope by inducing long-range conformational effects. Limited structural information in the form of computer-predicted representation of the folding of the polypeptide chain supports the possible role of conformational change in explaining the immunologic effects of amino acid substitutions at or near epitopes in the exchange of particular amino acid residues.

Functional Antigenic Map of ERA glycoprotein

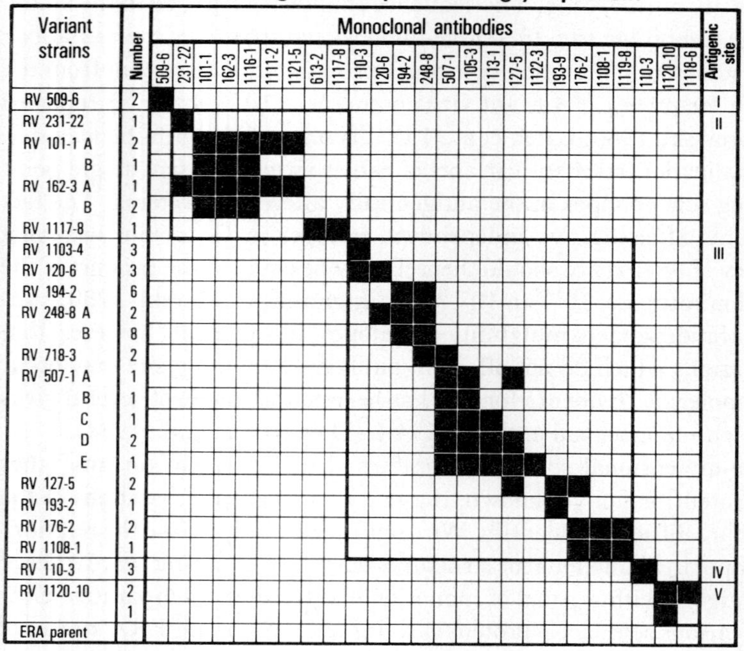

Functional Antigenic Map of CVS-11 glycoprotein

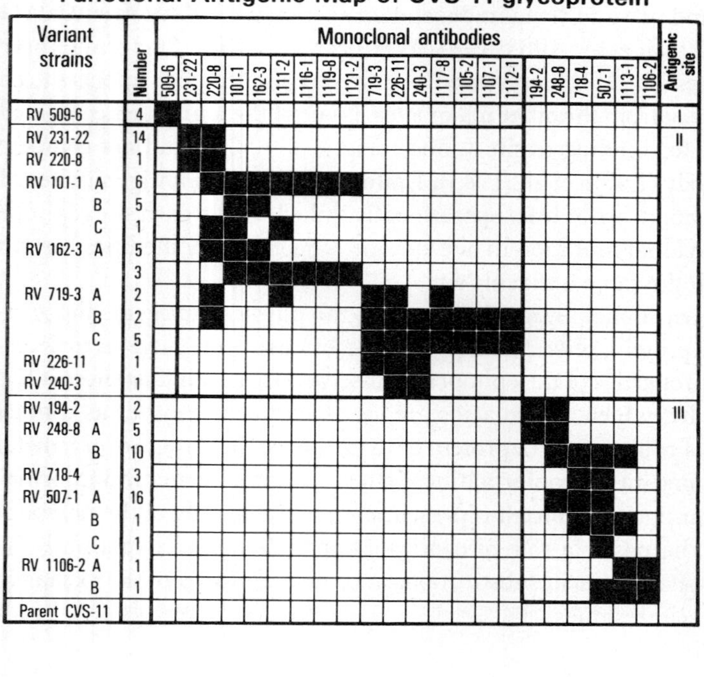

Direct nucleotide sequence analysis of the G protein gene of mutant viruses representing the operationally defined antigenic sites have been carried out by the dideoxy-chain termination sequencing procedure (Sanger et al., 1977), modified for analysis of RNA. This procedure has identified single base changes in the nucleotide sequence which predict amino acid substitutions in the G protein. All point mutations which have been detected in mutant virus glycoprotein genes from the ERA strain occur at positions which result in an amino acid substitution. Thus, a G to T change at position 1062 of the ERA strain of rabies variant (RV) 194-2 virus (the number refers to the monoclonal antibody used for the selection of a RV virus) or a G to A in CVS RV 194-2 virus, a G to T change at position 1062 of RV 248-8, a T to G change at position 1072 of RV 507-1, and a G to A change at position 1133 of RV 120-6 represent unique mutations characteristic of the different antigenic variants (Wunner et al., 1984). The amino acid change corresponding to each of these base mutations and the position occupied in the primary sequence (Fig. 20.4) is Arg-333 to Ile in RV 194-2 and RV 248-8 viruses, Arg-333 to Gln in CVS RV 194-2 virus, Asn-336 to Lys in RV 507-1 virus, and Gly-357 to Arg in RV 120-6 virus G proteins. From the locations of these four unique amino acid changes, it is clear that the viral epitopes which delineate site III in the operational antigenic map occupy positions close to each other in the G protein molecule.

The mapping of G protein epitopes of the ERA strain of rabies virus identified by monoclonal antibodies 194-2 and 248-8 to the same amino acid position, 333, was not totally surprising since both monoclonal antibodies selected antigenic variants which revealed a similar change in viral pathogenicity for adult mice (Coulson et al., 1982; Dietzschold et al., 1983b). Moreover, as expected, it was found that the change in pathogenicity of rabies virus CVS strain variants resistant to neutralization by the same monoclonal antibodies (194-2 and 248-8) also corresponded to an amino acid substitution at position 333 of the G protein molecule (Dietzschold et al., 1983b). The amino acid change in the CVS strain RV 194-2 virus G protein, however, was from arginine to glutamine (due to a G to A base change at position 1062 coding for the glutamine substitution (Wunner et al., 1984)), suggesting that arginine-333 in the parent molecule is essential both for the integrity of an antigenic determinant on G protein and for the ability of rabies viruses to produce a lethal infection in adult mice. Whether or not a similar structural correlate exists has to be determined to explain the same binding relationship of these two antibodies to the nonpathogenic Kelev and Flury HEP strains of rabies virus (Flamand et al., 1980b). Unlike the pathogenic parental ERA and CVS viruses, which cause a fatal disease in adult mice after intracerebral inoculation at doses between 10 and 10^6 plaque-forming units (PFU) per animal, the RV 194-2 viruses did not cause any

Fig. 20.3. Functional epitope maps of ERA and CVS-11 strain rabies virus glycoproteins. Neutralization-resistant variants (RV) were selected using monoclonal antibodies, and then tested for susceptibility (□) or resistance (■) to neutralization by antibodies in the panel.

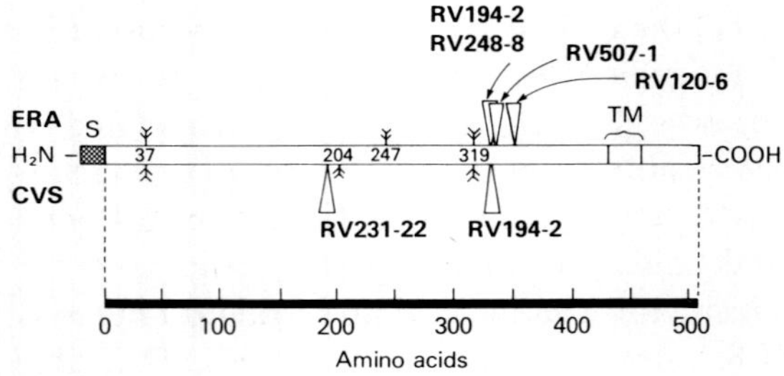

Fig. 20.4. Map location of altered amino acids in the glycoprotein sequence of rabies variant (RV) viruses. The diagram shows the location of mutations (darts) in antigenically altered glycoproteins and carbohydrate attachment sites (») at numbered asparagine residues in parental ERA (above) and CVS-11 (below) strain glycoproteins. The open bar represents the length of the rabies virus glycoprotein. The signal peptide (S) and transmembrane (TM) segments are also shown.

noticeable disease (Dietzschold et al., 1983b) and two out of thirteen CVS strain RV 248-8 viruses lost their virulence for adult mice (Coulson et al., 1982). Whether the other pathogenic CVS strain mutants resistant to antibody 248-8, but still neutralized by antibody 194-2 (Coulson et al., 1982), possess a different mutation close to position 333 remains to be determined. Such a change could reflect the altered reactivity with monoclonal antibodies 718-4, 507-1 and 1113-1 which delineate subregion IIIA of the operational epitope map (Lafon et al., 1983; Dietzschold et al., 1983b) or 613-2 derived from street rabies virus (Coulson et al., 1982).

4.3. Identification of glycosylation sites in G protein that influence conformation

The G proteins of CVS and PM strains of rabies virus characteristically exist in two forms which can be readily resolved by SDS-PAGE and are accordingly designated GI and GII (Dietzschold et al., 1979; Coslett et al., 1980). They give rise to identical leucine-containing tryptic peptides but differ in their carbohydrate content (Dietzschold et al., 1979). An antigenic variant of the CVS strain parental virus has been selected using monoclonal antibody 231-22 which contains G protein with a uniquely altered electrophoretic mobility in SDS-PAGE. The CVS strain variant, RV 231-22, which resisted neutralization by antibody 231-22 contained only one G protein species corresponding to the slow migrating GI form of the parent virus (Dietzschold et al., 1984).

Exhaustive digestion of the individual G protein species from CVS strain parent virus with trypsin and sequencing of tryptic glycopeptides to locate the attached carbohydrates in the deduced amino acid sequence of G protein, revealed that the number of tryptic glycopeptides varied significantly between GI and GII (Dietzschold et al., 1984). Two carbohydrate side chains were mapped by their glycopeptide residues to putative glycosylation sites II and III in the isolated slower migrating GI form, while only one glycopeptide mapping to glycosylation site III was identified from the faster migrating GII species. No glycopeptide containing the predicted amino acid sequence which includes glycosylation site I was obtained from any of the CVS strain viruses analyzed. The electrophoretic correspondence between G protein of RV 231-22 virus and the parent GI species was strengthened by the observation that the pattern of tryptic glycopeptides obtained from the single G protein of RV 231-22 virus also corresponded to that of the parent GI species. An explanation for this observation was then sought by comparing the sequences of the parent and RV 231-22 virus G proteins at the nucleotide level.

Nucleotide sequences of CVS strain genomic RNA determined by the dideoxy-chain termination sequencing method using synthetic complementary oligodeoxyribonucleotide (cDNA) primers, revealed the presence of all three potential glycosylation sites whether or not they appeared to be utilized in the various virion G protein molecules (Dietzschold et al., 1984; Wunner et al., 1985). Regarding glycosylation site I, for which no glycopeptides were obtained, the comparative nucleotide sequence analysis of the CVS parent and variant G protein genes revealed in both cases the predicted glycosylation signal (Asn-Leu-Ser) at Asn-37. In contrast, the signal sequence Asn-Lys-Thr for the predicted glycosylation site III at Asn 319 was utilized in all rabies viruses examined. More importantly, the difference in glycosylation at Asn-204 (site II) between GI (glycosylated) and GII (unglycosylated) had to be explained. The nucleotide sequences which included the predicted glycosylation signal at this site (Asn-Lys-Thr) in the parent and RV 231-22 viral transcripts were unambiguous, suggesting that the distinctive glycosylation patterns of these two virus G proteins were derived from the translation of two completely unique viral gene transcripts rather than from one G protein mRNA which may have been modified by a splicing event. However, comparative nucleotide sequence analysis revealed a mutation in the first position of the codon for residue 198 of RV 231-22 virus G protein (Fig. 20.4). This base change (A to G) predicts the substitution of glutamic acid in the variant for Lys-198 of the CVS parent G protein six residues from the putative glycosylation signal at Asn-204. As a result of this mutation, glycosylation occurs at Asn-204 in the presumptive GII species of RV 231-22 virus causing the observed comigration with GI in SDS-PAGE. It is reasonable to speculate that a change in protein conformation in the region of the point mutation (Lys to Glu-198) is responsible for the recognition of Asn-204 for carbohydrate attachment.

4.4. Variations in N and G proteins detected by monoclonal antibodies

Panels of monoclonal antibodies specific either for the nucleocapsid (NC) proteins or for G protein of rabies and rabies-related viruses have been used to differentiate viruses and detect antigenic variation among virus isolates (Wiktor and Koprowski, 1978; 1980; Wiktor et al., 1980). Of particular interest is the unclarified relationship of the rabies-related viruses Duvenhage, Mokola and Lagos bat to the other members of the Lyssavirus genus. A panel of four anti-NC monoclonal antibodies used in an immunofluorescence staining test permitted the rapid classification of isolates into rabies or rabies-related viruses and identified each member of the latter group (Wiktor et al., 1980). A practical application of this particular panel was recently demonstrated during an epidemic of rabies-like disease in dogs and cats in Zimbabwe. Eleven out of twenty isolates could be identified as Mokola viruses (Wiktor et al., 1984). Until now, rabies-related viruses have been little more than biological curiosities. This is no longer the case, since Mokola virus has been shown to be an agent responsible for an epidemic among domestic animals. Mokola virus infection of humans has been documented and current rabies vaccines do not protect against Mokola virus infection.

Mokola and Lagos bat viruses share a large number of antigenic determinants present on NC antigen of rabies virus, but no common antigenic determinant of the G protein has been detected. On the contrary, while Duvenhage virus shares as many NC determinants as Mokola and Lagos bat viruses, the analysis with the panel of anti-G monoclonal antibodies has demonstrated a strong cross-reactivity with rabies virus. Seven monoclonal antibodies that neutralize most rabies virus strains also neutralize Duvenhage virus (Flamand et al., 1980b; Wiktor, unpublished results). In addition, immunization of animals with a standard rabies virus vaccine confers a high degree of protection against challenge with Duvenhage virus but not against challenge with Mokola virus, suggesting a high correlation between rabies virus and Duvenhage virus. Thus the inclusion of Duvenhage virus in the group of rabies-related viruses may not be justified.

In several recent studies, field rabies virus isolates from animal species in Canadian (Charlton et al., 1982), Western European (Sureau et al., 1983), Pennsylvania (Rupprecht and Wiktor, 1984), Alabama, Kentucky and Missouri (Loretu et al., 1984) wildlife, and from human victims of rabies in the U.S.A. (Wiktor, 1984; Koprowski et al., 1984) have been typed by monoclonal antibodies. These experiments have revealed that not only can rabies virus be distinguished from rabies-related viruses, but also that field isolates of different geographic and species origin have different patterns of reactivity toward the antibodies. A panel of 36 monoclonal antibodies with anti-NC specificity and a panel of 44 monoclonal antibodies with anti-G specificity were used to analyse virus isolates described in the above studies. In one particular study, 42 rabies virus isolates were successfully cultured from 58 animal brains from approximately equal numbers of bats, raccoons and skunks. The virus isolates were analyzed with the panels of monoclonal antibodies and immunoreactivity patterns were compared. Two basic findings emerged from the antigenic analysis. Firstly, certain test combinations of anti-NC and anti-G monoclonal antibodies readily differentiated bat strains from both raccoon and skunk viruses. Raccoon and skunk isolates displayed a much greater similarity to each other than either of them did with respect to the bat isolates. Secondly, bat viruses displayed a heterogeneity, especially in the G immunoreactivity to monoclonal antibodies, not found in either skunks or raccoons. Bat rabies is clearly a major concern in many epizootics of rabies in terrestrial animals due to the greater mobility afforded by flight and its potential to interact during foraging and migratory activities (Baer, 1975). In addition, the antigenic complexity of bat rabies isolates and the considerable difference in antigenic structure of bat rabies virus G and NC proteins from rabies virus derived from terrestrial sources might suggest that the reservoire of rabies virus in bats is a constant threat to the spread of antigenic variant viruses. Nevertheless, the data suggest that the sylvatic pattern of insectivorous bat rabies is largely independent of the terrestrial cycle, and it is felt that the types of variants detected by monoclonal antibodies are fairly stable entities and that the two cycles retain their peculiar integrity (Rupprecht and Wiktor, 1984). Occasionally, however, terrestrial animals and human beings are infected by the bites of rabid bats. Cattle rabies, which is caused mainly by vampire bats feeding on cattle, is rarely transmitted from cattle before they die of the disease, or are killed as a practical means of disease control.

5. Prospects

Our present understanding of the antigenic relationships among rhabdoviruses, based on knowledge of the basic chemical structure and immunoreactivity of viral subunits and subunit fragments, is limited but increasing with the continued progress and application of modern molecular and immunological technologies. Much more has already been learned about antigenic relationships from the use of monoclonal antibodies. We recognize that antigenic variation among various rabies virus strains, for example, could be as great as between various strains of influenza virus (see Chapter 14) and that the dogma, which has been with us for over 100 years, that strains of virus isolated in different parts of the world and from different animal species are so much alike that they can be recognized by antibodies produced in animals exposed to any strain of rabies virus may not be the correct axiom to follow. We also recognize that the immune response to antiviral vaccine in animals comprises a large number of events including induction of VN antibodies and different subpopulations of T lymphocytes. The choice of an effective vaccine will therefore depend on the presence of those immunogenic domains which promote an optimal cross-protection by inducing all of these activities. Antigenic relatedness between virus strains by itself may not be sufficient to guarantee full protection, since antigenic determinants with similar structure may differ markedly in their immunogenic potential. Whether the vaccine of choice is whole virus, a component of the virus, or even synthetic analogues, the efficacy of the antiviral agent will depend upon meeting the above-mentioned requirements.

6. Acknowledgements

The authors' work is supported by Public Health Service Research Grants AI 18883 and AI 09706 from the National Institute of Allergy and Infectious Disease. We wish to thank Dr. Ellis Golub for his critical reading of the manuscript.

7. References

Anilionis, A., Wunner, W.H. and Curtis, P.J. (1981) *Nature* 294, 275–278.

Baer, G.M. (1975) In *The Natural History of Rabies* (G.M. Baer, Ed.) Academic Press, New York. pp. 78–97.

Ball, L.A. and White, C.N. (1976) *Proc. Natl. Acad. Sci. USA* 73, 442–446.

Ball, L.A. (1977) *J. Virol.* 21, 411–414.

Bauer, S.P. and Murphy, F.A. (1975) *Infect. Immun.* 12, 1157–1172.

Bishop, D.H.L. and Roy, P. (1972) *J. Virol.* 10, 234–243.

Brown, F., Bishop, D.H.L., Crick, J., Francki, R.I.B., Holland, J.J., Hull, R., Johnson, K., Martelli, G., Murphy, F.A., Obijeski, J.F., Peters, D., Pringe, C.R., Reichmann, M.E., Schneider, L.G., Shope, R.E., Simpson, D.I.H., Summers, D.F. and Wagner, R.R. (1979) *Intervirology* 12, 1–7.

Cantor, H. and Boyse, E.A. (1975) *J. Exp. Med.* 141, 1376–1389.

Cartwright, B., Smale, C.J. and Brown, F. (1970a) *J. Gen. Virol.* 7, 19–32.

Cartwright, B., Talbot, P. and Brown, F. (1970b) *J. Gen. Virol.* 7, 267–272.

Charlton, K.M., Casey, G.A., Boucher, D.W. and Wiktor, T.J. (1982) *Comp. Immun. Microbiol. Infect. Dis.* 5, 113–115.

Clinton, G.M., Burge, B.W. and Huang, A.S. (1978a) *J. Virol.* 27, 340–346.

Clinton, G.M., Little, S.P., Hagen, F.S. and Huang, A.S. (1978b) *Cell* 15, 1455–1462.

Coslett, G.D., Holloway, B.P. and Obijeski, J.F. (1980) *J. Gen. Virol.* 49, 161–180.

Coulson, P., Rollin, P., Aubert, M. and Flamand, A. (1982) *J. Gen. Virol.* 61, 97–100.

Cox, J.H., Dietzschold, B. and Schneider, L.G. (1977) *Infect. Immun.* 16, 754–759.

Cox, J.H., Weiland, F., Dietzschold, B. and Schneider, L.G. (1981) In *The Replication of Negative Strand Viruses* (D.H.L. Bishop and R.W. Compares, Eds.) Elsevier/North Holland, New York. pp. 639–645.

Crick, J. and Brown, F. (1969) *Nature* 222, 92.

Crick, J. and Brown, F. (1970) In *The Biology of the Large RNA Viruses* (R.D. Barry and B.W.J. Mahy Eds.) Academic Press, New York. pp. 133–140.

Dietzschold, B. (1977) *J. Virol.* 23, 286–293.

Dietzschold, B., Cox, J.H. and Schneider, L.G. (1979) *Virology* 98, 63–75.

Dietzschold, B., Wiktor, T.J., Macfarlan, R. and Varrichio, A. (1982) *J. Virol.* 44, 595–602.

Dietzschold, B., Wiktor, T.J., Wunner, W.H. and Varrichio, A. (1983a) *Virology* 124, 330–337.

Dietzschold, B., Wunner, W.H., Wiktor, T.J., Lopes, A.D., Lafon, M., Smith, C.L. and Koprowski, H. (1983b) *Proc. Natl. Acad. Sci. USA* 80, 70–74.

Dietzschold, B., Wunner, W.H., Lafon, M., Smith, C.L. and Verrichio, A. (1984) In *Nonsegmented Negative Strand Viruses* (D.H.L. Bishop and R.W. Compans, Eds.) Academic Press, San Diego. pp. 285–294.

Emerson, S.U. and Wagner, R.R. (1972) *J. Virol.* 10, 297–309.

Etchison, J.R. and Holland, J.J. (1974) *Proc. Natl. Acad. Sci. USA* 71, 4011–4014.

Evans, D., Pringle, C.R. and Szilagyi, J.F. (1979) *J. Virol.* 31, 325–333.

Flamand, A. and Delagneau, J.F. (1978) *J. Virol.* 28, 518–523.

Flamand, A., Wiktor, T.J. and Koprowski, H. (1980a) *J. Gen. Virol.* 48, 97–104.

Flamand, A., Wiktor, T.J. and Koprowski, H. (1980b) *J. Gen. Virol.* 48, 105–109.

Freeman, G.J., Rose, J.K., Clinton, G.M. and Huang, A.S. (1977) *J. Virol.* 21, 1094–1104.

Gallione, C.J., Greene, J.R., Iverson, L.E. and Rose, J.K. (1981) *J. Virol.* 39, 529–535.

Irving, R.A. and Ghosh, H.P. (1982) *J. Virol.* 42, 322–325.

Kingsford, L. and Emerson, S.U. (1980) *J. Virol.* 33, 1097–1105.

Knipe, D., Rose, J.K. and Lodish, H.F. (1975) *J. Virol.* 15, 1004–1011.

Knipe, D.M., Baltimore, D. and Lodish, H.F. (1977) *J. Virol.* 21, 1149–1158.

Knudsen, D.L. and MacLeod, R. (1972) *Virology* 47, 285–295.

Koprowski, H., Wiktor, T.J. and Abelseth, M.K. (1984) In *Rabies in the Tropics* (E. Kuwert, Ed.) Springer-Verlag, New York. In Press.

Lafon, M., Ideler, J. and Wunner, W.H. (1984) in *Development of Biological Standardization.* Joint WHO/AIBS Symposium on the Standardization of Rabies Virus. In press.

Lafon, M., Wiktor, T.J. and Macfarlan, R.I. (1983) *J. Gen. Virol.* 64, 843–851.

Lai, C.Y. and Dietzschold, B. (1981) *Biochem. Biophys. Res. Commun.* 103, 536–542.

Lefay, F. (1971) *J. Gen. Virol.* 13, 449–453.

Lingappa, V.R., Katz, F.N., Lodish, F. and Blobel, G. (1978) *J. Biol. Chem.* 253, 8667–8670.

Little, S.P. and Huang, A.S. (1978) *J. Virol.* 27, 330–339.

Loretu, K., Blanden, D.C. and Wiktor, T.J. (1984) In *North American Symposium on Rabies in Wildlife* (H. Fishman, Ed.) In press.

MacFarlan, R.I., Dietzschold, B., Koprowski, H. (1984a) *Cold Spring Harbor Quant. Symposium.* 139–143.

MacFarlan, R.I., Dietzschold, B., Wiktor, T.J., Kiel, M., Houghton, R., Lerner, R.A., Sutcliffe, J.G. and Koprowski, H. (1984b) *J. Immunol.* 133, 2748–2752.

Madore, H.P. and England, J.M. (1977) *J. Virol.* 22, 102–112.

Martinet, C., Lombard, A., Printz-Ané, C. and Printz, P. (1979) *J. Virol.* 29, 123–133.

McSharry, J.J., Ledda, C.A., Freiman, H.J. and Choppin, P.W. (1978) *Virology* 84, 183–188.

Mellon, M.G. and Emerson, S.U. (1978) In *Negative Strand Viruses and the Host Cell* (B.W.J. Mahy and R.D. Barry, Eds.) Vol. 7. Academic Press, London. pp. 293–300.

Morrison, T. and Lodish, H. (1975) *J. Biol. Chem.* 250, 6955–6962.

Morrison, T., Stempfer, M., Baltimore, D. and Lodish, H.F. (1974) *J. Virol.* 13, 62–72.

Murphy, F.A. and Harrison, A.K. (1979) In *Rhabdoviruses* (D.H.L. Bishop, Ed.) Vol. I. CRC Press, Boca Raton FL. pp. 65–106.

Neuberger, A., Gottschalk, A., Marshall, R.D. and Spiro, R.G. (1972) In *The Glycoproteins: Their Composition, Structure and Function* (A. Gottschalk, Ed.) Elsevier, Amsterdam. pp. 450–490.

Newcomb, W.W. and Brown, J.C. (1981) *J. Virol.* 29, 295–299.

Newcomb, W.W., Tobin, G.J., McGowan, J.J. and Brown, J.C. (1982) *J. Virol.* 41, 1055–1062.

Pennica, D., Holloway, B.P., Hayward, J.T. and Obijeski, J.F. (1980) *Virology* 103, 517–521.

Printz, P. and Wagner, R.R. (1971) *J. Virol.* 7, 651–662.

Rose, J.K. and Gallione, C.J. (1981) *J. Virol.* 39, 519–528.

Rose, J.K., Doolittle, R.F., Anilionis, A., Curtis, P.J. and Wunner, W.H. (1982) *J. Virol.* 43, 361–364.

Rupprecht, C.E. and Wiktor, T.J. (1984) In *North American Symposium on Rabies in Wildlife* (H. Fishman, Ed.) In press.

Sanger, F., Nicklen, S. and Coulson, A.R. (1977) *Proc. Natl. Acad. Sci. USA* 74, 5463–5467.

Schneider, L.G. and Diringer, J. (1976) *Curr. Top. Microbiol. Immunol.* 75, 153–180.

Schubert, M., Harmison, G.G. and Meier, E. (1984) *J. Virol.* 51, 505–514.

Shope, R.E. (1975) In *The Natural History of Rabies* (G.M. Baer, Ed.) Vol. 1, Academic Press, London. pp. 141–152.

Shope, R.E., Murphy, F.A., Harrison, A.K., Causey, O.R., Kemp, G.E., Simpson, D.I.H. and Moore, D.L. (1970) *J. Virol.* 6, 690–692.

Sokol, F. and Koprowski, H. (1975) *Proc. Natl. Acad. Sci. USA* 72, 933–936.
Sokol, F., Stancek, D. and Koprowski, H. (1971) *J. Virol.* 7, 241–249.
Sureau, P., Rollin, P. and Wiktor, T.J. (1983) *Am. J. Epidemiol.* 117, 605–609.
Szilagyi, J.F. and Pringle, C.R. (1979) *J. Virol.* 30, 692–700.
Tesh, R., Saidi, S., Javadian, E., Loh, P. and Nadin, A. (1977) *Am. J. Trop. Med. Hyg.* 26, 299–306.
Tesh, R.B., Travassos da Rosa, A.P.A. and Travassos da Rosa, J.S. (1983) *J. Gen. Virol.* 64, 169–176.
Tignor, G.H., Murphy, F.A., Clark, H.F., Shope, R.E., Madore, P., Bauer, S.P., Buckley, S.M. and Meredith, C.D. (1977) *J. Gen. Virol.* 37, 595–611.
Wagner, R.R., Prevec, L., Brown, F., Summers, D.F., Sokol, F. and MacLeod, R. (1972) *J. Virol.* 10, 1228–1230.
Wiktor, T.J. (1984) In *North American Symposium on Rabies in Wildlife* (H. Fishman, Ed.) In press.
Wiktor, T.J. and Koprowski, H. (1978) *Proc. Natl. Acad. Sci. USA* 75, 3938–3942.
Wiktor, T.J. and Koprowski, H. (1980) *J. Exp. Med.* 152, 99–112.
Wiktor, T.J., György, E., Schlumberger, H.D., Sokol, F. and Koprowski, H. (1973) *J. Immunol.* 110, 269–276.
Wiktor, T.J., Flamand, A. and Koprowski, H. (1980) *J. Virol. Meth.* 1, 33–46.
Wiktor, T.J., Lafon, M., Dietzschold, B. and Wunner, W.H. (1983) In *Advances in Gene Technology: Molecular Genetics of Plant and Animals* (K. Downey, R.W. Voellmy, F. Ahmad and J. Schultz, Eds.) Academic Press, New York. pp. 455–466.
Wiktor, T.J., Macfarlan, R.I., Foggin, C.M. and Koprowski, H. (1984) In *Development of Biological Standardization*. Joint WHO/AIBS Symposium on the Standardization of Rabies Virus. In press.
Wilson, T. and Lenard, J. (1981) *Biochemistry* 20, 1349–1354.
Wunner, W.H., Dietzschold, B., Curtis, P.J. and Wiktor, T.J. (1983) *J. Gen. Virol.* 64, 1649–1656.
Wunner, W.H. and Pringle, C.R. (1972) *J. Gen. Virol.* 16, 1–10.
Wunner, W.H., Smith, C.L., Lafon, M., Ideler, J. and Wiktor, T.J. (1984) in *Nonsegmented Negative Strand Viruses* (D.H.L. Bishop and R.W. Compans, Eds.) Academic Press, San Diego. pp. 279–284.
Wunner, W.H., Dietzschold, B., Smith, C.L., Lafon, M. and Golub, E. (1985) *Virology* 140, 1–12.
Yelverton, E., Norton, S., Obijeski, J.F. and Goeddel, D.V. (1983) *Science* 219, 614–620.
Zaidis, V.M., Krototova, L.I., Schinor, L.M., Schimor, M.A., Elbert, L.B. and Zhdanor, V.M. (1979) *J. Virol.* 29, 1226–1228.

Eds. M.H.V. van Regenmortel and A.R. Neurath
Immunochemistry of viruses. The basis for serodiagnosis and vaccines.

Antigenic structure of rotaviruses 21

MARY K. ESTES[1,2], DAVID Y. GRAHAM[1,2] and BETTY L. PETRIE[1]

Departments of [1]Virology and Epidemiology and [2]Medicine, Baylor College of Medicine, Houston, TX 77030, U.S.A.

The rotaviruses are ubiquitous human and veterinary pathogens that cause significant diarrheal disease (Estes et al., 1983b; Holmes, 1983). They constitute a genus of the family Reoviridae and viral particles are characterized by a 70 nm diameter double-shelled capsid surrounding an icosahedral core that contains 11 segments of double-stranded RNA. Epidemiologic and serologic surveys for rotavirus antigens and antibodies have demonstrated that these agents infect most, if not all, mammalian and avian species. The rotaviruses are primarily associated with gastroenteritis in the young, and the infections in children tend to be severe. In developing countries they are associated with high mortality, while in developed countries infections generally result in considerable morbidity but low mortality. The association of rotaviruses with severe gastroenteritis in adults (Echeverria et al., 1983) and with unusual clinical situations including immunosuppressed or immunodeficient patients (Halvorsrud and Orstavik, 1980; Holzel et al., 1980; Saulsbury et al., 1980; Yolken et al., 1982; Rotbart et al., 1983) emphasizes that our knowledge of the epidemiology, the factors that affect virulence and immunity to infection, and the mechanisms of transmission and evolution of strains is incomplete. In any event, sufficient epidemiologic data have been accumulated to indicate that methods of disease prevention are needed.

Vaccine programs have been proposed, although the best strategy for disease prevention is not currently known (Chanock, 1981; Vesikari et al., 1983a). The problem is complicated by the demonstration that the number of serotypes and virus host range may not be limited (i.e. serotypes defined by plaque-reduction neutralization assays contain viruses isolated from both human and animal species) (Table 21.1). It is possible that some of the observed genomic and antigenic variability

TABLE 21.1

Rotavirus serotypes

Serotype[a]	Isolated from:	Strains[b]	Subgroup[c]	RNA pattern
1	Humans	Wa, KU, K8, DB	2	long
2	Humans	DS-1, S2, KUN, HN-126, 390	1	short
3	Humans	M, P, YO, 14, 15, McM2, MO, Ito, Nemoto	2	long
	Animals	Simian MMU18006, SA11	1	long
		Canine CU-1	1	long
		Equine H2	NS	long
4	Humans	St. Thomas 4, Hochi, Hosokawa	2	long
	Animals	Porcine Gottfried, SB-1, SB-2		
5	Animals	Porcine OSU, EE, A580	1/NS	long
		Equine H-1	1	long
6	Animals	Bovine NCDV, UK, B641, B720, B14, II-2	1	long
7	Animals	Bovine B223	NT	long
8	Animals	Turkey Tyl, Ty3	NS	long
9	Animals	Chicken Ch1	NS	long

a Serotypes 1–4 as designated by Flewett et al. (1978), Wyatt et al. (1983b) and Beards and Flewett (1984). No uniform designation for the multiple serotypes from animals has yet been established.
b Compiled from Flewett et al. (1978), McNulty et al. (1980), Gaul et al. (1982), Sato et al. (1982), Thouless et al. (1982), Urasawa et al. (1982), Wyatt et al. (1983b), Hoshino et al. (1983b; c), Murakami et al. (1983), Woode et al. (1983), and Beards and Flewett (1984).
c NS denotes not able to be subgrouped; NT denotes not tested.

reflects the continual emergence of new virus strains which evolve through an exchange of genome segments (genetic reassortment) between two viruses that simultaneously infect a single host. Preliminary studies of natural isolates have supported this mechanism of virus variability (Clarke and McCrae, 1982; Street et al., 1982). Unequivocal demonstration of reassortment in nature will have a profound effect on the strategies for controlling rotavirus infections.

Rotaviruses that are morphologically identical, but that do not share common antigens with conventional rotaviruses, have also recently been isolated from several host species (McNulty et al., 1981; Bohl et al., 1982; Bridger et al., 1982; Rodger et al., 1982; Dimitrov et al., 1983; Nicolas et al., 1983). The clinical significance of these antigenically distinct rotaviruses is not known, but an outbreak of diarrhea in more than 12 000 adults in China has been attributed to these new rotaviruses (Hung et al., 1984). However, their discovery suggests that the rotavirus genus should be subdivided into several groups. It has been proposed to name the viruses studied extensively since 1969 as Group A rotaviruses and to classify the other viruses into Groups B, C, etc. as they become better characterized (Dimitrov et al., 1983; Pedley et al., 1983). This chapter will focus on the properties of the Group A rotaviruses because information on the viruses in the other groups is limited.

1. Virion structure

1.1. Morphology and physical properties of particles

Rotaviruses were named for their 'wheel-like' appearance in negatively stained preparations (Fig. 21.1A). The outer capsid of the virion (the rim) is a smooth layer about 5 nm thick (Fig. 21.1A). Examination of freeze dried particles revealed that this outer shell is perforated by numerous holes (probably 132) which correspond one to one with knob-like structures comprising the inner capsid layer (Roseto et al., 1979). These 'knobs' (seen in freeze dried preparations) form the spokes of the

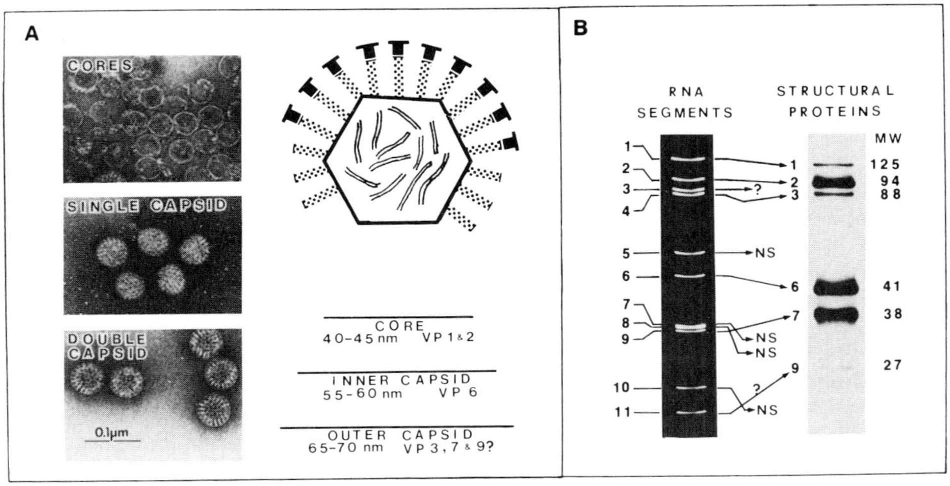

Fig. 21.1. Rotavirus virion structure. A. Electron micrographs of SA11 cores, single-capsid particles and double-capsid particles (stained with 10% saturated ammonium molybdate) and a schematic representation of rotavirus structure indicating the locations of the various structural proteins. B. Polyacrylamide gel electrophoresis patterns showing the eleven RNA segments (10% acrylamide) and the structural proteins (12% acrylamide) of double-capsid particles of SA11. Molecular mass (kDa) is indicated to the right of the photograph. Arrows indicate the gene coding assignments of each of the RNA segments. The indicated nonstructural proteins (NS) have molecular masses of 53, 35, 34 and 29 kDa. NS29 is a glycoprotein. Polypeptide nomenclature is reviewed in Holmes (1983) and Estes et al. (1983b).

wheel seen in negatively stained particles. When the outer capsid is removed, either naturally or by depletion of Ca^{2+} from the suspending medium, rotavirus particles take on a characteristic 'rough' appearance (Fig. 21.1A). Double and single-shelled particles are often characterized by their densities of 1.36 g/ml and 1.38 g/ml respectively, in cesium chloride (Newman et al., 1975). The inner capsid layer can be stripped from the particle by exposure to chaotropic agents such as sodium thiocyanate or 2 M calcium chloride (Almeida, 1979; Bican et al., 1982) leaving a smooth, hexagonal 'core' structure of 40–45 nm.

1.2. The genome and gene-coding assignments

The rotavirus genome is composed of eleven segments of double-stranded RNA that can be separated by electrophoresis in polyacrylamide gels. One interesting observation is that the relative migration of gene segment 11 may vary significantly in different strains, giving rise to 'long' and 'short' RNA patterns (Espejo et al., 1979; Rodger et al., 1981). Most animal and human rotavirus strains have 'long' electropherotypes, and viruses with 'long' RNA patterns are classified in more than one serotype (Beards, 1982). Therefore, at least 'long' RNA electropherotypes cannot be used to predict serotype.

Each of the 11 segments of rotavirus RNA codes for a single polypeptide. The simian rotavirus SA11 is the most extensively studied strain, and its genes, their corresponding proteins and the location and functions of the proteins are summarized in Fig. 21.1B and Table 21.2. Many of the primary gene products undergo post-translational modifications including glycosylation, trimming of oligosaccharides and proteolytic cleavage. Details of these novel features of rotavirus replication have been reviewed elsewhere (Estes et al., 1983b; Holmes, 1983). There is no consensus concerning the identification of the gene 3 product and whether the gene 10 and 11 proteins are structural or nonstructural polypeptides. Coding assignments with different group A virus strains generally agree except that inversions often occur among the 7, 8 and 9 segments and between segments 10 and 11 (Dyall-Smith and Holmes, 1981; McCrae and McCorquodale, 1982; Dyall-Smith et al., 1983; Both et al., 1983a).

1.3. Structural polypeptides

Six of the ten identified primary SA11 gene products are structural proteins. Definitive assignments of proteins to specific structures have not been made, although the proteins located in the different capsid layers have been identified (Fig. 21.1). Three proteins have been designated as outer shell components because they are not seen in purified single-capsid particles (Estes et al., 1981; Espejo et al., 1981; Mason et al., 1980; 1983). The outer capsid proteins include VP3, an 88-kDa trypsin-sensitive protein, and VP7, a 38-kDa glycosylated protein that contains

TABLE 21.2

SA11 rotavirus gene coding assignments and protein functions

Gene segment	Polypeptide[a]	Location in virus capsid[a]	Function[b]
1	VP1	Inner	–
2	VP2→VP3*,VP4*	Inner	–
3	?	?	–
4	VP3→VP5*,VP8*	Outer	Hemagglutinin, protease-enhanced infectivity, neutralization
5	NS53	Nonstructural	–
6	VP6	Inner	Subgroup antigen
7	NS34	Nonstructural	–
8	NS35	Nonstructural	RNA replication?
9	VP7	Outer glycoprotein	Major neutralization antigen, hemagglutination inhibiting activity (steric hindrance)
10	NS29	Nonstructural glycoprotein	Integral membrane protein, role in morphogenesis
11	VP9[c]	Outer?	–

* Cleavage products.

a Compiled from Mason et al. (1980), Arias et al. (1982), Ericson et al. (1982), Mason et al. (1983), Both et al. (1983a–c), Holmes (1983) and Kantharidis et al. (1983).

b From Estes et al. (1981), Both et al. (1982), Greenberg et al. (1983 a–c), Kalica et al. (1983), Ericson et al. (1983a, b), Matsuno and Inouye (1983), Killen and Dimmock (1982), Soler et al. (1983), Petrie et al. (1983) and unpublished data.

c The gene 11 product is a 26-kDa polypeptide; two related polypeptides are detected in infected cells. The designation of this gene product as a structural polypeptide remains tentative (Ericson et al., 1982; Estes et al., 1983b).

only high mannose oligosaccharides. A small amount of VP9, a 27-kDa protein, has been detected (Estes et al., 1981; Ericson et al., 1982), but is not reported by all investigators (Espejo et al., 1981; Holmes, 1983).

VP6, a 41-kDa protein, makes up the outer layer of the inner capsid and reportedly is a trimer in its native state (Novo and Esparza, 1981). The core particles contain VP1 and VP2 and possibly a third, high molecular mass polypeptide

(Bican et al., 1982; J. Cohen and M.K. Estes, unpublished results). Novo and Esparza (1981) estimated that VP6 comprises 80% of the protein mass of single-shelled particles.

The molecular masses of the polypeptides of different strains vary (Thouless, 1979; Offit et al., 1983b), and the apparent polypeptide pattern of particles depends on whether the particles have been exposed to proteolytic enzymes. Trypsin treatment of SA11 cleaves VP3 to yield a 60-kDa protein (VP5*) and a 28-kDa protein (VP8*). Prolonged exposure to proteases may also result in the appearance of additional cleavage products from VP2 (Espejo et al., 1981; Estes et al., 1981). It is not known if the post-translational modifications of these proteins affect the antigenicity of particles.

Heterogeneity in the structural glycoprotein has been demonstrated and this explains the discrepancy in studies that have shown protein profiles with one or two glycoproteins (Estes et al., 1982; Sabara and Babiuk, 1983). The oligosaccharides on VP7 have been found to range from $Man_5GlcNAc$ to $Man_9GlcNAc$ (Both et al., 1983b). A lack of glycosylation on VP7 does not affect virus morphology, growth, hemagglutination or stability to pH changes, protease digestion or elevated temperature (Estes et al., 1982; Petrie, 1983; Petrie et al., 1983).

2. Biologic properties of rotavirus structural polypeptides

Double-shelled, but not single-shelled particles are infectious (Elias, 1977b) and agglutinate red blood cells (Inaba et al., 1977), suggesting that the outer capsid proteins are necessary for infectivity and hemagglutination. The polypeptides responsible for virus neutralization, hemagglutination, protease-enhanced infectivity, and replication in tissue culture have been identified.

The structural glycoprotein VP7 is the major type-specific neutralizing antigen (Bastardo et al., 1981; Killen and Dimmock, 1982; Matsuno and Inouye, 1983; Greenberg et al., 1983c), and is probably important in eliciting a protective antibody response. It is not known if the heterogeneity in VP7 affects virus antigenicity or virulence. However, this heterogeneity suggests that rotaviruses may undergo genetic drift in a manner similar to influenza viruses.

Genetic studies and studies with monoclonal antibodies identified VP3 (the product of genome segment 4) as the hemagglutinin (Kalica et al., 1983; Greenberg et al., 1983a; 1983c). Studies with monoclonal antibodies and monospecific polyclonal antibodies initially yielded conflicting results in that antibodies directed at either VP3 or VP7 could inhibit hemagglutination. It is currently thought that the hemagglutination inhibiting activity of the antibodies to VP7 is due to steric hindrance (Greenberg et al., 1983c). Treatment of virus particles with reagents that break disulfide bonds destroys hemagglutinating activity.

It is likely, by analogy with other virus systems, that the rotavirus hemagglutinin is also the cell attachment protein, but limited data are available on the interaction of rotaviruses with cell receptors. It has been shown that SA11 interacts with carbohydrate moieties on the erythrocyte surface (Bastardo and Holmes, 1980). The presence of virus-specific receptors on enterocytes may influence age-dependent clinical disease by determining the extent or location of virus replication in the intestinal tract (Riepenhoff-Talty et al., 1982).

VP3 is also responsible for protease-enhanced plaque formation and the ability to replicate in tissue culture (Kalica et al., 1983). The cleavage of VP3 seems to be the key event in the enhancement of infectivity (Estes et al., 1981). It has been suggested that infection with particles having uncleaved VP3 may be blocked at the stage of viral uncoating (Clark et al., 1981).

The polypeptides responsible for virus attachment to cell receptors, for virus virulence, and for the enzymatic activities (RNA polymerase, poly A polymerase, guanyl transferase, methyl transferases, nucleotide phosphohydrolase) found in virus particles (Cohen, 1977; Gorziglia and Esparza, 1981) remain unknown. In a preliminary attempt to determine the proteins associated with virus virulence, Carpio et al. (1981) compared one avirulent and three virulent isolates of bovine rotavirus which differed in their ability to infect intestinal loops. The avirulent strain appeared to be more susceptible to cleavage of VP3. Whether polypeptide patterns will be correlated with the relative virulence of strains remains to be determined.

3. Antigenic properties of structural polypeptides

Several antigenic specificities have been associated with rotavirus particles. However, there is some difficulty in comparing results from different laboratories because diverse methods were used to quantitate serological relatedness. For example, different types of antisera (e.g. postinfection sera from conventional or gnotobiotic animals and human sera obtained either after natural infections or after experimental infections with virus given either orally or parenterally) have been used. In addition, virus antigens have often been uncharacterized and have included preparations made from stool samples and tissue culture lysates. This problem is not trivial because stool samples and uncloned stocks of tissue culture-adapted virus may contain multiple virus strains (Sabara and Babiuk, 1983; Hoshino et al., 1983c), as well as both structural and nonstructural proteins in unknown and, most likely, differing amounts. Despite these reservations, some conclusions can be made from existing data.

Group A rotaviruses contain antigen(s) that cross-react between viruses isolated from any mammalian or avian host as well as type-specific antigens that distinguish between virus strains (Woode et al., 1976; Kapikian et al., 1976; Thouless et al., 1977; Matsuno et al., 1977). The common antigen(s), designated the group antigens, are detectable in numerous serologic assays. Initial confusion in defining rotavirus antigens resulted from the assumption that different serologic tests (including enzyme-linked immunosorbent assays (ELISA) and neutralization assays) detected the same or related antigens and that they could define serotypes (Yolken et al., 1978; Zissis and Lambert, 1982). This was subsequently recognized to be incorrect (Kapikian et al., 1981), and serum neutralization tests are now acknowledged by all researchers as the correct assay to define serotypes.

Rotaviruses contain at least four major antigens: (*a*) a group antigen described above which has been associated with the 1.38 g/ml particles; (*b*) a subgroup antigen (possibly a part of the group antigen); (*c*) a type-specific neutralization antigen associated with the glycosylated polypeptide on the outer particles; and (*d*) a hemagglutinin associated with the outer capsid VP3. Hemagglutination probably reflects a type-specific reactivity, but has not been thoroughly characterized since most human rotavirus strains were thought to lack a hemagglutinin. Konno et al. (1983) have recently demonstrated that some human strains have detectable hemagglutinating activity if the virus is propagated in the absence of trypsin. Analysis of rotavirus strains by hemagglutination inhibition and neutralization assays confirms that these assays measure different antigens (Spence et al., 1978; Greenberg et al., 1983c; Hoshino et al., 1983a).

Bridger (1978) used immunoelectron microscopy to localize both common group and type-specific determinants on virus particles. Nonneutralizing sera made against heterologous rotavirus strains caused only single-shelled particles to agglutinate, which suggested that the common antigens are located on the inner capsid of the virus. More recently, all the structural polypeptides have been shown to contain common determinants (Offit et al., 1983a; Greenberg et al., 1983b; Estes, unpublished results). This result can be reconciled with Bridger's earlier observation if the common determinants on the polypeptides other than VP6 are not exposed on intact double or single-capsid particles. The common determinants found on the other proteins have only been demonstrated by assays where the antigens were denatured by detergents or with denaturing fixation procedures.

The presence of at least two antigenic determinants on VP6 (Kapikian et al., 1981; Thouless et al., 1982) has been confirmed with monoclonal antibodies. One determinant is common between all rotavirus strains and the second determinant reacts with viruses having one of the two well-characterized subgroup specificities. All animal strains that react in this assay, to date, are subgroup 1 viruses, while strains from humans have been found in both subgroups 1 and 2. A monoclonal antibody to VP6 that distinguishes between avian and mammalian rotavirus strains has also been described (Gary et al., 1982). Comparison of the nucleotide and deduced protein sequence data for VP6 from different strains and subgroups may help map these epitopes (Estes et al., 1984; Both et al., 1984b; Cohen et al., 1984).

Subgroups 1 and 2 have also been correlated with 'short' and 'long' electropherotypes, respectively (Kalica et al., 1981). Since the basis for this correlation is not clear, it may be unwise to rely on electropherotyping to classify viruses into subgroups.

The rotavirus type-specific determinants are detected in plaque-reduction or fluorescent focus-reduction neutralization assays (Beards et al., 1980; Estes and Graham, 1980; Wyatt et al., 1982). In the rotavirus system, hyperimmune antiserum, rather than immune or convalescent serum, is required to clearly differentiate serotypes (Estes and Graham, 1980; Gaul et al., 1982; Wyatt et al., 1982) as immune sera show broadly reactive and low titer neutralizing activities.

Although genetic studies showed that glycoprotein VP7 is the major neutralization antigen, studies with monospecific or monoclonal antibodies have indicated that other proteins (VP3, VP6 and a small molecular mass polypeptide) may also induce antibodies with neutralization activity (Bastardo et al., 1981; Killen and Dimmock, 1982; Greenberg et al., 1983c). The suggestion that gene segment 11 might code for a minor neutralization antigen was also supported by preliminary reassortant data of Matsuno et al. (1980). Further studies are needed to elucidate the relative importance of these other proteins in inducing neutralizing antibodies or in inducing protection from disease.

The fact that antibodies to both VP3 and VP7 can inhibit hemagglutination, and that both proteins may be involved in neutralization, must be considered in the interpretation of immune responses measured by hemagglutination inhibition or neutralization tests. In addition, this observation suggests that a complete understanding of the relationship between virus strains and serotypes may require the development of more specific reagents and characterization of numerous isolates. This will be important if the one-way antigenic cross-reactivities observed in neutralization assays reflect reactivities with VP3 rather than VP7.

4. The nucleotide sequence of the major neutralization antigen

The nucleotide sequence of the gene that codes for VP7 has been reported for two viruses (SA11 and UK bovine) that fall into distinct serotypes (Both et al., 1983a; Elleman et al., 1983). Comparison of the sequences of the VP7 genes reveals striking homology (88%) of the predicted amino acid sequences of VP7. Since no direct protein sequence data are available, these comparisons assume that translation of the proteins is initiated at the first AUG in both viruses and that the proteins are processed in a similar fashion.

Since VP7 is the major neutralizing antigen, it is of interest to analyze the deduced amino acid sequence data to determine whether distinct domains can be found that could represent the antigenic sites of VP7. Fig. 21.2 compares hydrophilicity plots of VP7 from the UK bovine and SA11 viruses. The profiles show several areas that are relatively hydrophilic, and the regions where divergence is seen in the gene sequence are highlighted. The hydrophilic regions that diverge between serotypes might be sites of interest for developing synthetic peptide vaccines for the rotaviruses. However, this approach may not be straightforward or successful, as neutral-

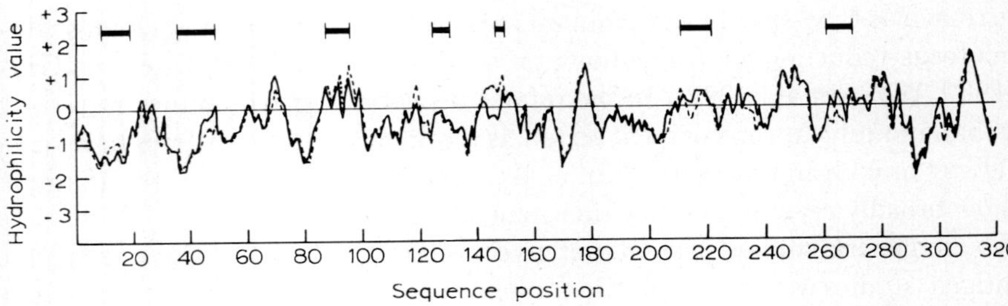

Fig. 21.2. Hydrophilicity plots for the deduced VP7 amino acid sequence of SA11 (——) and the bovine UK (- - - -) rotavirus strains. The average hydrophilicity for the deduced protein sequences from Both et al. (1983a) and Elleman et al. (1983) was calculated over an average of 6 amino acids according to Hopp and Woods (1981). The positive and negative values indicate hydrophilic and hydrophobic regions, respectively. The solid bars indicate regions where direct comparison of the sequences of these two viruses of different serotypes show divergence. Similar results were obtained by Both et al. (1984a) in a preliminary comparison using the method of Kyte and Doolittle (1982). The SA11 VP7 sequence has one glycosylation site at residues 68–70, while three potential glycosylation sites (at residues 68–70, 238–240, and 318–320) are found in the UK VP7.

ization titers of antisera produced to polypeptides reduced with dithiothreitol or 2-mercaptoethanol and purified from sodium dodecylsulfate gels are either low (Bastardo et al., 1981; Killen and Dimmock, 1982) or nondetectable (Estes, unpublished results). These results suggest that disulfide bonds may be important for antigenicity and that antigenic sites on VP7 may involve conformational determinants.

5. Immune responses to rotavirus infections

5.1. Humoral and cell-mediated immunity

A number of conclusions can be made from the available studies on the immune response to rotavirus infections: (*a*) Infection occurs frequently in the absence of diarrhea; (*b*) serum antibody (IgG) that indicates previous infection is not a reliable indicator of immunity to disease with another (or possibly the same) serotype; and (*c*) the presence of specific neutralizing intestinal antibody (acquired actively or passively) may provide protection from clinical disease but not protection from infection. Very high levels of intestinal antibody may prevent infection.

The development of intestinal and serum antibody after rotavirus infection has been studied in children and young animals (Sonza and Holmes, 1980; Riepenhoff-Talty et al., 1981; Vonderfecht and Osborn, 1982; Corthier and Vannier, 1983; Davidson et al., 1983; Sheridan et al., 1983). One problem in these studies, which evaluated the antibody responses to natural infections, is that the virus strains that initiated the infections were not cultivatable (or not available) and, therefore, heterologous viruses were used in the assays to measure virus-specific antibody isotypes and titers. The study conclusions were therefore based on the detection of common (not neutralizing) antigens on cross-reacting viruses. However, the results are in general agreement. Soon after infection, there is a short-lived increase in specific serum IgM which is followed by the appearance of IgA and IgG. In the intestine, a similar pattern of antibody development is seen (although an intestinal IgM response has not been demonstrated in mice). In the intestine, specific secretory IgA is the predominant immunoglobulin produced after infection.

The role of cell-mediated immunity is unclear, but it may function in recovery from infection as children with defects in their cell-mediated immune systems or immunosuppressed individuals may become chronic rotavirus excretors (Saulsbury et al., 1980; Yolken et al., 1982; Graham and Estes, unpublished results).

The rotaviruses are thought to infect only the mature enterocytes of the intestinal villi. This restricted site of infection makes it unlikely that serum antibody (resulting from previous infection) plays a major role in preventing disease. Protection from rotavirus disease must be mediated by mechanisms that prevent the virion from invading susceptible enterocytes or limit the spread of infection, or both. Although the local mucosal immune system probably plays the major role in immunity to rotavirus infection, many questions concerning the factors that affect host susceptibility remain to be answered. These include:

(*a*) What is the duration of active immunity and the frequency of reinfections?

(*b*) What is the specific immune response to infection with different serotypes?

(*c*) What is the specific serum and mucosal antibody response to specific rotaviral proteins?

(*d*) Does serum antibody play a role in modulating the extent or severity of disease?

5.2. Immune response to rotavirus polypeptides

A limited amount of information is available on the specific immune response to individual polypeptides during natural or experimentally induced infections. This is because a clear understanding of the rotavirus proteins has only recently been established and because a convenient animal model of rotavirus disease has been lacking.

Essentially all mammalian and avian species contain serum antibody directed against the Group A rotaviruses (Woode et al., 1976; Elias, 1977a). Radioimmunoprecipitation has been used to determine the polypeptides recognized by these antibodies acquired following natural infection. A caveat in the interpretation of these studies is that the antigens used in the radioimmunoprecipitation assays were not derived from the 'natural' infecting virus. However, mouse, rabbit and guinea-pig sera contain antibody directed against the inner capsid proteins (VP2 and VP6) of heterologous (human, simian and bovine) virus strains (Offit et al., 1983a; Estes, unpublished results). It is presumed that serum antibodies against the outer capsid VP3 and VP7 would also be detected if the infecting virus was used as the source of antigen for the assays. The polypeptides recognized by mucosal antibodies remain unstudied. Parenteral immunization of animals induces serum antibodies directed against both inner and outer capsid polypeptides and animals will reproducibly develop high titers of neutralizing antibody. This occurs despite the presence of pre-existing antibody (from natural infections) to antigens which share cross-reacting epitopes between heterologous rotaviruses.

5.3. Cross-neutralization between heterologous rotavirus strains

The observation that cross-reactive neutralizing antibodies are found in the serum of animals or children inoculated with heterologous rotavirus strains is provocative. Several studies have suggested that vaccination with heterologous virus strains may induce cross-protection, and the use of a bovine rotavirus or other animal rotavirus as a vaccine in humans is currently being tested (Wyatt et al., 1979; 1983a; Vesikari et al., 1983b; Kapikian et al., 1983). The molecular basis for this cross-protection is not understood, and cross-protection between heterotypic rotaviruses in experimental animals has been inconsistent. Some well-controlled studies have failed to demonstrate cross-protection or reduced virus shedding in animals unless the animals were immunized and challenged with viruses of the same serotype (Gaul et al., 1982; Woode et al., 1983). Since viruses from multiple host species can be in the same rotavirus serotype, it is possible that such 'heterologous cross-protection' reflects cross-protection of viruses in the same serotype. Cross-neutralization in vitro could also be due to nonantibody-mediated mechanisms such as nonspecific inhibitors (including protease inhibitors) present in serum (Graham and Estes, 1980). Cross-protection in vivo is multifaceted and may be mediated by humoral and cellular immune and other nonimmune mechanisms.

To fully understand immunity to rotavirus infection, factors such as (*a*) prior viral antigenic exposure of individuals or experimental animals, (*b*) the route, dose, preparation and serotype of the antigen administered, (*c*) the type of immune response and the time after virus administration for evaluation, and (*d*) the species and strain of animals used for study must be carefully considered.

5.4. Passive immunity

Reasonably clear results are available on the effectiveness of passive immunity in preventing disease. In ruminants and some monogastric animals (where the young do not receive immunoglobulins by transplacental transfer), administration of sufficient amounts of colostrum or milk containing high titers of antibody can reduce or postpone severe disease (Saif and Bohl, 1980; Saif et al., 1983). The role of colostrum or breast-feeding in preventing rotavirus disease in children is less clear cut.

Preliminary studies suggest that feeding bovine colostrum containing antibody to human rotavirus to babies may be useful in at least some situations (Ebina et al., 1983). The administration of human milk with rotavirus antibodies is reportedly effective in stopping the excretion of rotavirus in immunodeficient children (Saulsbury et al., 1980).

6. Antigenic determinants and diagnosis of rotavirus infections

Rotavirus infections are easily diagnosed by electron microscopy and by rapid immunologic methods such as enzyme-linked immunosorbent and radioimmunoassays. Methods that will detect rotavirus genetic information (dot blots) have been proposed as rapid tests to diagnose and also to determine the virus serotype (Flores et al., 1983; Estes et al., 1983a). Current immunological methods that use hyperimmune antisera will detect any of the Group A rotaviruses (if the virus particles are not coated with antibodies and no other interfering substances are present). False negative reactions can usually be minimized by treating the samples with low concentrations of chelating agents to remove the outer shell of the particles (together with the adherent antibody) so that single-shelled particles can be detected. False positive reactions remain a problem, especially in stool samples from younger children. In some cases, these can be eliminated by treating the samples with reducing agents (Yolken and Stopa, 1979), and confirmatory tests with pre-immune sera are needed for unequivocal interpretations of positive results (Beards et al., 1984). Monoclonal antibodies are needed to differentiate viruses into serotypes and to characterize virus strains by specific antigenic determinants encoded by individual gene segments. Once the important epitopes of the individual virus proteins are characterized, it should also be possible to use specific virus antigens produced in bacteria from cloned genes for seroepidemiologic studies or as vaccines for oral or parenteral immunization.

7. Conclusions

The last decade has seen a major advance in our knowledge of the rotaviruses as human and veterinary pathogens. The discovery of new groups of rotaviruses that do not share cross-reacting antigens with the conventional agents offers a challenge to preventing rotavirus disease by vaccination. The description of the rotavirus proteins was a beginning to understanding the precise immune and nonimmune mechanisms that mediate protection from disease. Information that will be accumulated in the next decade should determine whether rotavirus disease can be prevented by vaccination, whether disease can be effectively modulated by passive immunization or whether the rotaviruses will continue to challenge virologists and public health workers as the influenza viruses have for the past century.

8. Acknowledgements

We thank our numerous colleagues who shared their recent unpublished data with us. This laboratory is supported in part by grants from the National Institutes of Health, The Thrasher Research Fund, the Diarrheal Disease Program of the World Health Organization, and Vicks Health Care Division of Richardson-Vicks, Inc.

9. References

Almeida, J.D. (1979) In *Les colloques de l'INSERM* (F. Bricout and R. Scherrer, Eds.) Vol. 90. Institut National de la Santé et de la Recherche Médicale, Paris. pp. 379–392.
Arias, C.F., Lopez, S. and Espejo, R.T. (1982) *J. Virol.* 41, 42–50.
Bastardo, J.W. and Holmes, I.H. (1980) *Infect. Immun.* 29, 1134–1140.
Bastardo, J.W., McKimm-Greschkin, J.L., Sonza, S., Mercer, L.D. and Holmes, I.H. (1981) *Infect. Immun.* 34, 641–647.
Beards, G.M. (1982) *Arch. Virol.* 74, 65–70.
Beards, G.M. and Flewett, T.H. (1984) *Arch. Virol.* 80, 231–237.
Beards, G.M., Pilfold, J.N., Thouless, M.E. and Flewett, T.H. (1980) *J. Med. Virol.* 5, 231–237.
Beards, G.M., Campbell, A.D., Cottrell, N.R., Peiris, J.S.M., Rees, N., Sanders, R.C., Shirley, J.A., Wood, H.C. and Flewett, T.H. (1984) *J. Clin. Microbiol.* 19, 248–254.
Bican, P., Cohen, J., Charpilienne, A. and Scherrer, R. (1982) *J. Virol.* 43, 1113–1117.
Bohl, E.H., Saif, L.J., Theil, K.W., Agnes, A.G. and Cross, R.F. (1982) *J. Clin. Microbiol.* 15, 312–319.
Both, G.W., Bellamy, A.R., Street, J.E. and Siegman, L.J. (1982) *Nucl. Acids Res.* 10, 7075–7088.
Both, G.W., Mattick, A.R. and Bellamy, A.R. (1983a) *Proc. Natl. Acad. Sci. USA* 80, 3091–3095.
Both, G.W., Siegman, L.J., Bellamy, A.R. and Atkinson, P.H. (1983b) *J. Virol.* 48, 335–339.
Both, G.W., Mattick, J., Siegman, L., Atkinson, P.H., Weiss, S., Bellamy, A.R., Street, J.E. and Metcalf, P. (1983c) In *Double-stranded RNA Viruses* (R.W. Compans and D.H.L. Bishop, Eds.) Elsevier, New York. pp. 73–82.
Both, G.W., Siegman, L.J., Atkinson, P.H., Poruchynsky, M.S., Kabcenell, A.K., Street, J.E., Gunn, P.R., Sato, F., Powell, K.F.H. and Bellamy, A.R. (1984a) In *Modern Approaches to Vaccines* (R. Chanock and R. Lerner, Eds.) Cold Spring Harbor Laboratory, Cold Spring Harbor NY. pp. 53–58.
Both, G.W., Siegman, L.J., Bellamy, A.R., Ikegami, N., Shatkin, A.J. and Furuichi, Y. (1984b) *J. Virol.* 51, 97–101.
Bridger, J.C. (1978) *J. Clin. Microbiol.* 8, 625–628.
Bridger, J.C., Clarke, I.N. and McCrae, M.A. (1982) *Infect. Immun.* 35, 1058–1062.
Carpio, M., Bellamy, J.E.C. and Babiuk, L.A. (1981) *Can. J. Comp. Med.* 45, 38–42.
Chanock, R.M. (1981) *J. Infect. Dis.* 143, 364–374.
Clark, S.M., Roth, J.R., Clark, M.L., Barnett, B.B. and Spendlove, R.S. (1981) *J. Virol.* 39, 816–822.
Clarke, I.N. and McCrae, M.A. (1982) *Infect. Immun.* 36, 492–497.
Cohen, J. (1977) *J. Gen. Virol.* 36, 395–402.
Cohen, J., Lefevre, F., Estes, M.K. and Bremont, M. (1984) *Virology* 138, 178–182.
Corthier, G. and Vannier, P. (1983) *J. Infect. Dis.* 147, 293–296.
Davidson, G.P., Hogg, R.J. and Kirubakaran, C.P. (1983) *Infect. Immun.* 40, 447–452.

Dimitrov, D.H., Estes, M.K., Rangelova, S.M., Shindarov, L.M., Melnick, J.L. and Graham, D.Y. (1983) *Infect. Immun.* 41, 523–526.

Dyall-Smith, M.L. and Holmes, I.H. (1981) *J. Virol.* 38, 1099-1103.

Dyall-Smith, M.L., Azad, A.A. and Holmes, I.H. (1983) *J. Virol.* 46, 317–320.

Ebina, T., Sato, A., Umezu, K., Ishida, N., Ohyama, S., Ohizumi, A., Aikawa, K., Katagiri, S., Katsushima, N., Imai, A., Kitaoka, S., Suzuki, H. and Konno, T. (1983) *Lancet* ii, 1029–1030.

Echeverria, P., Blacklow, N.R., Cukor, G.C., Vibulbandhitkit, S., Changchawalit, S. and Boonthai, P. (1983) *J. Clin. Microbiol.* 18, 663–667.

Elias, M.M. (1977a) *J. Hyg.* 79, 365–372.

Elias, M.M. (1977b) *J. Gen. Virol.* 37, 191–194.

Elleman, T.C., Hoyne, P.A., Dyall-Smith, M.L., Holmes, I.H. and Azad, A.A. (1983) *Nucl. Acids Res.* 11, 4689–4701.

Ericson, B.L., Graham, D.Y., Mason, B.B. and Estes, M.K. (1982) *J. Virol.* 42, 825–839.

Ericson, B.L., Graham, D.Y., Mason, B.B., Hanssen, H.H. and Estes, M.K. (1983a) *Virology* 127, 320–332.

Ericson, B.L., Petrie, B.L., Graham, D.Y., Mason, B.B. and Estes, M.K. (1983b) *J. Cell. Biochem.* 22, 151–160.

Espejo, R.T., Calderón, E., Gonzalez, N., Salomon, A. Martuscelli, A. and Romero, P. (1979) *J. Infect. Dis.* 139, 474–477.

Espejo, R.T., Lopez, S. and Arias, C. (1981) *J. Virol.* 37, 156–160.

Estes, M.K. and Graham, D.Y. (1980) *Am. J. Vet. Res.* 41, 151–152.

Estes, M.K., Graham, D.Y. and Mason, B.B. (1981) *J. Virol.* 39, 879–888.

Estes, M.K., Graham, D.Y., Ramig, R.F. and Ericson, B.L. (1982) *Virology* 122, 8–14.

Estes, M.K., Graham, D.Y. and Dimitrov, D.H. (1983a) *Prog. Med. Virol.* 29, 1–22.

Estes, M.K., Palmer, E.L. and Obijeski, J.F. (1983b) *Curr. Topics Microbiol. Immunol.* 105, 124–184.

Estes, M.K., Mason, B.B., Crawford, S. and Cohen, J. (1984) *Nucl. Acids Res.* 12, 1875–1887.

Flewett, T.H., Thouless, M.E., Pilford, J.N., Bryden, A.S. and Candeias, J.A.N. (1978) *Lancet* ii, 632.

Flores, J., Boeggeman, E., Purcell, R.H., Sereno, M., Perez, I., White, L., Wyatt, R.G., Chanock, R.M. and Kapikian, A.Z. (1983) *Lancet* i, 555–559.

Gary, G.W., Humphrey, D., Black, R. and Palmer, E. (1982) *Arch. Virol.* 72, 223–227.

Gaul, S.K., Simpson, T.F., Woode, G.N. and Fulton, R.W. (1982) *J. Clin. Microbiol.* 16, 495–503.

Gorziglia, M. and Esparza, J. (1981) *J. Gen. Virol.* 53, 357–362.

Graham, D.Y. and Estes, M.K. (1980) *Virology* 101, 432–439.

Greenberg, H., Flores, J., Kalica, A., Wyatt, R. and Jones, R. (1983a) *J. Gen. Virol.* 64, 313–324.

Greenberg, H., McAuliffe, V., Valdesuso, J., Wyatt, R., Flores, J., Kalica, A., Hoshino, Y. and Singh, N. (1983b) *Infect. Immun.* 39, 91–99.

Greenberg, H.B., Valdesuso, J., VanWyke, K., Midthun, K., Walsh, M., McAuliffe, V., Wyatt, R.G., Kalica, A.R., Flores, J. and Hoshino, Y. (1983c) *J. Virol.* 47, 267–275.

Halvorsrud, J. and Orstavik, I. (1980) *Scand. J. Infect. Dis.* 12, 161–164.

Holmes, I.H. (1983) In *The Reoviridae* (W.K. Joklik, Ed.) Plenum Publishing Co., New York. pp. 359–423.

Holzel, H., Cubitt, D.W., McSwiggan, D.A., Sanderson, P.J. and Church, J. (1980) *J. Infect.* 2, 33–37.

Hopp, T.P. and Woods, K.R. (1981) *Proc. Natl. Acad. Sci. USA* 78, 3824–3828.

Hoshino, Y., Wyatt, R.G., Greenberg, H.B., Kalica, A.R., Flores, J. and Kapikian, A.Z. (1983a) *Infect. Immun.* 41, 169–173.

Hoshino, Y., Wyatt, R.G., Greenberg, H.B., Kalica, A.R., Flores, J. and Kapikian, A.Z. (1983b) *J. Clin. Microbiol.* 18, 585–591.

Hoshino, Y., Wyatt, R.G., Greenberg, H.B., Kalica, A.R., Flores, J. and Kapikian, A.Z. (1983c) *Infect. Immun.* 41, 1031–1037.

Hung, T., Chen, G., Wang, C., Yao, H., Fang, Z., Chao, T., Chou, Z., Ye, W., Chang, X., Den, S., Liong, X. and Chang, W. (1984) *Lancet* i, 1139–1142.

Inaba, Y., Sato, K., Takahashi, E., Kurogi, H., Satoda, K., Omori, T. and Matumoto, M. (1977) *Microbiol. Immunol.* 21, 531–534.

Kalica, A.R., Greenberg, H.B., Espejo, R.T., Flores, J., Wyatt, R.G., Kapikian, A.Z. and Chanock, R.M. (1981) *Infect. Immun.* 33, 958-961.

Kalica, A.R., Flores, J. and Greenberg, H.B. (1983) *Virology* 125, 194–205.

Kantharidis, P., Dyall-Smith, M.L. and Holmes, I.H. (1983) *J. Virol.* 48, 330–334.

Kapikian, A.Z., Cline, W.L., Kim, H.W., Kalica, A.R., Wyatt, R.G., VanKirk, D.H., Chanock, R.M., James, H.D. Jr. and Vaughn, A.L. (1976) *Proc. Soc. Exp. Biol. Med.* 152, 535–539.

Kapikian, A.Z., Cline, W.L., Greenberg, H.B., Wyatt, R.G., Kalica, A.R., Banks, C.E., James, H.D. Jr., Flores, J. and Chanock, R.M. (1981) *Infect. Immun.* 33, 415–425.

Kapikian, A.Z., Wyatt, R.G., Levine, M.M., Yolken, R.H., VanKirk, D.H., Dolin, R., Greenberg, H.B. and Chanock, R.M. (1983) *J. Infect. Dis.* 147, 95–106.

Killen, H.M. and Dimmock, N.J. (1982) *J. Gen. Virol.* 62, 297–311.

Konno, T., Kitaoka, S., Suzuki, H., Suto, T., Numazaki, Y. and Ishida, N. (1983) *Program of the Working Conference on Arboviruses, Viral Gastroenteritis, Rabies and Hemorrhagic Fever with Renal Syndrome* (U.S.-Japan Cooperative Medical Sciences Program, 7–9 September 1983).

Kyte, J. and Doolittle, R.F. (1982) *J. Mol. Biol.* 157, 105–132.

Mason, B.B., Graham, D.Y. and Estes, M.K. (1980) *J. Virol.* 33, 1111–1121.

Mason, B.B., Graham, D.Y. and Estes, M.K. (1983) *J. Virol.* 46, 413–423.

Matsuno, S. and Inouye, S. (1983) *Infect. Immun.* 39, 155–158.

Matsuno, S., Inouye, S. and Kono, R. (1977) *Infect. Immun.* 17, 661–662.

Matsuno, S., Hasegawa, A., Kalica, A.R. and Kono, R. (1980) *J. Gen. Virol.* 48, 253–256.

McCrae, M.A. and McCorquodale, J.G. (1982) *Virology* 117, 435–443.

McNulty, M.S., Todd, A.D., McFerran, J.B., McKillop, E.R., Collins, D.S. and McCracken, R.M. (1980) *Avian Pathol.* 9, 363–375.

McNulty, M.S., Allan, G.M., Todd, D., McFerran, J.B. and McCracken, R.M. (1981) *J. Gen. Virol.* 55, 405–413.

Murakami, Y., Nishioka, N., Hashiguchi, Y. and Kuniyasu, C. (1983) *Infect. Immun.* 40, 851–855.

Newman, J.F.E., Brown, F., Bridger, J.C. and Woode, G.N. (1975) *Nature* 258, 631–633.

Nicolas, J.C., Cohen, J., Fortier, B., Lourenco, M.H. and Bricout, F. (1983) *Virology* 124, 181–184.

Novo, E. and Esparza, J. (1981) *J. Gen. Virol.* 56, 325–335.

Offit, P., Clark, H.F. and Plotkin, S.A. (1983a) *Infect. Immun.* 42, 293–300.

Offit, P.A., Clark, H.F., Stroop, W.G., Twist, M. and Plotkin, S.A. (1983b) *J. Virol. Meth.* 7, 29–40.

Pedley, S., Bridger, J.C., Brown, J.F. and McCrae, M.A. (1983) *J. Gen. Virol.* 64, 2093–2101.

Petrie, B.L. (1983) In *Double-Stranded RNA Viruses* (R.W. Compans and D.H.L. Bishop, Eds.) Elsevier, New York. pp. 146–156.

Petrie, B.L., Estes, M.K. and Graham, D.Y. (1983) *J. Virol.* 46, 270–274.

Riepenhoff-Talty, M., Bogger-Goren, S., Li, P., Carmody, P.J., Barrett, H.J. and Ogra, P.L. (1981) *J. Med. Virol.* 8, 215–222.

Riepenhoff-Talty, M., Lee, P.-C., Carmody, P.J., Barrett, H.J. and Ogra, P.L. (1982) *Proc. Soc. Exp. Biol. Med.* 170, 146–154.

Rodger, S.M., Bishop, R.F., Birch, C., McLean, B. and Holmes, I.H. (1981). *J. Clin. Microbiol.* 13, 272–278.

Rodger, S.M., Bishop, R.F. and Holmes, I.H. (1982) *J. Clin. Microbiol.* 16, 724–726.

Roseto, A., Escaig, J., Delain, E., Cohen, J. and Scherrer, R. (1979) *Virology* 98, 471–475.

Rotbart, H.A., Levin, M.J., Yolken, R.H., Manchester, D.K. and Jantzen, J. (1983) *J. Pediatr.* 103, 454–459.

Sabara, M. and Babiuk, L.A. (1983) In *Double-stranded RNA Viruses* (R.W. Compans and D.H.L. Bishop, Eds.) Elsevier, New York. pp. 95–106.

Saif, L.J. and Bohl, E.H. (1980) In *Proc. 3rd Int. Symp. Neonatal Diarrhea* (S.D. Acres, A.J. Forman and H. Fast, Eds.) VIDO, Saskatoon. pp. 83–101.

Saif, L.J., Redman, D.R., Smith, K.L. and Theil, K.W. (1983) *Infect. Immun.* 41, 1118–1131.

Sato, K., Inaba, Y., Miura, Y., Tokuhisa, S. and Matumoto, M. (1982) *Arch. Virol.* 73, 45–50.

Saulsbury, F.T., Winkelstein, J.A. and Yolken, R.H. (1980) *J. Pediatr.* 97, 61–65.

Sheridan, J.F., Eydelloth, R.S., Vonderfecht, S.L. and Aurelian, L. (1983) *Infect. Immun.* 39, 917–927.

Soler, C., Musalem, C., Lorono, M. and Espejo, R.T. (1983) *J. Virol.* 44, 983–992.

Sonza, S. and Holmes, I.H. (1980) *Med. J. Aust.* 2, 496–499.

Spence, L., Fauvel, M., Petro, R. and Babiuk, L.A. (1978) *Can. J. Microbiol.* 24, 353–362.

Street, J.E., Croxson, M.C., Chadderton, W.F. and Bellamy, A.R. (1982) *J. Virol.* 43, 369–378.

Thouless, M.E. (1979) *J. Gen. Virol.* 44, 187–197.

Thouless, M.E., Bryden, A.S., Flewett, T.H., Woode, G.N., Bridger, J.C., Snodgrass, D.R and Herring, J.A. (1977) *Arch. Virol.* 53, 287–294.

Thouless, M.E., Beards, G.M. and Flewett, T.H. (1982) *Arch. Virol.* 73, 219–230.

Urasawa, S., Urasawa, T. and Taniguchi, K. (1982) *Infect. Immun.* 38, 781–784.

Vesikari, T., Maki, M. and Isolauri, E. (1983a) *Dev. Biol. Stand.* 53, 229–236.

Vesikari, T., Isolauri, E., Delem, A., D'Hondt, E., André, F.E. and Zissis, G. (1983b) *Lancet* ii, 807–811.

Vonderfecht, S.L. and Osburn, B.I. (1982) *J. Clin. Microbiol.* 16, 935–942.

Woode, G.N., Bridger, J.C., Jones, J.M., Flewett, T.H., Bryden, A.S., Davies, H.A. and White, G.B.B. (1976) *Infect. Immun.* 14, 804–810.

Woode, G.N., Kelso, N.E., Simpson, T.F., Gaul, S.K., Evans, L.E. and Babiuk, L. (1983) *J. Clin. Microbiol.* 18, 358–364.

Wyatt, R.G., Mebus, C.A., Yolken, R.H., Kalica, A.R., James, H.D., Kapikian, A.Z. and Chanock, R.M. (1979) *Science* 203, 548–550.

Wyatt, R.G., Greenberg, H.B., James, W.D., Pittman, A.L., Kalica, A.R., Flores, J., Chanock, R.M. and Kapikian, A.Z. (1982) *Infect. Immun.* 37, 110–115.

Wyatt, R.G., Kapikian, A.Z. and Mebus, C.A. (1983a) *J. Clin. Microbiol.* 18, 505–508.

Wyatt, R.G., James, H.D., Pittman, A.L., Hoshino, Y., Greenberg, H.B., Kalica, A.R., Flores, J. and Kapikian, A.Z. (1983b) *J. Clin. Microbiol.* 18, 310–317.

Yolken, R.H., Barbour, B., Wyatt, R.G., Kalica, A.R., Kapikian, A.Z. and Chanock, R.M. (1978) *Science* 201, 259–262.

Yolken, R.H. and Stopa, P.J. (1979) *J. Clin. Microbiol.* 10, 703–707.

Yolken, R.H., Bishop, C.A., Townsend, T.R., Bolyard, E.A., Bartlett, J., Santos, G.W. and Saral, R. (1982) *N. Engl. J. Med.* 306, 1009–1012.

Zissis, G. and Lambert, J.P. (1982) *J. Clin. Microbiol.* 11, 1–5.

Eds. M.H.V. van Regenmortel and A.R. Neurath
Immunochemistry of viruses. The basis for serodiagnosis and vaccines.

Antigenic structure of togaviruses 22

JAMES H. STRAUSS and ELLEN G. STRAUSS

Division of Biology, California Institute of Technology, Pasadena, CA 91125, U.S.A.

Togaviruses are defined as enveloped animal viruses whose genome is one linear molecule of infectious single-stranded RNA contained in an icosahedral nucleocapsid. The family Togaviridae currently contains four genera: Alphavirus, Flavivirus, Rubivirus and Pestivirus. Many of these viruses, including members of all four genera, are important human or veterinary pathogens (Matthews, 1982). Our knowledge of the structure and molecular biology of replication of the alphaviruses is now fairly detailed, and this chapter will focus primarily upon these viruses. The complete sequence of the RNA of the prototype alphavirus, Sindbis virus, is known (Strauss et al., 1984) and thus the primary sequence of all of the virus-encoded proteins is known. In addition, complete or partial sequences of the structural proteins of several of the alphaviruses, including Semliki Forest virus (Garoff et al., 1980a; 1980b), Ross River virus (Dalgarno et al., 1983), Western, Eastern and Venezuelan equine encephalitis viruses, Middelburg virus, and Barmah Forest virus (Bell et al., 1984a) are also known. For the flaviviruses, which include such important pathogens as yellow fever virus, dengue virus, Saint Louis encephalitis virus, Japanese encephalitis virus, and tick-borne encephalitis virus, knowledge of protein structure is only now emerging, while for the rubiviruses (Rubella virus) and pestiviruses (bovine viral diarrhea virus, hog cholera virus) little structural information is currently available.

Note added in proof: Since this chapter was completed, the flaviviruses have been removed from the family Togaviridae and classified as a separate family (Flaviviridae) based upon fundamental differences in the replication and genome organization of these two groups.

1. The alphaviruses

1.1. Structure of the alphaviruses

The genome of the alphaviruses is a positive sense single-stranded RNA approximately 11 700 nucleotides in length that is capped at the 5′ end and polyadenylated at the 3′ end (reviewed in Strauss and Strauss, 1983). This RNA is complexed with multiple copies of a single species of virus-encoded capsid (C) protein to form a nucleocapsid about 40 nm in diameter. The nucleocapsid is surrounded by a lipid bilayer in which are embedded two species of integral membrane glycoproteins, termed E1 and E2, such that the complete virus particle is about 70 nm in diameter (Strauss and Strauss, 1977). These glycoproteins are virus encoded and it is the external regions of the glycoproteins that interact with the host immune system to elicit neutralizing and protective antibodies. The envelope is acquired when the capsid buds through the plasma membrane, but there are no detectable host cell proteins in purified virions (Strauss, 1978).

1.2. Classification of the alphaviruses

1.2.1. Serological classification of the alphaviruses

Currently, viruses have been assigned to the genus Alphavirus (formerly known as the Group A Arboviruses) on the basis of serological cross-reactions with known members of the group, utilizing neutralization tests, complement fixation and hemagglutination inhibition. Approximately 25 alphaviruses are currently recognized of which many have geographical variants. These have been divided into three to six subgroups on the basis of cross-neutralization tests (see Porterfield, 1980; Calisher et al., 1980). Table 22.1 contains a partial list of alphaviruses with abbreviations to be used in this chapter; the grouping into complexes is modified from Calisher et al. (1980). Although grouping on the basis of serological cross-reactions has been a sensitive and usually reliable method of classification for the Togaviruses, the method has its limitations and protein and nucleic acid sequences now being obtained provide a more accurate basis for classification (see Section 1.2.2).

1.2.2. Classification of alphaviruses on the basis of protein sequences

Recently the N-terminal amino acid sequences of the two glycoproteins of eight different alphaviruses have been obtained (Bell et al., 1984a). These sequence data are shown in Fig.22.1 and clearly illustrate that these eight viruses, and presumably all alphaviruses, are closely related, having evolved from a common ancestor. Within this region the proteins are 32–83% homologous, depending upon which glycoproteins of which viruses are being compared. As the viruses diverged from one another, amino acid substitutions have occurred at many but not all positions within this sequence. All cysteines appear to have been conserved (although cysteine is difficult to recover in such sequencing experiments) and, in addition, many of the

TABLE 22.1

Alphavirus subgroups

Subgroup	Virus	Abbreviation
EEE	Eastern equine encephalitis	EEE
WEE	Western equine encephalitis	WEE
	Sindbis	SIN
	Highlands J	HJ
MID[a]	Middelburg	MID
NDU[a]	Ndumu	NDU
BF[b]	Barmah Forest	BF
VEE	Venezuelan equine encephalitis	VEE
	Everglades	EVE
SF[c]	Semliki Forest	SF
	Ross River	RR
	Chikungunya	CHIK
	O'nyong-nyong	ONN

[a]MID and NDU are grouped with the WEE complex by Porterfield (1980).
[b]BF was originally classified as a Bunyavirus (Marshall et al., 1982) on the basis of cross-reaction with Umbre virus. However, recent protein sequence data show that it is an alphavirus (see text). More extensive serological data will have to be obtained to determine if BF should be classified with one of the existing subgroups. The amino acid sequence data (Figs. 22.1 and 22.2) suggest that BF is most closely related to VEE, but in the absence of well-established serological cross-reactions we have placed BF in a separate subgroup.
[c]The SF complex is included in the VEE complex by Porterfield (1980).

prolines and aromatic amino acids as well as certain glycine residues are conserved. Since these amino acids are among the most important in determining the three dimensional configuration of proteins in solution it appears that the overall conformation of the glycoprotein is maintained and that within this framework a considerable amount of variation in primary sequence can be accommodated.

It is of interest that BF is clearly an alphavirus on the basis of protein sequence (Dalgarno et al., 1984). BF shows only limited serological cross-reaction with other alphaviruses tested and was originally classified as a Bunyavirus because of cross-reactions, presumably adventitious, with Umbre virus, a known Bunyavirus (Marshall et al., 1982). Serological cross-reactions examine only the epitopes responsible for antibody response and these can evolve to the point that these epitopes are serologically distinct while the core structure of the glycoprotein is essentially unchanged.

From the data in Fig. 22.1, an evolutionary tree can be derived which illustrates graphically the relationship of these viruses to one another (Fig. 22.2). This tree can be looked upon either as a measure of the evolutionary divergence of these viruses or as a measure of the relatedness of any pair of viruses. It is in reasonable agreement

GLYCOPROTEIN E1

```
                10            20            30            40            50            60
SIN    YEHATTVPNVPQIPYKALVERAGYAPLNLEITVMSSEVLPSTNQEYITCKFTTVVPSPKIKCCGSL
WEE    FEHATTVPNVPGIPYKALVERAGYAPLNLEITVVSSELTPSTNKEYVT?KFTTVIPSPQVK??GAS
VEE    YEHATTMPSQAGISYNTIVNRAGYAPLPISITPTKIKLIPTPNLEYVT?HYKTGV??PTY
BF     YEHSTTMPNQVGIPFKALIERPGYAALPLSLVVIKSELVPSLVQDYIT?NYKTVVPSPYI
EEE    YEHTAVMPNKVGIPYKALVERPGYAPVHLQIQLVNTRIIPITNLEYI??K
SF     YEHSTVMPNVVGFPYKAHIERPGYSPLTLQMQVVETSLEPTLNLEYITCEYKTVVPSPYVKCCGAS
MID    YEHSVTLPNAVGFPYRAHVDRPGFSPLTLHMEVV?TSLEPTLA??Y
RR     YEHTATIPNVVGFPYKAHIERNGFSPMTLQLEVVETSLEPTLNLEYITCEYKTVVPSPFIKCCGTS
```

GLYCOPROTEIN E2

```
                10            20            30            40            50            60
SIN    SVIDDFT    LTSPYLGTCSYCHHTEPCFSPVKIEQVWDEADDNTIRIQTSAQFGYDQSGAASANKYRY
WEE    SITDDFT    LTSPYLGFCPYCRHSAPCFSPIKIENVWDESDDGTIRIQVSAQFGYNQAGTADV?KFRY
VEE    STEELFKEYKLTRPYMARCIRCAVG SCHT?IAIEAVK
BF     SVA HFEAYKATRPYIGWCADCGLAGSCPSPVDIEHVWSD
EEE    DLDTHFTQYKLARPYIADCPNCMHS
SF     SVSQHFNVYKATRPYIAYCADCGAGHSCHSPVAIEAVRSEATDGMLKIQFSAQIGIDKSDNHDYTKIRY
MID    ?LTEDYEAYKLTKPYIAYCSDCGNGQFCYSPIAIETV?AEASDGMLKIQISAQIGLQVDEAHSWTKIRY
RR     SVTEHFNVYKATRPYLAYCADCGDGYFCYSPVAIEKIRDEAPDGMLKIQVSAQIGLDKAGTHAHTKIRY
```

Fig. 22.1. Protein sequences at the amino termini of the glycoproteins of eight alphaviruses. E1 and E2 were separated, purified and sequenced by automated Edman degradation in a spinning cup sequenator (Bell et al., 1978; 1983; 1984a; Dalgarno et al., 1984). In the case of SIN (Rice and Strauss, 1981), SF (Garoff et al., 1980b), MID (Bell et al., 1984a) and RR (Dalgarno et al., 1983) the data have been augmented with amino acid sequences derived from nucleotide sequencing (see Fig. 22.4). Underlining indicates some uncertainty in assignment; ? indicates that no assignment was made. Boxed residues are those of highest conservation. Virus abbreviations are the same as in Table 22.1.

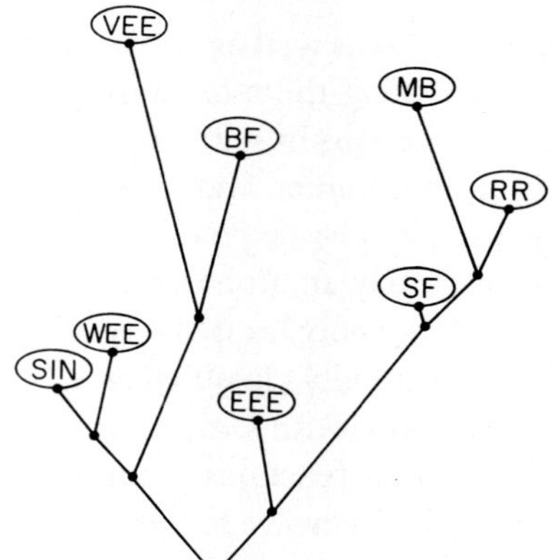

Fig. 22.2. An evolutionary tree of the alphaviruses. Using the protein sequence data for E1 from Fig. 22.1 a tree has been constructed. Branch lengths are proportional to the number of amino acid changes between any two virus E1s (Bell et al., 1984a).

with the classification based on serology (Table 22.1). However, in this tree, MID, SF and RR form a complex of closely related viruses which are well separated from the remaining alphaviruses, in contrast to their serological grouping.

1.3. Synthesis and processing of the structural proteins

Alphavirus structural proteins are produced as a polyprotein by translation of a subgenomic 26S mRNA and subsequently processed by proteolytic cleavage (Fig. 22.3; reviewed in Schlesinger and Kääriäinen, 1980; Strauss and Strauss, 1983). The capsid protein is first translated and then cleaved, probably autocatalytically, while still nascent. The two glycoproteins are translated on membrane-bound polysomes on the rough endoplasmic reticulum and cotranslationally inserted into the lumen of the endoplasmic reticulum. The N-terminus of the E2 precursor (PE2) serves as its signal sequence for insertion (Garoff et al., 1978), while the 6-kDa protein acts as an internal signal for insertion of E1 (Hashimoto et al., 1981). PE2 and E1, in the form of a complex, are transported to the smooth endoplasmic reticulum, the Golgi and finally to the plasma membrane of the cell. During this transport carbohydrate moieties are added and modified and fatty acid residues are attached to the hydrophobic anchors of both proteins using cellular pathways for the synthesis and transport of cell surface glycoproteins (reviewed in Schlesinger and Kääriäinen, 1980; Sabatini et al., 1982).

E1 is easily demonstrable on the cell surface by either antibody binding or lactoperoxidase iodination (reviewed in Brown, 1980). The situation with PE2 and E2 is less clear, and the location of the cleavage of PE2 to E2 is problematic. Because of

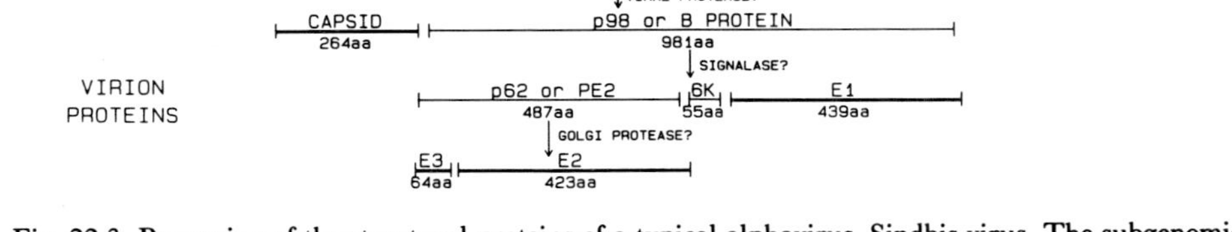

Fig. 22.3. Processing of the structural proteins of a typical alphavirus, Sindbis virus. The subgenomic region has been expanded. Untranslated regions of the subgenomic RNA are shown as single lines, and the translated region as an open box. The final protein products are shown as heavy lines. The open triangle is the initiation codon; the solid diamond is the termination codon. Data for construction of this figure are from Rice and Strauss (1981), Bell et al. (1978) and Ou et al. (1982).

```
      C
SIN   MNRGFFNML GRRPFPAPTAMWRPRRRRQAAPMPA        RNGLASQIQQLTTAVSALVIGQATRPQPPRPRPPPR QKKQAPKQPPKPKKPKTQEKKK  C ( 92)
SF    MNYIPTQTFYGRRWRPRPAARPWPL    QATPVAPVV     PDFQAQQMQQLISAVNALTMRQNAIAPARPPKPKKKKTTKPKPKTQPKKINGKTQQQKK    ( 93)
RR    MNYIPTQTFYGRRWRPRPAFRPWQVSM QPTPTMVTPMLQAPDLQAQQMQQLISAVSALTTKQNVKAPKGQRQKKQQKPKEKKENQKKKPTQKKKQQQKP    ( 99)

SIN   KQPA     KPKPGKRQRMALKLEADRLFDVKNEDGDVIGHALAMEGKVMKPLHVKGTIDHPVLSKLKFTKSSAYDMEFAQLPVNMRSEAFTYTSEHPEGF  C (188)
SF    KDKQADKKKKKPGKRERMCMKIENDCIFEVKHE GKVTGYACLVGDKVMKPAHVKGVIDNADLAKLAFKKSSKYDLECAQIPVHMRSDASKYTHEKPEGH    (192)
RR    KPQA    KKKKPGRRERMCMKIENDCIFEVKLD GKVTGYACLVGDKVMKPAHVKGTIDNPDLAKLTYKKSSKYDLECAQIPVHMKSDASKYTHEKPEGH    (195)
                                                                                      E3
SIN   YNWHHGAVQYSGGRFTIPRGVGGRGDSGRPIMDNSGRVVAIVLGGADEGTRTALSVVTWNSKGKTIKTTPEGTEEW SAAPLVTAMCLLGNVSFPCDRPP E3 ( 23)
SF    YNWHHGAVQYSGGRFTIPTGAGKPGDSGRPIFDNKGRVVAIVLGGANEGSRTALSVVTWN KDMVTRVTPEGSEEW SA PLITAMCVLANATFPCFQPP    ( 22)
RR    YNWHHGAVQYSGGRFTIPTGAGKPGDSGRPIFDNKGRVVAIVLGGANEGARTALSVVTWT KDMVTRVTPEGTEEW SA ALM  MCILANTSFPCSSPP    ( 20)
                                                    E2
SIN      TCYTREPSRALDILEENVNHEAYDTLLNAILRCGSSGRSKR SVIDGFT    LTSPYLGTCSYCHHTEPCFSPVKIEQVWDEADDNTIRIQTSAQFG E2 ( 52)
SF    CVPCCYENNAEATLRMLEDNVDRPGYYDLLQAALTCRNGTRHRR SVSQHFNVYKATRPYIAYCADCGAGHSCHSPVAIEAVRSEATDGMLKIQFSAQIG    ( 55)
RR    CYPCCYEKQPEQTLRMLEDNVNRPGYYELLEASMTCRNRSRHRR SVTEHFNVYKATRPYLAYCADCGDGYFCYSPVAIEKIRDEAPDGMLKIQVSAQIG    ( 55)

SIN   YDQSGAASANKYRYMSLKQDHTVKEGTMDDIKISTSGPCRRLSYKGYFLLAKCPPGDSVTVSIVSS NSATSCTLARKIKPKFVGREKYDLPPVHGKKIP E2 (151)
SF    IDKSDNHDYTKIRYA    DGHAIENAVRSSLKVATSGDCFVHGTMGHFILAKCPPGEFLQVSIQDTRNAVRACRIQYHHDPQPVGREKFTIRPHYGKEIP    (152)
RR    LDKAGTHAHTKIRYM    AGHDVQESKRDSLRVYTSAACSIHGTMGHFIVAHCPPGDYLKVSFEDADSHVKACKVQYKHDPLPVGREKFVVRPHFGVELP    (152)

SIN   CTVYDRLKETTAGYITMHRPRPHAYTSYLEESSGKVYAKPPSGKNITYECKCGDYKTGTVSTRTEITGCTAIKQCVAYKSDQTKWVFNSPDLIRHDDHTA E2 (251)
SF    CTTYQQTTAETVEEIDMHMPPDTPDRTLLSQQSGNV KITVGGKKVKYNCTCGTGNVGTTNSDMTINTC LIEQCHVSVTDHKKWQFNSPFVPRADEPAR    (250)
RR    CTSYQLTTAPTDEEIDMHTPPDIPDRTLLSQTAGNV KITAGGRTIRYNCTCGRDNVGTTSTDKTINTC KIDQCHAAVTSHDKWQFTSPFVPRADQTAR    (250)

SIN   QGKLHLPFKLIPSTCMVPVAHAPNVIHGFKHISLQLDTDHLTLLTTRRLGANPEPTTEWIVGKTVRNFTVDRDGLEYIWGNHEPVRVYAQESAPGDPHGW E2 (351)
SF    KGKVHIPFPLDNITCRVPMAREPTVIHGKREVTLHLHPDHPTLFSYRTLGEDPQYHEEWVTAAVERTIPVPVDGMEYHWGNNDPVRLWSQLTTEGKPHGW    (350)
RR    RGKVHVPFPLTNVTCRVPLARAPDVTYGKKEVTLRLHPDHPTLFSYRSLGAEPHPYEEWVDKFSERIIPVTEEGIEYQWGNNPPVRLWAQLTTEGKPHGW    (350)
                                                                                   6K
SIN   PHEIVQHYYHRHPVYTILAVASATVAMMIGVTVAVLCACKARRECLTPYALAPNAVIPTSLALLCCVRSANA ETFTETMSYLWSNSQPFFWVQL       6K ( 22)
SF    PHQIVQYYYGLYPAATVSAVVGMSLLALISIFASCYMLVAARSKCLTPYALTPGAAVPWTLGILCCAPRAHA ASVAETMAYLWDQNQALFWLEFAAPVA    ( 27)
RR    PHEIIQYYYGLYPAATIAAVSGASLMALLTLAATCCMLATARRKCLTPYALTPGAVVPLTLGLLCCAPRANA ASFAETMAYLWDENKTLFWMEFAAPAA    ( 27)
                                        E1
SIN   CIPLAAFIVLMRCCSCCLPFLVVAGAYLAKVDA YEHATTVPNVPQIPYKALVERAGYAPLNLEITVMSSEVLPSTNQEYITCKFTTVVPSPKIKCCGSL E1 ( 66)
SF    CILIITYCLRNVLCCCKSLSFLVLLSLGATARA YEHSTVMPNVVGFPYKAHIERPGYSPLTLQMQVVETSLEPTLNLEYITCEYKTVVPSPYVKCCGAS    ( 66)
RR    ALALLACCIKSLICCCKPFSFLVLLSLGASAKA YEHTATIPNVVGFPYKAHIERNGFSPMTLQLEVVETSLEPTLNLEYITCEYKTVVPSPFIKCCGTS    ( 66)

SIN   ECQPAAHADYTCKVFGGVYPFMWGGAQCFCDSENSQMSEAYVELSAVCASDHAQAIKVHTAAMKVGLRIVYGNTTSFLDVYVNGVTPGTSKDLKVIAGPI E1 (166)
SF    ECSTKEKPDYQCKVYTGVYPFMWGGAYCFCDSENTQLSEAYVDRSDVCRHDHASAYKAHTASLKAKVRVMYGNVNQTVDVYVNGDHAVTIGGTQFIFGPL    (166)
RR    ECSSKEQPDYQCKVYTGVYPFMWGGAYCFCDSENTQLSEAYVDRSDVCKHDHASAYKAHTASLKATIRISYGTINQTTEAFVNGEHAVNVGGSKFIFGPI    (166)

SIN   SASFTPFDHKVVIHRGLVYNYDFPEYGAMKPGAFGDIQATSLTSKDLIASTDIRLLKPSAKNVHVPYTQASSGFEMWKNNSGRPLQETAPFGCKIAVNPL E1 (266)
SF    SSAWTPFDNKIVVYKDEVFNQDFPPYGSGQPGRFGDIQSRTVESNDLYANTALKLARPSPGMVHVPYTQTPSGFKYWLKEKGTALNTKAPFGCQIKTNPV    (266)
RR    STAWSPFDNKIVVYKDDVYNQDFPPYGSGQPGRFGDIQSRTVESKDLYANTALKLSRPSPGVVHVPYTPTPSGFKYWLKEKGSSLNTKAPFGCKIKTNPV    (266)

SIN   RAVDCSYGNIPISIDIPNAAFIRTSDAPLVSTVKCEVSECTYSADFGGMATLQYVSDREGQCPVHSHSSTATLQESTVHVLEKGAVTVHFSTASPQANFI E1 (366)
SF    RAMNCAVGNIPVSMNLPDSAFTRIVEAPTIIDLTCTVATCTHSSDFGGVLTLTYKTNKNGDCSVHSHSNVATLQEATAKVKTAGKVTLHFSTASASPSFV    (366)
RR    RAMDCAVGSIPVSMDIPDSAFTRVVDAPAVTDLSCQVVVCTHSSDFGGVATLSYKTDKPGKCAVHSHSNVATLQEATVDVKEDGKVTVHFSTASASPAFK    (366)

SIN   VSLCGKKTTCNAECKPPADHIVSTPHKNDQEFQAAISKTSWSWLFALFGGASSLLIIGLMIFACSMMLTSTRR                             E1 (439)
SF    VSLCSARATCSASCEPPKDHIVPYAASHSNVVFPDMSGTALSWVQKISGGLGAFAIGAILVLVVVTC IGLRR                                 (438)
RR    VSVCDAKTTCTAACEPPKDHIVPYGASHNNQVFPDMSGTAMTWVQRLASGLGGLALIAVVVLVLVTC ITMRR                                 (438)
```

Fig. 22.4. Amino acid sequences of alphavirus structural proteins. The amino acid sequences of the structural proteins of SIN, SF and RR, deduced from the nucleotide sequences of their respective 26S subgenomic RNAs, are shown. Gaps have been introduced to align the proteins. Data are from Garoff (1980a; 1980b), Rice and Strauss (1981) and Dalgarno et al. (1983). Potential glycosylation sites are shaded and transmembrane domains are underlined with dots.

similarities in the cleavage of PE2 with those of proproteins and other virus glycoproteins (reviewed in Lodish, 1981), we believe that this cleavage is catalyzed by a Golgi protease during transport to the cell surface. In agreement with this PE2 cannot be iodinated at the surface of the cell, although E2 in virions is readily iodinated (Sefton et al., 1973). On the other hand the cleavage of PE2 is inhibited by anti-E1 antibodies added to intact cell monolayers, which suggests that the cleavage may not occur until the PE2-E1 complex reaches the cell surface (Bracha and Schlesinger, 1976). In any event there probably occurs a configurational rearrangement in a PE2-E1 heterodimer (Ziemiecki and Garoff, 1978) upon cleavage that could be important for studies showing that infected cells exhibit different epitopes from mature virions (see Section 1.5).

Maturation of the virus occurs when nucleocapsids, formed in the cytoplasm from genomic RNA and capsid protein, migrate to the cell surface, interact with the transmembranous roots of one or both glycoproteins and bud out through the plasmalemma (reviewed in Brown, 1980). Alphaviruses thus acquire their envelopes and are released into the extracellular space in the same event. On the surface of the virion, heterodimers of E1 and E2 (Ziemiecki and Garoff, 1978; Rice and Strauss, 1982) form the 'spikes' seen in electron micrographs. All analyses of either infected cells or virus must take into account the fact that these two glycoproteins form a tight complex and that antibodies to a site on one moiety may block the function of the other.

1.4. Domains of the structural proteins

The complete amino acid sequences of the structural proteins of three alphaviruses, SIN, SF and RR, have been obtained from the nucleotide sequence of the 26S RNA messenger which encodes these proteins. These amino acid sequences are shown in Fig. 22.4. The homologies among the virus proteins are illustrated graphically in Fig. 22.5 and a number of the features discussed below have been indicated in these figures.

The capsid proteins of the three viruses have two distinct domains. The N-terminal half shows little sequence homology among viruses but contains strings of arginine and lysine residues as well as large numbers of proline and glutamine residues. The lack of sequence conservation, as well as the high proportion of charged amino acids, suggests that this domain of the capsid protein interacts non-specifically and electrostatically with the genomic RNA, perhaps penetrating into the interior of the nucleocapsid in so doing. The C-terminal half of the capsid protein, in marked contrast, is highly conserved in all three viruses. This region is presumably responsible for the specific protein-protein and protein-RNA interactions required to form the nucleocapsid shell, for the interaction of the capsid with the glycoproteins during budding, and for the proteolytic cleavage which separates the capsid protein from the polyprotein precursor.

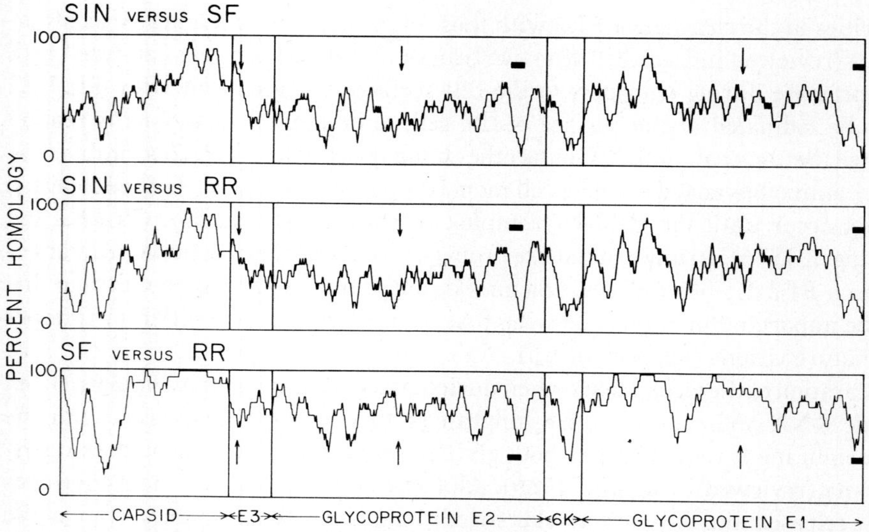

Fig. 22.5. Graphic homology plot of the structural polyprotein of SIN, SF and RR. Two amino acid sequences are compared over a string length of 20 and the degree of homology plotted as a moving average (100% homology means that all 20 amino acids are identical). Boundaries of the individual proteins are indicated. Features of the protein indicated are carbohydrate attachment sites (arrows) and transmembrane anchors (bars).

Glycoprotein E2 and its precursor PE2 contain a number of domains. As noted, the N-terminal 19 residues of PE2 form a signal sequence for membrane insertion during synthesis. This signal sequence is not cleaved during synthesis, but is eventually removed upon the cleavage of PE2 to form E3 and E2. E3 does not form a component of the virion in the case of most alphaviruses (although it does remain virus associated in the case of SF).

E2 has a membrane-spanning domain near the C-terminal end. In the case of Sindbis virus E2, this domain includes residues 365–390 (the total length of E2 is 423 residues; Fig. 22.3) (Rice et al., 1982). This domain shows very little conservation among alphaviruses (Figs. 22.4 and 22.5). The residues C-terminal to this are found in the cytoplasm or may possibly loop back through the bilayer. Conservation in the C-terminus is greater than in the transmembranous region; in particular cysteine and proline residues are conserved, implying that the secondary structure in this domain is important for function. This domain may interact with the capsid during budding.

Conservation of amino acid sequence throughout the rest of E2 is scattered, and no strongly conserved domains exist. The overall degree of conservation is less than that of E1 or C. Because of these observations and the fact that E2 carries the neutralization activity (see below), we have suggested that an evolutionary function of E2 is to generate strain diversity (Arias et al., 1983; Dalgarno et al., 1983).

Glycoprotein E1 also contains several important domains. There is a hydrophobic, highly conserved stretch of amino acids from residues 80 to 96 which may possess the fusion activity of the virus (Garoff et al., 1980b; Rice and Strauss, 1981; Dalgarno et al., 1983). This lies within a larger conserved region from residues 45 to 108 (Fig. 22.4). The available evidence suggests that, when the virus attaches to specific receptors it is pinocytosed and transported to lysosomes. Here the acidic pH is thought to trigger a conformational rearrangement of E1 leading to the fusion of the virus envelope with the lysosomal membrane and introducing the virus RNA into the cytoplasm (Helenius et al., 1980).

The membrane-spanning domain of E1 is at the extreme C-terminus and is not well conserved. A stretch of 25 uncharged amino acids spans the bilayer (residues 413–437) and is followed by two C-terminal arginine residues which are believed to be cytoplasmic in location (Rice et al., 1982).

The conservation of amino acid sequence observed in these structural proteins is of special interest because of the observation that the alphaviruses have evolved to the extent that the nucleic acid sequence encoding these proteins has diverged quite markedly and the codon used to encode any particular conserved amino acid has been essentially randomized (Rice and Strauss, 1981; Ou et al., 1982; Strauss et al., 1983). Thus any conservation of amino acid sequence observed is highly significant.

The E2s of SIN, SF and RR all have two attachment sites for asparagine-linked oligosaccharides. The first of these sites, Asn 196 for SIN or Asn 199 for SF and RR, is conserved. In the case of Sindbis virus a complex chain is known to be attached here (Mayne, 1983), and presumably this is so in all three alphaviruses. The second site is not conserved as to position. Sindbis E2 has a simple chain attached to Asn 318 (Mayne, 1983).

E1 has one or two carbohydrate chains attached, depending on the virus. SIN, SF and RR all have a complex carbohydrate chain attached to Asn 139 (SIN) or Asn 141 (SF, RR) and thus, as for E2, a complex chain site is conserved. SIN has a second, simple chain at Asn 245 (Mayne, 1983). Although both E1 and E2 are glycoproteins, all available evidence suggests that the carbohydrate chains per se do not serve as antigenic determinants.

1.5. The structural proteins as antigens

Both polyclonal and monoclonal antisera have been prepared which are specific for each of the three structural proteins of SIN and VEE, and these have been used to characterize these proteins (SIN: Dalrymple et al., 1976; 1981; Roehrig et al., 1980; 1982a; Chanas et al., 1982; Clegg et al., 1983; Schmaljohn et al., 1983; VEE: Pedersen and Eddy, 1974; France et al., 1979; Roehrig et al., 1982b; Mathews and Roehrig, 1982). In general antibodies to glycoprotein E1 or to the capsid protein are cross-reactive and nonneutralizing, whereas antibodies to E2 are type-specific and neutralizing.

Isolated E1 glycoproteins from SIN (Dalrymple et al., 1976) and SF (Helenius et al., 1976) have been shown to hemagglutinate. Furthermore, a temperature-sensitive mutant of SIN, ts23, which has been mapped to E1 (Arias et al., 1983; see also Section 1.6 below), has a temperature-sensitive hemagglutinin (Yin, 1969). These data suggest that E1 carries the physical domain primarily responsible for hemagglutination. Both polyclonal and monoclonal antibodies to SIN E1 which inhibit hemagglutination have been isolated (Dalrymple et al., 1976; Chanas et al., 1982). However, both polyclonal and monoclonal antibodies to E2 of VEE inhibit hemagglutination while antibodies to E1 do not (Pedersen and Eddy, 1974; France et al., 1979). Thus it may be appropriate to consider the E1-E2 heterodimer as being the viral hemagglutinin and antibodies to either E1 or E2 may be active in the hemagglutination inhibition assays used to classify alphaviruses.

Dalrymple et al. (1976) found that polyclonal antibodies to SIN E1 did not neutralize virus infectivity in a plaque assay, but this anti-E1 antiserum was broadly cross-reactive in precipitation experiments. More recent studies using monoclonal antibodies to E1 have in general confirmed these results. A series of E1 monoclonals studied by Dalrymple et al. (1981) were found to be nonneutralizing and cross-reactive. In other studies, two E1 monoclonals have been found to be weakly neutralizing (Chanas et al., 1982; Mathews and Roehrig, 1982; Schmaljohn et al., 1982). Despite being nonneutralizing, E1 monoclonals can confer cross-protection in animal experiments, apparently because they can lead to cytolysis of infected cells by a complement-mediated pathway (Schmaljohn et al., 1982; Cole et al., 1982).

Dalrymple et al. (1976) also found that polyclonal antibodies to E2 did neutralize infectivity and were virus specific in precipitation experiments. Monoclonal antisera to E2 confirm these results. Many anti-E2 monoclonals neutralize infectivity, and these antisera show much greater specificity for the virus strain used to elicit the antibody response than do E1 monoclonals (Chanas et al., 1982; Roehrig et al., 1982b). These E2 monoclonals can evidently confer cross-protection in animal experiments either by neutralizing infectivity or by complement-mediated cytolysis of infected cells (Schmaljohn et al., 1983; Cole et al., 1982).

Analysis of monoclonal antibodies to alphavirus proteins have illustrated that both E1 and E2 possess a number of different epitopes which have been shown to be spatially distinct by competitive binding assays. More epitopes have been identified on E1 than on E2 (Chanas et al., 1982; Schmaljohn et al., 1983; Mathews and Roehrig, 1982). Several of the sites appear unavailable to antibody interaction in mature virions and are only revealed after detergent disruption. One particularly interesting epitope was rendered reactive by lowering the pH and may be involved in the pH-dependent fusion of the viral envelope in lysosomes (Schmaljohn et al., 1983). Differences in the reactivity of E1 and E2 between infected cells and virions have illustrated the importance of three-dimensional conformation for the function of these proteins (Gates et al., 1982; Wolcott et al., 1984).

Antibodies to C are also cross-reactive and nonneutralizing. The C protein has been shown to be important in complement fixation assays used for alphavirus classification.

Animal protection experiments have shown that alphaviruses can cross-protect in a number of instances where there is no circulating antibody which can neutralize the challenge virus. Transplantation experiments have implicated cytotoxic T cells in this cross-protection (Wolcott et al., 1982). The recent experiments with animal protection induced by monoclonal antibodies (cited above) provide an alternative mechanism, namely complement-mediated cytolysis.

1.6. Mutants of alphaviruses

A large number of alphavirus mutants have been isolated and studied (reviewed in Strauss and Strauss, 1980), but little information is available on the effects of mutation on antigenic structure. It would be very useful to have sequence data on mutants selected to be resistant to monoclonal antibodies or on closely related strains of viruses which have been characterized antigenically in order to localize the antigenic epitopes in the primary sequence. However, there is one study of temperature sensitive (ts) mutants of SIN which relates mutation site to phenotype. In the case of two ts mutants, ts10 and ts23, glycoprotein E1 is synthesized at the nonpermissive temperature and is inserted into the endoplasmic reticulum; however, it then fails to be transported to the plasma membrane, instead accumulating in the endoplasmic reticulum and Golgi apparatus (Saraste et al., 1980). Sequencing of these mutants and of revertants revealed that ts23 was a double mutant and ts10 was a single mutant (Arias et al., 1983). These changes are widely spaced in E1 and involve an Ala to Thr mutation at residue 106 (ts23), a Lys to Gly change at residue 176 (ts10) and an Arg to Gly change at position 267 (ts23). These data suggest that the three-dimensional conformation of E1 is essential for its transport to the plasma membrane and its function.

1.7. Antibody-mediated enhancement of plaque formation by togaviruses

A number of antibodies to Togaviruses have been studied which, when added to a virus suspension at low concentration, gave greater numbers of plaques in tissue culture than virus alone, while the same antiserum in many cases was neutralizing at higher concentrations. The seeming contradiction was apparently resolved by the finding that soluble antibody-virus complexes could infect macrophage cultures which lacked virus receptors by binding to Fc receptors (Kliks and Halstead, 1983). Virus-antibody complexes, therefore, appear to be able to infect more of the cells of a given culture than virus alone, leading to antibody-mediated plaque enhancement observed with both alphaviruses and flaviviruses (Chanas et al., 1982; Symington et al., 1977).

2. The flaviviruses

2.1. Structure of the flaviviruses

The genome of the flaviviruses is a positive-sense single-stranded RNA whose size is of the order of 11 kilobases. The RNA is capped on the 5′ end, but the 3′ end is unmodified, i.e. there is no poly(A) tract (Wengler and Wengler, 1981). As in the alphaviruses, the flavivirus RNA is complexed with multiple copies of a single basic capsid protein, called C or V2, with a molecular mass of 13 kDa, to form the nucleocapsid (Westaway, 1980; Westaway et al., 1980). The nucleocapsid is enveloped in a lipid bilayer with which are associated two proteins, both of which are probably integral membrane proteins. One is a glycoprotein with a molecular mass of about 60 kDa, called E or V3. The second is a small (9 kDa) nonglycosylated protein which has been called M or V1. This M protein is probably derived from a 19-kDa precursor glycoprotein, NV2 (also known as GP19), by a cleavage similar to that undergone by the alphavirus PE2 (C.M. Rice, unpublished).

Flaviviruses mature in association with internal membranes but the process by which the nucleocapsid becomes enveloped is not understood, as no budding figures have been observed in infected cells. Similarly, the manner in which the assembled particles, of size about 50 nm, are released from the cell is not understood.

2.2. Antigenic classification of flaviviruses

The flavivirus group was formerly known as Group B arboviruses. Flaviviruses may be tick-borne, mosquito-borne, or may apparently lack an arthropod vector. A partial listing of flaviviruses arranged according to serological subgroups is given in Table 22.2. The mosquito-borne and tick-borne viruses cross-react but fall into distinct serological subgroups. These two groups can also be distinguished by the mobility of the M protein (Shapiro et al., 1972). The mosquito-borne viruses can be subdivided into several groups, three of which are given in Table 22.2. The flaviviruses which lack a vector represent an interesting set of viruses. Clearly closely related to the vector-borne flaviviruses by serological cross-reaction, they have evidently lost the requirement for vector transmission and can be transmitted in another fashion. These viruses are thus not arboviruses, since they lack an arthropod vector.

2.3. Relationships between flaviviruses established by amino acid sequence comparisons

Partial amino acid sequence data are now known for the structural proteins of several flaviviruses (Boege et al., 1983; Bell et al., 1984b: C.M. Rice, unpublished results). Sequence data for the glycoprotein of three mosquito-borne flaviviruses

TABLE 22.2

Classification of flaviviruses[a]

Subgroup	Virus	Abbreviation
Tick-borne	Tick-borne encephalitis	TBE
	Russian spring and summer encephalitis[b]	RSSE
Mosquito-borne		
YF group	Yellow fever	YF
JE group	Japanese encephalitis	JE
	Kunjin	KUN
	West Nile	WN
	St. Louis encephalitis	SLE
	Murray Valley encephalitis	MVE
DEN group	Dengue types 1–4	DEN

[a]See Porterfield (1980) for a more extensive catalog of flaviviruses.
[b]RSSE is a member of the TBE complex.

and for the nucleocapsid protein of two mosquito-borne and one tick-borne flavivirus are shown in Fig. 22.6. The relationships among the flaviviruses are analogous to those exhibited by the alphaviruses. The glycoproteins of YF, DEN and SLE,

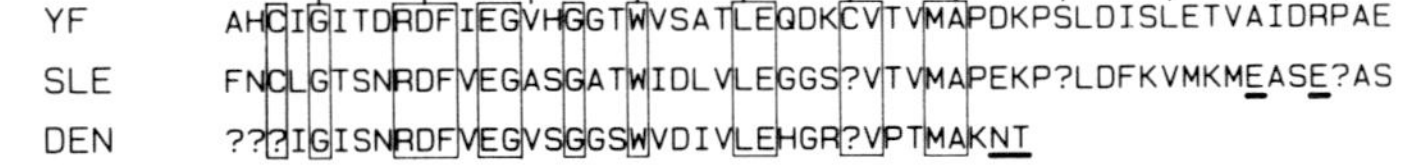

Fig. 22.6. N-terminal sequences of flavivirus proteins. Virus abbreviations are as in Table 22.2. Conventions are as in Fig. 22.1. Data are from Boege et al. (1983), Bell et al. (1984b), and C.M. Rice (unpublished).

which represent three different serotypes of the mosquito-borne flaviviruses (Table 22.2), are 52–60% homologous over the region sequenced and it appears that the three-dimensional conformation is conserved. The N-terminal domains of the nucleocapsid proteins are quite basic and the homology exhibited is less than that for the glycoproteins.

2.4. The flavivirus E protein as an antigen

Molecular characterization of flaviviruses has lagged behind that of alphaviruses for a number of reasons including the fact that flaviviruses grow poorly in culture and that most members are pathogenic. However, due to the importance of these agents as disease-causing organisms, the purely immunological investigations have been more aggressively pursued and there currently exist catalogs of polyclonal and monoclonal antibodies to DEN (Gentry et al., 1982), WN (Peiris et al., 1982), YF (Schlesinger et al., 1983), SLE (Roehrig et al., 1983), TBE (Heinz et al., 1983) and JE (Kimura-Kuroda and Yasui, 1983). All of the monoclonal antibodies are directed against the glycoprotein E. These were elicited either by immunization with intact virions or, more recently, with polymers of E protein isolated from virions after gentle disruption and delipidation by anionic detergents. Several classes of epitopes have been described, delineating the functions of the E protein, including neutralization and hemagglutination. Furthermore, the E protein appears to have regions of widely varying degrees of conservation among flaviviruses since some epitopes are group reactive (the most conserved), some subgroup reactive, and some are virus specific or even strain specific (the most divergent). Comparative protein sequence data of flavivirus E proteins will be of interest since this single polypeptide combines the functional and antigenic characteristics found on two separate glycoproteins in the alphaviruses.

2.5. Enhanced protection by sequential immunization with flaviviruses

Due to the seriousness of diseases caused by a number of flaviviruses there have been attempts to devise a vaccine which would protect against a number of these agents. One approach has been sequential immunization with 2 or 3 different flaviviruses, which leads to increased protection from challenge by the immunizing virus over that from a single immunization and also to the development of broadly cross-reactive circulating antibodies (Price et al., 1973). The antigenic basis for this response is not understood and is complicated by the fact that the efficacy of the procedure depends not only upon the particular viruses chosen, but also on the order of presentation. Although considered promising during the last decade, this approach has not been pursued, partially because of the severe nature of the disease caused by sequential natural infection with the dengue viruses (see next section).

2.6. Dengue hemorrhagic fever

One of the most perplexing of the disease syndromes associated with Togaviruses is dengue hemorrhagic fever/dengue shock syndrome. Uncomplicated dengue fever is an acute self-limiting febrile disease of children, but in a minority of cases it progresses to hemorrhagic fever, shock and death. The epidemiology of several recent outbreaks has been extensively reviewed by Halstead (1980). Most shock cases and mortality are associated with secondary dengue infections, usually when the primary infection was due to DEN type 1, 3 or 4 and the secondary agent was DEN 2. Moreover, the interval between the first and second exposures appears critical. This reaction is specific to sequential infection by dengue viruses since no excess morbidity occurs in populations first infected by JE or YF virus. The underlying causes of dengue shock syndrome are not understood and the analysis is complicated by the fact that DEN preferentially replicates in peripheral blood leukocytes and in other lymphatic tissues. Proliferation of these cells in an immune response to secondary infection could greatly expand the number of susceptible cells, increase viremia, and concomitantly enhance the severity of the disease. The severity of dengue hemorrhagic fever/dengue shock syndrome has led to extreme caution in initiating a vaccination program for any dengue virus.

3. Rubiviruses and pestiviruses

Only limited structural data on these two groups of Togaviruses are currently available (reviewed in Brinton, 1980). Rubella virus, the sole representative of the rubiviruses, has recently been found to exhibit properties suggesting that it is related to alphaviruses. The virion has two glycoproteins, E1 and E2 (molecular masses about 58 and 42–47 kDa, respectively), and a nucleocapsid protein C of 33 kDa. The sizes of all these proteins are similar to those of the corresponding proteins of alphaviruses (Trudel et al., 1982; Oker-Blom et al., 1983). Furthermore, a potential subgenomic RNA has been found which could conceivably encode these proteins (Oker-Blom et al., 1983; T. Frey, personal communication). No sequence data has been reported as yet, and thus the relationship to alphaviruses is unknown. Monoclonal antibodies to purified rubella virus have been prepared and one such antibody exhibited neutralizing activity without inhibiting hemagglutination, indicating that at least two epitopes are involved (Trudel et al., 1982).

4. Concluding remarks

The development of monoclonal antibodies and the development of protein and nucleic acid sequencing techniques have greatly increased our knowledge of the antigenic structure of Togaviruses in recent years and promise to expand greatly our future knowledge. Isolation of mutant strains resistant to monoclonal antibodies, together with sequence information on these and other antigenic variants, as well as studies using synthetic polypeptides as antigens, will be of importance in defining the immunologically active regions of the virion proteins. The finding that regions of alphavirus and flavivirus proteins are highly conserved may make it possible to develop group-reactive vaccines in a more systematic and successful fashion than the sequential immunization methodology tried to date (Section 2.5), using either recombinant DNA technology or synthetic peptide technology (reviewed in Lerner, 1982). As more sequence information on Togavirus genomes becomes available it will also increasingly define the evolutionary history of these viruses and the relationships among the various members of a genus and among the various genera of Togaviruses, knowledge that is important in the design and implementation of an effective strategy for the control of these viral diseases by vaccination.

5. Acknowledgements

We are grateful to T. Hunkapiller for computer programs used in construction of the figures, to L. Hood for use of computer facilities, and to J. Dalrymple and A. Schmaljohn for access to data prior to publication. Work of the authors is supported by Grants AI 10793 and AI 20612 from NIH and Grant PCM 83-16856 from NSF.

6. References

Arias, C., Bell, J.R., Lenches, E.M., Strauss, E.G. and Strauss, J.H. (1983) *J. Mol. Biol.* 168, 87–102.

Bell, J.R., Hunkapiller, M.W., Hood, L.E. and Strauss, J.H. (1978) *Proc. Natl. Acad. Sci. USA* 75, 2722–2726.

Bell, J.R., Bond, M.W., Hunkapiller, M.W., Strauss, E.G., Strauss, J.H., Yamamoto, K. and Simizu, B. (1983) *J. Virol.* 45, 708–714.

Bell, J.R., Kinney, R.W., Trent, D.W., Strauss, E.G. and Strauss, J.H. (1984a) *Proc. Natl. Acad. Sci. USA* 81, 4702–4706.

Bell, J.R., Kinney, R.W., Trent, D.W., Lenches, E.M., Dalgarno, L. and Strauss, J.H. (1984b) *Virology* In press.

Boege, U., Heinz, F.X., Wengler, G. and Kunz, C. (1983) *Virology* 126, 651–657.

Bracha, M. and Schlesinger, M.J. (1976) *Virology* 74, 441–449.

Brinton, M. (1980) In *The Togaviruses* (R.W. Schlesinger, Ed.) Academic Press, New York. pp. 623–666.

Brown, D.T. (1980) In *The Togaviruses* (R.W. Schlesinger, Ed.) Academic Press, New York. pp. 473–501.

Calisher, C.H., Shope, R.E., Brandt, W., Casals, J., Karabatsos, N., Murphy, F.A., Tesh, R.B. and Wiebe, M.E. (1980) *Intervirology* 14, 229–232.

Chanas, A.C., Gould, E.A., Clegg, J.C.S. and Varma, M.G.R. (1982) *J. Gen. Virol.* 58, 37–46.

Clegg, J.C.S., Chanas, A.C. and Gould, E.A. (1983) *J. Gen. Virol.* 64, 1121–1126.
Cole, G.A., Schmaljohn, A.L. and Dalrymple, J.M. (1982) In *Viral Diseases in Southwest Asia and the Western Pacific* (J.S. Mackensie, Ed.) Academic Press, Australia. pp. 541–545.
Dalgarno, L., Rice, C.M. and Strauss, J.H. (1983) *Virology* 129, 170–187.
Dalgarno, L., Short, N.J., Hardy, C.M., Bell, J.R., Strauss, J.H. and Marshall, I.D. (1984) *Virology* 133, 416–426.
Dalrymple, J.M., Schlesinger, S. and Russell, P.K. (1976) *Virology* 69, 93–103.
Dalrymple, J.M., Cole, G., Jahrling, P., Johnson, E. and Harrison, S. (1981) *Abstr. 5th Int. Congr. Virol.* 380.
France, J.K., Wyrick, B.C. and Trent, D.W. (1979) *J. Gen. Virol.* 44, 725–740.
Garoff, H., Simons, K. and Dobberstein, B. (1978) *J. Mol. Biol.* 124, 587–600.
Garoff, H., Frischauf, A.-M., Simons, K., Lehrach, H. and Delius, H. (1980a) *Proc. Natl. Acad. Sci. USA* 77, 6376–6380.
Garoff, H., Frischauf, A.-M., Simons, K., Lehrach, H. and Delius, H. (1980b) *Nature* 288, 235–241.
Gates, D., Brown, A. and Wust, C.J. (1982) *Infect. Immun.* 35, 248–255.
Gentry, M.K., Henchal, E.A., McCown, J.M., Brandt, W.E. and Dalrymple, J.M. (1982) *Am. J. Trop. Med. Hyg.* 31, 548–555.
Halstead, S.B. (1980) In *The Togaviruses* (R.W. Schlesinger, Ed.) Academic Press, New York. pp. 107–173.
Hashimoto, K., Erdei, S., Keränen, S., Saraste, J. and Kääriäinen, L. (1981) *J. Virol.* 38, 34–40.
Heinz, F.X., Berger, R., Tuma, W. and Kunz, C. (1983) *Virology* 126, 525–537.
Helenius, A., Fries, E., Garoff, H. and Simons, K. (1976) *Biochim. Biophys. Acta* 436, 319–334.
Helenius, A., Kartenbeck, J., Simons, K. and Fries, E. (1980) *J. Cell Biol.* 84, 404–420.
Kimura-Kuroda, J. and Yasui, K. (1983) *J. Virol.* 45, 124–132.
Kliks, S.C. and Halstead, S.B. (1983) *J. Virol.* 46, 394–404.
Lerner, R.A. (1982) *Nature* 299, 592–596.
Lodish, H.F. (1981) *Enzyme Microbiol. Technol.* 3, 178–188.
Marshall, I.D., Woodroofe, G.M. and Hirsch, S. (1982) *Aust. J. Exp. Biol. Med. Sci* 60, 457–470.
Mathews, J.H. and Roehrig, J.T. (1982) *J. Immunol.* 129, 2763–2767.
Matthews, R.E.F. (1982) *Intervirology* 17, 4–199.
Mayne, J.T. (1983) *PhD Thesis*, California Institute of Technology, Pasadena, CA.
Oker-Blom, C., Kalkkinen, N., Kääriäinen, L. and Pettersson, R.F. (1983) *J. Virol.* 46, 964–973.
Ou, J.-H., Rice, C.M., Dalgarno, L., Strauss, E.G. and Strauss, J.H. (1982) *Proc. Natl. Acad. Sci. USA* 79, 5235–5239.
Pedersen, C.E. and Eddy, G.A. (1974) *J. Virol.* 14, 740–744.
Peiris, J.S.M., Porterfield, J.S. and Roehrig, J.T. (1982) *J. Gen. Virol.* 58, 283–289.
Porterfield, J.S. (1980) In *The Togaviruses* (R.W. Schlesinger, Ed.) Academic Press, New York. pp. 13–46.
Price, W.H., Casals, J., Thind, I. and O'Leary, W. (1973) *Am. J. Trop. Med. Hyg.* 22, 509–523.
Rice, C.M. and Strauss, J.H. (1981) *Proc. Natl. Acad. Sci. USA* 78, 2062–2066.
Rice, C.M. and Strauss, J.H. (1982) *J. Mol. Biol.* 154, 325–348.
Rice, C.M., Bell, J.R., Hunkapiller, M.W., Strauss, E.G. and Strauss, J.H. (1982) *J. Mol. Biol.* 154, 355–378.
Roehrig, J.T., Corser, J.A. and Schlesinger, M.J. (1980) *Virology* 101, 41–49.
Roehrig, J.T., Gorski, D. and Schlesinger, M.J. (1982a) *J. Gen. Virol.* 59, 421–425.
Roehrig, J.T., Day, J.W. and Kinney, R.M. (1982b) *Virology* 118, 269–278.
Roehrig, J.T., Mathews, J.H. and Trent, D.W. (1983) *Virology* 128, 118–126.
Sabatini, D.D., Kreibich, G., Morimoto, T. and Adesnik, M. (1982) *J. Cell Biol.* 92, 1–22.
Saraste, J., von Bonsdorff, C.-H., Hashimoto, K., Kääriäinen, L. and Keränen, S. (1980) *Virology* 100, 229–245.

Schlesinger, M.J. and Kääriäinen, L. (1980) In *The Togaviruses* (R.W. Schlesinger, Ed.) Academic Press, New York. pp. 371–392.

Schlesinger, J.J., Brandriss, M.W. and Monath, T.P. (1983) *Virology* 125, 8–17.

Schmaljohn, A.L., Johnson, E.D., Dalrymple, J.M. and Cole, G.A. (1982) *Nature* 297, 70–72.

Schmaljohn, A.L., Kokubun, K.M. and Cole, G.A. (1983) *Virology* 130, 144–154.

Sefton, B.M., Wickus, G.G. and Burge, B.W. (1973) *J. Virol.* 11, 730–735.

Shapiro, D., Trent, D.W., Brandt, W.E. and Russell, P.K. (1972) *Infect. Immun.* 6, 206–209.

Strauss, E.G. (1978) *J. Virol.* 28, 466–474.

Strauss, J.H. and Strauss, E.G. (1977) In *The Molecular Biology of Animal Viruses* (D. Nayak, Ed.) Dekker, New York. pp. 111–166.

Strauss, E.G. and Strauss, J.H. (1980) In *The Togaviruses* (R.W. Schlesinger, Ed.) Academic Press, New York. pp. 393–426.

Strauss, E.G. and Strauss, J.H. (1983) *Curr. Top. Microbiol. Immun.* 105, 1–98.

Strauss, E.G., Rice, C.M. and Strauss, J.H. (1983) *Proc. Natl. Acad. Sci. USA* 80, 5271–5275.

Strauss, E.G., Rice, C.M. and Strauss, J.H. (1984) *Virology* 133, 92–110.

Symington, J., McCann, A.K. and Schlesinger, M.J. (1977) *Infect. Immun.* 15, 720–725.

Trudel, M., Nadon, F., Comtois, R., Payment, P., Bonneau, A.-M. and Lecomte, J. (1982) *J. Virol. Meth.* 5, 191–197.

Wengler, G. and Wengler, G. (1981) *Virology* 113, 544–555.

Westaway, E.G. (1980) In *The Togaviruses* (R.W. Schlesinger, Ed.) Academic Press, New York. pp. 531–581.

Westaway, E.G., Schlesinger, R.W., Dalrymple, J.M. and Trent, D.W. (1980) *Intervirology* 14, 114–117.

Wolcott, J.A., Wust, C.J. and Brown, A. (1982) *J. Immunol.* 129, 1267–1271.

Wolcott, J.A., Wust, C.J. and Brown, A. (1984) *J. Virol.* 49, 379–385.

Yin, F.H. (1969) *J. Virol.* 4, 547–548.

Ziemiecki, A. and Garoff, H. (1978) *J. Mol. Biol.* 122, 259–269.

Eds. M.H.V. van Regenmortel and A.R. Neurath
Immunochemistry of viruses. The basis for serodiagnosis and vaccines.

Antigenic structure of herpes simplex viruses

23

PATRICIA G. SPEAR

Department of Microbiology, The University of Chicago, Chicago, IL 60637, U.S.A.

Information is just becoming available about the identities and properties of herpes simplex virus (HSV) antigens that can induce protective immune responses, although the nature of protective immunity is not well understood. On cells infected with HSV and on virions, not only are viral antigens expressed but also viral receptors that have affinity for the Fc region of IgG or for the activated third component of complement. It has been suggested that these receptors may have immunoregulatory activity. The purpose of this chapter is to summarize and evaluate the data on properties of HSV gene products that interact with the immune system either as key immunogens or as putative regulators of immune responses.

1. Diseases associated with herpes simplex virus infection

The pathology, latency and epidemiology of HSV are discussed in depth in several reviews (Rawls, 1973; Nahmias and Roizman, 1973; Baringer, 1975; Stevens, 1977; Wildy et al., 1981; Klein, 1982; Corey et al., 1983; Hill, 1984; Whitley, 1984). As is true for most infectious agents, infection with HSV may be symptomatic or asymptomatic. The clinical manifestations most commonly associated with primary HSV infection are mucocutaneous lesions at the site of viral inoculation, usually in or near the mouth or on the genital mucosa. Replication and spread of virus may be largely restricted to cells in the lesion (and to nerves impinging on the lesion), although systemic symptoms of disease are evident in many individuals. In rare instances HSV causes encephalitis. Newborn infants have less natural resistance to the virus than do children or adults; perinatal infection may result in life-threatening disseminated disease in which all or most organ systems are affected. Immunodeficiencies can also predispose to severe herpetic disease.

Long after resolution of the primary infection, whether symptomatic or not, latent virus can be detected in sensory ganglia innervating the site of infection. Cocultivation of the nervous tissue with susceptible cells is usually necessary to reveal the presence of infectious virus, probably because activation of virus from a latent state is required. In some individuals there is periodic or sporadic activation of virus from the latent state in vivo, which may result in recurrences of lesions at or near the site of primary infection.

There are two serotypes of HSV (HSV-1 and HSV-2). Typing can be done in a variety of ways, the most reliable methods being neutralization tests and analysis of restriction endonuclease digests of the viral DNA. In many human populations the majority of virus isolates from genital lesions will be HSV-2 and the majority from facial lesions will be HSV-1. It appears that viruses of either serotype can infect at any inoculation site so that maintenance of the nonrandom distribution of serotypes with respect to anatomic site may depend primarily on the mode of transmission of virus from one person to another. Although lesions caused by HSV-1 and HSV-2 are indistinguishable, observed genetic differences between the serotypes may in fact determine some differences in pathogenicity. For example, it has been noted that primary genital infections with HSV-1 are usually less severe and less likely to recur than primary genital infections with HSV-2 (Reeves et al., 1981). In addition, encephalitis is usually caused by HSV-1 strains (Whitley, 1984) whereas meningitis is more likely to be associated with infection by HSV-2 than with infection by HSV-1 (Corey et al., 1983).

2. Feasibility of successful vaccination against herpes simplex virus

Feasibility of, and strategies for, development of HSV vaccines have also been discussed elsewhere (see e.g. Roizman et al., 1982; Allen and Rapp, 1982; Kit, 1982; Rouse and Lopez, 1984). A major aim of vaccination against HSV would presumably be to prevent (or reduce the frequency of) primary disease and establishment of latency. Conceivably, it might also be possible to interrupt the pattern of recurrent disease in already infected individuals.

To be considered successful, an HSV vaccine will have to fulfill requirements of efficacy and safety that may be challenging to meet. For many individuals who have no known immunodeficiency, natural infection with HSV fails to induce sufficient acquired immunity to prevent recurrent disease or even to prevent reinfection at adjacent or distant sites with endogenous or exogenous virus (Lazar, 1956; Buchman et al., 1979; Kit et al., 1983). Consequently, one requirement for an HSV vaccine may be that it induce more effective protective immunity than does natural infection in these individuals. Moreover, because HSV rarely causes life-threatening disease, except in infants or immunocompromised individuals, widespread vaccination of normal populations could not be considered unless the vaccine were shown to have very few harmful side effects.

A live virus vaccine that could be totally eliminated from the body by the induced immune response might conceivably be both safe and effective, although whether such a vaccine could induce more effective protective immunity than a natural infection is open to question. With sufficient knowledge of viral functions and by use of recombinant DNA technology, attenuation of HSV can be achieved by rational design; for example, introduction of appropriate mutations and deletions into the viral genome can abolish the ability of the virus to establish latency or to be reactivated from the latent state (Roizman et al., 1982). In addition, genes encoding important immunogens can be duplicated by inserting additional copies into the viral genome to yield, for example, viruses that express both the HSV-1 and HSV-2 forms of a particular glycoprotein (Gibson and Spear, 1983).

Alternatively, purified subunit vaccines, prepared by the developing methods of biotechnology, might also be both safe and effective, provided that the right subunits are chosen (from some half-dozen candidates) and that immunopathological reactions can be avoided. Whatever the strategy adopted for vaccine development, more work must be done to identify the antigens that contribute most to the induction of protective immunity.

3. Antigens expressed on the surfaces of virions and infected cells

Although many or all HSV proteins (from 50 to 100 in number) could induce the production of antibodies or other immune responses, only a fraction of the viral gene products are good candidates for elicitors of protective immunity. These candidates include all viral proteins exposed at the surface of the virion and any viral products that alter the antigenic structure of infected cell surfaces. To date only the membrane glycoproteins specified by HSV have been shown to belong to either category. HSV-1 specifies the synthesis of at least 4 glycoproteins and HSV-2 at least 5. This section will focus on the genomic locations of genes encoding HSV glycoproteins and on available information about structures and functions of the glycoproteins.

3.1. Genetic map and locations of genes encoding viral glycoproteins

Fig. 23.1 shows a map of the HSV genome and the locations of genes encoding HSV-1 and HSV-2 glycoproteins. The data summarized in this figure are based on a number of mapping studies and techniques (cited here and in the figure legend). In particular, the precise locations of the HSV-1 genes for gB-1, gC-1 and gD-1 are known from the use of cloned DNA fragments to select mRNAs (Lee et al., 1982a; 1982b; Frink et al., 1983; Holland et al., 1984; Rafield and Knipe, 1984) and also

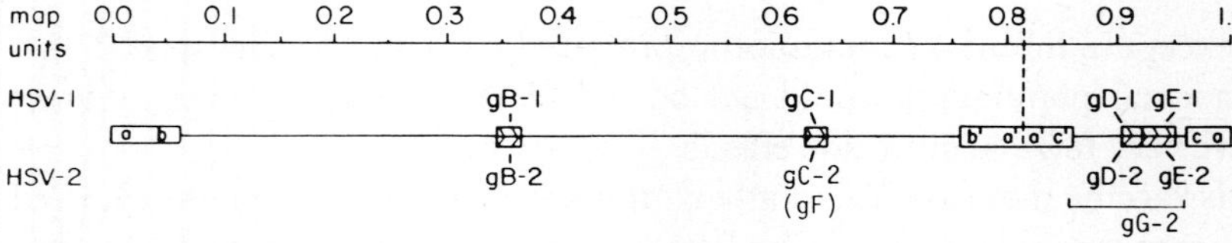

Fig. 23.1. Map of the HSV-1 and HSV-2 genomes showing the arrangement of repetitive (ab,ac) and unique sequences (see Roizman, 1979) and the locations of genes encoding viral glycoproteins. HSV-1 genes are shown above the line and HSV-2 genes below the line. Nomenclature of the HSV glycoproteins and recent changes in name are discussed elsewhere (Cohen et al., 1981; Spear, 1984). Note that the HSV-2 glycoprotein now designated gC-2 (Zweig et al., 1983; Para et al., 1983; Zezulak and Spear, 1983; 1984) was originally named gF (Balachandran et al., 1981; 1982a; 1982b). Also, the HSV-2 glycoprotein now designated gG-2 (Roizman et al., 1984) was originally called gC (Ruyechan et al., 1979). No species designated gA (Spear, 1976) is shown because gA proved to be a form of the gB gene product (Eberle and Courtney, 1980b; Pereira et al., 1981; Balachandran et al., 1982a). Approximate map locations for most of the glycoprotein genes were determined originally from analyses of intertypic recombinant viruses (Marsden et al., 1978; Ruyechan et al., 1979; Para et al., 1982b; 1983; Hope et al., 1982; Roizman et al., 1984). Experiments yielding more precise information on locations of the genes are discussed in the text.

from nucleotide sequence analysis (Watson et al., 1982; Frink et al., 1983; Bzik et al., 1984a). Location of the gene for gE-1 is based on the use of cloned DNA fragments to select gE-1 mRNA (Lee et al., 1982a). The HSV-2 genes for related glycoproteins (gB-2, gC-2, gD-2 and gE-2) are probably collinear with the equivalent HSV-1 genes, on the basis of analyses of intertypic recombinant viruses (Marsden et al., 1978; Ruyechan et al., 1979) and because mapping results and homology data obtained to date suggest that the HSV-1 and HSV-2 genomes are mostly if not entirely collinear. Except for gC-2 (Zezulak and Spear, 1984), however, these genes have not yet been independently mapped to the HSV-2 genome. The HSV-2 gene for gG-2 has been mapped to the S component of the genome (Roizman et al., 1984; Marsden et al., 1984) It is not yet known whether there is an HSV-1 equivalent of gG-2.

Although most of the glycosylated polypeptides detected by electrophoretic analyses of infected cell lysates can be accounted for as products of known glycoprotein genes, the possibility exists that other glycoproteins and their genes remain as yet undetected. In particular, it seems unlikely that an HSV-1 equivalent of gG-2 does not exist.

3.2. Structures of the glycoproteins

3.2.1. Polypeptide moieties

Partial amino acid sequences have been reported for gD-1 and gD-2 (Eisenberg et al., 1984). The only other information available about amino acid sequence has been inferred from nucleotide sequences published for gB-1 (Bzik et al., 1984a), gC-1 (Frink et al., 1983) and gD-1 (Watson et al., 1982). In each case, the open reading frame identified as encoding the glycoprotein is without introns. Evidence that the correct reading frame has been identified is based on construction of a hybrid gene and expression of a fusion protein antigenically related to the HSV glycoprotein, for gD-1 (Watson et al., 1982), or on inability to find any other open reading frame of appropriate characteristics consistent with mapping results summarized above, for the other glycoproteins. Fig. 23.2 shows the amino acid sequences for primary translation products, inferred from the published nucleotide sequences for the gB-1, gC-1 and gD-1 genes. What follows is a description of some of the more interesting features pointed out originally by the authors cited in the legend to Fig. 23.2 and below.

Signal sequences. Based on analogy with well-characterized secreted and membrane proteins, it seems likely that the hydrophobic N-terminal domains of the translation products for gB-1, gC-1 and gD-1 (underlined in Fig. 23.2) constitute part of a signal sequence (Blobel et al., 1979; Kreil, 1981) for each product. The features characteristic of signal sequences (a core of hydrophobic residues no less than 9 in length; a charged residue such as Arg near the N-terminus) are found in each case except that the N-terminal hydrophobic domain for gB-1 seems longer than usual. For gD-1 at least, the first 25 amino acids of the translation product (including the entire N-terminal hydrophobic domain) are known to be cleaved off during post-translational processing steps that yield the mature glycoprotein (Eisenberg et al., 1984). Post-translational cleavage occurs at the equivalent site in gD-2 (Eisenberg et al., 1984) but it has not yet been determined whether cleavages occur during processing of the gB-1 and gC-1 translation products. Experiments to test the hypothesis that the putative signal sequences are actually necessary for attachment of the translation products to membranes and vectorial transport have not yet been reported.

Potential sites for the addition of N-linked oligosaccharides. There are three potential signals (Asn-X-Ser or Asn-X-Thr) for the addition of N-linked oligosaccharides in the gD-1 translation product and nine each in gB-1 and gC-1 (Fig. 23.2). In the case of gD-1 (and gD-2) evidence has been presented for the existence of three N-linked oligosaccharide chains, indicating that all three potential sites can be used (Cohen et al., 1983; Matthews et al., 1983). The sites which are recognized for the addition of N-linked oligosaccharides in gB-1 and gC-1 have not yet been identified. Most or all of the HSV glycoproteins also contain O-linked oligosaccharides (see below).

Membrane-spanning domains and orientation in membranes. Based on locations of hydrophobic domains and positions of potential glycosylation sites in the HSV-1 glycoproteins (Fig. 23.2), it seems likely that gB-1, gC-1 and gD-1 will each be found to assume an orientation within membranes such that the small hydrophilic sequence at the C-terminus is in contact with the cytoplasm (or interior of the virion), the hydrophobic domain near the C-terminus spans the lipid bilayer and the

A. Predicted amino acid sequence of HSV-1(Patton) gD

1 MGGTAARLGAVILFVVIVGLHGVRGKYALADASLKMADPNRFRGKDLPVLDQLTDPPGVRRVYHIQAGLP

71 DPFQPPSLPITVYYAVLERACRSVLLNAPSEAPQIVRGASEDVRKQPYNLTIAWFRMGGNCAIPITVMEY

141 TECSYNKSLGACPIRTQPRWNYYDSFSAVSEDNLGFLMHAPAFETAGTYLRLVKINDWTEITQFILEHRA

211 KGSCKYALPLRIPPSACLSPQAYQQGVTVDSIGMLPRFIPENQRTVAVYSLKIAGWHGPKAPYTSTLLPP

281 ELSETPNATQPELAPEDPEDSALLEDPVGTVAPQIPPNWHIPSIQDAATPYHPPATPNNMGLIAGAVGGS

351 LLAALVICGIVYWMHRRTRKAPKRIRLPHIREDDQPSSHQPLFY

B. Predicted amino acid sequence of HSV-1(KOS) gC

1 MAPGRVGLAVVLWGLLWLGAGVAGGSETASTGPTITAGAVTNASEAPTSGSPGSAASPEVTPTSTPNPNN

71 VTQNKTTPTEPASPPTTPKPTSTPKSPPTSTPDPKPKNNTTPAKSGRPTKPPGPVWCDRRDPLARYGSRV

141 QIRCRFRNSTRMEFRLQIWRYSMGPSPPIAPAPDLEEVLTNITAPPGGLLVYDSAPNLTDPHVLWAEGAG

211 PGADPPLYSVTGPLPTQRLIIGEVTPATQGMYYLAWGRMDSPHEYGTWVRVRMFRPPSLTLQPHAVMEGQ

281 PFKATCTAAAYYPRNPVEFDWFEDDRQVFNPGQIDTQTHEHPDGFTTVSTVTSEAVGGQVPPRIFTCQMT

351 WHRDSVTFSRRNATGLALVLPRPTITMEFGVRHVVCTAGCVPEGVTFAWFLGDDPSPAAKSAVTAQESCD

421 HPGLATVRSTLPISYDYSEYICRLTGYPAGIPVLEHHGSHQPPPRDPTERQVIEAIEWVGIGIGVLAAGV

491 LVVTAIVYVVRTSQSRQLHRR

C. Predicted amino acid sequence of HSV-1(KOS) gB

1 MHQGAPSWGRRWFVVWALLGLTLGVLVASAAPSSPGTPGVARDPGGERGPCHSGAAALGAAPTGDPKPKK

71 NKKPKNPTPPRPAGDNATVAAGHATLREHLRDIKAENTDANFYVCPPPTGATVVQFEQPRRCPTRPEGQN

141 YTEGIAVVFKENIAPYKFKATMYYKDVTVSQVWFGHRYSQFMGIFEDRAPVPFEEVIDKINAKGVCRSTA

211 KYVRNNLETTAFHRDDHETDMELKPANAATRTSRGWHTTDLKYNPSRVEAFHRYGTTVNCIVEEVDARSV

281 YPYDEFVLATGDFVYMSPFYGYREGSHTEHTTYAADRFKQVDGFYARDLTTKARATAPTTRNLLTTPKFT

351 VAWDWVPKRPSVCTMTKWQEVDEMLRSEYGGSFRFSSDAISTTFTTNLTEYPLSRVDLGDCIGKDARDAM

421 DRIFARRYNATHIKVGQPQYYLANGGFLIAYQPLLSNTLAELYVREHLREQSRKPPNPTPPPPGASANAS

491 VERIKTTSSIEFARLQFTYNHIQRHVNDMLGRVAIAWCELQNHELTLWNEARKLNPNAIASVTVGRRVSA

561 KMLGDVMAVSTCVPVAADNVIVQNSMRISSRPGACYSRPLVSFRYEDQGPLVEGQLGENNELRLTRDAIE

631 PCTVGHRRYFTFGGGYVYFEEYAYSHQLSRADITTVSTFIDLNITMLEDHEFVPLEVYTRHEIKDSGLLD

701 YTEVQRRNQLHDLRFADIDTVIHADANAAMFAGLGAFFEGMGDLGRAVGKVVMGLVGGVVSAVSGVSSFM

771 SNPFGALAVGLLVLAGLAAAFFAFRYVMRLQSNPMKALYPLTTKELKNPTNPDASGEGEEGGDFDEAKLA

841 EAREMIRYMALVSAMERTEHKAKKKGTSRLLSAKVTDMVMRKRRNTNYTQVPNKDGDADEDDL

larger N-terminal domain, containing all oligosaccharide chains, is exposed to the exterior of the cell (or the virion). Evidence that this is the case has been presented for gD-1 and gD-2. Matthews et al. (1983) found that synthesis of the gD-1 or gD-2 translation product in vitro in the presence of dog pancreas microsomes resulted in N-glycosylation at all three potential sites and association of the glycosylated product with the microsomes such that the domain containing carbohydrate was protected from proteolysis by added trypsin, whereas a nonglycosylated fragment of 3 kDa was cleaved off by trypsin. They also showed that a type-specific antigenic determinant destroyed by cleavage of the 3-kDa fragment from gD-1 is inaccessible to antibody on intact infected cells expressing gD-1, whereas another antigenic determinant unaffected by proteolysis of the small fragment is accessible to antibody on intact infected cells. Taken together, these findings provide strong evidence for transmembrane orientation of gD-1 as described above.

Exploitation of the available nucleotide sequences and amino acid sequences to obtain information about three-dimensional structure and the locations of various functional domains on the glycoproteins has only just begun. For example, conditional-lethal ts mutations and mutations affecting epitopes recognized by neutralizing antibodies have been mapped to the N-terminal portion of the gB-1 gene and a mutation affecting cell fusion to very near the C-terminus (DeLuca et al., 1982; Holland et al., 1983; Kousoulas et al., 1984; Bzik et al., 1984a; 1984b). Between these two regions, a domain affecting rate of entry of virus has also been mapped (DeLuca et al., 1982). In addition, epitopes recognized by monoclonal antibodies have been mapped to proteolytic fragments of gD-1 and gD-2 (Eisenberg et al., 1982a; Matthews et al., 1983).

3.2.2. Prosthetic groups

N-Linked oligosaccharides. For all HSV glycoproteins examined to date (including gB, gC, gD, gE and gG-2), evidence of N-linked glycosylation has been obtained. Although most of the studies reported have been done with HSV-1, it seems likely that similar results would be obtained with HSV-2. It has been shown that tunicamycin partially or completely blocks glycosylation of the HSV glycoproteins mentioned above (Pizer et al., 1980; Olofsson and Lycke, 1980; Peake et al., 1982; Norrild and Pederson, 1982; Kousoulas et al., 1983; Hope and Marsden, 1983; Roizman et al., 1984). In addition, it has been shown that endo-β-*N*-acetyl glucos-

---

Fig. 23.2. Predicted amino acid sequence of primary translation products from the gD-1 (Patton), gC-1(KOS) and gB-1(KOS) genes, based on nucleotide sequences published by Watson et al. (1982) for gD-1, Frink et al. (1983) and Draper et al. (1984) for gC-1 and Bzik et al. (1984a) for gB-1. Hydrophobic domains at the N terminus and near the C terminus of each translation product are underlined. Potential signals for the addition of N-linked oligosaccharides are indicated by a double underlining. Determination of amino acid sequence at the N terminus of mature gD-1 showed that the first 25 amino acids of the primary translation product are cleaved off (at the arrow) during processing of the translation product (Eisenberg et al., 1984).

aminidase H (endo H) releases oligosaccharides from immature forms of the HSV glycoproteins, with a concomitant increase in electrophoretic mobility of the polypeptides indicative of decreased molecular weight, whereas all mature forms of the glycoproteins (except gB) are resistant to the action of endo H (Serafini-Cessi and Campadelli-Fiume, 1981; Wenske et al., 1982; Cohen et al., 1983; Johnson and Spear, 1983; Zezulak and Spear, 1983; Matthews et al., 1983). Because endo H selectively cleaves N-linked oligosaccharides of the high mannose type, but not the complex type (Tarentino and Maley, 1974), these results suggest that the high mannose-type oligosaccharides detected on immature forms of HSV glycoproteins are processed to yield complex type oligosaccharides in the mature forms, as has been shown to occur for many glycoproteins (Kornfeld and Kornfeld, 1980). In the case of gB, however, at least some of the high mannose-type N-linked oligosaccharides remain in a form that can be cleaved by endo H (Wenske et al., 1982; Johnson and Spear, 1983).

As mentioned above, the number of N-linked oligosaccharide chains added is three for gD-1 and gD-2 (Cohen et al., 1983; Matthews et al., 1983) and not known for the other glycoproteins. It is assumed for most of the HSV glycoproteins that the *same* number of N-linked oligosaccharide chains is added cotranslationally to each and every translation product of a single gene. Although this assumption will probably prove to be true for most of the glycoproteins, evidence to the contrary has been reported for gC-2. It was found that approximately equal amounts of two electrophoretically differentiable forms of gC-2 were detectable immediately after a short pulse-label and that these two discrete forms appeared to differ only in number of N-linked oligosaccharide chains (Zezulak and Spear, 1983). Although the significance of this finding is not known, further exploration of this phenomenon in relation to possible multiple pathways of processing of HSV glycoproteins is warranted.

O-Linked oligosaccharides. Evidence has been presented that HSV glycoproteins also contain O-linked oligosaccharides, although for at least one glycoprotein (gC-1) contradictory conclusions have been published. Results reported from three laboratories (Olofsson et al., 1981b; 1983a; Johnson and Spear, 1983; Wenske and Courtney, 1983) indicate that oligosaccharides can be released from mature forms of gC-1 by β-elimination (treatment with alkaline borohydride under conditions selective for release of O-linked oligosaccharides (Spiro, 1966; Marshall and Neuberger, 1977)) whereas negative results were obtained for such experiments done in a fourth laboratory (Kumarasamy and Blough, 1982). There is no obvious explanation at present for the contradictory findings. Other observations supporting the presence of O-linked oligosaccharides on gC-1, however, include (*a*) the affinity of a fraction of gC-1 for *Helix pomatia* lectin or soybean lectin (Olofsson et al., 1981a; 1983a), both of which bind to *N*-acetylgalactosamine (GalNAc) residues (Hammarström et al., 1977) moieties found in O-linked but not in N-linked oligosaccharides (Kornfeld and Kornfeld, 1980); (*b*) the reduced but detectable incorporation of glucosamine into oligosaccharides of gC-1 in the presence of amounts of tunicamycin sufficient to block N-glycosylation (Wenske and Courtney, 1983; Olofsson et al., 1983b); and (*c*) release of oligosaccharides from gC-1 by GalNAc oligosaccharidase (Johnson and Spear, 1983), an enzyme that cleaves O-linked oligosaccharides (Huang and Aminoff, 1972; Pomato and Aminoff, 1978; Johnson and Spear, 1983).

Presence of O-linked oligosaccharides on mature forms of other HSV glycoproteins is indicated by the release of oligosaccharides from gD-1 due to treatment with alkaline borohydride (Johnson and Spear, 1983); by shifts (increases) in the electrophoretic mobilities of mature forms of gB-1, gC-1, gC-2, gD-1 and gE-1 after treatment with GalNAc oligosaccharidase (Johnson and Spear, 1983; Zezulak and Spear, 1983); and by acceptor activity of immature forms, but not mature forms, of gB-1, gC-1 and gD-1 for the enzymatic transfer in vitro of GalNAc from UDP-GalNAc (Serafini-Cessi et al., 1983). It should be noted that although O-glycosylation of gC-1, and perhaps also of gE (Hope and Marsden, 1983), can apparently occur in the presence of tunicamycin, glycosylation of other HSV glycoproteins in the presence of this drug has not yet been reported. O-Glycosylation of other HSV glycoproteins (in the absence of drug) is not ruled out by these negative results, however, on the basis of the following considerations:

(*a*) The effects of inhibiting N-glycosylation are different for different membrane proteins; in some cases the translation product remains trapped in the rough endoplasmic reticulum, whereas in others, the translation product can be transported to other compartments of the cell and to the cell surface (Gibson et al., 1980).

(*b*) The addition or elongation of O-linked oligosaccharides, at least for HSV glycoproteins, appears to occur in the Golgi apparatus (Johnson and Spear, 1983).

(*c*) As a consequence of (*a*) and (*b*), an HSV glycoprotein could be O-glycosylated in the presence of tunicamycin only if it is transported from the rough endoplasmic reticulum. Consistent with the notion that selective O-glycosylation of gC-1 in the presence of tunicamycin might be due simply to selective transport of gC-1 under these conditions is the finding that antigenic determinants of gC-1, but not of other HSV-1 glycoproteins, could be detected on the surfaces of infected cells incubated in the presence of tunicamycin (Norrild and Pederson, 1982).

Sulfation. Inorganic sulfate is incorporated into all the major HSV-1 and HSV-2 glycoproteins, particularly gE-1 and gE-2 (Hope et al., 1982; Hope and Marsden, 1983). The nature of the linkage between sulfate and the glycoproteins is not known and, as pointed out by these authors, several possibilities exist. Sulfations of N-linked oligosaccharides (Prehm et al., 1979), of O-linked oligosaccharides (reviewed by Kornfeld and Kornfeld, 1980) and of tyrosine residues in proteins (Huttner, 1982) have all been described. Tunicamycin was shown to inhibit the incorporation of radioactive sulfate into HSV-1 glycoproteins, although reduced amounts of sulfate could be detected in an abnormal form of gE-1 (Hope and

Marsden, 1983). As noted by these authors, this result indicates that at least some of the sulfate on gE-1 is incorporated into O-linked oligosaccharides or into the polypeptide backbone. They favored the idea that, for the other glycoproteins, sulfate is incorporated into the N-linked oligosaccharides, but recognized the possibility that attachment of O-linked oligosaccharides (and sulfation of these oligosaccharides if it occurs) may be dependent on N-glycosylation.

Fatty acylation. It has been shown that ^{3}H from palmitate becomes incorporated into gE-1 (Johnson and Spear, 1983), indicating that fatty acid may become covalently attached to this protein, as has been documented for other membrane glycoproteins (reviewed by Schmidt, 1983) and for other proteins such as membrane-bound forms of SV40 T antigen (Henning and Lange-Mutschler, 1983). The physiological significance of fatty acylation of membrane proteins is not yet understood.

3.2.3. Relatedness of the HSV-1 and HSV-2 glycoproteins

The HSV-1 and HSV-2 glycoproteins assigned the same alphabetic designations are antigenically related (Cohen et al., 1972; Sim and Watson, 1973; Norrild et al., 1978; Vestergaard and Norrild, 1979; Pereira et al., 1980; 1981; Killington et al., 1981; Showalter et al., 1981; Eisenberg et al., 1982b; Para et al., 1982b; Zweig et al., 1983; Zezulak and Spear, 1983) and are generally assumed to be also functionally related. This assumption is supported by evidence in only a few instances and will be discussed in Section 3.3.

Antigenic determinants that are specific for each serotype can be detected on all the glycoproteins as well as determinants that are shared by members of both serotypes. A particular virus strain may not be true to type by every available criterion, however. For example, a very few HSV-2 strains among many tested were shown to react with an anti-gC-1-specific monoclonal antibody (Pereira et al., 1982). Also, a single HSV-1 isolate was shown to react with a particular anti-gD-2-specific monoclonal antibody but not with any other type 2-specific antibodies tested; nucleotide sequence analysis revealed that the gD gene of this virus isolate was identical to the gD-1(Patton) gene except for 3 nucleotide differences, one of which caused an amino acid substitution that could account for the presence of the gD-2-specific epitope (Rawls et al., 1984).

Determinations of nucleotide and amino acid sequences, which are not yet available, are necessary to assess the extent of structural similarity between the antigenically related HSV-1 and HSV-2 glycoproteins. Other kinds of evidence indicate that gC-1 and gC-2 are probably the least closely related of the HSV-1 and HSV-2 glycoproteins. The gene for gC-2 encodes a glycoprotein whose immature and mature forms are all significantly smaller in size than those of gC-1 (Balachandran et al., 1981; Zezulak and Spear, 1983). In addition, purified forms of gC-1 and gC-2 can each induce what appear to be type-specific immune responses (Vestergaard and Norrild, 1978; Eberle and Courtney, 1980a; Schrier et al., 1983; Zezulak and

Spear, 1983). When first described, the product of the gC-2 gene appeared, on the basis of size and antigenic structure, to be distinct from all known HSV-1 and HSV-2 glycoproteins and was originally designated gF (Balachandran et al., 1981; 1982a). With appropriate antisera and monoclonal antibodies, however, it has been possible to show that gC-2 (gF) is antigenically related to gC-1 (Zweig et al., 1983; Zezulak and Spear, 1983). Furthermore, the genes for gC-1 and gC-2 have been mapped to collinear partially homologous regions of the HSV-1 and HSV-2 genome, respectively (Fig. 23.1).

3.3. Functions of the glycoproteins

None of the HSV glycoproteins has been well-defined with regard to its function. There are some indications, however, as to which species may be assigned roles in virion infectivity, in virus-induced cell fusion and in the expression of Fc and C3b receptors, as summarized below and in Table 23.1. This subject has been reviewed in more depth elsewhere (Courtney, 1984; Spear, 1984).

3.3.1. Virion infectivity

HSV has a broad host range and is capable of infecting many cell types from many species. Consequently, the cell structures and cell surface constituents necessary for the adsorption and penetration of HSV must be nearly ubiquitous on vertebrate cells. The properties of cell receptors for HSV are not known.

As to virion constituents required for infectivity, to date only gB and gD have been implicated in the adsorption/penetration process. Several lines of evidence suggest that gB-1 may be multifunctional and may be required both for the initial attachment of virus to cells and for subsequent steps leading to penetration. With

TABLE 23.1

Activities associated with HSV-1 glycoproteins[a]

Glycoprotein	Process or activity			
	Infectivity	Cell fusion	Fc receptor	C3b receptor
gB-1	×	×		
gC-1				×
gD-1	×	×		
gE-1			×[b]	

[a] Evidence for these assignments of activity is summarized in the text. For any of the processes listed, glycoproteins other than those indicated could participate, although evidence for such participation has not yet been obtained.

[b] Direct evidence has been obtained that gE-2 also has Fc-binding activity (Para et al., 1982b).

regard to attachment, results suggest that the ability of heparin to block the binding of HSV to cells (Vaheri and Cantrell, 1963; Takemoto and Fabish, 1964; Nahmias and Kibrick, 1964; Hochberg and Becker, 1968) may be due to interaction of heparin with gB in virions. It was found that both intact HSV-1 virions and solubilized gB-1 bound quantitatively to heparin-Sepharose whereas other solubilized virion envelope constituents failed to bind or partitioned between the bound and unbound fractions (P.G. Spear and M.L. Parish, unpublished). In addition, experiments done with liposomes containing a mixture of virion envelope proteins (virosomes) indicate that gB-1 may mediate binding of these liposomes to cells (Johnson et al., 1984). Specifically (*a*) virosomes bound to cells were greatly enriched for gB-1 and depleted of the other envelope glycoproteins in comparison with virosomes in the unbound fraction; (*b*) anti-gB antibodies inhibited the binding of virosomes to cells whereas anti-gD antibodies were less effective and anti-gC antibodies were without effect, and (*c*) virosomes depleted of gB-1 and, to a lesser extent, virosomes depleted of gD-1 had reduced cell-binding activity.

Results implicating gB-1 in viral penetration have come from the study of two temperature-sensitive mutants (tsB5 and tsJ12) that were independently isolated but belong to the same complementation group (Schaffer et al., 1978). These strains have gB-1 gene mutations (Ruyechan et al., 1979; Little et al., 1981; DeLuca et al., 1982; Holland et al., 1983; Kousoulas et al., 1984; Bzik et al., 1984a; b) that prevent normal post-translational processing of the gB-1 polypeptide at nonpermissive temperature (Manservigi et al., 1977; Sarmiento et al., 1979; Little et al., 1981). At this temperature, although virions are produced, these virions are deficient in the normal mature form of gB-1, probably containing instead an aberrant form of gB, and are noninfectious. The block to infectivity appears to be at the level of penetration because the noninfectious virions can bind to cells but do not initiate infection unless the virion-cell complexes are treated with polyethylene glycol (Sarmiento et al., 1979; Little et al., 1981). These results and others mentioned below suggest that gB-1 may be necessary for virus-induced membrane fusion.

gD is also believed to have a role in the adsorption/penetration process. Experimental support for this hypothesis comes principally from observations that this glycoprotein can elicit the production of neutralizing antibodies (Cohen et al., 1972; Sim and Watson, 1973; Eisenberg et al., 1982b). Much attention has been focused on gD as a target of neutralizing antibodies despite the fact that other HSV glycoproteins can also be targets of neutralizing antibodies (Powell et al., 1974; Vestergaard and Norrild, 1979; Eberle and Courtney, 1980a; Pereira et al., 1980; 1981; Killington et al., 1981; Showalter et al., 1981; Para et al., 1982a; Zezulak and Spear, 1983). For a variety of reasons it is of interest to know which of the isolated HSV glycoproteins can, when used as immunogens under similar conditions, induce the highest titers of neutralizing antibodies. Unfortunately, such comparative experiments have not been reported. Even if gD proves to be the inducer of the most potent neutralizing antibodies, this finding would shed little light on the precise role of gD in infectivity as the mechanism of neutralization is unknown. Recent studies with mouse monoclonal antibodies suggest that several potent neutralizing anti-gD monoclonal antibodies block infection at some stage subsequent to the adsorption of virus to cells (M.F. Para, M.L. Parish, A.G. Noble and P.G. Spear; A.O. Fuller and P.G. Spear, submitted for publication).

No roles have been proposed for gC, gE or gG in infectivity. The fact that HSV-1 and HSV-2 mutants producing no recognizable form of gC-1 or gC-2, respectively, are viable in cell culture (Heine et al., 1974; Manservigi et al., 1977; Zezulak and Spear, 1984) suggests that gC may have no essential role in infectivity, at least not for the kinds of cells usually grown in culture. It remains to be proven, however, that these mutants fail to produce gC-related products that could substitute for gC for some essential role in infectivity.

3.3.2. HSV-induced cell fusion

Whether or not HSV induces cell fusion is determined not only by the cell type infected but also by the products of several viral genes (reviewed by Spear, 1984). The process of HSV-induced cell fusion may be analogous in some respects to the process of virion-cell fusion necessary to initiate infection. There is evidence at least that the two HSV glycoproteins implicated in infectivity (gB and gD) also have roles in cell fusion.

For one of the gB-1 mutants mentioned above (tsB5), it has been shown that temperature sensitivity for production of functional gB-1 correlates with temperature sensitivity in ability of the mutant to induce cell fusion (Manservigi et al., 1977; Haffey and Spear, 1980). Evidence that gD-1 also has some role in cell fusion comes from the finding that anti-gD monoclonal antibodies (9 of 9 tested) blocked HSV-1-induced cell fusion, whereas monoclonal antibodies directed against other viral glycoproteins did not (Noble et al., 1983).

3.3.3. Fc-Binding and C3b-binding receptors

Cells infected with HSV express new receptors, one kind of which has affinity for the Fc region of IgG and another for the activated third component of complement. Evidence summarized below indicates that these receptors are viral glycoproteins or, more precisely, that specific viral glycoproteins form at least part of each receptor. The physiological significance of these receptors for HSV replication and for HSV pathogenesis is not understood, nor is it known whether the ligands used to detect the receptors (IgG, C3b) are actually the physiologically important ligands.

It is particularly intriguing, of course, that receptor activities often found together on certain cells of the lymphoreticular system can be detected on the surfaces of other cell types infected by HSV (Fc receptors are also induced by infections with human cytomegalovirus (Keller et al., 1976; Rahman et al., 1976; Westmoreland et al., 1976) and with varicella-zoster virus (Ogata and Shigeta, 1979)). It has been speculated that such receptors may interfere with or modulate immune responses to

HSV-infected cells (Costa and Rabson, 1975; Lehner et al., 1975; Adler et al., 1978) or that binding of appropriate ligands to the receptors might influence what happens inside the infected cell (Westmoreland and Watkins, 1974; Lehner et al., 1975; Costa et al., 1977). Although experiments have shown that aggregated non-immune IgG can interfere with immune cytolysis in vitro (Adler et al., 1978), it remains to be seen whether such phenomena are demonstrable or important in vivo. In addition, although it was reported that high concentrations of nonimmune IgG can reduce the yields of infectious HSV produced in cell culture (Costa et al., 1977), the effect was not large and it remains to be determined whether this effect was actually due to binding of IgG to the Fc receptors.

The observation that antibody-sensitized erythrocytes could bind to HSV-infected cells (Watkins, 1964) led to the demonstration that both HSV-1 and HSV-2 induce in many cell types the expression of receptors with affinity for the Fc region of IgG (reviewed by Spear, 1984). Evidence that gE-1 and gE-2 are the HSV-1 and HSV-2 Fc-binding receptors, respectively, comes in part from findings that the solubilized forms of these glycoproteins selectively bind to antigen-antibody complexes on affinity columns, provided that the Fc regions of the antibodies are intact (Baucke and Spear, 1979; Para et al., 1982b). It has also been shown that $F(ab')_2$ fragments of anti-gE-1 antibodies can partially block the Fc-binding activity of infected cells (Para et al., 1982a).

It was recently shown that endothelial cells infected with HSV-1 develop receptors for the C3b component of complement as well as Fc receptors (Cines et al., 1982; Friedman et al., 1984). Evidence that gC-1 forms part of, or is, the C3b receptor comes from findings that a gC^- viral mutant is also C3b-receptor negative and that monoclonal antibodies specific for gC-1, but not for other viral glycoproteins, block C3b binding activity (Friedman et al., 1984). Viability of the gC^-, C3b-receptor negative mutant indicates that this receptor activity is not essential for HSV-1 replication in cell culture. In light of differences observed between gC-1 and gC-2, as described above, it is of interest that C3b receptors could *not* be detected on endothelial cells infected with HSV-2 (Friedman et al., 1984). Other cell types or conditions may be required for detection of HSV-2 C3b receptors, should they exist. Although the study of HSV-1 C3b receptors has so far been restricted to infected endothelial cells, it seems likely that these receptors will also be detected on other infected cell types.

4. Progress in identifying the inducers of protective immunity

Studies on the immunobiology of HSV infections have yielded considerable information about the kinds of natural and acquired immune responses that can be detected after infection and that might be protective (reviewed in a recent book edited by Rouse and Lopez, 1984). It seems reasonable to assume that *protective* immune responses induced by infection are directed principally against viral antigens exposed on the surface membranes of infected cells and virions, but it is extremely difficult to obtain experimental support for this assumption by studying the immune responses of infected animals or humans, or to determine whether one viral membrane antigen is any more important than another as a target for a protective response. The difficulty is in proving that an immune effector mechanism demonstrable in vitro actually contributes to immune protection in vivo.

It is possible, however, to empirically determine whether purified viral antigens, alone or in combination, can induce protection in suitable experimental animals against challenge with infectious virus. Isolated viral antigens can be compared as vaccines with other immunogens such as attenuated or inactive virus. Experimental work of this kind is just beginning to be reported. Use of monoclonal and monospecific antibodies and cloned T cells in tests for passive transfer of immunity could also conceivably provide information about the identities of antigens important for protection.

In this section attention will be focused on results obtained using the experimental approaches mentioned above. The animals used for most of the studies to be described were mice. In different studies protocols for the administration of antigen and challenge virus varied, as did the criteria for protection (ranging from survival after a lethal dose of virus to reduction in severity of localized cutaneous lesions after peripheral inoculation of nonlethal doses). Only empirical tests can determine whether results obtained in such experimental animals will also apply to humans.

4.1. Passive transfer of immunity with antibodies of defined specificity

The passive transfer of either anti-HSV antibodies or HSV-primed lymphocytes can confer protection against challenge with virus under appropriate conditions (see e.g. Worthington et al., 1980; Kapoor et al., 1982; Sethi et al., 1983, in addition to papers cited below). The discussion here will be limited to studies done with antibodies because only with antibodies have the specificities been known. The tentative conclusions that emerge from several studies are that antibodies specific for any of the HSV glycoproteins can confer protection by passive transfer and that multiple mechanisms may have to be invoked to account for the protection observed.

Kapoor et al. (1982) found that only one of two different anti-gD monoclonal antibodies blocked the spread of infectious virus in nude mice from ear to ganglia and other organs after subcutaneous inoculation of virus into the ear. They correlated the protective activity with in vitro neutralizing activity of the antibody.

Dix et al. (1981) and Balachandran et al. (1982b) tested the ability of mouse monoclonal antibodies injected intraperitoneally to protect mice from lethal neurological disease occurring after footpad inoculation. In the former study, an antibody specific for gC-1, and another specific for gD-1 and gD-2, were shown to confer protection against several HSV strains, whether given 2 hours before virus or 24 hours after, provided that the challenge virus expressed the relevant determinant recognized by the antibody. In the latter study antibodies specific for gB-2, gC-2 (gF), gD-2, gE-2 and gG-2 (gC) were all shown to confer protection, whereas a second anti-gG-2 antibody did not. It was also reported by Balachandran et al. (1982b) that the passive transfer of immunity was as effective for complement (C5)-deficient mice as for normal mice and that protection did not correlate with neutralizing activities of the antibodies. A correlation was noted, however, between levels of protection and antibody-dependent cell-mediated cytolysis (ADCC) measured in vitro. In contrast, Rector et al. (1982) found that antibodies specific for gB-1, gC-1, gD-1 and gE-1 protected mice against ocular HSV-1 infection and that the level of protection did not correlate with any of three in vitro measures of antibody activity (neutralization, ADCC and complement-mediated cell lysis).

With regard to the mechanisms by which antibodies can confer passive protection, the results of several investigations done with polyspecific anti-HSV sera should be noted. Firstly, anti-HSV sera raised in several species can protect mice against HSV-induced disease (Oakes and Rosemond-Hornbeak, 1979; McKendall et al., 1979; Oakes et al., 1980; Oakes and Lausch, 1981; Hayashida et al., 1982). Secondly, under certain conditions the Fc domains of the antibodies are necessary for the protection observed. For example, intact human antibodies, but not $F(ab')_2$ fragments, delayed the onset of lesions and time of death after intracutaneous inoculation of HSV into nude mice; the $F(ab')_2$ fragments were ineffective despite repeated injections to maintain high levels in the face of more rapid turnover (Hayashida et al., 1982). In another study (Oakes and Lausch, 1981), $F(ab')_2$ fragments from rabbit antibodies were as effective as intact IgG in protecting mice from disease after ocular or subcutaneous inoculation, but only if the antibodies or fragments were administered before or simultaneously with virus. If administration was delayed until 8 hours after virus inoculation, then only intact IgG was effective. Thirdly, treatment with antithymocyte serum or sublethal irradiation of recipient mice abrogated the immune protection conferred by anti-HSV antibodies (Davis et al., 1979; Oakes et al., 1980) and, under certain conditions, nude mice were not protected by antibody unless reconstituted with syngeneic thymus cells (Oakes et al., 1980). Finally, complement-deficient mice can be as effectively protected by anti-HSV antibodies as complement-sufficient mice (Oakes and Lausch, 1981; Hayashida et al., 1982).

These observations taken together suggest that, among the effector mechanisms recognized for anti-HSV antibodies, neutralization and complement-dependent neutralization or cell lysis probably contribute only marginally to protection conferred by passive transfer of antibodies. ADCC may contribute more importantly. Another possibility is raised, however, by the findings (*a*) that both Fc regions of the antibodies and radio-sensitive cells (including T lymphocytes) of the recipient animals are necessary for optimal protection conferred by antibodies, and (*b*) that antibodies specific for any of the glycoproteins, regardless in some instances of in vitro activities, can confer protection. It may be that the injected antibodies are more important as regulators and enhancers of the recipient animal's own immune response than as direct effectors for elimination of virus or virus-infected cells. If so, there should be demonstrable differences between antibody-treated and control animals in the immune response induced by challenge virus.

Whatever the mechanisms by which antibodies confer protection after passive transfer, it is clear that the use of monospecific antibodies has not yet provided any clues as to whether one HSV glycoprotein is any more important than another as a target for protective immune responses.

4.2. Immunization with purified antigens

That subunits of HSV can induce protective immunity in laboratory animals has been demonstrated by injection of material extracted with detergents from partially purified virions or infected cells (Cappel, 1976; Kitces et al., 1977; Cappel et al., 1980; Ohashi et al., 1980; Slichtova et al., 1980; Rajcani et al., 1980; Zweerink et al., 1981; Hilleman et al., 1981; Klein et al., 1981). Where comparisons were done, the crude subunit vaccine used, under appropriate conditions, could be as effective as live attenuated or inactivated virus in inducing protection against virulent challenge virus (Cappel, 1976; Ohashi et al., 1980; Zweerink et al., 1981; Klein et al., 1981). These results are promising when one considers that much can probably be done to improve the efficacy of HSV subunit vaccines by optimizing vaccine composition, conformation and presentation of antigens, choice of adjuvants and routes and timing of administration.

Information as to which components of the crude subunit vaccines are most important for efficacy is lacking. Moreover, only limited information is available as to which isolated HSV antigens can induce protective immunity under various conditions. It has been shown that immunization with purified gC-1, gD-1 or gD-2 can protect mice from lethal intraperitoneal challenge with virus (Schrier et al., 1983; Long et al., 1984). The immunity induced by gC-1 was type specific (Schrier et al., 1983) whereas both gD-1 and gD-2 induced protection against challenge with either HSV-1 or HSV-2 (Long et al., 1984). It has also been shown that a recombinant vaccinia virus which carries and expresses the gD-1 gene can be used as a live vaccine to protect mice against challenge with HSV-1 (Paoletti et al., 1984). Because neutralizing antibodies can be produced by rabbits in response to a synthetic peptide corresponding to residues 33–48 of the gD-1 translation product (Cohen et al., 1984) (Fig. 23.2) or in response to a chimeric gD-related polypeptide produced in bacteria (Weis et al., 1983), it will be of interest to learn whether these gD-related products can also induce protective immunity.

4.3. CONCLUDING REMARKS

The possibility remains that each of the HSV glycoproteins (and perhaps other products as well) may be able to induce some level of protective immunity. Considerably more experimental work must be done to determine which of the HSV antigens, alone or in combination, are the most effective inducers of protective immunity against both HSV-1 and HSV-2 and which membrane antigens can be ignored with impunity in the development of vaccines. In addition, investigation is needed to determine the most effective way of presenting HSV antigens to the immune system. The ideal vaccine may prove to be a synthetic hybrid product carrying key determinants from several of the HSV glycoproteins, and perhaps other viral components as well.

5. References

Adler, R., Glorioso, J.C., Cossman, J. and Levine, M. (1978) *Infect. Immun.* 21, 442–447.

Allen, W.P. and Rapp, F. (1982) *J. Infect. Dis.* 145, 413–421.

Balachandran, N., Harnish, D., Killington, R.A., Bacchetti, S. and Rawls, W.E. (1981) *J. Virol.* 39, 438–446.

Balachandran, N., Harnish, D., Rawls, W.E. and Bacchetti, S. (1982a) *J. Virol.* 44, 344–355.

Balachandran, N., Bacchetti, S. and Rawls, W.E. (1982b) *Infect. Immun.* 37, 1132–1137.

Baringer, J.R. (1975) *Prog. Med. Virol.* 20, 1–26.

Baucke, R.B. and Spear, P.G. (1979) *J. Virol.* 32, 779–789.

Blobel, G., Walter, P., Chang, C.N., Goldman, B.M., Erikson, A.H. and Lingappa, V.R. (1979) *Symp. Soc. Exp. Biol.* 33, 9–36.

Buchman, T.G., Roizman, B. and Nahmias, A.J. (1979) *J. Infect. Dis.* 140, 295–304.

Bzik, D.J., Fox, B.A., DeLuca, N.A. and Person, S. (1984a) *Virology* 133, 301–314.

Bzik, D.J., Fox, B.A., DeLuca, N.A. and Person, S. (1984b) *Virology* 137, 185–190.

Cappel, R. (1976) *Arch. Virol.* 52, 29–35.

Cappel, R., de Cuyper, F. and Rickaert, F. (1980) *Arch. Virol.* 65, 15–23.

Cines, D.B., Lyss, A.P., Bina, M., Corkey, R., Klialidis, N.A. and Friedman, H.M. (1982) *J. Clin. Invest.* 69, 123–128.

Cohen, G.H., Ponce de Leon, M. and Nichols, C. (1972) *J. Virol.* 10, 1021–1030.

Cohen, G., Halliburton, I. and Eisenberg, R. (1981) In *The Human Herpesviruses* (A.J. Nahmias, W.R. Dowdle and R.F. Schinazi, Eds.) Elsevier, New York. pp. 549–554.

Cohen, G.H., Long, D., Matthews, J.T., May, M. and Eisenberg, R. (1983) *J. Virol.* 46, 679–689.

Cohen, G.H., Dietzschold, B., Ponce de Leon, M., Long, D., Golub, E., Varrichio, A., Pereira, L. and Eisenberg, R.J. (1984) *J. Virol.* 49, 102–108.

Corey, L., Adams, H.G., Brown, Z.A. and Holmes, K.K. (1983) *Ann. Intern. Med.* 98, 958–972.

Costa, J.C. and Rabson, A.S. (1975) *Lancet* i, 77–78.

Costa, J., Rabson, A.S., Yee, C. and Tralka, T.S. (1977) *Nature* 269, 251–252.

Courtney, R.J. (1984) In *Immunobiology of Herpes Simplex Virus Infection* (B.T. Rouse and C. Lopez, Eds.) CRC Press, Boca Raton. pp. 33–44.

Davis, W.B., Taylor, J.A. and Oakes, J.E. (1979) *J. Infect. Dis.* 140, 534–540.

DeLuca, N., Bzik, D.J., Bond, V.C., Person, S. and Snipes, W. (1982) *Virology* 122, 411–423.

Dix, R.D., Pereira, L. and Baringer, J.R. (1981) *Infect. Immun.* 34, 192–199.

Draper, K.G., Costa, R.H., Lee, G.T.-Y., Spear, P.G. and Wagner, E.K. (1984) *J. Virol.* 51, 578–585.

Eberle, R. and Courtney, R.J. (1980a) *J. Virol.* 35, 902–917.
Eberle, R. and Courtney, R.J. (1980b) *J. Virol.* 36, 665–675.
Eisenberg, R.J., Long, D., Pereira, L., Hampar, B., Zweig, M. and Cohen, G.H. (1982a) *J. Virol.* 41, 478–488.
Eisenberg, R.J., Ponce de Leon, M., Pereira, L., Long, D. and Cohen, G.H. (1982b) *J. Virol.* 41, 1099–1104.
Eisenberg, R.J., Long, D., Hogue-Angeletti, R. and Cohen, G. (1984) *J. Virol.* 49, 265–268.
Friedman, H.M., Cohen, G.H., Eisenberg, R.J., Seidel, C.A. and Cines, D.B. (1984) *Nature* 309, 633–635.
Frink, R.J., Eisenberg, R., Cohen, G. and Wagner, E.K. (1983) *J. Virol.* 45, 634–647.
Gibson, M.G. and Spear, P.G. (1983) *J. Virol.* 48, 396–404.
Gibson, R., Kornfeld, S. and Schlesinger, S. (1980) *Trends Biochem. Sci.* 5, 290–293.
Haffey, M.L. and Spear, P.G. (1980) *J. Virol.* 35, 114–128.
Hammarström, S., Murphy, L.A., Goldstein, I.J. and Etzler, M.E. (1977) *Biochemistry* 16, 2750–2755.
Hayashida, I., Nagafuchi, S., Hayashi, Y., Kino, Y., Mori, R., Oda, H., Ohtomo, N. and Tashiro, A. (1982) *Microbiol. Immunol.* 26, 497–509.
Heine, J.W., Honess, R.W., Cassai, E. and Roizman, B. (1974) *J. Virol.* 14, 640–651.
Henning, R. and Lange-Mutschler, J. (1983) *Nature* 305, 736–738.
Hill, T.J. (1984) In *The Herpesviruses* (B. Roizman, Ed.) Vol. 3. Plenum Press, New York. In Press.
Hilleman, M.R., Larsen, V.M., Lehman, E.D., Salerno, R.A., Conard, P.G. and McLean, A.A. (1981) In *The Human Herpesviruses* (A.J. Nahmias, W.R. Dowdle and R.F. Schinazi, Eds.) Elsevier, New York. pp. 503–506.
Hochberg, E. and Becker, Y. (1968) *J. Gen. Virol.* 2, 231–241.
Holland, T.C., Sandri-Goldin, R.M., Holland, L.E., Marlin, S.D., Levine, M. and Glorioso, J.C. (1983) *J. Virol.* 46, 649–652.
Holland, L.E., Sandri-Golden, R.M., Goldin, A.L., Glorioso, J.C. and Levine, M. (1984) *J. Virol.* 49, 947–959.
Hope, R.G. and Marsden, H.S. (1983) *J. Gen. Virol.* 64, 1943–1953.
Hope, R.G., Palfreyman, J., Suh, M. and Marsden, H.S. (1982) *J. Gen. Virol.* 58, 399–415.
Huang, C.C. and Aminoff, D. (1972) *J. Biol. Chem.* 247, 6737–6742.
Huttner, W.B. (1982) *Nature* 299, 273–276.
Johnson, D.C. and Spear, P.G. (1983) *Cell* 32, 987–997.
Johnson, D.C., Wittels, M. and Spear, P.G. (1984) *J. Virol.* 52, 238–247.
Kapoor, A.K., Nash, A.A., Wildy, P., Phelan, J., McLean, C.S. and Field, H.J. (1982) *J. Gen. Virol.* 60, 225–233.
Keller, R., Peitchel, R., Goldman, J.N. and Goldman, M. (1976) *J. Immunol.* 116, 772–777.
Killington, R.A., Newhook, L., Balachandran, N., Rawls, W.E. and Bacchetti, S. (1981) *J. Virol. Methods* 2, 223–236.
Kit, S. (1982) *Infect. Dis.* 12, 4–9; 18–19.
Kit, S., Trkula, D., Qavi, H., Dreesman, G., Kennedy, R.C., Adler-Storthz, K., Kaufman, R. and Adam, E. (1983) *Sex. Trans. Dis.* 10, 67–71.
Kitces, E.N., Morahan, P.S., Twe, J.G. and Murray, B.K. (1977) *Infect. Immun.* 16, 955–960.
Klein, R.J. (1982) *Arch. Virol.* 72, 143–168.
Klein, R.J., Buimovici-Klein, E., Moser, H., Moucha, R. and Hilfenhaus, J. (1981) *Arch. Virol.* 68, 73–80.
Kornfeld, R. and Kornfeld, S. (1980) In *The Biochemistry of Glycoproteins and Proteoglycans* (W.J. Lennarz, Ed.) Plenum Press, New York. pp. 1–34.
Kousoulas, K.G., Bzik, D.J., DeLuca, N. and Person, S. (1983) *Virology* 125, 468–474.

Kousoulas, K.G., Pellet, P.E., Pereira, L. and Roizman, B. (1984) *Virology* 135, 379–394.
Kreil, G. (1981) *Annu. Rev. Biochem.* 50, 317–348.
Kumarasamy, R. and Blough, H.A. (1982) *Biochem. Biophys. Res. Comm.* 109, 1108–1115.
Lazar, M.P. (1956) *Arch. Dermatol.* 73, 70–71.
Lee, G.T.-Y., Para, M.F. and Spear, P.G. (1982a) *J. Virol.* 43, 41–49.
Lee, G.T.-Y., Pogue-Geile, K.L., Pereira, L. and Spear, P.G. (1982b) *Proc. Natl. Acad. Sci. USA* 79, 6612–6616.
Lehner, T., Wilton, J.M.A. and Shillitoe, E.J. (1975) *Lancet* ii, 60–62.
Little, S.P., Jofre, J.T., Courtney, R.J. and Schaffer, P.A. (1981) *Virology* 115, 149–160.
Long, D., Madara, T.J., Ponce de Leon, M., Cohen, G.H., Montgomery, P.C. and Eisenberg, R.J. (1984) *Infect. Immun.* 37, 761–764.
Manservigi, R., Spear, P.G. and Buchan, A. (1977) *Proc. Natl. Acad. Sci. USA* 74, 3913–3917.
Marsden, H.S., Stow, N.D., Preston, V.G., Timbury, M.C. and Wilkie, N.M. (1978) *J. Virol.* 28, 624–642.
Marsden, H.S., Buckmaster, A., Palfreyman, J.W., Hope, R.G. and Minson, A.C. (1984) *J. Virol.* 50, 547–554.
Marshall, R.C. and Neuberger, A. (1977) *Adv. Carbohydrate Chem. Biochem.* 25, 407–478.
Matthews, J.T., Cohen, G.H. and Eisenberg, R.J. (1983) *J. Virol.* 48, 521–533.
McKendall, R.R., Klassen, T. and Baringer, J.R. (1979) *Infect. Immun.* 23, 305–311.
Nahmias, A.J. and Kibrick, S. (1964) *J. Bacteriol.* 87, 1060–1066.
Nahmias, A.J. and Roizman, B. (1973) *N. Engl. J. Med.* 289, 667–674; 719–725; 781–789.
Noble, A.G., Lee, G.T.-Y. and Spear, P.G. (1983) *Virology* 129, 218–224.
Norrild, B. and Pedersen, B. (1982) *J. Virol.* 43, 395–402.
Norrild, B., Ludwig, H. and Rott, R. (1978) *J. Virol.* 26, 712–717.
Oakes, J.E. and Lausch, R.N. (1981) *Infect. Immun.* 33, 109–114.
Oakes, J.E. and Rosemond-Hornbeak, H. (1979) *Infect. Immun.* 21, 489–495.
Oakes, J.E., Davis, W.B., Taylor, J.A. and Weppner, W.A. (1980) *Infect. Immun.* 29, 642–649.
Ogata, M. and Shigeta, S. (1979) *Infect. Immun.* 26, 770–774.
Ohashi, Y., Sakaue, Y., Kato, S., Wada, T. and Sato, K. (1980) *Biken J.* 23, 199–204.
Oloffson, S. and Lycke, E. (1980) *Arch. Virol.* 65, 201–209.
Oloffson, S., Jeansson, S. and Lycke, E. (1981a) *J. Virol.* 38, 564–570.
Oloffson, S., Blomberg, J. and Lycke, E. (1981b) *Arch. Virol.* 70, 321–329.
Oloffson, S., Norrild, B., Andersen, A.B., Pereira, L., Jeansson, S. and Lycke, E. (1983a) *Arch. Virol.* 76, 25–38.
Oloffson, S., Sjöblom, I., Lundström, M., Jeansson, S. and Lycke, E. (1983b) *J. Gen. Virol.* 64, 2735–2747.
Paoletti, E., Lipinskas, B.R., Samsonoff, C., Mercer, S. and Panicali, D. (1984) *Proc. Natl. Acad. Sci. USA* 81, 193–197.
Para, M.F., Baucke, R.B. and Spear, P.G. (1982a) *J. Virol.* 41, 129–136.
Para, M.F., Goldstein, L. and Spear, P.G. (1982b) *J. Virol.* 41, 137–144.
Para, M.F., Zezulak, K.M., Conley, A.J., Weinberger, M., Snitzer, K. and Spear, P.G. (1983) *J. Virol.* 45, 1223–1227.
Peake, M.L., Nystrom, P. and Pizer, L.I. (1982) *J. Virol.* 42, 678–690.
Pereira, L., Klassen, T. and Baringer, J.R. (1980) *Infect. Immun.* 29, 724–732.
Pereira, L., Dondero, D., Norrild, B. and Roizman, B. (1981) *Proc. Natl. Acad. Sci. USA* 78, 5202–5206.
Pereira, L., Dondero, D.V., Gallo, D., Devlin, V. and Woodie, J.D. (1982) *Infect. Immun.* 35, 363–367.
Pizer, L.I., Cohen, G.H. and Eisenberg, R.J. (1980) *J. Virol.* 34, 142–153.
Pomato, N. and Aminoff, D. (1978) *Fed. Proc.* 37, 1602.
Powell, K.L., Buchan, A., Sim, C. and Watson, D.H. (1974) *Nature* 249, 360–361.

Prehm, P., Scheid, A. and Choppin, P.W. (1979) *J. Biol. Chem.* 254, 9669–9677.
Rafield, L.F. and Knipe, D.M. (1984) *J. Virol.* 49, 960–969.
Rahman, A.A., Teschner, M., Sethi, K.K. and Brandis, H. (1976) *J. Immunol.* 117, 253–258.
Rajcani, J., Kutinova, L. and Vonka, V. (1980) *Acta Virol.* 24, 183–193.
Rawls, W.E. (1973) In *The Herpesviruses* (A.S. Kaplan, Ed.) Academic Press, New York. pp. 291–325.
Rawls, W.E., Balachandran, N., Sisson, G. and Watson, R.J. (1984) *J. Virol.* 51, 263–265.
Rector, J.T., Lausch, R.N. and Oakes, J.E. (1982) *Infect. Immun.* 38, 168–174.
Reeves, W.C., Corey, L., Adams, H.G., Vontver, L.A. and Holmes, K.K. (1981) *N. Engl. J. Med.* 305, 315–319.
Roizman, B. (1979) *Cell* 16, 481–494.
Roizman, B., Warren, J., Thuning, C.A., Fanshaw, M.S., Norrild, B. and Meignier, B. (1982) *Dev. Biol. Stand.* 52, 287–304.
Roizman, B., Norrild, B., Chan, C. and Pereira, L. (1984) *Virology* 133, 242–247.
Rouse, B.T. and Lopez, C. (1984) In *Immunobiology of Herpes Simplex Virus Infection* (B.T. Rouse and C. Lopez, Eds.) CRC Press, Boca Raton. pp. 145–155.
Ruyechan, W.T., Morse, L.S., Knipe, D.M. and Roizman, B. (1979) *J. Virol.* 29, 677–697.
Sarmiento, M., Haffey, M. and Spear, P.G. (1979) *J. Virol.* 29, 1149–1158.
Schaffer, P.A., Carter, V.C. and Timbury, M.C. (1978) *J. Virol.* 27, 490–504.
Schmidt, M.F.G. (1983) *Curr. Top. Microbiol. Immunol.* 102, 101–129.
Schrier, R.D., Pizer, L.I. and Moorhead, J.W. (1983) *J. Immunol.* 130, 1413–1418.
Serafini-Cessi, F. and Campadelli-Fiume, G. (1981) *Arch. Virol.* 70, 331–343.
Serafini-Cessi, F., Dall'Olio, F., Scannavini, M., Costanzo, F. and Campadelli-Fiume, G. (1983) *J. Virol.* 48, 325–329.
Sethi, K.K., Omata, Y. and Schneweis, K.E. (1983) *J. Gen. Virol.* 64, 443–447.
Showalter, S.D., Zweig, M. and Hampar, B. (1981) *Infect. Immun.* 34, 684–692.
Sim, C. and Watson, D.H. (1973) *J. Gen. Virol.* 19, 217–233.
Slichtova, V., Kutinova, L. and Vonka, V. (1980) *Arch. Virol.* 66, 207–214.
Spear, P.G. (1976) *J. Virol.* 17, 991–1008.
Spear, P.G. (1984) In *The Herpesviruses* (B. Roizman, Ed.) Vol. 3. Plenum Press, New York. In press.
Spiro, R.G. (1966) *Meth. Enzymol.* 8, 26–52.
Stevens, J.G. (1977) *Adv. Cancer Res.* 22, 227–256.
Takemoto, K.K. and Fabish, P. (1964) *Proc. Soc. Exp. Biol. Med.* 116, 140–144.
Tarentino, A.L. and Maley, F. (1974) *J. Biol. Chem.* 249, 811–817.
Vaheri, A. and Cantrell, K. (1963) *Virology* 21, 661–662.
Vestergaard, B.F. and Norrild, B. (1978) *J. Infect. Dis.* 138, 639–643.
Vestergaard, B.F. and Norrild, B. (1979) *I.A.R.C. Sci. Publ.* 24, 225–234.
Watkins, J.F. (1964) *Nature* 202, 1364–1365.
Watson, R.J., Weis, J.H., Salstrom, J.S. and Enquist, L.W. (1982) *Science* 218, 381–384.
Wenske, E.A. and Courtney, R.J. (1983) *J. Virol.* 46, 297–301.
Wenske, E.A., Bratton, M.W. and Courtney, R.J. (1982) *J. Virol.* 44, 241–248.
Weis, J.H., Enquist, L.W., Salstrom, J.S. and Watson, R.J. (1983) *Nature* 302, 72–74.
Westmoreland, D. and Watkins, J.F. (1974) *J. Gen. Virol.* 24, 167–178.
Westmoreland, D., St. Jeor, S. and Rapp, F. (1976) *J. Immunol.* 116, 1566–1570.
Whitley, R.J. (1984) In *The Herpesviruses* (B. Roizman, Ed.) Vol. 3. Plenum Press, New York. In press.
Wildy, P., Field, H.J. and Nash, A.A. (1981) In *Virus Persistence* (B.W.J. Mahy, A.C. Minson and G.K. Darby, Eds.) Cambridge University Press, Cambridge. pp. 133–167.

Worthington, M., Conliffe, M.A. and Baron, S. (1980) *J. Infect. Dis.* 142, 163–174.
Zezulak, K.M. and Spear, P.G. (1983) *J. Virol.* 47, 553–562.
Zezulak, K.M. and Spear, P.G. (1984) *J. Virol.* 49, 741–747.
Zweerink, H.J., Martinez, D., Lynch, R.J. and Stanton, L.W. (1981) *Infect. Immun.* 31, 267–275.
Zweig, M., Showalter, S.D., Bladen, S.V., Heilman, C.J. Jr. and Hampar, B. (1983) *J. Virol.* 47, 185–192.

Eds. M.H.V. van Regenmortel and A.R. Neurath
Immunochemistry of viruses. The basis for serodiagnosis and vaccines.

Antigenic structure of retroviruses 24

STEPHEN OROSZLAN[1] and RAYMOND V. GILDEN[2]

[1] *Laboratory of Molecular Virology and Carcinogenesis, LBI-Basic Research Program and*
[2] *Program Resources, Inc., NCI-Frederick Cancer Research Facility, Frederick, MD 21701, U.S.A.*

Retroviruses constitute a large family of enveloped RNA viruses united by the common property of replication via a DNA intermediate generated by the RNA-dependent DNA polymerase called reverse transcriptase. Shortly after infection, the newly formed DNA may integrate into the host cell genome. Members of this family can exist both as endogenous inherited genetic elements in their natural hosts and as conventional horizontally transmitted viruses. The pathogenic members are associated with a variety of diseases including malignancies, neurological disorders, anemia and arthritis in infrahuman species, and are strongly implicated in leukemia lymphoma, acquired immune deficiency syndrome (AIDS) and related diseases in man. The family Retroviridae is divided into three subfamilies:

(*a*) Oncovirinae, including all the oncogenic and many other closely related non-oncogenic members.

(*b*) Lentivirinae, the slow viruses responsible for neurological diseases such as visna virus (maedi).

(*c*) Spumavirinae, the foamy virus group capable of inducing persistent infections without known disease.

On the basis of their morphology, oncovirinae can be classified into four categories: type A, B, C and D particles. The morphology and taxonomy of these viruses was summarized by Teich (1982). Type A particles, which do not have any infectivity, occur intracellularly as intracisternal or intracytoplasmic forms. Intracytoplasmic A particles are precursor forms to type B particles, which have an eccentrically placed nucleoid within the envelope bearing prominent spikes. Mouse mammary tumor virus (MMTV) is the prototype type B virus. Type C particles have a centrally located nucleoid and start virion assembly at the plasma membrane. The envelope spikes are not as well defined as on type B particles. The majority of retroviruses and all leukemia and sarcoma viruses are of the C-type morphology. Type D particles, which are somewhat larger than type C viruses, have bullet-shaped nucleoids which are eccentrically located within the envelope.

Retroviruses are either endogenous or exogenous to the host. Murine leukemia viruses are classified as ecotropic, xenotropic or amphotropic viruses according to their infectivity for cells of murine, nonmurine or both murine and nonmurine origin.

Before we describe the immunochemical properties of retroviral structural proteins, we will first discuss briefly their biosynthesis and processing. Next we will consider a newly proposed nomenclature for retroviral proteins. We then discuss in more detail the chemical and antigenic structure of the protein components of mature virions. Emphasis will be placed on the structural proteins of replication-competent nonacute type C leukemia viruses. The discussion of oncogene products of acute transforming viruses is outside the scope of this brief review.

1. Biosynthesis and processing of retroviral gene products

The 60S–70S genomic RNA of all replication-competent retroviruses consists of two identical subunits of single-stranded 35S RNA approximately 9.5 kilobases in length that, like all mRNAs, has a methylated cap structure at the 5′ end and is polyadenylated at the 3′ end. The 35S RNA of all replication-competent retroviruses is divided into three genes encoding the viral structural proteins: the *gag* (group-specific antigen) gene codes for the internal structural proteins, the *pol* (polymerase) gene codes for the enzymes protease, reverse transcriptase and endonuclease and the *env* (envelope) gene codes for the proteins of viral envelope. These genes are arranged in the order of 5′-*gag*-*pol*-*env*-3′, as shown in Fig. 24.1. The replication-competent avian sarcoma viruses (ASV) have in addition a *src* gene, located 3′ to *env*, which codes for the pp60 transforming protein (Brugge and Erikson, 1977). In murine leukemia viruses (MuLV), the primary translational product of the *gag* gene is a polyprotein designated $Pr65^{gag}$ and shown to be the proximal precursor to the internal proteins designated p15, p12, p30 and p10 (Fig. 24.1) which occur in this order from the NH_2 to COOH-terminus of $Pr65^{gag}$ (for review see Dickson et al., 1982). This gene order was first determined by immunochemical methods utilizing antibodies to the purified individual protein components in radioimmunoassays and conditional lethal mutants of Rauscher (Ra)-MuLV having temperature-sensitive lesions in post-translational processing of $Pr65^{gag}$ (for review see Stephenson, 1980). The complete primary structure of MuLV $Pr65^{gag}$, as determined by protein and DNA sequencing (Henderson et al., 1982; Schinnick et al., 1981; Miller and Verma, 1984), is shown in Fig. 24.2. The peptide chain of $Pr65^{gag}$, which consists of 538 residues, is more hydrophobic in the region extending

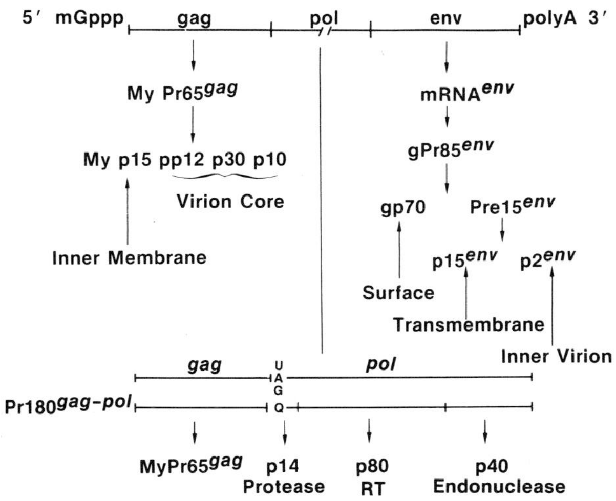

Fig. 24.1. Schematic representation of MuLV protein synthesis and processing of *gag*, *pol* and *env* gene products.

from the NH_2-terminus through approximately 400 residues than in the COOH-terminal region, which is highly polar. $Pr65^{gag}$ is phosphorylated before or during assembly with RNA, after which it is processed into the structural proteins by proteolytic cleavage inside the budding virions. Glycosylated proteins derived from a distinct form of *gag* gene product have also been found. These are present in the infected cells only and do not occur in virions. The nature of retroviral proteins as

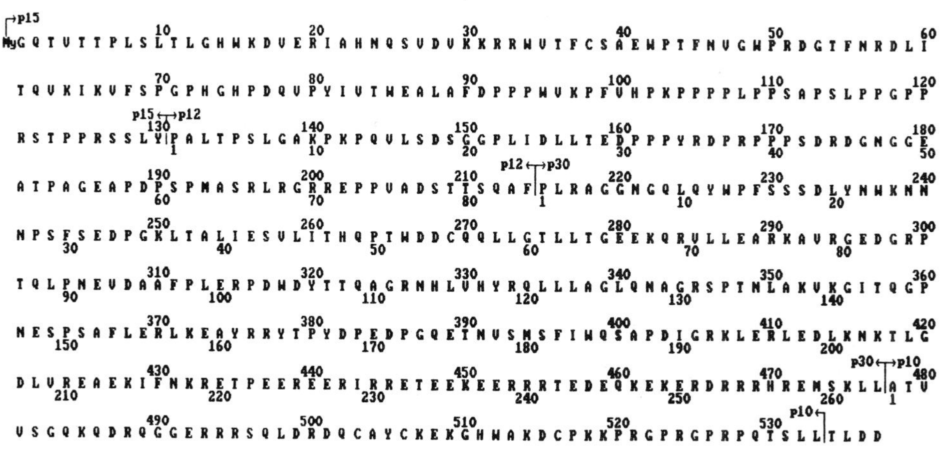

Fig. 24.2. Complete amino acid sequence of Mo-MuLV $Pr65^{gag}$. The cleavage sites are indicated by arrows.

cellular antigens has recently been reviewed by Fleissner and Snyder (1982).

Translation of 35S RNA, which has an amber termination codon at the end of the *gag* gene, also yields a larger precursor designated Pr180$^{gag\text{-}pol}$. This polyprotein is synthesized in 20–50-fold lower amounts than Pr65^{gag} (Jamjoon et al., 1977). In addition to all *gag* gene products, it consists of the viral proteolytic enzyme, reverse transcriptase and endonuclease. The map order of the *pol* gene is also shown in Fig. 24.1. This order has been established by the results of recent experiments in which both the protease and reverse transcriptase have been purified and their NH_2- and COOH-terminal sequences determined (Yoshinaka et al., 1985; Copeland et al., 1985a). The structural studies provided for the first time definitive evidence that the protease responsible for processing both precursor polyproteins in mammalian retroviruses is virally coded and that the synthesis of Pr180$^{gag\text{-}pol}$ occurs through a read-through mechanism whereby Gln is inserted at the UAG codon by tRNA^{Gln}.

The translation of the *env* gene is independent of *gag* and *pol* and preceded by splicing of the viral genome to form a subgenomic-size (28S) mRNA. In Ra- and Moloney (Mo)-MuLV the primary translational products are Pr68^{env} and Pr60^{env} respectively, both of which are post-translationally modified to yield the corresponding glycosylated precursor polyproteins gPr85^{env} and gPr80^{env}. The glycosylated precursor is cleaved during or after transport to the plasma membrane to gp70, the surface glycoprotein, and Pr15^{env}, which remain anchored to the membrane (Dickson et al., 1982). In MuLV *env* processing an additional morphogenetic cleavage takes place. Inside the budding virion a short COOH-terminal segment (16 residues) is cleaved off to generate the transmembrane protein, p15^{env}, and p2^{env} (Henderson et al., 1984). It is important to note that in several retroviruses such as ASV (Dickson et al., 1982), MMTV (Henderson et al., 1983), and bovine leukemia virus (BLV) (Schultz et al., 1984) the transmembrane *env* proteins are also glycosylated but are not cleaved at their carboxyl end. The complete amino acid sequence of Mo-MuLV gPr80^{env} is shown in Fig. 24.3.

2. Nomenclature

By convention (August et al., 1974) the retroviral gene products have been annotated by prefixing p, pp, gp or Pr (for protein, phosphoprotein, glycoprotein or precursor polyprotein) to the apparent molecular weight in thousands (see Fig. 24.1). While this nomenclature has been useful, the differences in the molecular weights (apparent or real) of analogous proteins from viruses of different species and variations in the results by the techniques required to resolve the proteins led to confusion. One of us (S. Oroszlan) has recently proposed a preferable systematic nomenclature which takes into account the known structural and biochemical properties of individual viral structural proteins as well as their genetic origin. Accordingly, in the following parts of the chapter a nomenclature will be used as follows.

→gp70

M A R S T L S K P L K N K V N P R G P L I P L I L L M L R G V S T | A S P G S S P H Q V Y N I T W E V T N G D R E T V W A

T S G N H P L W T W W P D L T P D L C M L A H H G P S Y W G L E Y Q S P F S S P P G P P C C S G G S S P G C S R D C E E

P L T S L T P R C N T A W N R L K L D Q T T H K S N E G F Y V C P G P H R P R E S K S C G G P D S F Y C A Y W G C E T T

G R A Y W K P S S S W D F I T V N N N L T S D Q A V Q V C K D N K W C N P L V I R F T D A G R R V T S W T T G H Y W G L

R L Y V S G Q D P G L T F G I R L R Y Q N L G P R V P I G P N P V L A D Q Q P L S K P K P V K S P S V T K P P S G T P L

S P T Q L P P A G T E N R L L N L V D G A Y Q A L N L T S P D K T Q E C W L C L V A G P P Y Y E G V A V L G T Y S N H T

S A P A N C S V A S Q H K L T L S E V T G Q G L C I G A V P K T H Q A L C N T T Q T S S R G S Y Y L V A P T G T M W A C

gp70←→p15E

S T G L T P C I S T T I L N L T T D Y C V L V E L W P R V T Y H S P S Y V Y G L F E R S N R H K R | E P V S L T L A L L L

G G L T M G G I A A G I G T G T T A L M A T Q Q F Q Q L Q A A V Q D D L R E V E K S I S N L E K S L T S L S E V V L Q N

R R G L D L L F L K E G G L C A A L K E E C C F Y A D H T G L V R D S M A K L R E R L N Q R Q K L F E S T Q G W F E G L

p15E←→p2E

F N R S P W F T T L I S T I M G P L I V L L M I L L F G P C I L N R L V Q F V K D R I S V V Q A L | V L T Q Q Y H Q L K P

p2E←

I E Y E P

Fig. 24.3. Complete amino acid sequence of Mo-MuLV gPr80^{env}. The cleavage sites are indicated by arrows.

Since the gene designations *gag* and *env*, which code for the structural proteins, are unique for retroviruses, we name all *gag* gene-coded proteins of mature virions *gagtins*. Similarly, the *env* gene-coded proteins are called *envins*. The structural protein derived from the NH_2-terminal region of the *gag* precursor polyprotein of mammalian type C and also B and D viruses is always myristylated (Henderson et al., 1983; Schultz and Oroszlan, 1983; 1984) at its NH_2-terminus. We name it *myrgagtin*. Large amounts of structural data have accumulated showing that the proteins which are adjacent to myrgagtin in the *gag* precursor are the most proline rich *gag* proteins in a given retrovirus and will therefore be called *prolgagtins*. The protein previously designated p24, p26, p30, etc. and frequently called 'group-specific antigen' or 'major internal protein' is always the largest *gag* gene-coded retroviral mature protein. It is to be named *largagtin*. The usually smallest *gag* gene product (p10–p15) within a single virus is rich in basic amino acids and has been invariably shown to be the protein which binds unspecifically to single-stranded RNA or DNA. This nucleic acid-binding *gag* protein will be called *nugagtin*. The location of the *env* gene-coded structural proteins on the virus has been identified. These are the surface glycoprotein and the transmembrane protein which, by the same convention, should be named *surenvin* and *tramenvin*. The new nomenclature for the *gag* and *env* gene-coded proteins of mammalian type C viruses, MMTV, the prototype type B virus, Mason-Pfizer monkey virus (MPMV), the prototype type D virus, and human T-cell leukemia virus (HTLV) is shown in Table 24.1 and is compared with the designations used previously.

TABLE 24.1

Nomenclature for mammalian retrovirus structural proteins based on their common properties

New nomenclature	Old nomenclature				Major antigenic determinant[a]	Location	Other characteristics
	Type C	Type B	Type D	HTLV			
Myrgagtin	p15, p12	p10	p10	p19	G	Inner membrane	Hydrophobic; N-terminus: myristyl-Gly
Prolgagtin	pp12, p15	p23	p14	—[b]	T	Not determined	Proline rich; acidic p*I*; phosphorylated
Largagtin	p24, p26, p27, p30	p27	p27	p24	G[c]	Core shell	Largest gagtin; neutral p*I*; N-terminus: Pro
Nugagtin	p10, p12	p14	p12	p15	G[c]	Inner core	Rich in basic amino acids; alkaline p*I*; nucleic acid binding
Surenvin	gp60, gp70	gp52	gp70	gp46–51	T, G	External surface	Glycosylated; neutralizing antigen
Tramenvin	p15E, p12E, gp30	gp36	gp20	p21, gp20	G[c]	Transmembrane	Most hydrophobic gagtin

[a]G, group-specific; T, type-specific.
[b]HTLV and BLV *gag* genes do not code for complete prolgagtin sequences at their 5′ end. In these viruses, the gene coding for prolgagtin-like protein is located 3′ to nugagtin (S. Oroszlan, presented at the 15th Int. Symp. Princess Takamaišu Cancer Res. Fund, Tokyo, 13–15 November 1984).
[c]Type C also has interspecies-specific antigenic activity.

3. Chemical properties of structural proteins

3.1. Myrgagtin

The myrgagtins of mammalian retroviruses are internal structural proteins which are coded for by the extreme 5′ region of the *gag* gene and thus are the NH_2-terminal polypeptides in the primary translational products, the precursor polyproteins such as the prototype $Pr65^{gag}$ of MuLVs. The NH_2-terminus is always glycine, to which the myristyl group is linked in amide linkage (Henderson et al., 1983; Schultz and Oroszlan, 1983; 1984). The complete amino acid sequence for Mo-MuLV myrgagtin (a prototype) is shown in Fig. 24.2. It encompasses residues 1–130 in the sequence of $Pr65^{gag}$ and is designated p15. In general hydrophobic amino acids are the dominant residues, but the myristate contributes greatly to the overall hydrophobicity of these proteins. Myrgagtins have been shown to be attached to the inner membrane of the viral envelope (Barbacid and Aaronson, 1978; Pepinsky and Vogt, 1979). The hydrophobic fatty acid moiety should have a major role in membrane binding. Myrgagtins have at least one cysteine, which is responsible for the dimerization of $Pr65^{gag}$ through disulfide linkage (Yoshinaka et al., 1984).

3.2. Prolgagtin

Retroviral prolgagtins are proline-rich *gag* gene-encoded proteins, which among the structural proteins of a given virus invariably contain the highest percentage of proline; amounting to 21–31% for the various mammalian type C viruses of the four subgroups (see Fig. 24.4). Prolgagtins are usually phosphorylated and consequently have acidic isoelectric points (Dickson et al., 1982). However, they may not be the only phosphorylated proteins within a single virus. Phosphorylation converts the hydrophobic peptide chain into a rather hydrophilic protein. The previous designations pp12 for MuLV, pp15 for baboon endogenous virus (BaEV) and pp18 for reticuloendotheliosis virus (REV-A) were made on the basis of apparent molecular mass. A comparison of these values with the calculated M_r of the peptide chains, which range from 6.7 to 11.5 kDa, may indicate post-translational modification other than phosphorylation itself. Nucleotidylation has been suggested (Versteegen et al., 1982). The complete amino acid sequence of Mo-MuLV prolgagtin (pp12), as deduced from the nucleotide sequence of proviral DNA (Shinnick et al., 1981) and also determined by protein sequencing (Versteegen et al., 1982), is shown in Fig. 24.2. It is designated p12 and encompasses residues 131–214 in the continuous sequence of $Pr65^{gag}$. The complete amino acid sequence of Ra-MuLV prolgagtin has also been determined by direct sequencing while those of feline leukemia virus (FeLV), BaEV and spleen necrosis virus (SNV) were obtained from the nucleotide sequences (Laprevotte, 1984; Tamura, 1983; O'Rear and Temin, 1982) and partial NH_2- and COOH-terminal amino acid sequences (Copeland et al., 1981; Tsai et

```
SUBGROUP I     1        5              10              15             20             25             30

  RA-MLV               LEU
  MO-MLV  ProLeuArgAlaGlyGly - - - - - AsnGly - - GlnLeuGlnTyrTrpProPheSerSerSerAspLeuTyrAsnTrpLys
  FELV                 Glu   Pro                  Asn      ArgPro                    Ala

SUBGROUP II

  BAEV                                                         Ile
          ProLeuArgThrVal - - - - - - AsnArg - - Thr   GlnTyrTrpProPheSerAlaSerAspLeuTyrAsnTrpLys
  RD-114                                                        Val

SUBGROUP III

  GALV
          ProLeuArgAlaIleGlyProProAlaGluProAsnGlyLeuValProLeuGlnTyrTrpProPheSer X AlaAspLeuTyr
  SSAV

SUBGROUP IV

  MMC-1                                           His
  MAC-1   ProLeuArgGluIleGlySerLeuAspAspThr - GlyLeuSerArgLeuMetTyrTrpProPheSerThrSerAspLeuTyrAsnTrpLys
  CPC-1                                       AlaProPro

COMMON    ProLeuArg                                          TyrTrpProPheSer     AspLeuTyrAsnTrpLys
__________

  REV-A   ProLeuArgGluThrGlyGluArgAspMetGly - GlyArgProMetArgThrTyrValProPheThrThrSerAspLeuTyrAsnTrpLys
```

Fig. 24.1. Schematic representation of MuLV protein synthesis and processing of *gag*, *pol* and *env* gene products.

al., 1984). Ra-MuLV and Mo-MuLV prolgagtins are of identical length: each 84 residues. The total number of amino acids is 70 for FeLV Subgroup I, and ranges from 68 to 106 for prolgagtins of viruses in the other three subgroups (see Fig. 24.4). None of the prolgagtins of mammalian type C viruses for which the complete structure is known contains either cysteine or tryptophan, another unique property for these mammalian *gag* gene products. A comparison of Ra-MuLV and Mo-MuLV prolgagtin sequences reveals the existence of two variable regions and two highly conserved regions within the two molecules. Residues 18–37 are identical, with the exception of residue 24 (valine in Ra-MuLV, isoleucine in Mo-MuLV). A similar conserved sequence has been assigned in FeLV prolgagtin running from the corresponding residue 22 to 34 of Mo-MuLV p12 and ending with Pro-Pro-Pro-Tyr. The other conserved sequence runs from residue 61 to 84 with only one substitution at residue 64 (valine in Ra-MuLV, alanine in Mo-MuLV). This segment is also conserved in FeLV. The two variable regions, comprising residues 1–17 and 38–60 of p12, contain 7 and 11 amino acid changes respectively, corresponding to a variability of 41% in the N-terminal variable region and 49% in the internal variable region. These two regions account for the immunological type specificity in mouse type C virus p12 (unpublished results).

The phosphorylation of p12 has been implicated in nucleic acid binding, but there is no information as to the location of binding sites or phosphorylated residues within the molecules. A similarity between both the NH_2- and COOH-terminal amino acid sequences of murine retrovirus p12s and avian histone H_5 has been reported (Henderson et al., 1981; Versteegen et al., 1982). The N-terminal region of histone H_5 has been implicated in phosphorylation-mediated nucleic acid binding, a property which may be shared by retroviral prolgagtins. The observed homology involving basic amino acids in the COOH-terminal region may also have some similar functional significance.

3.3. Largagtin

This is the structural protein commonly called the major internal protein or group-specific antigen. The molecular masses of largagtins of the various retroviruses show notable size differences in the range of 24–30 kDa, but their size is always the largest among the *gag* gene-encoded proteins of mature infectious virions, hence the name. The isoelectric point of largagtins is close to neutral but can vary from slightly acidic to slightly alkaline pH and corresponds well to the number of basic and acidic amino acids present in the molecule. Nevertheless, substantial charge heterogeneity has been observed. Largagtins readily self-associate in solution at low ionic strength and in the absence of urea or other dissociating agents. The amino acid compositions of type C virus largagtins are strikingly similar, including the total number of polar and nonpolar amino acids. The complete primary sequence of Mo-MuLV largagtin is shown in Fig. 24.2 designated p30 and encompassing residues 215–477 in the continuous sequence of the precursor polyprotein $Pr65^{gag}$. All the largagtins of type C viruses have a common NH_2-terminal sequence (Pro-Leu-Arg) and their sequences are highly conserved, with average homology of 60–80% among various species and higher within a single species, viz. mouse. The initial 2/3 of the molecule is fairly hydrophobic and may form globular structures, but the carboxy-terminal 1/3 is highly polar, having clusters of charged amino acids. In this respect, this region somewhat resembles nuclear histones. It is noted that the highly polar carboxy-terminal domain is not present in largagtins of avian, bovine and human type C viruses.

The mammalian p30s had shown cross-reactive (interspecies) determinants (for review see Oroszlan and Gilden, 1978) as established by both immunodiffusion and radioimmunoassays, and have been found to be compositionally similar. Sequence analysis provided support for the immunologic findings and showed that the NH_2-terminal regions of the p30s of mammalian type C viruses are highly conserved (Oroszlan et al., 1975). We have identified four subgroups based on alignments of NH_2-terminal sequences of p30s. The classification is shown in Fig. 24.4. All p30s initiate with a common tripeptide (Pro-Leu-Arg) and contain a highly homologous (conserved) region beginning with glutamine (position 11 in the alignment), except in Subgroup IV where glutamine is substituted with methionine (a few additional substitutions also occur in this region). While in the first subgroup (mouse, rat and cat leukemia viruses) residues 4–10 are variable and essentially species specific, the p30s in Subgroup II (baboon endogenous virus and cat endogenous virus RD-114) differ only in position 17 and both have a deletion in position 6. Gibbon ape leukemia virus (GaLV) and simian sarcoma-associated virus (SSAV) in Subgroup III have identical sequences throughout the known first 28 residues, and both have two insertions involving a total of seven residues in the variable region of the mammalian p30s. In Subgroup IV, the p30s of the recently isolated endogenous rhesus monkey (*Macaca mulatta*) virus, MMC-1 (Rabin et al., 1979) and that of *Macaca arctoides,* MAC-1 (Todaro et al., 1978), are also identical but again have two insertions. These sequences have the first residue in the conserved region (glutamine) substituted with methionine. As already mentioned, the p30 of REV-A belongs to this group, which is more closely related to the p30s of Subgroup IV than to those of other subgroups (Fig. 24.4).

3.4. Nugagtin

Usually this is the smallest (lowest M_r) *gag* gene-encoded internal structural protein of a given retrovirus. It is uniquely characterized by the highest relative content of basic amino acids (lysine and arginine), and as a consequence has a high isoelectric point (pH 9.5–10.4). This protein, which has been identified as the component of the ribonucleoprotein complex present in the viral nucleoid (Fleissner and Tress, 1973), has been shown to bind strongly to single-stranded RNA as well as DNA in vitro. The complete sequence of Mo-MuLV nugagtin, which occupies the COOH-terminal position in $Pr65^{gag}$, is shown in Fig. 24.2. Nugagtins have been shown to possess strong group- and interspecies-reactive antigenicity and lack type-specific determinants. This is consistent with the complete amino acid sequences, which are highly conserved not only among type C viruses but also in BLV and human T-cell leukemia viruses (HTLV) and possess the putative nucleic acid binding sites each involving three cysteines (Henderson et al., 1981; Copeland et al., 1983a; 1983b; 1984).

3.5. Surenvin

Surenvins are without exception glycosylated retroviral proteins, located on the viral envelope surface as demonstrated by surface labeling techniques. They have been shown to be involved in recognition functions relating to virus-host range and to serve as primary targets for antibody-mediated virus neutralization, as will be discussed later. Surenvins are found on the surface of viral envelopes in association with tramenvins, probably through noncovalent hydrophobic bonds and/or covalent disulfide linkages. Both these proteins are generated before virus maturation by cleavage of the envelope precursor protein. This protein is found only on the surface of infected cells and not on viruses. The complete primary structure of the Mo-MuLV *env* gene translational product is shown in Fig. 23.3. Surenvin is designated as gp70 in the figure.

3.6. Tramenvin

Tramenvins are the transmembrane proteins of retroviruses. They contain two exceptionally hydrophobic regions, one located near the NH_2-terminus and the other in the COOH-terminal region. This latter segment has an α-helical structure and is believed to anchor the protein to the membrane. Tramenvins of mammalian type C viruses are usually not glycosylated but tramenvins of ASV (gp37), MMTV (gp36) and BLV (gp30) were shown to be glycoproteins. The transmembrane proteins have been implicated in immunosuppression by RNA tumor viruses (see Cerny and Essex, 1979; Cianciolo et al., 1980). They have in their amino acid sequence a highly conserved region (Fig. 24.5) encompassing residues 68–95 in MuLV p15E (Fig. 24.3). It may be that these structures can be responsible for or involved in common biological functions resulting in immune deficiency.

4. Antigenic determinants of structural proteins

As discussed in the preceding chapters of this volume, protein molecules usually possess multiple antigenic determinants including both conformational and sequential determinants, i.e. discontinuous and continuous antigenic sites. It is also realized that, in general, the majority of antibodies elicited by globular proteins are directed toward conformational rather than sequential antigenic determinants (Sela et al., 1967; Crumpton, 1974). The protein components of retroviruses are not an exception. They have been shown to contain type-specific, group(species)-specific and interspecies-specific antigenic determinants (for review see Gilden, 1977; Oroszlan and Gilden, 1978; Stephenson et al., 1980). Chemical modification of proteins combined with appropriate serological tests and quantitative immunoassays, and other approaches, including protein fragmentation, have been shown to provide an effective method for the immunological characterization of protein antigens. Such studies have led to the determination of the entire antigenic structure of myoglobin and lysozyme (Atassi, 1975; 1978).

Some time ago we investigated the relationship among the antigenic determinants in p30s (now called largagtins). Using native, unmodified and citraconylated pro-

```
                1        10        20        30
A-RSV gp 37     VLQNRAAIDFLLLAHGH**GCEDVAGMCCF
Mo-MuLV p15E    VLQNRRGLDLLFLKEGG**LCAALKEECCF
BLV gp30E       VAQNRRGLDWLYIRLGFQSLCPTINEPCCF
HTLV-I p21      AAQNRRGLDLLFWEQGG**LCKALQEQCRF
BaEV p15E       VLQNRRGLDLLTAEQGG**ICLALQEKCCF
```

Fig. 24.5. Alignment of conserved regions of avian and mammalian type C virus tramenvins. A-RSV, avian Rous sarcoma virus (Schwartz et al., 1983); Mo-MuLV (Shinnick et al., 1981); BLV (Rice et al., 1984); HTLV-I (Seiki et al., 1983); BaEV (Copeland et al., 1985b; M. Cohen, unpublished).

teins and a peptide fragment encompassing the NH_2-terminal one-third of the molecule we demonstrated by the double immunodiffusion technique that mammalian largagtins contain multiple antigenic determinants of different categories and were able to distinguish at least three species-specific and three interspecies-specific determinants (Davis et al., 1973; 1974; 1975). In addition, type-specific chemical differences were shown by peptide mapping (Oroszlan et al., 1974), providing the basis for and confirming previous immunological data on type-specific determinants (Strand and August, 1974). More recently, we investigated the effects of protein modification on the antigenic determinants of p30 and gp70 of type C retroviruses by using solid-phase competition radioimmunoassays. Proteins were modified by reduction with 2-mercaptoethanol and subsequent carboxamidomethylation of SH groups with iodoacetamide or by amidination of α and ϵ amino groups with methylacetimidate. The type-specific determinants of gp70 were found to be conformational in nature, as they were destroyed by these chemical modifications. However, group- and interspecies-specific determinants of gp70 antigens detectable by the antisera used appeared to be sequential and did not involve residues susceptible to the chemical reagents. Conformation-dependent type-specific determinants of p30 were affected only by methylacetimidate. Group- and interspecies-specific determinants of p30 were similar to those of gp70 in that they also appeared to be sequential antigenic sites. Therefore, the broadly reactive group- and interspecies-specific determinants of gp70 and p30 could be found in small peptides. Accordingly, a cyanogen bromide cleavage fragment derived from the COOH-terminal one-third of Ra-MuLV p30 was found to carry group-specific determinants but no detectable interspecies-specific determinants. In contrast, a peptide obtained by limited trypsin cleavage of p30, which was derived from the NH_2-terminal region of the protein, contained at least one of the interspecies determinants shared with FeLV p27. Studies with a synthetic undecapeptide (Oroszlan and Gilden, 1978), representing the NH_2-terminal conserved region of mammalian largagtins (see Fig. 24.4), and with an octaeicosapeptide (Sallay et al., 1979), corresponding to an NH_2-terminal segment of GaLV largagtin which contains sequences unique to both GaLV/SSV and to the conserved undecapeptide, showed that the antiserum to the octaeicosapeptide contained GaLV/SSAV type-specific as well as interspecific antibodies and that both peptides were recognized by antibodies to the native protein molecule (A.M. Schultz, T.D. Copeland and S. Oroszlan, unpublished results), although neither of them is the immunodominant determinant. We have also identified several antigenically active peptides from surenvin of Ra-MuLV. Some of these, derived from the COOH-terminal region, were found to carry interspecies reactive determinants and represented common continuous antigenic sites (Versteegen and Oroszlan, 1980). In general, methods to elucidate in detail the relationship between the primary structure and the nature and location of natural epitopes of type C viral proteins are essential for the development of specific synthetic viral antigens (immunogens) for eventual use as vaccines.

Similar approaches in the mouse system led to the demonstration that the gp70 of AKV, an ecotropic virus from the AKR inbred mouse strain which recombines in the *env* gene with endogenous sequences to form the dualtropic mink cell focus-forming MCF-247 virus, and the gp70 of the recombinant virus, contain highly conserved COOH-terminal domains but unique NH_2-terminal sequences (Pinter and Honnen, 1983). It has been shown previously that a monoclonal antibody against an epitope located in the COOH-terminal domain of AKV gp70 has potent neutralizing activity for both AKV and MCF-247 MuLVs (Nowinski et al., 1981). These findings are in full agreement with the complete amino acid sequences of surenvins (Khan, 1984). Although the antigenic mapping of mouse retroviral surenvins is far from complete, the availability of a large number of highly specific monoclonal antibodies (O'Donnell and Nowinski, 1980; Chesebro et al., 1983) already permits the differentiation and localization in different domains of several unique epitopes. Furthermore, recognition by monoclonal antibodies of native tertiary protein structure or functional sites of biologically active protein complexes makes these reagents superior to antibodies against short synthetic peptides.

Most recently, an elegant study by Nunberg et al. (1984) succeeded in localizing a determinant of virus neutralization on FeLV surenvin (the viral surface protein gp70). These authors used virus-neutralizing monoclonal antibody to screen recombinant phage libraries expressing, as fusion proteins, short random fragments of DNA sequences encoding FeLV gp70, which were molecularly cloned into the β-galactosidase gene of phage λ Charon 16. They mapped the relevant antibody binding to a 14-residue amino acid segment located in the NH_2-terminal half of gp70. The sequence of this immunogenic determinant and flanking sequences of the viral protein are compared with the corresponding regions of MuLV surenvins in Fig. 24.6. The hydropathic plot, also shown in the figure, indicates that this region probably resides, as expected, on the outside surface of FeLV gp70, as do the corresponding regions in the mouse retroviral surenvins. This high resolution method can be used to map antigenic determinants recognized by neutralizing monoclonal antibodies. Its advantage over conventional methods, such as protein fragmentation, is high specificity combined with speed and simplicity of analysis. Therefore, it can potentially be extremely useful in attempts to develop (bio)synthetic vaccines for the prevention of retrovirus-associated malignancies including human leukemia and related diseases.

5. Cell receptors – viral infection and neutralization

As described above, the retroviral envelope gene product is processed into two major polypeptides, one of which is always glycosylated and projects from the virion surface, and the second of which may be considered a transmembrane protein. The apparent molecular weight of these two components may vary from virus to virus

```
GA-FeLV-B   GYDPIALFSVSRQVMTITPPQA MGPNLVLPDQKPPS RQSQIESRVTPHHSQGNGGTPGIT
                                    ... .. .. ...
Mo-MuLV     GIDPVTRFSLTRQVLNIGPRVS IGPNPVITDQLPPS RPVQINLPRPPQPPPPGAASIVPE
                                    ... .  .. ...
Mo-MCF      GQDPGLTFGIRLRYQNLGPRVP IGPNPVLADQQPLS KPKPVKSPSVTKPPSGTPLSPTQL
                                    ... .. .. . .
Fr-MuLV     GRDPGLTFGIRLRYQNLGPRVP IGPNPVLADQLSLP RPNPLPKPAKSPPASNSTPTLISP
                                    ... .. ..
AKV         GHDPGLIFGIRLKITDSGPRVP IGPNPVLSDRRPPS RPRPTRSPPPSNSTPTETPLTLPE
                                    ... .. .  ...
```

Fig. 24.6. The neutralization epitope (MGPNLVLPDQKPPS) of the Gardner (GA)-FeLV strain B gp70 is shown with the neighboring sequences, together with a hydrophilicity plot and alignment with homologous Mo-MuLV, Mo-MCF, Friend (Fr)-MuLV and AKV-MuLV surenvins (gp70) sequences. The dots indicate identities. (courtesy of J.H. Nunberg; see Nunberg et al., 1984).

depending somewhat on the extent of glycosylation, but also on the length of the polypeptide chain. The amino-terminal *env* gene product, surenvin (gp70 in murine type C viruses, gp85 in avian type C viruses, gp52 in mouse mammary tumor virus, etc.), is the determining virion component for virus entry into target cells. This has been shown in a variety of ways, most directly by the well-documented ability of monospecific and monoclonal antibodies to surenvins to neutralize virus infectivity both in vitro and in vivo. Antibodies to other virion proteins do not neutralize directly, although antibodies to the transmembrane protein, tramenvin (p15E in murine viruses), may do so in the presence of complement. This neutralization is based on the phenomenon of virolysis (Oroszlan and Gilden, 1970) in which 'holes' are made in the viral membrane by the action of complement on the immune complexes.

The entry of retroviruses into cells is based on specific virus receptors. Such receptors may be distributed in species, tissue and stage of differentiation-specific fashion, and their expression is highly relevant to disease induction. Interference phenomena, commonly used for subtyping retroviruses, occur when cells expressing a given type are resistant to all others of that type but not to viruses of other types (Steck and Rubin, 1966; Vogt and Ishizaki, 1966; Rein, 1982). The interference operates at the level of surenvin receptor interaction and, for example, may be simulated by addition of purified surenvin to appropriate target cells or abolished by inhibition of surenvin synthesis (Rein et al., 1982). Certain retroviruses are able to induce syncytia in appropriate host cells and this has provided the basis for in vitro infectivity and neutralization assays (Rowe et al., 1970; Rand et al., 1974). Direct cell fusion has been shown with intact (but noninfectious) virions and this phenomenon can be inhibited by purified surenvin. Retrovirus-cell membrane fusion appears to be a pH-dependent process (Redmond et al., 1984) resembling that seen with other enveloped viruses. The virus is thought to bind via the specific surenvin receptor, after which it is internalized by endocytosis through coated pits and, finally, the viral core is released into the cytoplasm. The specific interaction with receptors may help to explain the complex interactions of chronic leukemia viruses with their hosts leading to disease after long latent periods. In the mouse, at least three distinct gp70 receptor interactions have been described (Rein, 1982). Thus, based on interference properties, these viruses can be divided into amphotropic, ecotropic and recombinant mink cell-focus forming (MCF) dualtropic viruses. In the well-studied AKR mouse, it appears to be the MCF variants arising late in life which are responsible for disease (the mechanism is not yet established), while the ecotropic virus of early life is an essential component but not the proximal carcinogen. Recent data indicates that MCF viruses are recombinants between ecotropic and endogenous sequences involving the 5′ region of the *env* gene and specific long terminal repeat (LTR) sequences (Lung et al., 1983). These viruses acquire a new property of being biologically active in thymocytes. Neither ecotropic or xenotropic viruses have this property. In this complex system, neutralization of the ecotropic virus early in life by in vivo administration of neutralizing antibody (Huebner et al., 1976) or antibody monospecific to gp70 reduces virus disease, presumably by reducing the availability of one of the parental components in generation of MCF viruses.

Given that availability of unsaturated receptors for a pathogenic virus is crucial for disease induction, it has been a frequent speculation that endogenous retroviral sequences may express *env* products, reacting with receptors and thus preventing infection by horizontally transmitted pathogenic viruses. While there are several situations in which this apparently has been found (Bassin et al., 1982), there is currently no substantial evidence for the generality of the phenomenon. The theory that endogenous viral genomes endow their hosts with a selective advantage by providing a resistance factor against pathogenic variants and, in this manner, ensure their own survival is only one of several explanations for the multiplicity of endogenous genomes in some species. We previously suggested that these genomes represent a prime example of 'selfish DNA' (Gilden, 1981) and that biological manifestations of their presence are likely to be accidental. The occurrence of surenvin-like molecules on cell surfaces and their role in natural biology remains an important topic for investigation in man.

6. Humoral and cellular immune responses

Humoral immune responses to retroviral infection are highly variable depending on virus type and host interactions. In certain cases where prolonged viremia does not occur, serum antibody provides a most valuable tool for establishing virus prevalence. In the case of bovine leukemia, virtually all infected animals develop antibody to *gag* and *env* products of BLV and diseased animals actually show the highest titers (Levy et al., 1977). In any given herd, the existence of even a single case of bovine leukosis will indicate a very high incidence of infected antibody-positive individuals. In herds where no disease is observed, animals are negative for antibody. In this system, freshly isolated lymphocytes, when placed in vitro, generally show no evidence of virus production; this requires mitogenic stimulation or cocultivation with susceptible target cells. In parallel fashion, the detection of HTLV in vitro was highly dependent on the ability to grow T cells in vitro (requiring interleukin 2) (Gallo and Wong-Staal, 1982). Virus expression in vivo, as for BLV (HTLV and BLV are distant relatives), tends to be relatively low. Thus, in both of these systems, the presence of antibody indicates the likelihood of isolating infectious virus by appropriate in vitro methods. In both cattle infected with BLV and humans with HTLV, disease occurs only in a small fraction of infected individuals. This may reflect the low probability of eventual integration of virus promoter or enhancer sequences so as to activate resident 'oncogenes'. Alternatively (or in addition), disease resistance may result from cellular immunity directed against virus-producing cells. This remains to be established.

In contrast to the restricted cell tropism of viruses like BLV and HTLV, FeLV replicates extensively in many cell types in susceptible cats, the end result being an extensive and life-long viremia (Hardy, 1980). The pathogenic endpoint may be either malignancy induction or a spectrum of diseases resulting from immunosuppressive effects of the virus (McClelland et al., 1980). Diagnosis of virus infection in cats thus depends on actual assays for infectious virus (or viral antigen) (Hardy, 1980) in blood smears or saliva (the source of transmitted virus), and antibody to viral proteins will not ordinarily be found. Not all cats exposed to FeLV develop viremia; a significant percentage develop neutralizing antibody and thus are resistant to superinfection. Such cats also frequently have antibody to viral *gag* gene products (Charman et al., 1976).

A complicating factor in malignancy induction results from immune responses to an antigen designated FOCMA (feline oncornavirus cell membrane antigen) which occurs on both virus-positive and virus-negative lymphosarcomas (Hardy, 1980). Antibody to this protein even in the presence of viremia appears protective against malignant conversion. FOCMA appears to be highly related to FeLV subgroup C surenvin and the most recent evidence suggests analogy to the MCF category of murine viruses, i.e. it is a product of an endogenous *env* gene or a recombinant between such sequences and subgroup C ones (Besmer et al., 1983). Since FOCMA is found on virus-free lymphosarcomas, the endogenous activation model may be most logical. The suggestion here seems to be that FOCMA is a target for antibody-mediated cytolysis, thus eliminating potentially malignant cells. Gibbons infected with GaLV likewise either become virus-negative and antibody-negative or viremic and antibody-positive (Kawakami et al., 1973). Here the antibody-positive individuals (as in the cat situation) may remain healthy while the viremic animals will develop disease.

The above cases represent contrasting situations between presence of antibody and disease prediction. In the antibody-negative relationships, it is possible that antibody is actually made but removed from the circulation by the tremendous excess of antigen. In addition to the restricted cell tropism of viruses like BLV and HTLV, the newly described HTLV-III (evidently the etiologic agent of AIDS) (Gallo et al., 1984; Sarngadharan et al., 1984) is also cytolytic for the T4 subpopulation of T cells; thus again reducing virus mass to low levels allowing for production of circulating antibody. In this instance, as for HTLV-I (the etiologic agent of the adult T-cell leukemia syndrome), the presence of antibody is not restricted to individuals with overt disease.

In most retrovirus-host interactions, the importance of cellular immunity has not been critically evaluated. Equally important may be both protection against and augmentation of the disease. In one model (Ihle et al., 1984), the cellular immune response to viral antigen actually induces the production of lymphokines which may increase the cell population at risk for the proximal carcinogenic event. Supportive evidence for this comes from studies in mice with immune dysfunction in which high levels of virus are found in the absence of disease (Ihle et al., 1984). Cellular immune responses may thus have conflicting effects, the extremes being elimination of transformed cells bearing virus-specific target molecules and the provision of an expanded target cell population for newly generated transforming viruses or activated cellular elements, i.e. oncogenes.

It should be evident that host-virus interactions are not readily generalized in the case of retroviruses. Rather, each system must be analyzed in detail with respect to the variables outlined above.

7. Acknowledgements

Research was sponsored in part by the National Cancer Institute, DHHS, under contract no. N01-CO-23909 with Litton Bionetics, Inc. The contents of this publication do not necessarily reflect the views or policies of the Department of Health and Human Services, nor does mention of trade names, commercial products, or organizations imply endorsement by the US Government.

8. References

Atassi, M.Z. (1975) *Immunochemistry* 12, 423–438.

Atassi, M.Z. (1978) *Immunochemistry* 15, 909–936.

August, J.T., Bolognesi, D.P., Fleissner, E., Gilden, R.V. and Nowinski, R.C. (1974) *Virology* 60, 595–601.

Barbacid, M. and Aaronson, S.A. (1978) *J. Biol. Chem.* 253, 1408–1414.

Bassin, R.H., Ruscetti, S., Al, I., Haapala, D.K. and Rein, A. (1982) *Virology* 123, 139–151.

Besmer, P., Snyder, H.W. Jr., Murphy, J.E., Hardy, W.D. Jr. and Parodi, A. (1983) *J. Virol.* 46, 606–613.

Brugge, J.S. and Erikson, R.L. (1977) *Nature* 269, 346–348.

Cerny, J. and Essex, M. (1979) In *Naturally Occurring Biological Immunosuppressive Factors and Their Relationship to Disease* (R.H. Neubauer, Ed.) CRC Press, Boca Raton FL. pp. 233–256.

Charman, H.P., Kim, N., Gilden, R.V., Hardy, W.D. Jr. and Essex, M. (1976) *J. Natl. Cancer Inst.* 56, 859.

Chesebro, B., Britt, W., Evans, L., Wehrly, K., Nishio, J. and Cloyd, M. (1983) *Virology* 127, 134–148.

Cianciolo, G.J., Matthews, T.J., Bolognesi, D.P. and Snyderman, R. (1980) *J. Immunol.* 124, 2900–2905.

Copeland, T.D., Henderson, L.E., Vanlaningham-Miller, E.S., Stephenson, J.R., Smythers, G.W. and Oroszlan, S. (1981) *Virology* 109, 13–24.

Copeland, T.D., Morgan, M.A. and Oroszlan, S. (1983a) *FEBS Lett.* 156, 37–40.

Copeland, T.D., Oroszlan, S., Kalyanaraman, V.S., Sarngadharan, M.G. and Gallo, R.C. (1983b) *FEBS Lett.* 162, 390–395.

Copeland, T.D., Morgan, M.A. and Oroszlan, S. (1984) *Virology* 133, 137–145.

Copeland, T.D., Gerard, G., Hixson, C.G. and Oroszlan, S. (1985a) *Virology* In press.

Copeland, T.D., Tsai, W.P. and Oroszlan, S. (1985b) *Biochem. Biophys. Res. Comm.* In press.

Crumpton, M.J. (1974) In *The Antigens* (M. Sela, Ed.) Vol. II. Academic Press, New York. pp. 1–80.

Davis, J., Gilden, R.V. and Oroszlan, S. (1973) *Virology* 56, 411–415.

Davis, J., Charman, H.P., Oroszlan, S. and Gilden, R.V. (1973/74) *Intervirology* 2, 326–336.

Davis, J., Gilden, R.V. and Oroszlan, S. (1975) *Immunochemistry* 12, 67–72.

Dickson, C., Eisenman, R., Fan, H., Hunter, E. and Teich, N. (1982) In *RNA Tumor Viruses* (R. Weiss, N. Teich, H. Varmus and J. Coffin, Eds.) Cold Spring Harbor, New York. pp. 513–648.

Fleissner, E. and Snyder, H.W. Jr. (1982) In *Current Topics in Microbiology and Immunology* (E. Wecker and I. Horak, Eds.) Vol. 98. Springer-Verlag, New York. pp. 37–80.

Fleissner, E. and Tress, E. (1973) *J. Virol.* 12, 1612–1615.

Gallo, R.C. and Wong-Staal, F. (1982) *Blood* 60, 545–557.

Gallo, R.C., Salahuddin, S.Z., Popovic, M., Shearer, G.M., Kaplan, M., Haynes, B.F., Palker, T., Redfield, R., Oleske, J., Safai, B., White, G., Foster, P. and Markham, P.D. (1984) *Science* 224, 500–503.

Gilden, R.V. (1977) In *Biology of RNA Tumor Viruses* (P.D. Nayak, Ed.) Marcel Dekker, New York.

Gilden, R.V., Oroszlan, S., Young, H.A., Rice, N.R., Gonda, M.A., Cohen, M. and Rein, A.R. (1981) In *Frontiers in Immunogenetics* (W.H. Hildemann, Ed.) Elsevier/North Holland, Amsterdam. pp. 191–223.

Hardy, W.D. Jr. (1980) In *Feline Leukemia Virus* (W.D. Hardy Jr. et al., Eds.) Elsevier/North Holland, New York. pp. 33–78.

Henderson, L.E., Copeland, T.D., Sowder, R.C., Smythers, G.W. and Oroszlan, S. (1981) *J. Biol. Chem.* 256, 8400–8406.

Henderson, L.E., Copeland, T.D., Smythers, G.W., Sowder, R.C. and Oroszlan, S. (1982) In *Molecular Biology of Tumor Viruses: RNA Tumor Viruses* (R.A. Weiss, N.M. Teich, H.E. Varmus and J.M. Coffin, Eds.) Cold Spring Harbor, New York. pp. 1378–1382.

Henderson, L.E., Krutzsch, H.C. and Oroszlan, S. (1983a) *Proc. Natl. Acad. Sci. USA* 80, 339–343.

Henderson, L.E., Sowder, R., Smythers, G.W. and Oroszlan, S. (1983b) *J. Virol.* 48, 314–319.

Henderson, L.E., Sowder, R., Copeland, T.D., Smythers, G. and Oroszlan, S. (1984) *J. Virol.* 52, In press.

Huebner, R.J., Gilden, R.V., Toni, R., Hill, R.W., Trimmer, R.W., Fish, D.C. and Sass, B. (1976) *Proc. Natl. Acad. Sci. USA* 73, 4633–4635.

Ihle, J.N., Rein, A. and Mural, R. (1984) In *Advances in Viral Oncology* (G. Klein, Ed.) Vol. 4. Raven Press, New York. pp. 95–137.

Jamjoon, G.A., Naso, R.B. and Arlinghaus, R.B. (1977) *Virology* 78, 11–34.

Kawakami, T.G., Buckley, P.M., McDowell, T.S. and DePaoli, A. (1973) *Nature* 246, 105–107.

Khan, A.S. (1984) *J. Virol.* 50, 864–871.

Laprevotte, I., Hampe, A., Sherr, C.J. and Galibert, F. (1984) *J. Virol.* 50, 884–894.

Levy, D., Deshayes, L., Paroid, A.L., Levy, J.P., Stephenson, J.R., Devare, S.G. and Gilden, R.V. (1977) *Int. J. Cancer* 20, 543.

Lung, M.L., Hartley, J.W., Row, W.P. and Hopkins, N.H. (1983) *J. Virol.* 45, 275–290.

McClelland, A.J., Hardy, W.D. Jr. and Zuckerman, E.E. (1980) In *Feline Leukemia Virus* (W.D. Hardy Jr. et al., Eds.) Elsevier/North Holland, New York. pp. 121–126.

Miller, A.D. and Verma, I.M. (1984) *J. Virol.* 49, 214–222.

Nowinski, R.C., Pickering, R., O'Donnell, P.V., Pinter, A. and Hammerling, U. (1981) *Virology* 111, 84–92.

Nunberg, J.H., Rodgers, G., Gilbert, J.H. and Snead, R.M. (1984) *Proc. Natl. Acad. Sci. USA* 81, 3675–3679.

O'Donnell, P.V. and Nowinski, R.C. (1980) *Virology* 107, 81–88.

O'Rear, J.J. and Temin, H.H. (1982) *Proc. Natl. Acad. Sci. USA* 79, 1230–1234.

Oroszlan, S. and Copeland, T.D. (1984) In *Current Topics in Microbiology and Immunology* (P.K. Vogt, Ed.) Springer-Verlag, New York. pp. 221–233.

Oroszlan, S. and Gilden, R.V. (1970) *Science* 168, 1478.

Oroszlan, S. and Gilden, R.V. (1978) *Scand. J. Immunol.* 7, 81–96.

Oroszlan, S. and Gilden, R.V. (1980) In *Molecular Biology of RNA Tumor Viruses* (J.R. Stephenson, Ed.) Academic Press, New York. pp. 299–344.

Oroszlan, S., Summers, M.R., Foreman, C. and Gilden, R.V. (1974) *J. Virol.* 14, 1559–1574.

Oroszlan, S., Copeland, T., Summers, M.R., Smythers, G. and Gilden, R.V. (1975) *J. Biol. Chem.* 250, 6232–6239.

Oroszlan, S., Henderson, L.E., Copeland, T.D., Schultz, A.M. and Rabin, E.M. (1980) In *Biosynthesis, Modification, and Processing of Cellular and Viral Polyproteins* (G. Koch and D. Richter, Eds.) Academic Press, New York. pp. 219–233.

Pepinsky, R.B., and Vogt, V.M. (1979) *J. Mol. Biol.* 131, 819–837.

Pinter, A. and Honnen, W.J. (1983) *Virology* 129, 40–50.

Rabin, H., Benton, C.V., Tainsky, M.A., Rice, N.R. and Gilden, R.V. (1979) *Science* 204, 841–842.

Rand, K., Long, C., Wei, T. and Gilden, R.V. (1974) *J. Natl. Cancer Inst.* 53, 449.

Redmond, S., Peters, G. and Dickson, C. (1984) *Virology* 133, 393–402.

Rein, A. (1982) *Virology* 120, 251–257.

Rein, A., Schultz, A.M., Bader, J.P. and Bassin, R.H. (1982) *Virology* 119, 185–192.

Rice, N.R., Stephens, R.M., Couez, D., Deschamps, J., Kettmann, R., Burny, A. and Gilden, R.V. (1984) *Virology* 138, 82–93.

Rowe, W.P., Puch, W.D. and Hartley, J.W. (1970) *Virology* 42, 1136–1139.

Sallay, S., Srivastava, K.S.L., Oroszlan, S. and Gilden, R.V. (1979) In *Peptides* (E. Gross and J. Meienhoefer, Eds.) Pierce Chemical Co., Rockford IL. pp. 377–380.

Sarngadharan, M.G., Popovic, M., Bruch, L., Schupbach, J. and Gallo, R.C. (1984) *Science* 224, 506–508.

Satake, M. and Luftig, R.B. (1983) *Virology* 124, 259–273.

Schultz, A.M. and Oroszlan, S. (1983) *J. Virol.* 46, 355–361.

Schultz, A.M. and Oroszlan, S. (1984) *Virology* 133, 431–437.

Schultz, A.M., Copeland, T.D. and Oroszlan, S. (1984) *Virology* 135, 417–427.

Schwartz, D.E., Tizard, R. and Gilbert, W. (1983) *Cell* 32, 853–869.

Seiki, M., Hattori, S., Hirayama, Y. and Yoshida, M. (1983) *Proc. Natl. Acad. Sci. USA* 80, 3618–3622.

Sela, M., Schechter, B., Schechter, I. and Borek, F. (1967) *Cold Spring Harbor Symp. Quant. Biol.* 32, 537.

Shinnick, T.M., Lerner, R.A. and Sutcliffe, J.G. (1981) *Nature* 293, 543–548.

Steck, F.T. and Rubin, H. (1966) *Virology* 29, 628–641.

Stephenson, J.R. (1980) In *Molecular Biology of RNA Tumor Viruses* (J.R. Stephenson, Ed.) Academic Press, New York. pp. 245–297.

Strand, M. and August, J.T. (1974) *J. Virol.* 13, 171–180.

Tamura, T.-A. (1983) *J. Virol.* 47, 137–145.

Teich, N. (1982) In *RNA Tumor Viruses* (R. Weiss, N. Teich, H. Varmus and J. Coffin, Eds.) Cold Spring Harbor, New York. p. 25.

Todaro, G.J., Benveniste, R.E., Sherwin, S.A. and Sherr, C.J. (1978) *Cell* 13, 775–782.

Todaro, G.J. (1980) In *Molecular Biology of RNA Tumor Viruses* (J.R. Stephenson, Ed.) Academic Press, New York. pp. 47–76.

Tsai, W.-P., Copeland, T.D. and Oroszlan, S. (1985) *Virology* 140, 289–312.

Versteegen, R.J. and Oroszlan, S. (1980) *J. Virol.* 33, 983–992.

Versteegen, R.J. Copeland, T.D. and Oroszlan, S. (1982) *J. Biol. Chem.* 257, 3007–3013.

Vogt, P.K. and Ishizaki, R. (1966) *Virology* 30, 368–374.

Yoshinaka, Y., Katoh, I., Copeland, T.D. and Oroszlan, S. (1985) *Proc. Nat. Acad. Sci. USA* In press.

Yoshinaka, Y., Katoh, I. and Luftig, R.B. (1984b) *Virology* 136, 274–281.

Eds. M.H.V. van Regenmortel and A.R. Neurath
Immunochemistry of viruses. The basis for serodiagnosis and vaccines.

Antigenic structure of plant viruses 25

M.H.V. VAN REGENMORTEL

IBMC, 15 rue René Descartes, 67084 Strasbourg, France

Although plant viruses do not give rise to an immune response when they infect their plant hosts, they are nevertheless good immunogens when administered to standard laboratory animals. The behaviour of plant viruses as antigens is of interest mainly because serological procedures have been found to provide the most reliable way to identify viruses. Sensitive serological tests such as the enzyme-linked immunosorbent assay (ELISA) have become widely used for the diagnosis of plant viruses and for selecting healthy propagating plant material (Clark, 1981).

Another reason for studying the antigenicity of plant viruses is that specific antibodies represent particularly sensitive tools for unraveling fine details of virus structure. As a result, serological methods are often used for assessing the degree of similarity between plant viruses and for studying the surface topography of viral capsids and protein subunits (Van Regenmortel, 1978).

As the antigenic properties of plant viruses have been described in considerable detail in a recent monograph (Van Regenmortel, 1982) the present chapter will be restricted to a survey of the developments that have occurred during the last three years. Some of the new methods that have been introduced to study the antigenic properties of plant viruses will be reviewed, and the antigenic structure of two well-studied plant viruses will be described.

1. New methodological advances

A comprehensive list of the 262 plant virus species that could be identified in 1980 by serological techniques has been published (Van Regenmortel, 1982). Since then about 40 additional virus species have been characterized by serology, which brings the number of viruses that can be distinguished by their antigenic properties to more than 300 (Van Regenmortel, 1984a).

1.1. Monoclonal antibodies

During the last three years, monoclonal antibodies have been prepared against more than twenty different plant viruses (Hsu et al., 1984; Van Regenmortel, 1984b). The results that have been obtained show that the hybridoma technology will be as useful in the case of plant viruses as in other fields of virology (see Chapter 3). Since only very small quantities of viral antigens are needed for immunizing mice, the hybridoma method is particularly useful for viruses such as the luteoviruses that reach only a low concentration in plants. Another advantage of using monoclonal antibodies for the diagnosis of plant virus diseases lies in the fact that large quantities of standard antibody preparations are obtainable; this ensures that uniform results will be obtained in all laboratories interested in detecting a particular virus (Halk et al., 1984). The increased specificity of monoclonal antibodies compared to polyclonal antisera also ensures that viral strains can now be identified in a reproducible manner (Briand et al., 1982). It should be stressed, however, that monoclonal antibodies are sometimes too specific in that they may recognize only the immunizing serotype and not other serotypes likely to be found in the field. This difficulty can be overcome by selecting monoclonal antibodies specific for the common epitopes of different strains of a virus (Gugerli and Fries, 1983) or by working with the right mixture of monoclonal antibodies.

The antigenic structure of only one plant virus, tobacco mosaic virus (TMV), has been investigated so far by monoclonal antibodies. Fifteen mutants of TMV showing single or double amino acid substitutions in the coat protein as well as several TMV strains showing as much as 50% difference in sequence were compared by means of monoclonal antibodies elicited against wild type virus (Al Moudallal et al., 1982; Briand et al., 1982). Many amino acid substitutions occurring at the outer surface of the viral capsid were detectable by monoclonal antibodies, but in addition, certain internal substitutions located 2–5 nm from the viral surface also altered the binding of some antibodies. The ability of monoclonal antibodies directed against TMV to recognize small conformational changes in TMV protein was also demonstrated by the fact that about half the antibodies did not bind to the monomeric viral protein subunit (Al Moudallal et al., 1984). The superior discrimination achieved by monoclonal antibodies compared to polyclonal antisera was shown by the fact that certain strains which had previously been considered to be serologically identical with TMV were found to be distinguishable by certain monoclonal antibodies. On the other hand, two virus species such as TMV and tomato mosaic virus that are always distinguishable by conventional antisera may appear to be antigenically identical when tested with a monoclonal antibody directed against a common epitope in the two viruses (Briand et al., 1982).

Several instances of heterospecificity were encountered among the TMV monoclonal antibodies. This phenomenon arises from the ability of an antibody to bind to many different related epitopes. As a result of this multispecificity, an antibody can often be shown to bind more weakly to the immunogen than to other antigens (Al Moudallal et al., 1982). Clearly, clonal selection of antibody-producing cells does not guarantee that the best possible fit between epitope and paratope is always actualized with the molecule used for immunization. This state of affairs has a parallel in the human selection of marital partners which also does not necessarily lead to the formation of couples endowed with the highest degree of fit (Van Regenmortel, 1984c). Calling antibodies by the antigen's name is no more guarantee of a superior degree of fit than bearing the partner's name achieves in marriage. In both situations a prolonged search often reveals the existence of better matched individuals.

Heterospecificity also occurs with polyclonal antisera (Van Regenmortel, 1982) although it is then less conspicuous than in the case of monoclonal antibodies. This phenomenon is of considerable importance in hybridoma work since it allows the selection of clones that possess a high affinity not only for the immunogen but also for other antigens. Since heterospecific activity present in clones may go undetected if only the immunogen is used for selection, it is often profitable to screen hybridoma clones against a variety of related viral antigens. In this way, monoclonal antibodies can be obtained that react weakly (or not at all in a particular type of assay) with the virus strain used for immunization, but which have a high degree of specificity for other strains.

1.2. Solid-phase immunoassays

The enzyme-linked immunosorbent assay (ELISA) has become the most popular diagnostic procedure in plant virology. Most workers use the direct double-antibody sandwich method of ELISA which requires the preparation of a different antibody conjugate for each virus to be tested (Clark, 1981). This method has a rather narrow strain specificity (Van Regenmortel and Burckard, 1980; Koenig, 1981) but has the advantage that it requires only one antiserum prepared in a single animal species. Indirect ELISA methods, on the other hand, necessitate the use of two antisera prepared in different animal species, but a single enzyme conjugate can be used for all virus systems. The indirect methods detect a broader range of serologically related viruses and are usually more sensitive than the direct methods (Van Regenmortel, 1982; Koenig and Paul, 1982). If only a single virus antiserum is available, the indirect method can also be applied by utilizing specific $F(ab')_2$ immunoglobulin fragments to trap the virus on the microtiter plate (Barbara and Clark, 1982). Trapped virus is then detected with the whole viral antibody followed by a conjugate reacting only with the Fc portion of the immunoglobulin.

Avian antibodies extracted from the egg yolk of hens immunized with plant viruses have been found very useful in indirect ELISA (Devergne et al., 1981; Van Regenmortel, 1982). There is no cross-reactivity between mammalian and avian immunoglobulins, and chicken antibody can therefore be used in multilayered sandwich procedures without any concomitant high background reading resulting from the cross-reactions between globulins of different species. In a comparison of seven ELISA procedures using monoclonal antibodies specific for TMV, the most sensitive methods were found to be those that involved five successive layers, one of which was avian antibody (Al Moudallal et al., 1984).

Radioimmunoassays have only rarely been used in plant virology. Recently a number of solid-phase radioimmunoassays that have the same sensitivity as ELISA have been applied to the detection of plant viruses (Bryant et al., 1983; Hill et al., 1984). There is little evidence, however, that these tests, which require expensive counting equipment, are in any way superior to the more commonly used ELISA procedures.

1.3. Immunoelectron microscopy

The technique introduced by Derrick (1973) of trapping viruses onto electron microscope grids coated with specific antiserum has become widely used in plant virology (Van Regenmortel, 1982; Roberts et al., 1982). Because of its similarity with solid phase immunoassays, the method has been called immunosorbent electron microscopy (ISEM). When the grids are pretreated with protein A, the antibody molecules can be anchored more strongly to the support film. Another variation of the method consists of trapping virus particles at the surface of protein A-containing *Staphylococcus aureus* cells coated with a layer of viral antibodies (Nicolaïeff et al., 1982).

The specificity of trapping on grids can be demonstrated by a subsequent decoration step, which consists of coating the virions with a layer of antibody molecules (Milne and Luisoni, 1977). A double decoration with a second layer of antiglobulin molecules may further facilitate the detection of a positive reaction (Kerlan et al., 1982).

The main problems encountered with ISEM are caused by the variability of support films, the pH sensitivity and the breakage of virus particles (Cohen et al., 1982; Pares and Whitecross, 1983) and by the fact that the trapping of virions can be blocked by viral subunits when these are present in the sample (see also Chapter 6).

1.4. New analytical methods

Immunoprecipitation and immunoblotting techniques have been increasingly used for analyzing the virus-encoded polypeptides produced in infected cells (see Chapter 7) and they have found many applications in plant virology (Dougherty and Hiebert, 1980; Rybicki and Von Wechmar, 1982). Rocket immunoelectrophoresis has been found highly suitable for quantitative assays of plant viruses in small samples of infected tissue (Tobias et al., 1979; Hagen et al., 1982).

Another recent development concerns the use of ELISA inhibition tests which have been shown to be a powerful technique for demonstrating the presence of antigenic activity on short peptide fragments of viral proteins (Altschuh and Van Regenmortel, 1982).

Refinements in analytical techniques have demonstrated that the serological properties of plant viruses can be altered significantly by formalinization or swelling of the capsids (Rybicki and Von Wechmar, 1981) as well as by proteolytic degradation of the viral protein (Du Plessis and Von Wechmar, 1980; Hiebert et al., 1984).

2. Classification of plant viruses by serology

The usefulness of serology for classifying plant viruses is based on the fact that viruses that are serologically related always share most of their other properties. Serological cross-reactions are found only between the members of each of the 26 recognized plant virus groups, and the existence of a serological cross-reaction thus constitutes prima facie evidence for including a newly described virus within an existing group (Van Regenmortel, 1982). The most striking exception to the above rule concerned the alleged serological cross-reaction between the rod-shaped tobacco mosaic virus and the isometric cocksfoot mild mosaic virus (Bercks and Querfurth, 1971). However, when this unusual antigenic relationship was reinvestigated by ELISA, no cross-reaction between the two viruses could be demonstrated (Altschuh and Van Regenmortel, 1983).

The validity of serological data for classifying viruses has sometimes been questioned on the basis that the antigenically relevant part of viral coat proteins corresponds to only a small percentage of the total coding capacity of the viral genome. The fallacy of this argument is demonstrated by the success achieved in reconstructing phylogenetic trees of higher organisms on the basis of the degree of antigenic similarity found in a small number of their proteins (Wilson et al., 1977). Studies with several protein families have shown that the degree of sequence difference correlates with the antigenic distance between members. In the tobamovirus group, for instance, the degree of serological relatedness among members also agrees closely with the degree of similarity revealed by chemical analysis of the different coat proteins. The correlation between sequence difference and antigenic distance between several tobamoviruses is illustrated in Fig. 25.1. The degree of antigenic relatedness between these viruses is expressed by a serological differentiation index (SDI) which corresponds to the average number of two-fold dilution steps separating homologous from heterologous precipitin titers (Van Regenmortel, 1975).

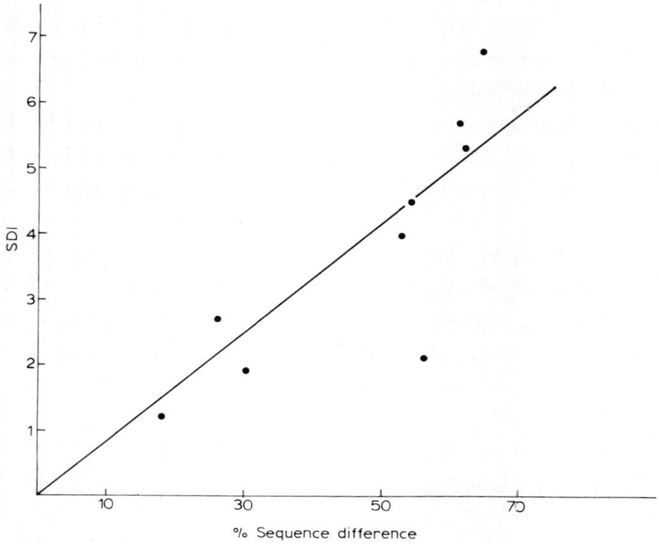

Fig. 25.1. Correlation between sequence difference in the coat protein and antigenic difference as measured by serological differentiation index (SDI) in the tobamovirus group (Van Regenmortel, 1975; 1982).

When a large number of virus strains are examined serologically, it is always found that they lie on a continuum of decreasing serological relatedness. There is therefore no simple rule for deciding whether two related viruses should be considered strains of the same virus or separate virus species. It has been suggested that when two virus isolates have different host ranges and differ antigenically by a SDI greater than 3, it would be justified to consider them as separate virus species (Hamilton et al., 1981). This is not a firm rule, however, as many viruses that differ only by a SDI of 1–2 are currently considered to be distinct virus species (Chu and Francki, 1983; Thomas, 1984; Tamada et al., 1984). As more virus groups are analyzed with respect to the extent of serological relationships between their members, it seems that all of them are made up of clusters of serologically more or less closely related viruses for which a clear-cut distinction between strain and separate species is not possible. This has been shown to be the case in the tobamo-, tymo-, tombus-, como-, poty-, potex-, ilar-, gemini- and luteovirus groups (Lin et al., 1981; Sequeira and Harrison, 1982; Van Regenmortel, 1982; Gracia et al., 1983; Uyeda and Mink, 1983; Cohen et al., 1983; Stein et al., 1983; Thomas, 1984).

The impossibility of drawing a precise borderline between clusters of virus strains and different virus species has made many plant virologists reluctant to use the taxons 'species' and 'genus' in virus classification. Instead of these terms, the labels 'virus' and 'group' are commonly used in plant virology. However, much confusion can arise when the same word 'virus' is used both for individual strains and for taxonomic entities consisting of all the strains of a species. Although there are no hard rules for deciding when a newly described pathogen qualifies as a new virus species, it seems pointless to deny that the category 'species' is useful for referring to taxonomic entities made up of individual viruses that seem to form a cluster to a human observer. Although the distinction between strain and species may be arbitrary, human minds need conceptual divisions not found in nature, i.e. taxons, in order to cope with the bewildering variety of the viral world. Furthermore, all viruses should be classified in a uniform manner and there seems to be little justification for restricting the concept of virus species to animal virology.

3. Antigenic structure of tobacco mosaic virus

Tobacco mosaic virus (TMV) is a rod-shaped particle, 300 nm long, which consists of 2130 identical protein subunits arranged as a helix around a single RNA molecule. The TMV protein subunit contains 158 amino acid residues and its three-dimensional structure has been established by x-ray crystallography. A detailed account of the antigenic structure of TMV as it was known in 1980 has been published (Van Regenmortel, 1982), and the present section will concentrate on new information acquired in recent years.

Cleavage peptides and synthetic peptides corresponding to various regions of TMV protein have been used in complement fixation and ELISA inhibition experiments (Altschuh and Van Regenmortel, 1982) to locate the position of epitopes in the dissociated subunit. Seven epitopes were localized in the vicinity of residues 1–10, 34–39, 55–61, 62–68, 80–90, 105–112 and 153–158 (Altschuh et al., 1983). The position of these seven epitopes in the subunit is shown in the stereo pair of the TMV protein backbone illustrated in Fig. 25.2.

All seven epitopes were found to correspond to those regions in the protein, as determined by x-ray crystallography, which possess high temperature factors and a high segmental mobility (Westhof et al., 1984). As is the case in many proteins (Thornton and Sibanda, 1983), the N- and C-termini of TMV protein are flexible and accessible at the surface of the molecule, and are also antigenic determinants. It was striking, however, that all seven epitopes identified by inhibition tests with short peptides belonged to regions with higher than average mobility. This seems to indicate that flexibility in the main chain atoms is an essential ingredient for the activity of continuous epitopes in proteins (see Chapter 1). Clearly, continuous epitopes which are made up of a series of amino acid residues in peptide linkage are the only type of epitope that can be identified by means of short peptides. Flexible segments in proteins are more likely to mimic the conformation of peptides in solution and thus to be recognized as antigenic determinants in inhibition experiments. This dependence on mobility may also explain the success that has been achieved in raising antipeptide antibodies able to recognize native proteins (Shinnick et al., 1983).

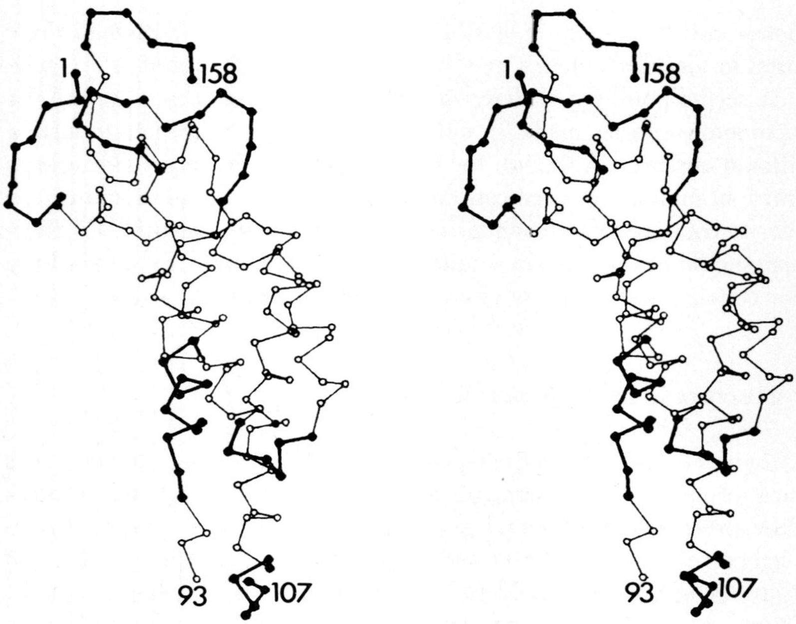

Fig. 25.2. Stereo drawing of the alpha-carbon backbone of the TMVP subunit. Residues 94–106 are omitted because this region has no specific conformation in the disk (Bloomer et al., 1978). The locations of the seven continuous antigenic determinants are indicated by thick lines.

Discontinuous and topographic epitopes cannot be identified by studying short linear peptides and their presence in TMV protein was ascertained only when monoclonal antibodies were used to compare a panel of TMV strains and mutants. The majority of monoclonal antibodies raised against TMV were able to detect many amino acid substitutions located at the outer surface of the virion, but some of the antibodies also recognized conformational alterations induced by exchanges occurring deep inside the viral subunit (Al Moudallal et al., 1982). Some of the TMV monoclonal antibodies reacted with viral strains that showed up to 18–56% difference in amino acid sequence compared to TMV (Briand et al., 1982; Altschuh, 1984) and a few were able to bind to both the C- and N-terminal tryptic peptides of TMV protein. This latter finding clearly demonstrated the presence in TMV of discontinuous epitopes made up of residues in the regions 1–10 and 142–158 that become contiguous through the folding of the polypeptide chain. The surface of globular proteins is only to a very limited degree constituted of linear arrays of residues and it should come as no surprise that many monoclonal antibodies appear to be specific for discontinuous epitopes (Van Regenmortel, 1984c).

About half the monoclonal antibodies raised against TMV were unable to react with dissociated viral subunits and were thus specific for the neotopes of the virus.

Neotopes are antigenic determinants that arise from the quaternary structure of the viral protein (Van Regenmortel, 1966; 1982); they owe their existence to conformational changes in the protein induced by intersubunit bonds and may involve juxtaposing residues from neighboring subunits. Monoclonal antibodies specific for neotopes of TMV were not reactive in ELISA procedures in which the virus was directly adsorbed to the plastic surface of microtiter plates (Al Moudallal et al., 1984). These antibodies only recognized the virus when it was attached to the plastic by an intermediary layer of antibodies. Adsorption to the plastic seems to alter the conformation of the virus and to render it similar to that found in dissociated subunits, since only those monoclonal antibodies that also reacted with the depolymerized viral protein were able to bind to virus adsorbed directly to the solid phase. These results demonstrate that monoclonal antibodies are useful tools for studying the conformational changes that occur when viral subunits associate into capsids.

4. Antigenic structure of turnip yellow mosaic virus

Turnip yellow mosaic virus (TYMV) is the type member of a large group of isometric plant viruses known as tymoviruses (Koenig, 1976). Some of the members of this group are closely related serologically while others, which appear to be unrelated in comparative serological tests, can be interconnected via other members that are more closely related. The degree of serological cross-reactivity between tymoviruses did not correlate with similarities in amino acid composition of their proteins, in contrast to the situation found in the tobamovirus group (Paul et al., 1980).

The icosahedral capsid of TYMV consists of 180 20-kDa protein subunits arranged as 32 capsomeres. The protein subunit (TYMVP) has been sequenced (Peter et al., 1972) but there is little information available on its conformation and topography.

The first study aimed at locating the epitopes of TYMVP made use of a number of peptides obtained by cleavage of the protein with trypsin or cyanogen bromide (Pratt et al., 1980). Four antigenic regions were identified in the protein by means of inhibition of complement fixation tests. The reaction between TYMVP and its antiserum was inhibited most strongly by peptides corresponding to residues 1–12 and 46–67, whereas the reaction between the virus and its specific antibodies was inhibited only by the peptide corresponding to residues 46–67. This led to the conclusion that at least part of this peptide was located at the surface of the subunit that is exposed at the periphery of the virion. Since cross-linking studies between the coat protein and the viral RNA seemed to indicate that the region 1–12 was linked to the RNA (Ehresmann et al., 1980), it was concluded that this antigenic region of the protein was located at the surface of the subunit that faces the central hole in the capsid (Pratt et al., 1980).

In a subsequent study of the antigenic activity of the N-terminal region of TYMVP, it was discovered by solid-phase immunoassays that the region 1–12 was in fact exposed at the external surface of the assembled capsid. It seems that polymeric viral antigens fix complement so efficiently that it becomes very difficult to demonstrate the inhibitory capacity of a peptide in a complement fixation inhibition test. In contrast, inhibition of solid-phase immunoassays using the heterologous systems virus-antisubunit serum and subunit-antivirus serum were much more sensitive for revealing the presence of antigenic activity in short peptides (Quesniaux et al., 1983a). In the reaction between antiserum specific for protein subunits and viral capsids, only antibodies that recognize the outer surface of the capsid participate in the heterologous reaction; the binding of peptides to this subpopulation of antibodies, therefore, will increase the extent of measurable inhibitory activity. The antibodies found in a TYMVP antiserum that are directed against the hidden surfaces of the monomers will not participate in the reaction with viral capsids and will not be able to mask the inhibitory activity of a peptide that binds to antibodies specific for the external surface of the virus.

Using a series of synthetic peptides corresponding to overlapping regions of residues 1–12 of TYMVP, it was found that all twelve residues contribute to an antigenic site of the virus. Clearly, if the N-terminal residues are facing outwards in the capsid, they cannot also be bound to the internal RNA. The earlier cross-linking data which had suggested such an orientation are in fact suspect, for they were based on negative and conflicting evidence (Quesniaux et al., 1983a).

The antigenic activity of the C-terminal region of TYMVP was also investigated by complement fixation, ELISA and radioimmunoassay procedures. Synthetic peptides corresponding to residues 187–189 and 183–189 of TYMVP were found to inhibit the reaction between whole virions and antibodies as measured in these different assays (Quesniaux et al., 1983b). The heptapeptide was always more active than the tripeptide, suggesting that all seven terminal residues probably contribute to the structure of an epitope of the virus.

The inhibitory activity of peptide 183–189 was higher in the protein-antiprotein system than in the virus-antivirus system. This finding suggests that the conformation of the synthetic peptide is more similar to that of the corresponding region in the protein than in the virus.

5. Conclusions

Viruses such as tobacco mosaic and turnip yellow mosaic viruses represent some of the best-studied macromolecular structures in the whole of molecular biology. A great deal of information is available on the structural and physicochemical properties of these viruses, and our understanding of the antigenic structure of TMV is probably as comprehensive as that of myoglobin and lysozyme. Although plant viruses have been used as models for a variety of immunological investigations (Slaoui et al., 1983; Rennick et al., 1983) their antigenic properties have been studied mainly because serological techniques are extremely useful for quantitative virus assays, for routine diagnosis and for assessing the degree of similarity between viruses. The in vitro serological reactions of all viral antigens are very similar and much useful knowledge has emerged from studies performed with all groups of viruses, irrespective of whether the host is a bacterium, an animal or a plant.

6. References

Al Moudallal, Z., Briand, J.P. and Van Regenmortel, M.H.V. (1982) *EMBO J.* 1, 1005–1010.

Al Moudallal, Z., Altschuh, D., Briand, J.P. and Van Regenmortel, M.H.V. (1984) *J. Immunol. Methods* 68, 35–43.

Altschuh, D. (1984) *PhD thesis.* University of Strasbourg.

Altschuh, D. and Van Regenmortel, M.H.V. (1982) *J. Immunol. Methods* 50, 99–108.

Altschuh, D. and Van Regenmortel, M.H.V. (1983) *Int. J. Microbiol.* 1, 13–17.

Altschuh, D., Hartmann, D., Reinbolt, J. and Van Regenmortel, M.H.V. (1983) *Molec. Immunol.* 20, 271–278.

Barbara, D.J. and Clark, M.F. (1982) *J. Gen. Virol.* 58, 315–322.

Bercks, R. and Querfurth, G. (1971) *J. Gen. Virol.* 12, 25–32.

Briand, J.P., Al Moudallal, Z. and Van Regenmortel, M.H.V. (1982) *J. Virol. Methods* 5, 293–300.

Bryant, G.R., Durand, D.P. and Hill, J.H. (1983) *Phytopathology* 73, 623–629.

Chu, P.W.G. and Francki, R.I.B. (1983) *Virology* 129, 350–356.

Clark, M.F. (1981) *Annu. Rev. Phytopathol.* 19, 83–106.

Cohen, J., Loebenstein, G. and Milne, R.G. (1982) *J. Virol. Methods* 4, 323–330.

Cohen, S., Duffus, J.E., Larsen, R.C., Liu, H.Y. and Flock, R.A. (1983) *Phytopathology* 73, 1669–1673.

Derrick, K.S. (1973) *Virology* 56, 652–653.

Devergne, J.C., Cardin, L., Burckard, J. and Van Regenmortel, M.H.V. (1981) *J. Virol. Methods* 3, 193–200.

Dougherty, W.G. and Hiebert, E. (1980) *Virology* 104, 183–194.

Du Plessis, D.H. and Von Wechmar, M.B. (1980) *Virology* 107, 298–301.

Ehresmann, B., Briand, J.P., Reinbolt, J. and Witz, J. (1980) *Eur. J. Biochem.* 108, 123–129.

Gracia, O., Koenig, R. and Lesemann, D.E. (1983) *Phytopathology* 73, 1488–1492.

Gugerli, P. and Fries, P. (1983) *J. Gen. Virol.* 64, 2471–2477.

Hagen, T.J., Taylor, D.B. and Meagher, R.B. (1982) *Phytopathology* 72, 239–242.

Halk, E.L., Hsu, H.T., Aebig, J. and Franke, J. (1984) *Phytopathology* 74, 367–372.

Hamilton, R.I., Edwardson, J.R., Francki, R.I.B., Hsu, H.T., Hull, R., Koenig, R. and Milne, R.G. (1981) *J. Gen. Virol.* 54, 223–241.

Hiebert, E., Tremaine, J.H. and Ronald, W.P. (1984) *Phytopathology* 74, 411–416.

Hill, E.K., Hill, J.H. and Durand, D.P. (1984) *J. Gen. Virol.* 65, 525–532.

Hsu, H.T., Jordan, R.L. and Lawson, R.H. (1984) *Am. Soc. Microbiol. News* 50, 99–102.

Kerlan, C., Mille, B., Detienne, G. and Dunez, J. (1982) *Ann. Virol. Inst. Pasteur* 133E, 3–14.

Koenig, R. (1976) *Virology* 72, 1–5.

Koenig, R. (1981) *J. Gen. Virol.* 55, 53–62.

Koenig, R. and Paul, H.L. (1982) *J. Virol. Methods* 5, 113–125.

Lin, M.T., Anjos, J.R.N. and Rios, G.P. (1981) *Phytopathology* 71, 435–438.

Milne, R.G. and Luisoni, E. (1977) In *Methods in Virology* (K. Maramorosch and H. Koprowski, Eds.) Vol. VI. Academic Press, New York. pp. 265–281.
Nicolaïeff, A., Katz, D. and Van Regenmortel, M.H.V. (1982) *J. Virol. Methods* 4, 155–166.
Pares, R.D. and Whitecross, M.I. (1983) *J. Virol. Methods* 7, 241–250.
Paul, H.L., Gibbs, A. and Wittmann-Liebold, B. (1980) *Intervirology* 13, 99–109.
Peter, R., Stéhelin, D., Reinbolt, J., Collot, D. and Duranton, H. (1972) *Virology* 49, 615–617.
Pratt, D., Briand, J.P. and Van Regenmortel, M.H.V. (1980) *Molec. Immunol.* 17, 1167–1171.
Quesniaux, V., Briand, J.P. and Van Regenmortel, M.H.V. (1983a) *Molec. Immunol.* 20, 179–185.
Quesniaux, V., Jaegle, M. and Van Regenmortel, M.H.V. (1983b) *Biochim. Biophys. Acta* 743, 226–231.
Rennick, D.M., Morrow, P.R. and Benjamini, E. (1983) *J. Immunol.* 131, 567–571.
Roberts, I.M., Milne, R.G. and Van Regenmortel, M.H.V. (1982) *Intervirology* 18, 147–149.
Rybicki, E.P. and Von Wechmar, M.B. (1981) *Virology* 109, 391–402.
Rybicki, E.P. and Von Wechmar, M.B. (1982) *J. Virol. Methods* 5, 267–278.
Sequeira, J.C. and Harrison, B.D. (1982) *Ann. Appl. Biol.* 101, 33–42.
Shinnick, T.M., Sutcliffe, J.G., Green, N. and Lerner, R.A. (1983) *Annu. Rev. Microbiol.* 37, 425–446.
Slaoui, M., Franssen, J.-D., Hérion, P. and Urbain, J. (1983) *Molec. Immunol.* 20, 1073–1080.
Stein, V.E., Coutts, R.H.A. and Buck, K.W. (1983) *J. Gen. Virol.* 64, 2493–2498.
Tamada, T., Harrison, B.D. and Roberts, I.M. (1984) *Ann. Appl. Biol.* 104, 107–116.
Thomas, J.E. (1984) *Ann. Appl. Biol.* 104, 79–86.
Thornton, J.M. and Sibanda, B.L. (1983) *J. Mol. Biol.* 167, 443–460,
Tobias, I., Hornok, L. and Balàzs, E. (1979) *Acta Phytopathol. Acad. Sci. Hung.* 14, 231–237.
Uyeda, I. and Mink, G.I. (1983) *Phytopathology* 73, 47–50.
Van Regenmortel, M.H.V. (1966) *Adv. Virus Res.* 12, 207–271.
Van Regenmortel, M.H.V. (1975) *Virology* 64, 415–420.
Van Regenmortel, M.H.V. (1978) *Annu. Rev. Phytopathol.* 16, 57–81.
Van Regenmortel, M.H.V. (1982) *Serology and Immunochemistry of Plant Viruses.* Academic Press, New York.
Van Regenmortel, M.H.V. (1984a) *Recent Advances in Immunodiagnosis of Viral Diseases in Crops. Applied Virology* (E. Kurstak, Ed.) Vol. 1. Academic Press, New York. pp. 463–477.
Van Regenmortel, M.H.V. (1984b) *Trends Microbiol. Sci.* 1, 73–78.
Van Regenmortel, M.H.V. (1984c) In *Hybridoma Technology in Agricultural and Veterinary Research* (N.J. Stern and H.R. Gamble, Eds.) Rowman and Allenheld, Totowa NJ. pp. 42–80.
Van Regenmortel, M.H.V. and Burckard, J. (1980) *Virology* 106, 327–334.
Westhof, E., Altschuh, D., Moras, D., Bloomer, A.C., Mondragon, A., Klug, A. and Van Regenmortel, M.H.V. (1984) *Nature* 311, 123–126.
Wilson, A.C., Carlson, S.S. and White, T.J. (1977) *Annu. Rev. Biochem.* 46, 573–639.

Eds. M.H.V. van Regenmortel and A.R. Neurath
Immunochemistry of viruses. The basis for serodiagnosis and vaccines.

Antigenic structure of insect viruses 26

T.W. TINSLEY

Natural Environment Research Council, Institute of Virology, Mansfield Road, Oxford OX1 3SR, U.K.

The number of insects in the Animal Kingdom in commonly underestimated and it must be emphasized that the Class Insecta comprises almost three quarters of the known species. As an illustration, there are many more species of beetles (500 000) than mammals (a mere 3 000). Therefore, in turn, it is very probable that insects harbor a larger number of viruses of all types and groups. Ten distinct groups of viruses have been described from insects to date but this has to be regarded very much as a minimum number. The reasons for this situation are quite clear; there are few virologists concerned with the isolation and description of new viruses from insects and most attention has been paid, for good and compelling reasons, to those species having pest status in Agriculture, Forestry and Public Health. It is understandable that attention naturally centers on large insects that are easily caught and can be reared in the laboratory whereas those species that live in soil, in water or are less than 1 cm in length are usually ignored. This means that a very restricted sample of the total insect population has been selected for virological study.

The picture is complicated further by the fact that several of the extraction methods in current use may in fact eliminate certain viruses on initial treatment. For example, organic solvents such as carbon tetrachloride and/or butanol/chloroform mixtures are commonly used in the primary extraction procedures. Such formulations would readily destroy viruses containing lipid in the outer layers of the nucleocapsid. This may explain why no viruses of the family Togaviridae have been described from defoliating insects. This is very surprising when one considers the close association that viruses of this family have with insects. Indeed, a convincing argument can be made that yellow fever is basically a virus of certain mosquito species and that primates and man are only secondary and accidental hosts. Further, many viruses readily aggregate on release from the infected tissue of insects and, unless special precautions are taken, it is very easy to lose such viruses in the initial cycle of centrifugation. Therefore, a warning note must be sounded that many of the methods in current use for the extraction and purification of insect pathogenic viruses may be exerting their own selection pressure to a varying and unknown degree. Many of these problems are avoided in the isolation of viruses from mammals by the use of a wide range of permissive cell systems. Unfortunately, there are not yet enough insect-derived cells for the direct recovery of viruses from diseased material to be used in routine procedures. However, this deficiency is widely recognized and strenuous efforts are being made to remedy the situation.

The groups of viruses presently known from insects are shown in Table 26.1, which is a condensation of the taxonomic system presented by Matthews (1982). The similarities between viruses that cause disease in insects and those found in vertebrates are principally at the gross biochemical level. Considering the few comparisons that have been made, it will not be surprising if future experimentation reveals relationships between insect and mammalian viruses, and perhaps even cross-infectivity. In this context, it is interesting to note that Nodamura viruses will replicate in suckling mice, BHK_{21} and mouse L cells as well as in honeybees, various mosquito species and the greater wax moth (Scherer and Hurlbut, 1967; Scherer et al., 1967; Bailey and Scott, 1973).

Reference has already been made to the apparent absence of Arboviruses (Alphavirus, Flavivirus: Togaviridae) from such orders of nonhemophagous insects as the Lepidoptera, Hymenoptera and Coleoptera. Nevertheless, various workers have shown that many members of the Togaviridae can infect and replicate in larvae of such species if the viruses are injected directly into the hemocoele and it is significant that this route avoids the oral entry point and subsequent digestive processes (Hurlbut, 1956; Hurlbut and Thomas, 1960; 1969). On occasion, larvae of certain parasitic species of flies (Diptera) can acquire arboviruses through feeding on infected tissues (Aitken et al., 1958) but again, although the adult forms are not parasitic they can be infected by injection. Thus, specificity in these examples seems to be determined principally by the entry route of the virus particles and probably involves a degree of tissue tropism.

Invertebrate animals do not possess an immune system analogous to that of vertebrates and therefore do not have circulating immunoglobulins. However, they do have other systems of defence that are obviously effective (Tinsley, 1975). Therefore it must be appreciated that our knowledge of the antigenic structure of insect viruses has been derived entirely from reactions with antibodies produced by test animals in a hyperimmune state (guinea pigs, rabbits and mice). This means that throughout this chapter we will be concerned with the effects of insect virus antigens on the antibody responses in warm-blooded animals taxonomically far-removed from the original hosts. Classical antibody pressure on virions invading the insect hosts obviously does not occur, so insect viruses are free from this type of induced variation. However, we should not ignore the possibility that preformed virus inhibitors may have a role in virus selection pressure in the invertebrate animal.

1. Groups of insect pathogenic viruses and their antigenic relationships

1.1. Baculoviruses

Baculoviruses have been recorded from the following Natural Orders of insects: Neuroptera, Trichoptera, Lepidoptera, Diptera, Hymenoptera and Coleoptera. This list of families again must be regarded as a minimum because few searches for these viruses have been made in more primitive Orders such as Collembola, Psoptera, Mallophaga or Thysanoptera. Also, there is a tendency to avoid either collecting diseased larvae of aquatic species of the Ephemeroptera and Odonata or attempting to catch the actively flying adults.

Baculoviruses are thought to be confined to invertebrate animals. This is a reasonable assumption in that no cross-infectivity has been recorded between invertebrate and vertebrate animals or between invertebrates and plants. This holds true both for whole animals and for the tissue culture systems derived from them. Furthermore, such large and complex virions and the associated inclusion body would surely have been seen by the numerous virologists and pathologists engaged in the examination of animal and plant tissues.

Baculoviruses have rod-shaped particles (40–70 nm $\times$ 250–400 nm) with a lipoprotein envelope and are usually occluded within an inclusion body composed of a face-centered cubic lattice of protein. The properties of the viruses in the four subgroups (Table 26.1) are quite similar in their basic properties and this has contributed to a logical grouping. Only subgroups A and B are associated with inclusion bodies and the associated proteins are referred to as polyhedrin and granulin, respectively. These inclusion body proteins have a single polypeptide with a molecular weight of 25–33 $\times 10^3$. The rod-shaped virus particles have an envelope of lipoprotein which surrounds a protein/DNA core or nucleocapsid. The virions have a varying number of polypeptides (11–25) with molecular weights of 10–160 $\times 10^3$ (Payne, 1974; Harrap et al., 1977; Brown et al., 1977; Summers and Smith, 1978; Harrap and Payne, 1979). Approximately 50% of the polypeptides are associated with the nucleocapsid and the other half with the capsid envelope.

Baculovirus isolates are usually differentiated either by comparisons of inclusion body or virion protein patterns as produced in polyacrylamide gel systems (Harrap et al., 1977; Summers and Smith, 1978; Maskos and Miltenburger, 1981), or by similarities in restriction enzyme profiles (Miller and Dawes, 1978a; 1978b; Smith and Summers, 1978). This is in marked contrast with most groups of vertebrate viruses where identification follows from the use of a range of serological techniques of varying sophistication, which has led to the development of a considerable body of knowledge of the antigenic structure and function of such viruses. This is sadly

TABLE 26.1

Taxonomic arrangement of viruses pathogenic to insects

Family	Genus	Subgroups	Prototype species	Nucleic acid	Particle	Comparative data and similarities
Baculoviridae	Baculovirus	(A) Nuclear polyhedrosis virus (B) Granulosis virus (C) (Proposed) Non-occluded rod-shaped nuclear viruses (D) (Proposed) Non-occluded nuclear viruses	(A) *Autographa californica* NPV (B) *Trichoplusia ni* GV (C) *Oryctes rhinoceros* virus (D1) *Hyposoter exiguae* virus (D2) *Apanteles melanoscelus* virus	DNA ds	Rod	None to any other groups
Poxviridae, entomopox-viridae	A	—	*Melolontha melolontha* entomopox virus	DNA ds	Ovoid	Poxviruses of vertebrates
	B	—	*Amsacta moorei* entomopox virus			
	C	—	*Chironomus luridus* entomopox virus			
Reoviridae	Cytoplasmic polyhedrosis virus	—	*Bombyx mori* CPV	RNA ds	Isometric	Reoviruses from plants and vertebrates

Iridoviridae	Iridovirus	—	*Tipula paludosa* iridescent virus	DNA ds	Isometric (120 nm)	Iridoviruses of amphibia and fish, possibly African swine fever virus
	Chloriridovirus		Mosquito iridescent virus	DNA ds	Isometric (180 nm)	
Parvoviridae	Densovirus		*Galleria mellonella* densovirus	DNA ss	Isometric	Parvoviruses
Picornaviridae	Unassigned	—	Unassigned members: Cricket paralysis virus, *Drosophila* C virus, *Gonometa* virus	RNA ss	Isometric	Picornaviruses
Caliciviridae	Unassigned	—	Possible member *Amyelois* virus	RNA ss	Isometric	Caliciviruses
Nodaviridae	Nodavirus	—	Nodamura virus	RNA ss (2 molecules)	Isometric	Nodamura virus will grow in vertebrate cell systems and suckling mice
Nudaurelia virus group	Unassigned	—	*Nudaurelia* β virus	RNA ss	Isometric	Not known
Rhabdoviridae	Unassigned	—	Sigmavirus (probable)	RNA ss	Bullet-shaped	Rhabdoviruses of vertebrates and plants

lacking in our present knowledge of the Baculoviruses and, indeed, of insect pathogenic viruses in general.

Nevertheless, the multiple antigen systems presented by virions with complex proteins, which in turn are embedded within a proteinaceous inclusion body, presents perhaps a unique system for serological studies. Furthermore, the inclusion body protein would at first sight appear to be the ideal material for serodiagnostic studies. It is produced in large quantities in infected larvae and in tissue culture cells and it is not difficult to prepare in purified form. However, it is highly insoluble at neutral pH and seems not to be particularly tractable when used in complement fixation and immunodiffusion tests. Furthermore, there is considerable evidence to suggest that the inclusion bodies of most Baculoviruses share common antigens (Harrap and Payne, 1979; Pritchett et al., 1979; Smith and Summers, 1981). These relationships have led to considerable confusion in the literature as published results were frequently misleading in that it was obvious that inclusion body proteins (IBP) were present as contaminants in preparations of virions. Later work confirmed these reservations and showed that extensive precautions had to be taken to ensure that preparations of virus particles were completely free of IBP (Harrap et al., 1977; McCarthy and Lambiase, 1979; Smith and Summers, 1981). The relationships between nucleocapsids of different viruses have not been studied in detail. Harrap et al. (1977) showed that certain nucleocapsid antigens were specific for individual isolates from three species of Spodoptera, whereas proteins that were specific to the capsid envelope revealed antigens in common.

Smith and Summers (1981) used the Western blot radioimmunoassay technique to study possible relationships between various baculovirus structural proteins. Their work suggested that the capsid polypeptides provided sites that were more covered than the determinants found in the viral envelope proteins. The antisera used recognized common antigenic determinants in all the samples of polyhedrin and granulin tested. The development of techniques to prepare monoclonal antibodies (Köhler and Milstein, 1975) has provided more sensitive tools for serological analysis of antigens. Roberts and Naser (1982) successfully produced monoclonal antibodies against the *Autographa californica* nuclear polyhedrosis virus (AcNPV) and at least one clone was specific for this AcNPV. Naser and Miltenberger (1983) used this specific monoclonal antibody with a Western blotting-ELISA technique to test the degree of relationship between a range of Baculoviruses. The specificity of the particular antibody was confirmed, and it detected AcNPV together with a NPV from *Galleria mellonella* (GmNPV). This confirmed the earlier suggestion of Smith and Summers (1979) that AcNPV and GmNPV were closely related. The technique has also been used to detect virus in infected larvae before any visual signs of infection appeared (Volkman and Falcon, 1982). The stage is now set for the development of monoclonal antibodies specific for other Baculoviruses and, when these are available, the inter-relationships of this large and complex group of viruses will become amenable to immunochemical investigation. The establishment of criteria for unequivocal determination of relatedness would have several applications of practical value in the field of biological control. Most workers in this field are of the opinion that many of the collections of Baculoviruses, ostensibly of distinct viruses, may in fact represent only a few distinct viruses together with isolates of identical viruses from other hosts and possibly with strains of differing host ranges. To be able to state categorically that virus isolate 'A' from a new pest species is the same as a virus already known, tested and registered as a biological insecticide, would have many attractions, both scientifically and commercially.

1.2. Poxviridae – entomopoxviridae

Entomopoxviruses occur in the Lepidoptera, Orthoptera, Coleoptera and Diptera (Kurstak and Garzon, 1977). This group shares many morphological and biochemical properties with similar viruses found in vertebrates. The virions are large, brick-shaped or ovoid, 170–250 × 300–400 nm and contain similar enzymes to those found in vertebrate poxviruses. No comparative serological studies have been undertaken as yet, although no serological relationship apparently exists between two of the viruses found in insects and vaccinia virus (Bergoin and Dales, 1971).

1.3. Reoviridae – cytoplasmic polyhedrosis virus

These viruses have been found in more than 200 species of insects and are important natural pathogens (Payne and Kelly, 1981). Morphologically similar viruses have been recorded in a crustacean, *Simocephalus expinosus* (Federici and Hazard, 1975). This insect virus group has similarities to other animal and plant viruses grouped into the family Reoviridae. The genome comprises 10 segments of a dsRNA of molecular weight ranging from 0.34 to 2.59×10^6 (Harrap and Payne, 1979). It can be assumed that each segment represents a single gene if one accepts the analogy with Reoviruses. A preliminary classification has been made possible on the basis of size differences between the RNA segments of some 40 cytoplasmic polyhedrosis viruses (Payne and Rivers, 1976; Payne et al., 1977).

The differentiation of these viruses by gel-profiles of the RNA segments has perhaps over-shadowed the use of serological analysis. Furthermore, serological analyses are complicated by the fact that antibodies are not only produced to the virus structural proteins but also to the virus dsRNA. The antibodies that are specific to the dsRNA will cross-react with dsRNA from quite distinct viruses and so the results mean little more than that the viruses had the same type of nucleic acid. The problem can be overcome by removal of the cross-reacting antibodies by absorption of the sera with dsRNA (Payne and Churchill, 1977), or by using 'fixed' or intact particles as antigen so that the dsRNA is not exposed to the antibody-forming system (Payne and Kalmakoff, 1974). Cytoplasmic polyhedrosis viruses have icosahedral symmetry and range in size from 50 to 65 nm. The particles bear large projections or spikes at the vertices of icosahedra (Hosaka and Aizawa, 1964; Payne and Harrap, 1977). These distinct 20 nm-long spikes appear to be hollow and to originate within the virion core (Miura et al., 1969). It would be of considerable interest to investigate the antigenic structure of the coat proteins and of the spike material. This is particularly so as the virus particles will agglutinate vertebrate erythrocytes (Miyajima and Kawase, 1969). Asai et al. (1972) described terminal knobs on the apical spikes, which they associated with the hemagglutinin activity.

1.4. Iridoviridae – iridovirus

Iridescent viruses are very common in invertebrates and have been recorded from most natural orders in the Class Insecta. These DNA viruses are large, icosahedral, and replicate in the cytoplasm of cells (Kelly and Robertson, 1973). Two major groups of iridescent viruses have been described; large (about 180–210 nm in diameter) and small (about 150 nm in diameter). Very little work has been done on the antigenic structure of the large viruses although two variants of a virus isolated from the mosquito *Aedes taeniorhynchus* have been shown to be serologically identical (Wagner et al., 1974).

The fact that the small iridescent viruses apparently have very wide host ranges suggests that the possibility is high of the same virus infecting different species living within the same ecosystem (Kalmakoff et al., 1972). The serological relationships of at least 14 types of the small viruses have been studied and all but two were related, the unrelated viruses occurring in the bee, *Apis cerana*, and the moth, *Chilo suppressalis* (Kelly et al., 1979). ELISA has been used to detect and to discriminate between 4 different isolates of iridescent viruses and virus could be detected in the presence of an excess of larval tissue (Kelly et al., 1978). It seems that at present the ELISA technique is the method of choice for analyzing iridescent viruses. Immunodiffusion and counterelectrophoresis are of no value as the viruses are very large and, in the intact state, do not diffuse in agar or cellulose acetate.

As a detailed study of the serological relationships has yet to be made, the present situation is somewhat confused. The apparent morphological similarity of the viruses is misleading and there is undoubtedly real diversity in their properties; whether this is sufficiently marked or not to call into question the present systematic groupings is not known. The confusing number of apparently distinct strains being described in the literature led Tinsley and Kelly (1970) to suggest a system of interim nomenclature analogous to the one used for adenoviruses. This list was extended by Kelly and Robertson (1973) and a list of some 27 types was provided by Tinsley and Harrap (1978). A list of types does not contribute in any way to answering the question of relationships, but at least it provides a degree of order until definitive studies are undertaken.

1.5. Parvoviridae – densovirus

Densoviruses, or Densonucleosis viruses as they are more usually known, are some of the smallest viruses found in insects. Only the viruses isolated from the greater wax-moth *Galleria mellonella*, DV1 (Meynadier et al., 1964), and from the buck-eye caterpillar, *Junonia coenia*, DV2 (Rivers and Longworth, 1972) and the Limacodiid *Casphalia extranea* (Fediere, 1983) have been investigated in any detail. Four other probable densoviruses have been described from the silkworm *Bombyx mori* (Watanabe et al., 1976) the blackfly *Simulium vittatum* (Federici, 1976), the sugar cane moth *Diatrea saccharalis* (Meynadier et al., 1977) and the silver spotted flambeau *Agraulis vanillae* (Kelly et al., 1980b). It is very likely that this last virus has sufficient properties in common with DV1 and DV2 to be classed as DV3.

The viruses are small (circa 22 nm) and contain single-stranded DNA. The complementary DNA strands are packaged in separate virions. The virion size and nucleic acid content are sufficient to distinguish them from the small RNA-containing viruses found in insects, often in association with the Densoviruses. DV1, DV2 and DV3 have been shown to be closely related by immunodiffusion tests (Tinsley and Longworth, 1973; Spilling, 1970; Kelly et al., 1980a; 1980b). Fediere (1983) found that the densonucleosis virus from *Casphalia extranea* was not related to DV1 or DV2 and this virus should perhaps now be accorded the type number of DV4. Hoggan (1971) could not demonstrate any serological relationship between DV1 of insects and Parvoviruses isolated from mammals.

DV1 and DV2 are very closely related in serological terms and yet there are profound differences in their biological behaviour. DV1 will only infect *G. mellonella* as far as is known, whereas DV2 will infect several species in the families Noctuidae and Nymphalidae in the Lepidoptera but not *G. mellonella*.

1.6. Picornaviridae

It has been convenient until recently to place all the small single-stranded RNA viruses found in insects into this family. Whether all the viruses so included will prove to be Picornaviruses, as defined by the International Committee on the Taxonomy of Viruses (ICTV), can only be determined by further work.

Various workers have attempted to bring some order to the confused picture of these particular viruses (Longworth, 1978; Payne and Kelly, 1981). To date, at least two groups have now been assigned to definite virus families (Nodaviridae and the *Nudaurelia* virus group). In the latest report of the ICTV (Matthews, 1982), viruses such as cricket paralysis virus (CrPV), *Drosophila* C virus and *Gonometa* virus are classed as members of the Picornaviridae but not yet assigned to genera, whereas some 30 others are simply noted as ‘unclassified small RNA viruses of invertebrates’.

Cricket paralysis virus is related to *Drosophila* C virus in immunodiffusion tests

(Reinganum and Scotti, 1976) but is not identical (Plus et al., 1978). *Drosophila* C must therefore be regarded as a strain of CrPV. Many isolations have been made of viruses from a wide range of insect species showing relationship to CrPV, and it is clear that many serotypes of this virus must exist in nature (Scotti et al., 1981). Scotti and Longworth (1980) made the significant observation that mammalian sera taken from numerous cattle, one pig and one horse, contained both precipitating and neutralizing antibodies to CrPV. The class of antibody was the IgM. It was thought that cattle had been exposed to the insect virus as it was present in the cricket populations in the pastures but there was no indication that cross infections had occurred. The structural proteins of CrPV are formed from high molecular weight precursors by multiple proteolytic cleavages (Moore et al., 1980). The virus-induced proteins can thus be divided into three major groups, (*a*) precursor proteins, (*b*) viral structural proteins, and (*c*) apparently nonfunctional and wasted low molecular weight proteins arising from the multiple proteolysis cleavage. The production of the latter group of proteins would indicate that virus infection and replication had occurred. No such proteins were found in the sera of sheep and horses that gave immune precipitates in gel-diffusion plates with CrPV antigen (Moore et al., 1981). These authors concluded that in these instances, antibodies had been produced against preformed virus introduced from another source. However, these experiments do not rule out the possibility that the antibodies in mammals were as a result of an infecting virus related to but not identical with CrPV. Tinsley et al. (1984) have shown that CrPV is a strain of encephalomyocarditis virus (EMCV) by immunodiffusion, counter immunoelectrophoresis and neutralization techniques. These results show that CrPV is a member of the genus Cardiovirus in the Picornaviridae.

The virus from *Gonometa podocarpi* was regarded by Longworth (1978) to be sufficiently different from the accepted vertebrate picornavirus standards to not be regarded as an Enterovirus. He therefore grouped it with several viruses from *Drosophila*, the honey bee and the silkworm. Interestingly enough, Longworth et al. (1973) showed that this virus and a similar virus from *Darna trima* also reacted with IgM antibodies present in sera from a wide range of wild and domestic animals. These authors concluded that the most plausible explanation was that the vertebrate animals had been subjected to a low but persistent exposure to the insect virus antigens. In view of the reported relationship between CrPV and EMCV it may well be that the mammalian antibodies were produced to an as yet unknown virus which was nevertheless related to the *Gonometa/Darna* viruses.

1.7. Nudaurelia β group

These viruses are 38 nm in diameter with a single polypeptide of about 66 kDa. The serological relatedness among the various members has been reported by Reinganum et al. (1978) and Greenwood and Moore (1981). The virus (*Nudaurelia* β), isolated from the pine emperor moth, *Nudaurelia cytherea capensis* (Juckes, 1970) is regarded as the prototype species and the other virus isolates, though related, are not identical. Therefore, here again, a complex of virus strains exists, occurring in a wide range of insect hosts in many different geographical areas. Some superficial resemblances to members of the Caliciviridae have been noted, but there are fundamental differences that clearly separate the two families. These have been presented in detail by Reinganum et al. (1978).

It is interesting that members of the *Nudaurelia* group also react with antibodies found in the sera of wild and domestic animals and, most significantly, with antibodies in many samples of human sera (MacCallum et al., 1979). The nature of the virus or virus strains stimulating the production of these cross-reacting antibodies in mammalian sera is not known, although MacCallum et al. (1979) observed virus-like particles in one of the sheep sera involved in these reactions. The particles were similar in size to the insect-virus antigen being used in the experiments and formed a line of identity in immunodiffusion tests when exposed to the insect virus antiserum.

1.8. Nodaviridae

This family is noted for its members containing two single-stranded segments of RNA. Thus, they are commonly known as the 'divided-genome RNA viruses'. Four virus isolates have been described to date, from *Culex tritaeniorhynchus* (Nodamura virus), *Heteronynchus arator* (black beetle virus), *Apis mellifera* (Arkansas bee virus) and *Lymantria ninayi*. All four viruses are reported to be distantly related as measured by immunodiffusion and ELISA tests (Longworth, 1978; L.K. Greenwood and N.F. Moore, personal communication). Nodamura virus is the only strain known to replicate in both insects and vertebrates and it is interesting that high levels of neutralizing antibody were found in pigs and herons (Scherer and Hurlbut, 1967).

1.9. Caliciviridae

One possible member of this family has been described from the navel orangeworm *Amyelois transitella* (Kellen and Hoffman, 1981; Hillman et al., 1982). Preliminary serological studies did not indicate any relationship to mammalian caliciviruses.

1.10. Rhabdoviridae – sigmavirus

No serological studies have been made on this virus and indeed only very recently has it been successfully grown in tissue culture (Richard-Molard et al., 1984). It is possible that, once sufficient quantities of pure virus are obtained, some knowledge of the antigenic structure of Sigmavirus can be obtained and so enable comparisons to be made with other members of this pathologically important virus family.

2. Conclusions

It is clear from the literature dealing with the antigenic structure and serological reactions of insect pathogenic viruses that much remains to be investigated. Most attention has been paid to the Baculoviridae and this is not surprising as these viruses are not only the most numerous in the Class Insecta but also are of the greatest practical importance as biological control agents. Another area that has received perhaps disproportionate attention is beneficial insects, and the silkworm, *B. mori*, in particular. It has already been noted that only a tiny fraction of the possible number of insect viruses has been described and the use of serological methods will remain a vital step in the diagnosis and identification in all future work.

3. References

Aitken, T.H.G., Downs, W.G. and Anderson, C.R. (1958) *Proc. Soc. Exp. Biol. Med.* 99, 635–637.

Asai, J., Kawamoto, F. and Kawase, S. (1972) *J. Invertebr. Pathol.* 19, 279–280.

Bailey, L. and Scott, H.A. (1973) *Nature* 241, 545.

Bergoin, M. and Dales, S. (1971) In *Comparative Virology* (K. Maramorosch and E. Kurstak, Eds.) Academic Press, New York. pp. 169–205.

Brown, D.A., Bud, H.M. and Kelly, D.C. (1977) *Virology* 81, 317–327.

Federici, B.A. (1976) *Proc. 1st Inter. Coll. on Insect Path.* Kingston, Ontario, Canada. pp. 341–342.

Federici, B.A. and Hazard, E.I. (1975) *Nature* 254, 327–328.

Fediere, G. (1983) *PhD Thesis.* Presented to Université des Sciences et Techniques du Languedoc, Montpellier.

Greenwood, L.K. and Moore, N.F. (1981) *Microbiologica* 4, 271–280.

Harrap, K.A. and Payne, C.C. (1979) *Adv. Virus Res.* 25, 273–355.

Harrap, K.A., Payne, C.C. and Robertson, J.S. (1977) *Virology* 79, 14–31.

Hillman, B., Morris, T.J., Kellen, W.R., Hoffman, D. and Schlegel, D.E. (1982) *J. Gen. Virol.* 60, 115–123.

Hoggan, M.D. (1971) In *Comparative Virology* (K. Maramorosch and E. Kurstak, Eds.) Academic Press, New York. pp. 43–79.

Hosaka, Y. and Aizawa, K. (1964) *J. Insect Pathol.* 6, 53–77.

Hurlbut, H.S. (1956) *Am. J. Trop. Med. Hyg.* 5, 76–85.

Hurlbut, H.S. and Thomas, J.I. (1960) *Virology* 12, 381–407.

Hurlbut, H.S. and Thomas, J.I. (1969) *J. Med. Entomol.* 6, 423–427.

Juckes, I.R.M. (1970) *Bull. S.A. Soc. Plant Pathol. Microbiol.* 4, 18.

Kalmakoff, J., Moore, S. and Pottinger, R.P. (1972) *J. Invertebr. Pathol.* 20, 70–76.

Kellen, W.R. and Hoffman, D.F. (1981) *J. Invertebr. Pathol.* 38, 52–66.

Kelly, D.C. and Robertson, J.S. (1973) *J. Gen. Virol.* 20, 17–41.

Kelly, D.C., Edwards, M.L. and Robertson, J.S. (1978) *Am. J. Appl. Biol.* 90, 369–374.

Kelly, D.C., Ayres, M.D., Lescott, T., Robertson, J.S. and Happ, G.M. (1979) *J. Gen. Virol.* 42, 95–105.

Kelly, D.C., Moore, N.F., Spilling, C.R., Barwise, A.H. and Walker, I.O. (1980a) *J. Virol.* 36, 224–235.

Kelly, D.C., Ayres, M.D., Spencer, L.K. and Rivers, C.F. (1980b) *Microbiologica* 3, 455–460.

Köhler, G. and Milstein, C. (1975) *Nature* 256, 495–497.

Kurstak, E. and Garzon, S. (1977). In *The Atlas of Insect and Plant Viruses* (K. Maramorosch, Ed.) Academic Press, New York. pp. 29–66.

Longworth, J.F. (1978) *Adv. Virus Research* 23, 103–157.

Longworth, J.F., Robertson, J.S., Tinsley, T.W., Rowlands, D.J. and Brown, F. (1973) *Nature* 242, 314–316.

MacCallum, F.O., Brown, G. and Tinsley, T.W. (1979) *Intervirology* 11, 234–237.

McCarthy, W.J. and Lambiase, J.T. (1979) *J. Invertebr. Pathol.* 34, 170–177.

Maskos, C.B. and Miltenburger, H.G. (1981) *J. Invertebr. Pathol.* 37, 174–180.

Matthews, R.E.F. (1982) *Intervirology* 17, 1–200.

Meynadier, G., Vago, C., Plantevin, G. and Atger, P. (1964) *Rev. Zool. Ecol. Animale* 2, 79–91.

Meynadier, G., Galichet, P.F., Veyrunes, J.C. and Armagier, A. (1977) *Entomophaga* 22, 115–120.

Miller, L.K. and Dawes, K.P. (1978a) *Appl. Environ. Microbiol.* 35, 411–421.

Miller, L.K. and Dawes, K.P. (1978b) *Appl. Environ. Microbiol.* 35, 1206–1210.

Miura, K., Fujii-Kawata, I., Iwata, H. and Kawase, S. (1969) *J. Invertebr. Pathol.* 14, 262–265.

Miyajima, S. and Kawase, S. (1969) *Virology* 39, 347–348.

Moore, N.F., Kearns, A. and Pullin, J.S.K. (1980) *J. Virol.* 33, 1–9.

Moore, N.F., McKnight, L. and Tinsley, T.W. (1981) *Infect. Immun.* 31, 825–827.

Naser, W.L. and Miltenberger, H.G. (1983) *J. Gen. Virol.* 64, 639–647.

Payne, C.C. (1974). *J. Gen. Virol.* 25, 105–116.

Payne, C.C. and Churchill, M.P. (1977) *Virology* 79, 251–258.

Payne, C.C. and Harrap, K.A. (1977) In *The Atlas of Insect and Plant Viruses* (K. Maramorosch, Ed.) Academic Press, New York. pp. 105–129.

Payne, C.C. and Kalmakoff, J. (1974) *Intervirology* 4, 365–368.

Payne, C.C. and Kelly, D.C. (1981) In *Microbial Control of Pests and Plant Diseases 1970–1980* (H.D. Burges, Ed.) Academic Press, London. pp. 61–91.

Payne, C.C. and Rivers, C.F. (1976) *J. Gen. Virol.* 33, 71–85.

Payne, C.C., Piasecka-Serafin, M. and Pilley, B. (1977) *Intervirology* 3, 155–163.

Plus, N., Crozier, G., Reinganum, C. and Scotti, P.D. (1978) *J. Invertebr. Pathol.* 31, 296–302.

Pritchett, D.W., Scott, H.A. and Young, S.Y. (1979) *J. Invertebr. Pathol.* 33, 193–188.

Reinganum, C. and Scotti, P.D. (1976) *J. Gen. Virol.* 31, 131–134.

Reinganum, C., Robertson, J.S. and Tinsley, T.W. (1978) *J. Gen. Virol.* 40, 195–202.

Richard-Molard, C., Blondel, D., Wyers, F. and Dezelee, S. (1984) *J. Gen. Virol.* 65, 91–99.

Rivers, C.F. and Longworth, J.F. (1972) *J. Invertebr. Pathol.* 20, 369–370.

Roberts, P.L. and Naser, W. (1982) *Virology* 122, 424–430.

Scherer, W.F. and Hurlbut, H.S. (1967) *Am. J. Epidemiol.* 86, 271–285.

Scherer, W.F., Verna, J.E. and Richter, G.W. (1967) *Am. J. Trop. Med. Hyg.* 17, 120–128.

Scotti, P.D. and Longworth, J.F. (1980) *Intervirology* 13, 186–191.

Scotti, P.D., Longworth, J.F., Plus, N., Crozier, G. and Reinganum, C. (1981) *Adv. Virus Res.* 26, 117–143.

Smith, G.E. and Summers, M.D. (1978) *Virology* 89, 517–527.

Smith, G.E. and Summers, M.D. (1979) *J. Virol.* 30, 828–838.

Smith, G.E. and Summers, M.D. (1981) *J. Virol.* 39, 125–137.

Spilling, C.R. (1970) *PhD Thesis*. University of Oxford.

Summers, M.D. and Smith, G.E. (1978) *Virology* 84, 390–402.

Tinsley, T.W. (1975) In *Invertebrate Immunity – Mechanisms of Invertebrate Vectors-Parasite Relations.* Academic Press, New York. pp. 55–63.
Tinsley, T.W. and Harrap, K.A. (1978) In *Comprehensive Virology* (H. Fraenkel-Conrat and R.R. Wagner, Eds.) Vol. 12. Plenum Press, New York. pp. 1–101.
Tinsley, T.W. and Kelly, D.C. (1970) *J. Invertebr. Pathol.* 16, 470–472.
Tinsley, T.W. and Longworth, J.F. (1973) *J. Gen. Virol.* 20, 71–75.
Tinsley, T.W., MacCallum, F.O., Robertson, J.S. and Brown, F. (1984) *Intervirology* 21, 181–186.
Volkman, L.E. and Falcon, L.A. (1982) *J. Econ. Entomol.* 75, 868–871.
Wagner, G.W., Paschke, J.D., Champbell, W.R. and Webb, S.R. (1974) *Intervirology* 3, 97–105.
Watanabe, H., Maeda, S., Matsui, M. and Shimizu, T. (1976) *J. Sericulture Sci. Jpn.* 45, 29–34.

Subject index